ASSESSING CLIMATE CHANGE

Members of the Model Evaluation Consortium
for Climate Assessment (MECCA)

Electric Power Research Institute (EPRI), USA

University Corporation for Atmospheric Research (UCAR)

Electricité de France (EdF)

Italian Agency for New Technologies, Energy and Environment (ENEA)

Central Research Institute of Electric Power Industry (CRIEPI), Japan

N.V. KEMA, The Netherlands

Southern California Edison Company (SCE)

National Supercomputing Center for Energy and the Environment
at the University of Nevada at Las Vegas (UNLV)

US Environmental Protection Agency

ASSESSING CLIMATE CHANGE

Results from the Model Evaluation Consortium for Climate Assessment

Edited by

Wendy Howe
Centre for Natural Resources
Department of Land and Water Conservation
Parramatta, Australia

Ann Henderson-Sellers
Deputy Vice-Chancellor (Research and Development)
Royal Melbourne Institute of Technology
Melbourne, Australia

Gordon and Breach Science Publishers

Australia • Canada • China • France • Germany • India • Japan
Luxembourg • Malaysia • The Netherlands • Russia • Singapore
Switzerland • Thailand • United Kingdom

Amsteldijk 166
1st Floor
1079 LH Amsterdam
The Netherlands

British Library Cataloguing in Publication Data

Assessing climate change : results from the model
   evaluation consortium for climate assessment
   1. Climatic changes 2. Climatic changes – Forecasting
   I. Howe, Wendy II. Henderson-Sellers, Ann
   551.6

ISBN: 90-5699-067-5

Cover Photography: John Cleasby, LAPS

CONTENTS

FOREWORD

The Model Evaluation Consortium for Climate Assessment (MECCA) was a unique international global climate research consortium embracing universities, government and industry. Established in 1991, MECCA had a stated goal of providing insight about uncertainties inherent in climate model projections of greenhouse gas-induced climate changes. MECCA was created to (1) accelerate studies that would attempt to quantify estimates of uncertainty in predicting climate change, and (2) assist the public to understand the nature and limitations of predictions from climate models. MECCA's research was designed to improve understanding of the models being used to forecast global climate change. By joining in the assessment of existing model capabilities, the sponsors hoped that MECCA would facilitate a more detailed analysis of theoretical predictions of climate change than otherwise would have been possible.

MECCA resulted from a merger of needs foreseen by two communities. First, scientists recognized the need to expand their efforts to analyze climate forcing from anthropogenic sources. Second, representatives of the private sector became increasingly concerned about the reliability of climate predictions that were affecting international energy policy. By 1990, supercomputer technology and the climate science community had advanced the capabilities of theoretical models to a stage where true experimentation could be undertaken on simulated atmospheres. At about the same time, broad public awareness of the potential for human activities to interfere with climate increased dramatically. The reliability of the scientific foundations for this assertion was questioned by some. Many investigators believed that the uncertainties in prediction of climate change could be defined, and ultimately reduced, through accelerated development and experimentation with climate models.

Through the 1980s, the electric utilities became increasingly concerned about the potential for a peremptory public response to the prospect of interference with climate that would have implications for fossil-fuel-based energy production. Since decisions for major changes depend on the quality of information used for intervention, industry expressed concern and even skepticism about the reliability of the projections of climate change. Representing the US electric utility collaborative research, the Electric Power Research Institute (EPRI) president, Richard Balzhiser talked with Robert White, the president of the US National Academy of Engineering, about the climate change issue. Dr White agreed that an effort should be made to accelerate the development and evaluation of climate models to insure that their projections were defensible to the public and private sectors. Their conversation appears to have led to the conception in 1989 of the concept for a collaborative research effort between industry and the government on climate modeling. About the same time, the need for accelerated model development and evaluation was articulated by a group of well-known scientists at the Marshall Institute (Seitz et al. 1989). This group proposed that a fully dedicated supercomputer facility could facilitate a substantial improvement in climate models within five years.

Through Dr Balzhiser's suggestion, Dr George Hidy, vice president and director of EPRI's Environment Division, initiated a program in February of 1990 to coalesce industry, government and university support for the concept. Dr Peter K. Mueller, program manager for Atmospheric Sciences in EPRI's Environment Division, was asked to pursue the formation of, and provide the framework for, funding the project. This activity represented the major component of a broader EPRI program aimed at sponsoring climate and climate effects science that would lead to capabilities for undertaking meaningful risk assessment of climate change.

Informal discussions with several supercomputer manufacturers, supercomputing centers, and other industries culminated with the submittal in the summer of 1990 of a proposal from the University Corporation for Atmospheric Research (UCAR) for the National Center for Atmospheric Research to house and operate a dedicated supercomputer for climate modeling at NCAR's facilities in Boulder, Colorado, with Gary Jensen of the NCAR Scientific Computing Division serving as the on-site manager. The National Science Foundation agreed to support NCAR's participation in this endeavor. Cray Research provided the principal hardware for the supercomputer facility and became an active partner in the venture by providing an attractive lease proposal for a Y-MP2/16 supercomputer. Discussions with other potential financial sponsors proceeded in parallel with the technical review of the UCAR proposal. At the time of the official launch of the facility at Boulder in June of 1991, financial sponsorship had expanded to include EPRI, UCAR, the Central Research Institute of the Electric Power Industry of Japan (CRIEPI), Electricité de France (EdF) and the Italian Agency for New Technologies, Energy and Environment (ENEA). Later, the dedicated supercomputer was augmented substantially with computing hours obtained in collaboration with the National Supercomputing Center for Energy and Environment (NSCEE) at the University of Nevada, Las Vegas, and Southern California Edison (SCE) and the US Environmental Protection Agency (EPA) added their sponsorship. Dr Peter K. Mueller and Dr Chuck Hakkarinen of EPRI were selected to serve as the Chair of the MECCA Policy Committee and the MECCA Project Manager, respectively, for the duration of the Consortium.

The overall aim of MECCA was to help quantify the reliability of climate change forecasts that are based on increasing atmospheric concentrations of greenhouse gases. MECCA's research plan included two key elements:

1. a set of carefully selected model experiments designed to address major model uncertainties, supplementing ongoing work of other groups; and
2. an analysis strategy that focused on the social and environmental impact-oriented needs of policy makers in the public and private sectors.

Modeling experiments were to be carried out in two phases. Phase 1 concentrated on experiments of relatively short simulation periods (decades or less) to clarify parameters in models to which the results were most sensitive. Some of the parameters of interest studied in Phase 1 included model sensitivities to atmospheric CO_2, other greenhouse gas and aerosol concentrations, heat transfer between atmosphere and oceans, changes in vegetation characteristics, and the spatial resolution of the model's computational grid. Phase 2 experiments would emphasize fewer, but longer-term (decades to centuries) simulations to examine climate effects in models forced by more realistic representations of greenhouse gas increases. The experiments were scheduled to begin in mid-1991 and continue for approximately three years.

The MECCA Consortium supported research by offering the use of the dedicated facility free of charge to qualifying experimenters, chosen from the worldwide community of climate scientists. Proposals for experiments and analysis of results were solicited from the international scientific community, and evaluated and recommended through peer review by a scientific advisory committee nominated by the sponsoring organizations. Through a separate, informal competitive review, a leader for an Analysis Team was selected (Professor Ann Henderson-Sellers, at that time at Macquarie University in Sydney) to conduct comparative analyses among selected individual experiments, and to prepare a combined data archive that could be used in future impacts-related assessment studies.

The uniqueness of the MECCA project is summarized for scientists in this book. Most of the chapters have been contributed, on invitation, by principal investigators of selected MECCA-sponsored experiments. These contributions are reviewed in groups in the form of chapter commentaries. The work is organized around three major technical areas of uncertainty which, in the 1980s, were believed to be crucial to establishing the reliability of climate forecasts. These focused on the model designs to handle cloud and aerosol processes, ocean and atmosphere interactions, and sensitivity to surface-atmosphere interactions. The main chapters of the book have been grouped under these topics.

The book is divided into six parts with Part 1 setting the scene for the rest of the book. Parts 2, 3 and 4 describe tasks under MECCA's first goal—the careful selection of model experiments designed to supplement ongoing work by other modeling groups. Commentary chapters are intended to place the MECCA results into a broader context as a review of the usefulness and insight contained in the MECCA results. These chapters, together with Parts 5 and 6, address MECCA's second goal: an analysis strategy focusing on the impact-oriented needs of policy makers.

The editors have chosen to include not only a synthesis of important results bearing on climate science, but also a commentary on the 'social' experience of the MECCA community. The consortium brought together a diverse group of investors and scientists in fields of science previously unaccustomed to working toward common goals. The MECCA Analysis Team (MAT), after initiation of the projects, was requested to synthesize the wide variety of results in a form that would communicate the 'whole' as much greater than 'the sum of its parts'. As noted by the editors, this posed a formidable challenge, given the international character of the consortium and the diverse backgrounds of the participants.

The MECCA experience, viewed as a 'social experiment', represents one attempt to translate high quality, esoteric science into information useful and relevant to the public-at-large. In such an ambitious undertaking, there are bound to be perceived deficiencies in design and implementation. The editors find value in exploring these from their experiences in hopes of improving future ventures of this kind. The mixture of physical science discussion with social science critique for 'getting things done credibly' adds an important dimension to this work, especially for those who seek to continue such consortia to address major science issues in the future.

From the beginning, the consortium members sought to create a bridge between the science community, whose goal is to advance the state of scientific knowledge, and the public, whose interest is in obtaining information for developing improved strategies to protect the environment. As an experiment in timely investigation of an important environmental question, and in bridging scientific knowledge with the international 'user' community, MECCA can certainly be considered a success. The consortium produced a significant amount of new information on uncertainties inherent in climate models and their ability to predict future conditions. It also demonstrated that collaborative actions by the public and private sectors can stimulate scientific effort to inform the public in a timely manner. These included many references to research conducted under MECCA sponsorship in the 1992 and 1995 assessments of the Intergovernmental Panel on Climate Change (IPCC). Particularly prominent in the 1995 Assessment are the MECCA findings that highlight the importance of sulfur aerosols in moderating global warming estimated from greenhouse gases on a regional basis, a concept that is now recognized in all climate model simulations and international discussions on future climate change.

MAT has stated that this book represents the final product of its efforts. The MECCA goals, however, are not quite complete in that the extensive scientific results described in this book remain to be synthesized and integrated into laypersons' language. The discussion of these results in the broad context of their implications to strategy development for environmental and energy policy in the next century remains a final challenge for participants in the MECCA experiment.

George M. Hidy
Department of Civil and Environmental Engineering
University of Alabama at Birmingham

Reference

Seitz, F., Bendelson, K., Jastrow, R. and Nierenberg, W.A. (1989) *Scientific Perspectives on the Greenhouse Problem: Executive Summary.* Washington, DC: Marshall Institute.

CONTRIBUTORS

Richard A. Anthes
University Corporation for Atmospheric Research
Boulder, Colorado, USA

Vincenzo Artale
Ente per le Nouve Tecnologie
l'Energia l'Ambiente
Rome, Italy

Jean-Yves Caneill
Electricité de France (EdF)
Châtou Cedex, France

Robert M. Chervin
National Center for Atmospheric Research
Boulder, Colorado, USA

Catherine Ciret
Climatic Impacts Centre
Macquarie University
North Ryde, New South Wales, Australia

Robert A. Colman
Bureau of Meteorology Research Centre
Melbourne, Victoria, Australia

Anthony P. Craig
National Center for Atmospheric Research
Boulder, Colorado, USA

Robert E. Dickinson
Institute of Atmospheric Physics
University of Arizona
Tucson, Arizona, USA

W.L. Gates
Lawrence Livermore National Laboratory
Livermore, California, USA

Andrea N. Hahmann
Institute of Atmospheric Physics
University of Arizona
Tucson, Arizona, USA

Akira Hasegawa
Institute of Geoscience
University of Tsukuba
Tsukuba, Japan

Matthew Hecht
Institute for Geophysics and Planetary Physics
Los Alamos National Laboratory
Los Alamos, New Mexico, USA

Ann Henderson-Sellers
Deputy Vice-Chancellor (Research & Development)
Royal Melbourne Institute of Technology
Bundoora, Victoria, Australia

Neil Holbrook
Climatic Impacts Centre
Macquarie University
North Ryde, New South Wales, Australia

Gregory Holland
Bureau of Meteorology Research Centre
Melbourne, Victoria, Australia

William Holland
National Center for Atmospheric Research
Boulder, Colorado, USA

Wendy Howe
Centre for Natural Resources
Department of Land and Water Conservation
Parramatta, New South Wales, Australia

Peter Jaumann
National Center for Atmospheric Research
Boulder, Colorado, USA

Christopher Landsea
National Oceanic and Atmospheric Administration
AOML/Hurricane Research Division
Miami, Florida, USA

Shuhua Li
Climatic Impacts Centre
Macquarie University
North Ryde, New South Wales, Australia

Bryant J. McAvaney
Bureau of Meteorology Research Centre
Melbourne, Victoria, Australia

Kendal McGuffie
Bureau of Meteorology Research Centre
Melbourne, Victoria, Australia

Linda O. Mearns
National Center for Atmospheric Research
Boulder, Colorado, USA

Gerald A. Meehl
National Center for Atmospheric Research
Boulder, Colorado, USA

Bahram Nassersharif
National Supercomputing Center for Energy and
   the Environment
University of Las Vegas
Las Vegas, Nevada, USA

Shaw Nishinomiya
Central Research Institute for Electric Power Industry
Komae Research Laboratory
Tokyo, Japan

Joyce E. Penner
Atmospheric, Oceanic and Space Sciences
University of Michigan
Ann Arbor, Michigan, USA

Nadia Pinardi
IMGA-CNR
Modena, Italy

Andrew J. Pitman
Climatic Impacts Centre
Macquarie University
North Ryde, New South Wales, Australia

David Pollard
Interdisciplinary Climate Systems Section
Climate and Global Dynamics Division
National Center for Atmospheric Research
Boulder, Colorado, USA

Cynthia Rosenzweig
Goddard Institute for Space Studies
   and Colombia University
New York, USA

Walter Ruijgrok
N.V. KEMA
Arnhem, The Netherlands

Ben D. Santer
Program for Climate Model Intercomparison and
   Diagnostics
Lawrence Livermore National Laboratory
Livermore, California, USA

Sascha Schubert
School of Mathematics
University of New South Wales
Kensington, New South Wales, Australia

Maurizio Sciortino
Italian Agency for New Technologies, Energy and
   Environment
C.R.E. Casaccia, Rome, Italy

Albert J. Semtner Jnr.
Naval Postgraduate School
Monterey, California, USA

Robert Serafin
National Center for Atmospheric Research
Boulder, Colorado, USA

Qingqiu Shao
Institute of Atmospheric Physics
University of Arizona
Tucson, Arizona, USA

Hiroshi L. Tanaka
Institute of Geoscience
University of Tsukuba
Tsukuba, Japan

Karl E. Taylor
Atmospheric Sciences Division and
Program for Climate Model Intercomparison and
   Diagnostics
Lawrence Livermore National Laboratory
Livermore, California, USA

Starley L. Thompson
Interdisciplinary Climate Systems Section
Climate and Global Dynamics Division
National Center for Atmospheric Research
Boulder, Colorado, USA

Heather Tonkin
Climatic Impacts Centre
Macquarie University
North Ryde, New South Wales, Australia

Warren M. Washington
National Center for Atmospheric Research
Boulder, Colorado, USA

Tom M.L. Wigley
National Center for Atmospheric Research
Boulder, Colorado, USA

James Young
Research and Development
Southern California Edison Company
Rosemead, California, USA

Huqiang Zhang
Climatic Impacts Centre
Macquarie University
North Ryde, New South Wales, Australia

ACKNOWLEDGMENTS

This book represents the final 'product' from the MECCA Analysis Team. We would especially like to thank Ann-Maree, Cathy, Chris and Chris, Craig, Heather, Hong, Jay, Jeljer, Kendal, Margaret, Margriet, Neil, Olga, Sascha, Shuhua and Zav, who have all been friends and colleagues during this MECCA experience. We would also like to thank Dr Chuck Hakkarinen for the bibliography.

ABBREVIATIONS

AAP	Atmospheric Analysis and Prediction
ABL	Atmospheric Boundary Layer
ACARP/ESAA	Australian Coal Association Research Program/Electricity Supply Association of Australia
ACAS	Advisory Committee on Atmospheric Sciences
ACC	Antarctic Circumpolar Current
ACD	Atmospheric Chemistry Division
AGCM	Atmospheric General Circulation Model
AMIP	Atmospheric Model Intercomparison Project
ARM	Atmospheric Radiation Measurements
ATM	Atmospheric Science Division
BATS	Biosphere–Atmosphere Transfer Scheme
BATSle	Biosphere–Atmosphere Transfer Scheme Version le
BEST	Bare Essentials for Surface Transfer
BMRC	Bureau of Meteorology Research Centre (Australia)
CAP	Corporate Affiliates Program
CCC	Canadian Climate Centre
CCM	Community Climate Model
CCM0	Community Climate Model Version 0
CCM1	Community Climate Model Version 1
CCM1-OZ	Community Climate Model Version 1 in Australia
CCM1W	Community Climate Model Version 1 of Wang
CCM2	Community Climate Model Version 2
CCS	Coupled Climate System
CEES	Committee on Earth and Environmental Science
CES	Committee on Earth Sciences
CGD	Climate and Global Dynamics
CHAMMP	Computer Hardware, Advanced Mathematics and Model Physics
CIG	Climate Impacts Group
CLIVAR	Climate Variability and Predicability Programme
CMAP	Climate Modeling and Analysis Program
CME	Community Modeling Effort
COADS	Comprehensive Oceanic Atmospheric Data Set
CRIEPI	Central Research Institute of the Electric Power Industry (Japan)

CSL	Climate Simulation Laboratory
CSM	Climate System Model
CSMAC	Climate System Model Advisory Committee
CSMC	Climate System Modeling Core
CSMI	Climate System Modeling Initiative
CSMIG	Climate System Model Investigator Group
CSMP	Climate System Modeling Program
CSMP SAC	Climate System Modeling Program Scientific Advisory Committee
DOE	Department of Energy
DOE/FE	Department of Energy/Fossil Energy Research
DRS	Data Retrieval and Storage System
ECMWF	European Centre for Medium-Range Weather Forecasts
EdF	Electricité de France
ENEA	Italian Agency for New Technologies, Energy and the Environment
ENSO	El Niño-Southern Oscillation
EPA	Environmental Protection Agency
EPRI	Electric Power Research Institute
ERBE	Earth Radiation Budget Experiment
ET	Evapotranspiration
FANGIO	Feedback Analysis of GCMs and Intercomparison with Observations
FGGE	First GARP Global Experiment
FIFE	First ISLSCP Field Experiment
GARP	Global Atmospheric Research Program
GCM	General Circulation Model or Global Climate Model
GENESIS	Global Environmental and Ecological Simulation of Interactive Systems
GFDL	Geophysical Fluid Dynamics Laboratory
GISS	Goddard Institute for Space Studies
GTCP	Global Tropospheric Chemistry Program
HDP	Human Dimensions Programme
ICMAP	Integrated Climate Modeling and Analysis Program
ICRCCM	Intercomparison of Radiation Codes for Climate Models
IGBP	International Geosphere–Biosphere Programme
IMAP	Integrated Modeling Analysis and Prediction
IPCC	Intergovernmental Panel on Climate Change
IRM	Institute for Resource Management
ISCCP	International Satellite Cloud Climatology Project
ISLSCP	International Satellite Land Surface Climatology Project
ITCZ	Inter-Tropical Convergence Zone
JAM	Joint Atlantic Mediterranean
JCSC	Joint CSMI Steering Committee
JMA	Japanese Meteorological Agency

JOI	Joint Oceanographic Institutions
LAI	Leaf Area Index
LAM	Limited Area Model
LLNL	Lawrence Livermore National Laboratory
LSX	Land-Surface-Exchange Scheme
MADD	MECCA Archive of Daily Data
MAT	MECCA Analysis Team
MCA	Moist Convective Adjustment
MECCA	Model Evaluation Consortium for Climate Assessment
MIT	Massachusetts Institute of Technology
MOS	Model Output Statistics
MPI	Maximum Potential Intensity
NADWF	North Atlantic Deep Water Formation
NASA	National Aeronautics and Space Administration
NCAR	National Center for Atmospheric Research
NCDC	National Climate Data Center
NMC	National Meteorological Center
NOAA	National Oceanic and Atmospheric Administration
NRL	Naval Research Laboratory
NSCEE	National Supercomputing Center for Energy and the Environment
NSF	National Science Foundation
N.V. KEMA	Royal Energy Company Anhem (The Netherlands)
OAGCM	Ocean–Atmosphere Global Climate Model
OGCM	Ocean General Circulation Model
OIES	Office of Interdisciplinary Earth Sciences
OLR	Outgoing Longwave Radiation
OMB	Office of Management and Budget
PCA	Principal Component Analysis
PCMDI	Program for Climate Diagnosis and Intercomparison
PI	Principal Investigator
PILPS	Project for Intercomparison of Landsurface Parameterization Schemes
P-M	Penman-Monteith
POCM	Parallel Ocean Climate Model
RegGCM	Regional GCM
SAC	Scientific Advisory Committee
SCD	Science Computing Division
SCE	Southern California Edison
SERDP	Strategic Environmental Research and Development Program
SPCZ	South Pacific Convergence Zone
SST	Sea Surface Temperature
TOA	Top of the Atmosphere
TOGA	Tropical Ocean Global Atmosphere
UCAR	University Corporation for Atmospheric Research

UNCED	United Nations Conference on Environment and Development
USGCRP	US Global Change Research Program
VEMAP	Vegetative Mapping and Analysis Program
WWW	World Wide Web
WCRP	World Climate Research Programme
WGNE	Working Group on Numerical Experimentation

PART 1

ABOUT THE MECCA PROJECT

CHAPTER 1

ORIGINS AND ESTABLISHMENT OF THE MECCA PROJECT

Richard A. Anthes

1.1 Origins of the MECCA project

The MECCA project became a reality when a two-processor CRAY Y-MP-2D, named 'Castle', was dedicated in a ceremony at the National Center for Atmospheric Research (NCAR) on 16 July 1991. MECCA had its roots in a number of institutions. Chief among these were NCAR, its parent body University Corporation for Atmospheric Research (UCAR) and the Electric Power Research Institute (EPRI). This chapter reviews the origins and history of MECCA from these institutions' point of view.

From the NCAR and UCAR points of view, the evolution of MECCA was dependent upon and closely connected to other NCAR and UCAR initiatives and programs, namely the NCAR Coupled Climate System (CCS) initiative, the UCAR Climate System Modeling Initiative (CSMI), which later became the Climate System Modeling Program (CSMP), and finally, the NCAR Climate System Model (CSM) project and the Climate Simulation Laboratory (CSL). MECCA, in turn, heavily influenced the development of the CSL by proving the value of a numerical laboratory dedicated to climate modeling research. Scientific studies from MECCA and CSMP were often closely connected. For example, climate simulations supported by MECCA were used by CSMP scientists Timothy Kittel and David Schimel for input into the Vegetative Mapping and Analysis Program (VEMAP) which compared terrestrial vegetation and ecosystem process models for climates with present-day and doubled concentrations of carbon dioxide. Because these programs were so closely related and were developed in a synergistic way, this chapter describes the history of all of these initiatives and programs.

1.1.1 The 1980s: The rise of global climate change as a national and international issue and role of climate system models in understanding global change

The political context within which MECCA and related activities were developed was one marked by broad interest in the potential impacts on climate of human activities, especially those associated with increasing rates of emissions of carbon dioxide and other radiatively active trace gases (greenhouse gases). The mid-1980s was a period of many national and international planning activities at the highest level in the general area of *earth system science* or *global change*. Major international initiatives included the World Climate Research Program (WCRP) and the International Geosphere–Biosphere Programme (IGBP). In the United States, NASA was developing its Earth System Science Program (NASA 1986), and the National Science Foundation (NSF) was developing its Global Geosciences initiative.

In the late 1980s, the various national initiatives in the general area of global change coalesced around the US Global Change Research Program (USGCRP), an ambitious, multi-agency supported national program (Pielke 1995). In *Our Changing Planet: A U.S. Strategy for Global Change Research*, a publication prepared as a supplement to the US President's 1990 budget, the Committee on Earth Sciences (CES) describes the goal of the USGCRP (Committee on Earth Sciences 1989):

> to provide a sound scientific basis for developing national and international policy on global change issues.

The USGCRP had three major research objectives:

1. Establish an integrated comprehensive monitoring program for earth system measurements on a global scale;
2. Conduct a program of focused studies to improve our understanding of the physical, chemical, and biological processes that influence earth system changes and trends on global and regional scales; and
3. Develop integrated conceptual and predictive earth system models.

In the NSF–UCAR atmospheric sciences community, a blue-ribbon long-range planning committee, chaired by John Dutton (Pennsylvania State University), was appointed in 1986 by NSF and UCAR to examine the major community initiatives and scientific opportunities and combine them into a coherent and balanced program of research in the atmospheric sciences. In the Dutton Committee report *The Atmospheric Sciences: A Vision for 1989–1994*, published on 1 July 1987 (UCAR 1987), the committee recommended four major community science initiatives, five facility and organizational/infrastructure enhancements, and eight new scientific initiatives. The highest priority in the last category was an initiative from NCAR called the Coupled Climate System (CCS).

1.1.1.1   *The NCAR Coupled Climate System (CCS) initiative*

In the early 1980s, NCAR Senior Scientist John Firor introduced the concept of a CCS initiative at NCAR. This led to the formation of a CCS Committee at NCAR composed of myself, Stephen Schneider, Robert Dickinson, Ralph Cicerone, Robert Serafin, and Edward Zipser, with Firor as the Chairman. Under the early leadership of Schneider and Firor, the CCS project was envisioned to be interdisciplinary and aimed at studying the Earth's climate in a holistic way—as a coupled, interactive system involving the atmosphere, the oceans, the land and ice masses, the biosphere, and human activities. In a memo to the NCAR Directors' Committee of 5 November 1985, Firor argued:

> ... we are talking about some of the grander intellectual challenges ever set before a group of scientists—what caused the warm Cretaceous, the ice ages, the climatic optimum and the Little Ice Age? Does the climate control life or does life control the climate? How will the climate respond in the future to the combination of its own internal processes, external forces of nature, and the forcings produced by the industrial and agricultural activities of people?

Firor gave five reasons why the time was right for a new focus on climate:

1. A whole new ball game has opened up with the realization that the coupling of the air to the rest of the earth is not only through the exchange of heat, momentum and moisture at the lower boundary but also by the radiative effects of trace gases emitted by the lower boundary or created within the air itself; and, the trace gases that are radiatively active are frequently also highly interactive with living things.
2. A new force has emerged as important within our lifetimes: there are now so many people and each person on the average commands so much energy that the influence of people on the climate must in the future be explicitly allowed for.
3. The development of satellite techniques means that the questions can now be asked about the workings of the global climate system that would have been impossible some years ago.
4. A host of new techniques are appearing for the reconstruction of past climates, holding out the possibility of verifying climate models against real data sets, some of which exhibit large changes from the present climate.
5. The capability of today's and tomorrow's computers allows the coupling of models of two or more complex physical, and hopefully, biological, systems.

Because of the intellectual excitement, scientific challenge, societal relevance and emerging observational and computational technologies as described by Firor, the CCS concept became NCAR's highest priority in developing its plans in 1985–87. In the first draft of the CCS plan, dated 20 November 1985, I, then Director of the Atmospheric Analysis and Prediction (AAP) Division, argued that NCAR was an ideal place to develop such a program because of its discipline-oriented studies directed toward understanding the elements of the climate system, including the sun and the global atmosphere, atmospheric and ocean dynamics, radiation, clouds, precipitation and atmospheric chemistry. At this time, the NCAR Community Climate Model (CCM) involved more than 20 NCAR scientists and 40 scientists from the university community. Even at this early stage, the CCM was being used to address such diverse problems as the climate of the Cretaceous period (about 100 million years ago) and the last ice age (18 000 years ago), the response of the climate to increasing CO_2 and other trace gases, and short-term climate phenomena involving atmosphere–ocean interactions such as the El Niño-Southern Oscillation (ENSO). In summary, I wrote:

The experience gained with these studies of coupled atmospheric systems, the strong foundation in basic mathematics, physics and chemistry, the availability of an on-site supercomputer, and the archive of extensive atmospheric data establish the foundation for NCAR to make major advances in the understanding of CCS.

The central role of three-dimensional models to studies of the CCS and the requirement for the fastest, most powerful computers to run these models were recognized from the start. As stated in the 20 November 1985 draft CCS plan:

> Because of the inherently three-dimensional nature of atmospheric forcing at the lower boundary and the strong zonal, meridional and vertical variations of the atmosphere, the coupled models must be three-dimensional and global in coverage. Research versions of the CCM already exist that include mutual interactions between the atmosphere and the ocean, although these models are limited by their coarse horizontal and vertical resolution. Conceptual and numerical models of atmospheric chemistry, ice, and surface processes including vegetation already exist, but completely interacting global models incorporating all of these processes have yet to be developed. With the arrival of the CRAY X-MP 4800 at NCAR in 1986, it will be possible to increase the resolution of the existing atmospheric-ocean models and to include simple representations of atmospheric chemistry, ice, and vegetation in a truly coupled model.

NCAR's proposed CCS project would directly address the third research objective of the new USGCRP. Given this relevance, and the high priority put upon the CCS by the Dutton Committee, one might think that the CCS would quickly receive the support required to begin. However, this was not the case in the restricted budget climate of the late 1980s, and budget shortfalls in 1988 and 1989 resulted in an actual *decrease* of resources directed toward climate modeling at NCAR.

Budgetary limitations were not the only problems facing the CCS. In spite of the high priority put upon global change at the national and international level, and the good marks given CCS by the Dutton Committee, not all scientists at NCAR or in other institutions wholeheartedly supported the notion of global change/earth system science in general or the CCS in particular. The two major objections were related to *scientific readiness* and *competition*. Some NCAR scientists, for example, argued that it was scientifically premature to spend a lot of effort developing coupled models. In a memo dated 20 December 1985, two AAP scientists stated:

> The large uncertainties in our current uncoupled models imply that our highest priority should not be coupling models together. Work along these lines has been progressing at its current rate not because of a shortage of manpower or interest in the subject, but because the science itself has not warranted a more rapid development. For example, the oceanography section has been reticent to devote much effort to coupled ocean-atmosphere systems because ocean modeling is not far enough along due to insufficient computer power. Similar and more damning criticisms can be made of the other models to be coupled. In sum, uncertainties coupled to additional uncertainties likely lead to (uncertainty)**2, not the type of solid science a national center ought to be pursuing.

Competition and fear of new initiatives in a zero-sum-game budget climate were also major factors in many people's concerns about the CCS. At NCAR, there were serious proposals to reprogram existing resources into the CCS initiative, even in a declining overall NCAR budget. Given the lack of complete agreement that CCS should be NCAR's highest priority, this led to resistance on the part of some to the CCS concept, as exemplified by one NCAR Division Director's memo to the NCAR Director on 26 March 1986:

> I have just been informed that I was cited as a party to an agreement on the part of the Coupled Climate Systems Committee to add money to the NCAR/NSF base budget for two positions for CCS, at the expense of various NCAR Divisions. I must respond immediately to assert that I recall no such agreement, and to request that the 1987 target figures not include cross-divisional taxation for CCS.

The skepticism about the value of planning, competition from other global change initiatives and fears of a zero-sum-game were in fact verified when the first global change funds came to NCAR in the 1987 fiscal year. These funds (US$1.5 million) were all designated for an earlier established program, the Global Tropospheric Chemistry Program (GTCP), a top NSF Atmospheric Science Division (ATM) priority. In addition, after taking out the US$1.5 million, the total NSF budget for NCAR decreased by US$0.8 million, requiring cuts in other parts of the program, including climate research and computational facilities that were crucial to the CCS.

The budget shortfalls and reductions in program for NCAR continued in the 1988 fiscal year. Another broader problem then became evident with the USGCRP. Even though incremental funds were being identified for high priority scientific research as part of the USGCRP, no funds were identified for the observational and computational facilities required to support the research, and in fact resources allocated to the facilities divisions at NCAR were being reduced in order to support the mandated increases in other scientific efforts. Specific to the CCS project, the NCAR XMP-4800, a computer serving all disciplines of atmospheric and related sciences, had been saturated since its acquisition in 1986, severely limiting the rate of progress on climate modeling at NCAR and in the NSF-sponsored university climate modeling community. Other climate modeling centres in the United States were experiencing similar resource limitations, particularly in the availability of computational capability.

1.1.1.2   The UCAR Climate System Modeling Initiative (CSMI)
Frustrations related to the gradual decrease of support to climate modeling at NCAR and the universities and limited computational facilities led myself, Dickinson and others to explore other ways than through the NCAR budgetary and planning process to increase resources to the climate system modeling community, broadly defined. In September 1988, UCAR, NCAR, the UCAR Office of Interdisciplinary Earth Studies (OIES), and the NSF-ATM began discussing the concept of a community Climate System Modeling Initiative (CSMI). In a memorandum to NCAR Senior Scientists Warren Washington and Robert Dickinson dated 29 September 1988, I wrote:

As you know, we have been discussing the possibility of a true community initiative to make a major advance in climate modeling, as part of the global change program. One of the greatest things the NSF scientific community could do for global change (in my view) would be to create a 'Manhattan Project' to make a quantum advance in climate modeling, with emphasis on climate prediction resulting from greenhouse gas warming over the next 25, 50 and 100 years. The problem is simple to state:

1. Numerical models, with coupled oceans, atmosphere, land surface and ice processes and atmospheric chemistry are the primary tools for quantifying global climate change and making estimates of future climates based on various assumptions of human activities.
2. The present models are inadequate for the task. Their resolution is too coarse and they use physics that are in some cases 20 years old. Bob Dickinson has estimated that there are fewer real resources going into climate modeling now than 20 years ago. Climate modelers are forced, by inadequate computer resources, to do 'safe' experiments with low-resolution models. The high-risk, innovative experiments that are required to make progress on fundamental issues such as cloud-radiation feedbacks are not being done.
3. The NSF is in an ideal position to create and support this project, making use of the best climate scientists in the universities and in NCAR who have had extensive experience in climate modeling, but are relatively unorganized and computer resource limited at this time.

Key to the proposed project was a

dedicated supercomputer, of at least the CRAY 3 class. A significant fraction of this computer would be devoted to high-risk, innovative experiments, selected on the basis of peer review.

The idea at this time was that NSF, with perhaps some support from industry for the computer, would supply the incremental funds (estimated to be a maximum of US$25 million per year).

The proposed project soon became known as the Climate System Modeling Initiative (CSMI). In November 1988, I, as UCAR President, invited the Joint Oceanographic Institutions (JOI) to join UCAR in supporting the CSMI and helping obtain NSF funds for it. President of the Joint Oceanographic Institutions (JOI), D. James (Jim) Baker, was enthusiastic about the idea, and JOI participated in the early stages of the planning process.

During the fall of 1988, I advanced the CSMI concept, discussing it with the UCAR Board of Trustees in October and the UCAR University Relations Committee in November. Both supported the concept strongly, and on 15 December 1988 the Executive Committee of the UCAR Board of Trustees made the following statement:

The Executive Committee of the UCAR Board of Trustees notes with pleasure the progress made by UCAR, NCAR, and the OIES in initiating planning for a program aimed at improving climate change models. The program concept requires:

- a focused goal of developing improved global and regional climate systems models;
- involvement of scientists from a broad spectrum of disciplines, including atmospheric sciences, oceanography, biology, chemistry and other related fields;
- major participation by the universities; and
- dedicated computer capabilities.

We believe that this modeling development activity is an essential contribution to evolving global change programs, and we strongly endorse continuation of the planning efforts currently underway.

Former UCAR President and then President of the National Academy of Engineering, Robert M. White also supported the CSMI; in a letter dated 15 November 1989 White states:

I think the climate system modeling project is long overdue and is badly needed to improve the scientific basis for policy formulation and to reduce the uncertainty in our climate projections.

The next step in the planning process was the appointment of the Joint CSMI Steering Committee, which consisted of Eric Barron (Pennsylvania State University), Kirk Bryan (NOAA Geophysical Fluid Dynamics Laboratory), Lawrence Gates (Lawrence Livermore National Laboratory and Co-Chair), Berrien Moore (University of New Hampshire), V. Ramanathan (University of Chicago), Guy Brasseur (NCAR), Robert Dickinson (NCAR), Michael Ghil (University of California at Los Angeles), John Pastor (University of Minnesota) and Stephen Schneider (NCAR and Co-Chair). This committee met on 19–20 December and developed a consensus statement that strongly supported the CSMI (Appendix 1.1).

The final recommendation from the steering committee was that UCAR sponsor a community planning workshop for 25–27 January 1989, and, despite the extremely short notice, more than 100 scientists from around the country participated in the First CSMI Workshop. (Colour plate 1.1, land use categories for North America at four horizontal resolutions (480 km, 240 km, 120 km and 60 km), was one of the colour figures used in the First CSMI Workshop.)

The participants included leading scientists from the atmospheric, oceanic, ecological, hydrologic and related disciplines, and included senior leaders of the atmospheric and global change communities, such as Baker, Francis Bretherton, Jack Eddy (UCAR Office for Interdisciplinary Earth Studies), J. Michael Hall (NOAA Office of Global Programs), Tom Malone (who gave a rousing 'call to action' keynote address), Walter Orr Roberts, and Verner Suomi (University of Wisconsin). The participants at the workshop reached a strong consensus in support of the CSMI, and adopted the overall scientific goal:

to accelerate progress toward reliable predictions of global and regional climate changes in the decades ahead, with particular attention to the effects of increasing greenhouse gases.

The workshop results and conclusions were published in spring of 1989 in a report *The UCAR Climate System Modeling Initiative: Report of the First CSMI Workshop* (UCAR 1989). By this time, a much broader participation and base of support than just NSF was envisioned, including other federal agencies and industry:

> The CSMI will solicit, through UCAR's Corporate Affiliates Program, the support and participation of the private sector on the basis of their general and specific interests in the effects of regional and global climate change.

The requirement for a dedicated supercomputer was again strongly emphasized:

> Support will be sought from federal, other government, and private sources to obtain and operate a next-generation supercomputer dedicated to the CSMI.

CSMI was always intended to be more than a basic research project; it was envisioned as providing useful information to policy makers in the government and in industry and to reducing uncertainties in future climate model projections. One path aimed at this societally relevant side of climate modeling was developed with Robert Redford's Institute for Resource Management (IRM), which was developing a project called CLIMAX ('climate action'). As the name implied, the proposed CLIMAX project was more action than research oriented. According to an August 1989 document prepared jointly between UCAR and IRM, the purpose of CLIMAX was:

> ... to assemble and focus the best of the world's intellectual, scientific, technological, cultural, and political resources to help guide the implementation of a strategy to aid political, economic, and social institutions to accommodate successfully to global climate change that may prove inevitable, and possibly, to slow the rate of change.

Three key components of CLIMAX were research, public policy and education, and corporate involvement. A number of UCAR, NCAR and IRM staff participated in a conference on global change held in Sundance, Utah in 1989.

Perhaps because the communities represented by UCAR and IRM were too far apart culturally and politically, the joint aspects of CLIMAX did not develop. However, the three *components* of the proposed CLIMAX project survived in CSMI and MECCA.

Although the UCAR Corporate Affiliates Program (CAP) did not play the large role that was anticipated by the participants of the First CSMI Workshop, CAP was responsible for initiating several contacts that led to MECCA and related climate modeling activities. Chief among these was an early contact with EPRI's Ralph Perhac, and it was through this contact that EPRI was invited to participate in the Second CSMI Workshop held in April 1990. In addition UCAR Vice President Harriet Crowe, CAP Director Robert Bunting and I visited EPRI on 8 November 1989 to discuss the CSMI concept.

In October 1989, Baker and I appointed a multidisciplinary scientific advisory committee. Members included Barron, William Bendel (Digital Equipment Corporation), Bretherton, Moustafa Chahine (Jet Propulsion Laboratory), Stanley Changnon (Illinois Water Survey),

Ralph Cicerone (NCAR), Lawrence Gates, Michael Glantz (NCAR), Klaus Hasselmann (Max Planck Institut für Meteorologie), William Jenkins (Woods Hole Institute for Oceanography), Margaret LeMone (NCAR), Syukuro Manabe (NOAA GFDL), Gordon McBean (University of British Columbia), James McWilliams (NCAR), Ramanathan, Paola Rizzoli (MIT), David Schimel (Colorado State University), Steven Schneider, Albert Semtner, Jr. (Naval Postgraduate School), Herman Shugart (University of Virginia), Joanne Simpson (NASA Goddard), and Susan Soloman (NOAA, Aeronomy Laboratory). Ghil and Brasseur were designated co-Principal Investigators of the effort to write a proposal for a Phase 7 CSMI effort. UCAR Trustee Richard Somerville (Scripps Institution of Oceanography) was appointed UCAR Trustee Liaison.

Further community support for the CSMI came from NSF's Advisory Committee on Atmospheric Sciences (ACAS), which at their meeting of 23–24 October 1989 stated:

> ACAS supports the UCAR Climate System Modeling Initiative and its goal to accelerate progress toward reliable predictions of global and regional climate changes in the years ahead.

However, not all organizations supported the CSMI. JOI, which had given CSMI its initial support, withdrew in a letter dated 15 January 1990. Concern about too much of CSMI being centered at NCAR was cited as a major factor in their decision, as the JOI Board of Governors preferred a more distributed national research effort. JOI also expressed concerns that the Scientific Advisory Committee was too heavily weighted toward atmospheric sciences. Finally, competition for limited federal funds was a factor in JOI's decision.

There was also some perceived competition for the CSMI from related global change programs being developed in the NOAA and the Department of Energy (DOE). In 1989, NOAA was developing a 'Centers Initiative' for climate analysis, modeling and prediction. The Centers Initiative did not overlap directly with the CSMI because it focused on the seasonal to interannual time scales rather than the decadal time scales that were central to the CSMI. Nevertheless, the two initiatives were viewed as competitors for incremental global change funds.

In early 1990, DOE was developing two major programs, ARM (Atmospheric Radiation Measurements) and CHAMMP (Computer Hardware, Advanced Mathematics and Model Physics). Although DOE participated in the planning of CSMI and provided some support, ARM and CHAMMP were higher DOE priorities. (Both ARM and CHAMMP became highly successful programs involving many scientists in the university community, and made major scientific contributions to the USGCRP.)

1.1.1.3 *Collaboration with Japan*

Collaboration with Japan became a very important part of MECCA and related climate modeling activities. In 1989, NCAR Senior Scientist Akira Kasahara was communicating with University of Tokyo Professor Taroh Matsuno, who was interested in developing in Japan a climate modeling program similar to CSMI. In a letter dated 17 February 1989, Matsuno wrote:

I agree with you that it is now the most important mission of the world's climate science community to answer questions how the earth's climate will change in response to the future loading of greenhouse gases, and for this purpose we need to concentrate our efforts to grading up of computer models of the climate system.

On 23 January 1989, the *Daily Yomiuri* reported:

In the United States, the University Corporation for Atmospheric Research has been heading a project to establish a large-scale center for simulating climatic changes.

 The phenomenon of the earth becoming warmer will not be solved unless many countries put their brains and resources together. It is hoped that this second 'Manhattan Project' for the sake of peace will prove a success.

During this time period, EPRI and the Central Research Institute of the Electric Power Industry (CRIEPI) in Japan had been discussing increased collaboration in several scientific areas. In September 1989, Charles (Chuck) Hakkarinen and other EPRI staff visited CRIEPI to discuss enhanced collaboration in climate research. A year later, EPRI Vice President George Hidy invited his counterpart, Hiromasa Amano, to join MECCA, and that invitation was quickly accepted by CRIEPI.

Kasahara's contacts in Japan and the joining of CRIEPI with the MECCA Consortium eventually led to an offer by Fujitsu Ltd to donate a VP-2600 supercomputer to MECCA. Fujitsu made the offer after Amano approached them for computer support. However, political pressures from the US Congress and Cray Research caused Fujitsu to withdraw the offer in late 1991 (*Supercomputing Review*, January 1992).

1.1.2 1990: The UCAR Climate System Modeling Program (née CSMI) is developed and the MECCA concept is formed

The year 1990 turned out to be the defining year for the Climate System Modeling Program (the CSMI was renamed CSMP early in 1990) and for MECCA.

In 1989, EPRI President Richard Balzhiser promoted the use of supercomputers to explore and visualize solutions to environmental problems. Balzhiser encouraged the redirection of EPRI resources into climate change issues. In response, on 24–26 January 1990, EPRI conducted a workshop in Boulder, Colorado to explore ways of using supercomputing capability already in place or rapidly emerging to contribute in a meaningful way to the more than $1 billion per year US Government global change research effort and the substantial research efforts in Europe, Russia, China, Japan, Australia, India and elsewhere (Electric Power Research Institute 1990). From this workshop, it became clear that scientists active in climate model development, testing and evaluation had such limited access to supercomputing capability that completion of some numerical experiments would literally take years. A dedicated numerical laboratory would enable simulations to be completed within weeks. In April 1990, I met with Hakkarinen and proposed a cost-effective option of locating a numerical laboratory dedicated to climate modeling within the Scientific

Computing Division (SCD) of NCAR. This laboratory would take advantage of the computing infrastructure already in place, the scientific data sets, and the climate models already running at NCAR. Hakkarinen supported this proposal with enthusiasm and began working closely with UCAR and NCAR to make this dream come true.

During the early part of 1990, Ghil and Brasseur developed a proposal for CSMP. Phase 1 was to be aimed at developing the specific research and engineering plans for Phase 2, including an assessment of the present state of the art of climate system modeling and publication of that assessment as a monograph. This assessment was proposed to be used as a text for a summer school for about forty graduate students in 1991. On 2 March 1990, UCAR submitted a proposal to EPRI requesting US$200 000 to support Phase 1.

In spring of 1990, NSF-ATM, under the leadership of Jay Fein, was developing its own climate modeling initiative, originally called the Integrated Climate Modeling and Analysis Program (ICMAP). Fein had been working closely with UCAR on the CSMI/CSMP from the start. Fein viewed ICMAP, which eventually became the Climate Modeling and Analysis Program (CMAP) and a part of the USGCRP, as the appropriate NSF response to the CSMP community initiative.

A second CSMP workshop, followed by a meeting of the CSMP Scientific Advisory Committee (SAC), was held on the 11–12 April 1990 under the sponsorship of NSF, NOAA, NASA and DOE. Representation was good, with participants from the sponsoring agencies, universities, national laboratories and industry. International representatives from the United Kingdom, Japan, the (then) Soviet Union, Australia and Canada also participated. The workshop reached consensus on seven important areas:

1. A major organized, coordinated national effort to advance climate modeling is needed urgently.
2. We are scientifically and technologically ready to pursue this accelerated effort in Climate Systems Modeling.
3. At least two organized efforts are needed, one for the monthly to interannual time scales, one for the decadal to 100 year time scale. These two must be developed in a coordinated and harmonious fashion.
4. UCAR is the appropriate mechanism to initiate and facilitate the organization of CSMP.
5. CSMP should concentrate on the longer time scales, but should support and be coordinated with the shorter time scale effort.
6. NCAR, the universities and other research institutions should be major contributors to the science and development of models for the CSMP.
7. We should begin CSMP as soon as possible rather than have a separate two year Phase 1 effort.

The CSMP SAC then met and agreed to prepare a short prospectus describing the goals and implementation of the CSMP, for forwarding to the seven CES agencies.

A search committee was formed to identify a pool of world-class scientists for scientific director of the CSMP Project Office. A solicitation from the community for nominations for the scientific director of CSMP began immediately. After considering several nominations and in close consultation with the community, in early May 1990, I approached Francis Bretherton at the University of Wisconsin-Madison to become the director of CSMP. Bretherton accepted this position on a half-time basis. I then appointed David Schimel (Colorado State University) to be the full-time CSMP Project Scientist, working with Bretherton from Boulder. With the appointment of Bretherton and Schimel, Ghil and Brasseur stepped down as co-Principle Investigators (PIs) and became co-chairs of the CSMP SAC. With the promise of first year funding of the CSMP Project Office from NSF and the immediate transfer of US$100 000 to UCAR from NSF, CSMP was off and running.

During this very busy period, on 17 May 1990, Serafin, Ginger Caldwell (NCAR SCD), Warren Washington (NCAR CGD Director) and Robert Street (Stanford University and Chairman of the UCAR Board of Trustees) and I visited EPRI to see about possible EPRI support for the CSMP monograph and to continue operating the XMP-48 and dedicating it to CSMP when the new computer, a Cray Y-MP8/864 was acquired on 1 October 1990. This was a very encouraging meeting, as EPRI agreed to provide US$160 000 to sponsor development of the monograph and an additional US$25 000 to help support the 1991 summer school. (Although these two activities were technically separate from MECCA, they did represent important components of the overall climate system modeling effort envisioned by UCAR.) EPRI also agreed to continue talking about further cooperation between EPRI and UCAR in climate system modeling, including possible support of a dedicated computer.

With support for the monograph in hand, I asked NCAR Senior Scientist Kevin Trenberth to be chief editor, design the content of the book, and recruit expert scientists to author each chapter. Trenberth accepted this giant task, and eventually produced the definitive text *Climate System Modeling*, which was published by Cambridge University Press in 1992. With the publication costs subsidized by EPRI, this affordable 788-page book, which contained 276 two-color figures and 16 full-color plates, became very popular with scientists and students around the world.

I then approached David Houghton at the University of Wisconsin-Madison, to host the 1991 summer school and Houghton readily agreed.

On 1 June 1990, UCAR submitted a CSMP Prospectus, written by Bretherton and Schimel, to NSF proposing,

> to undertake a collaborative, community-based, interdisciplinary Climate System Modeling Program to accelerate progress toward reliable predictions of global and regional climate changes. The program will be a collaborative effort among the universities, national laboratories and private industry.

This prospectus included a schematic of an ideal 'coupled climate model' (Figure 1.1).

NCAR played a major role in developing the CSMP during these months. In letters from Serafin (3 May 1990) and Washington (14 June 1990), NCAR offered to firstly,

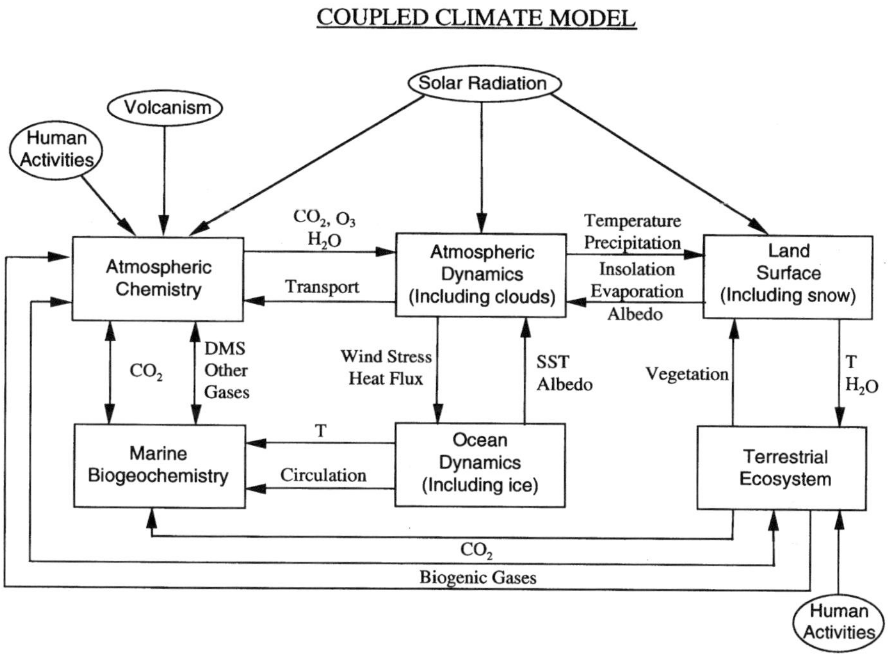

<u>COUPLED CLIMATE MODEL</u>

Figure 1.1 Schematic Diagram of Coupled Climate Model

lead the development of the next generation of climate models, building upon experience with the CCM, secondly, provide the dedicated supercomputer facilities through SCD, and thirdly, provide global data sets for model verification and other climate research studies. The Climate and Global Dynamics Division (CGD) and the Atmospheric Chemistry Division (ACD) offered to contribute in the areas of atmospheric and oceanic dynamics, atmospheric chemistry, land surface processes, marine biogeochemistry, and terrestrial ecosystems.

On 27 July 1990, UCAR submitted a proposal to EPRI for the *incremental* funding for the XMP-48 for three years, beginning 1 October 1990. The incremental funding required was US$13 million; NCAR's contributions through the infrastructure already in place and supported was estimated to be US$8.25 million over the three-year period. Discussions with Mueller and Hakkarinen of EPRI during the summer revealed that a major issue was cost; the requested US$13 million was too much for EPRI to commit to alone. Hence the concept of seeking other industrial partners was born.

Another concern of EPRI was that the proposal, although strong scientifically, did not address in sufficient detail how the computer would be used to improve our assessment capability and define the limits of our ability to predict climate change, thereby contributing to the policy arena.

Other contacts were being made in an effort to obtain private sector support for CSMP. Discussions with Burlington Northern Foundation led to a contribution to support a specific study on possible drying in the central part of the United States. On 27 August 1990, I sent a letter to colleague Jean-Yves Caneill at Electricité de France (EdF) asking if EdF was interested in becoming an industrial partner with EPRI to support the XMP-48. After further discussions between Antoine Bastin at EdF and Mueller at EPRI, EdF became a member of MECCA later in the year.

Federal agencies were also approached to be partners with EPRI in supporting the XMP-48. In a meeting with the CES agency representatives on 18 September, Bretherton and I received a positive response on the CSMP in general, but a negative response on the immediate support of the XMP. One reason given was that DOE had recently offered free time on a CRAY X-MP for climate modeling and had received a minimal response from the community. Lack of CES support for the XMP was a disappointment, not only because the funds were needed, but also because joint public and private sector support of the computer would have sent a strong message of cooperation on a project of international interest to all.

On 28 September 1990, the EPRI Environmental Division Committee passed favorably on the UCAR proposal for XMP-48 funding, clearing the last hurdle in the EPRI review and approval process. EPRI then began seeking industrial partners to help raise the US$13 million support. At about this time, representatives from the Italian Agency for New Technologies, Energy and the Environment (ENEA) indicated that they would be willing to contribute US$1.5 million to EPRI in return for becoming a partner in the three-year project. Serafin visited ENEA on 20 November 1990 and met with ENEA President Umberto Colombo, Gian Felice Clemente, Director of Environmental Programs, and Vincenzo Ferrara, Head of the climate program and delegate to the Intergovernmental Panel on Climate Change (IPCC). ENEA was particularly interested in the regional climate modeling work of NCAR scientist Filippo Giorgi.

On 18 October 1990, EPRI convened representatives from eighteen industries to a meeting in Boulder to discuss the possibility of forming a consortium to support the XMP project. Participating in this Climate Modeling Assessment Industry Liaison meeting were representatives from General Motors, Southern Company services, Ford Motor Company, Hughes Aircraft, CRAY Research, IBM Scientific Center, American Petroleum Institute, Western Fuels Association and MITRE in addition to UCAR, NCAR and EPRI management. There was strong consensus from this group that the project was necessary and an industry subgroup was formed to write a management plan. Soon thereafter CRIEPI, in a 13 November 1990 letter to EPRI Vice President George Hidy, pledged US$1 million of support over the three-year period. One of CRIEPI's interests in the project was to develop a strong collaborative relationship between NCAR and CRIEPI scientists and, to this end, CRIEPI supported Hiromaru Hirakuchi to visit NCAR for a period of two years beginning in April 1992.

It was at the 18 October Climate Modeling Assessment Industry Liaison meeting that Hakkarinen coined the name MECCA for the developing project and the consortium of sponsors.

In the emerging CSMP vision, it was recognized that a 'core group' of modelers was needed to provide, at a minimum, centralized leadership and support for developing the community models. Although NCAR was a logical place for this core group in many people's minds, there was a significant number of others who thought that the location and institution hosting the core group should be open to competition, and CSMP remained neutral on this issue. However, NCAR did express a strong interest in providing leadership within the CSMP context. Stephen Schneider developed a major proposal for a 'climate system modeling core' (CSMC) group to support the UCAR CSMP. In a draft plan, dated 25 October 1990, Schneider proposed:

> The CSMC will attempt to concentrate the existing skills and resources of NCAR in computing, climate modeling, and related scientific activities into a focused core group that would support the Climate System Modeling Program. In addition, it will be proposed to augment the traditional atmospheric science staff with the addition of several researchers from the related fields of ecology, glaciology, hydrology, etc. needed to round out a critical mass of talent in climate systems modeling at NCAR. It is our belief that a strong interdisciplinary center that can focus the talents and skills of the university/NCAR community or 'UCAR community', in short, is essential to the national goal of accelerating progress in forecasting the climatic effects of increase in trace gases on regional climates in the decades ahead.

In the somewhat controversial proposal, Schneider argued that the CSMC should be involved in the policy as well as the scientific side of CSMP:

> In addition, state-of-the-art scientific assessments needed by the policy community should be issued on a regular schedule, presumably after review by external groups such as the Committee on Global Change from the National Academy of Sciences.

1.1.3 Community concerns with CSMP

During the summer and fall of 1990, Bretherton and Schimel led the effort to produce an overall science strategy for CSMP. This draft was circulated widely to the community and received excellent reviews on the science. Most concerns centered around funding and management issues, and even whether an organized national effort was really needed. Concerns included competition with other programs, whether NCAR was the right institution to take the leadership role, whether CSMP was going to draw away resources from other climate modeling groups in the United States, whether CSMP was paying enough attention to what other groups were doing—many of the age-old issues of 'big' versus 'little' science. These concerns, including the competition issue, received national attention when *Science* Editor Richard Kerr summarized them in a Research News article titled 'Climatologists Debate How to Model the World' (*Science*, 23 November 1990). From these concerns came the philosophical approach that CSMP should have 'minimum central control but maximum achievable coordination', a phrase contributed by Bretherton.

Concerns in the community about NCAR's role presented NCAR with a major dilemma. On the one hand, there were many who believed that, based on the large scientific group at NCAR working on climate modeling, NCAR's long record of support to the climate modeling community through the Community Climate Model, and the excellent supercomputing facilities provided by SCD, NCAR should take a strong leadership role in the CSMP. On the other hand, there were those who felt that CSMP should be a widely distributed program and were concerned that NCAR should not grow at the expense of the community and dominate CSMP. NCAR recognized explicitly this dilemma; in a letter to Bretherton dated 9 November 1990, NCAR Director Serafin offered NCAR's strong support of CSMP and outlined how NCAR could contribute, whilst recognizing the need for balance:

> It is essential that NCAR not overwhelm the program or monopolize all the resources. Therefore, at any level of funding, I strongly support a broad community involvement as decided upon by the CSMP Directorate.

The political issues raised by the community were explicitly addressed by Bretherton in a briefing to the Committee on Earth and Environmental Sciences (CEES) on 10 January 1991. In this briefing, Bretherton listed four key scientific issues impeding progress in developing improved climate models:

1. the treatment of clouds;
2. the role of terrestrial ecosystems in the carbon and hydrologic cycles;
3. unsatisfactory coupling of ocean and atmosphere GCMs; and
4. inadequate computer power for high resolution models to predict regional climate.

Bretherton also addressed the central issues being raised by the community, namely:

1. Is CSMP really necessary?
2. Is the science ready?
3. Is CSMP unique?
4. Will CSMP detract from existing efforts?
5. Location of the CSMP core.

In Bretherton's view, the answers to issues one and two were a definite 'yes'. Issue three was addressed by the proposed CSMP being agency neutral, involvement of national labs and the broad academic community, and admitting some constructive overlap with CHAMMP and the NOAA Centers initiative. Issue four, perhaps the most sensitive one politically, was addressed by the concept of a distributed research program with an active core, moving tasks to people rather than people to tasks, a science team type of management, and funding coordinated by CEES. The final sensitive issue, location of the CSMP core, was to be resolved by minimizing the size of the core and having an open competition for its institutional location.

On 5 February 1991, Bretherton and Schimel sent a letter to approximately 200 members of the community asking for their input on the development of a core computing and modeling group to support the distributed science effort of CSMP. They envisioned the core group as a small group of scientists and a team of programmers whose task would be to facilitate the integration of the submodels and parameterizations being developed by the CSMP projects and other groups into the CSMP climate system model or models and to support the use of the model or models for validation, sensitivity analysis and projections. The letter asked for expressions of interest in hosting the core group or for suggestions of other institutions that would be suitable candidates. In response, CSMP received several expressions of interest.

However, perhaps because the implementation of a major national CSMP effort seemed to be imminent, some of the old concerns about competition, centralized versus distributed effort etc., were raised again, and agency support for CSMP as an organized, national program began to fade. With DOE, its highest priorities continuing to be the development of better understanding of cloud radiation interactions and numerical and computational techniques throughout ARM and CHAMMP; with NASA supporting the existing paradigm for a national climate modeling effort of coordination of existing modeling activities among agencies; and with NOAA focusing on its interannual prediction centers; CSMP did not have the strong support of three of the major CEES agencies involved in climate modeling. Furthermore, without such strong support, CSMP could not become the centerpiece of the USGCRP's multi-agency national climate system modeling effort. Furthermore, Congressional interest in the USGCRP appeared to be waning, and the optimistic budget futures of recent years were being replaced by more pessimistic outlooks, creating a conservative view in the agencies toward new programs.

1.1.4 CSMP, the NCAR Climate System Model and the Climate Simulation Laboratory

Because of these national factors, at the 18–19 October 1991 meeting of the CSMP Scientific Advisory Committee, the SAC and CSMP management agreed that CSMP should begin its activities as an NSF-sponsored collaborative project between university and NCAR scientists, in contrast to the earlier vision of a full CEES-sponsored national project. It would remain open to and welcome collaboration with other agencies of the CEES, but would not wait for nor depend upon other agency support. CSMP planned to continue its focused workshops, the postdoctoral fellowship program and, in addition, begin fostering research collaborations in three focused areas: (1) interdecadal variability and predictability; (2) regional systems (atmosphere/soils/biota/chemistry); and (3) the science of assessment.

In spite of the reticence of some of the principal agencies interested in climate system modeling to fully embrace CSMP, CSMP carried out several valuable efforts in 1991, including establishment of a postdoctoral visiting scientist program and hosting, or co-hosting several workshops. The first of these workshops was on reanalysis, jointly sponsored by CSMP and the National Meteorological Center (NMC) on 25–26 April. This workshop led to a reanalysis project that produced forty years of global analyses—a data set of great importance in climate studies and model verification (Kalnay et al., 1996). On 20–21 May, CSMP sponsored a workshop titled Natural Variability, Model Validation and Climate Change Detection, in Boulder. CSMP and DOE's CHAMMP program jointly sponsored a workshop titled Datasystems for Parallel Climate Models on 15–16 July at Argonne National Laboratory. The final workshops was one on Global Climate Change and Regional Processes held 21–23 October in Pingree Park, Colorado.

In addition to the CSMP workshops a two-week summer school on climate system science was held on the 14–27 July 1991 at the University of Wisconsin-Madison, chaired by David Houghton. Supported by EPRI and NSF, this summer school was a joint CSMP–MECCA project and involved approximately forty students from around the world. The primary source materials for the summer school were the first drafts of the climate system monograph edited by Trenberth.

A highly successful follow-on to the July 1991 summer school on climate system modeling was held at Macquarie University in Sydney on the 8–19 February 1993 (Giambelluca and Henderson-Sellers 1996). Developed under the leadership of Ann Henderson-Sellers, the theme of this summer school was 'Coupled Climate System Modeling: A Southern Hemisphere Perspective'. Sixteen nations were represented at the summer school: Australia, Brazil, Canada, China, France, Germany, India, Kenya, Malawi, The Netherlands, New Zealand, Tonga, United Kingdom, United States of America, Vietnam and Zimbabwe. Thirty-five world-class lecturers and teachers and fifty-seven graduate students participated, representing a wide variety of disciplines including meteorology, oceanography, geography, zoology, physics, geology, engineering, mathematics and resource management. Trenberth's newly finished text *Climate System Modeling* was provided to all participants and was the basis for many of the lectures. Although not officially part of CSMP or MECCA, this workshop involved many leaders and participants in these programs and directly supported the objectives of CSMP and MECCA.

Following the CSMP Science Advisory Committee meeting in October 1991, CSMP and NCAR began working closely with the NSF CMAP program to realize the overall goals of CSMP. A very important development beginning about this time was the formation of a group of NCAR scientists who began to develop the concept of an NCAR-led effort to model the climate system. On 18 November 1991, NCAR, led by CGD, held a workshop to develop NCAR's goals and priorities for responding to the goals of CMAP. This workshop led to the concept of CSM and CSL at NCAR.

In February 1992, Serafin appointed the Climate System Model Advisory Committee (CSMAC), comprised of NCAR scientists, to advise him on how NCAR should proceed with the development of the CSM. Under the leadership of Jim McWilliams, CSMAC was organized around six groups:

1. Ocean–Atmosphere–Sea Ice (Peter Gent and Joe Tribbia)
2. Cloud Systems (Jeff Kiehl and Mitch Moncrieff)
3. Chemical Cycles (David Erickson, Sasha Madronich, Phil Rasch and Dave Schimel)
4. Planetary Boundary Layers (McWilliams and Chin-Hoh Moeng)
5. Terrestrial Surface Processes (Moncrieff, Schimel, Starley Thompson and Dave Williamson)
6. Upper Atmosphere (Byron Boville, Brasseur and Ray Roble).

The CSM effort became one of the three components of the NSF CMAP program, which was announced as a new program, part of the USGCRP, in the 1992 fiscal year with a budget of US$2 million. The other two components of CMAP were university projects, led by individual PIs, and CSMP. As CMAP and the CSM developed, the CSMP Scientific Advisory Committee assumed a broader role. After discussions with Fein, Schimel and Serafin, in a memo dated 22 October 1993, I defined the terms of reference for the new CMAP Scientific Advisory Committee which were to:

> provide broad scientific advice to NSF-ATM, UCAR, and NCAR Management regarding CMAP programs, including the UCAR Climate System Modeling Program (CSMP) and the NCAR Climate System Model (CSM) effort. The CMAP SAC will also provide advice on other general CMAP issues upon request by NSF-ATM.

With the core model of the climate system being developed by NCAR, with collaboration by the university community, the CSMP role became one of outreach, promoting collaboration between modelers and data analysts in different disciplines. In 1992, CSMP made five postdoctoral appointments, (NCAR, GFDL, University of Washington, University of Ohio and Colorado State), collaborated with MECCA on analyses of climate change and ecosystem impacts, and sponsored interdisciplinary workshops on *The Science of Assessment* at the Pennsylvania State University and the *Role of Ecological Modeling in Earth System Modeling* in Boulder. CSMP also joined with NASA and the Environmental Protection Authority (EPA) in sponsoring a workshop on terrestrial-atmosphere interactions in Columbia, Maryland (June 1992) and the National Research Council on a workshop in interdecadal variability of the climate system.

Another important development in late 1993 was discussions between UCAR, NCAR and NSF about developing a new supercomputing facility at NCAR to serve the needs of the USGCRP, particularly the development of Earth system models. In a letter dated 1 November to Corell, Serafin used the MECCA experience as justification for the proposed Climate Simulation Laboratory:

> For the past two years NCAR has operated, with support from industry, a numerical laboratory for the Model Evaluation Consortium for Climate Assessment (MECCA). A dedicated supercomputer, for carrying out a small number of climate simulations requiring large computer allocations, has been the core of the numerical laboratory. By dedicating a supercomputer to a small number of simulations, progress was greatly accelerated. Although MECCA was funded by an International Consortium of non-profit organizations, computer time was allocated to scientists around the world, based on scientific merit via a peer-reviewed proposal process. The success of MECCA, the importance of global and regional climate change, and NCAR's links to the international environmental research community have led NCAR and UCAR to consider establishing a leading edge computer simulation facility at NCAR that would support US climate system research, as well as an international network of researchers in other countries.

In a document titled 'UCAR/NCAR's Plan for a Climate Simulation Laboratory' dated 7 January 1994, Bill Buzbee, Wayne Shiver and Warren Washington wrote, in words strongly reminiscent of the First CSMI Workshop report in 1989:

> Building on its international collaborations in atmospheric and oceanic sciences and its initiatives in supercomputing, the University Corporation for Atmospheric Research (UCAR) plans to establish a Climate Simulation Laboratory (CSL) at the National Center for Atmospheric Research (NCAR). The context for this will be cooperation with other organizations as partners in a research initiative of mutual interest—global and regional climate change. There will be strong university, industry and government participation in the program.

In this 7 January 1994 plan, MECCA was again cited as a paradigm for the success of the CSL:

> MECCA is an example of how an international project with strong interactive industry, academic and government participation can make a significant contribution to an issue of global importance. The diversity of both opinions and concerns among its membership has been a strength of the project and has contributed to the vitality of the scientific discussion around the early research results.
>
> The success of MECCA leads to optimism that a larger scale climate research program could succeed.

In January 1994, Serafin appointed Byron Boville and Bill Holland co-chairmen of the NCAR CSM project. Under their leadership, a group of NCAR scientists (the Climate System Model Investigator Group (CSMIG), developed a comprehensive plan titled the *NCAR Climate System Plan*. At their meeting of 20–21 May, the CMAP SAC reviewed the 25 April version of this plan and gave it very high marks:

> The SAC notes with great satisfaction the strong, coherent plan for model development put forward by CSMIG and its co-chairs B. Boville and W. Holland, as well as the impressive progress made in five months of the program's two-year start-up phase.

In April 1994, Serafin and Buzbee presented the concept of the CSL to the Integrated Modeling Analysis and Prediction (IMAP) Subcommittee of the USGCRP and received a favorable response.

This response represented the convergence of two independent activities that were underway in 1993 and 1994—the planning activities for a Climate Simulation Laboratory as described above and a multiagency activity to enhance research in climate modelling and prediction as part of the USGCRP. In January 1994 the President's request to Congress included a US$12 million increment for enhanced USGCRP funding for climate modelling and prediction for the 1995 fiscal year. The Office of Management and Budget (OMB) put this request (along with US$43 million of other USGCRP increments) in the NSF budget, with the understanding that NSF would work with the other agencies in deciding how to best use these funds. With respect to the climate modeling increment, the agency representatives to the IMAP decided to use the full amount to augment the computing resources available to the broad climate modeling community. Because of the high reputation of the Scientific Computing Division at NCAR and the NCAR scientific staff, IMAP decided to place the interagency computing facility at NCAR, a decision coincident with and supportive of the CSL planning.

During the rest of 1994, NCAR continued to make significant progress toward developing a family of prototype coupled models of the climate system, based on the second generation version of the Community Climate Model (CCM2). Key to the developing CSM was the flux coupler, a code that serves as an interface among the atmosphere, ocean/sea ice, land/biology/hydrology, atmospheric chemistry, and upper atmosphere components of the CSM.

At the 2–3 March 1995 meeting of the CMAP SAC, the SAC reviewed the CSMP activities over the past year, focusing on the Interdecadal Variability and Regional-Global Interactions projects. The SAC proposed that both projects lead to major community workshops, one on land surface-atmosphere interactions and another on interdecadal-to-decadal variability in instrumental and proxy records. The SAC also received a report from Fein on the generally favorable peer reviews of the CSM plan and on progress on the CSM.

Throughout the latter part of 1994 and early 1995, Buzbee and others at UCAR and NCAR worked closely with Fein, Cliff Jacobs and ATM Director Dick Greenfield of NSF to develop the detailed plans and justification for the CSL In spring of 1995, the National Science Board approved the funding of NCAR's Climate Simulation Laboratory and on 17 March 1995, NSF transferred US$7.5 million to NCAR for support of the CSL. Thus, just as CSMP played a major role in establishing MECCA, the success of MECCA in turn was a major factor in establishing the CSL at NCAR.

1.2   Establishment of the MECCA project, 1991

Meanwhile, more or less independently of CSMP activities, in late 1990 and early 1991, MECCA was making rapid progress. In December 1990, EPRI obtained authorization from its Board of Directors and established the MECCA consortium of sponsors and the MECCA Policy Committee, which consisted of Peter Mueller (Chairman), Hiromaso Amano (CRIEPI), Jacques Delcambre (EdF), Vincenzo Ferrara (ENEA), Clifford Jacobs (NSF), Robert Serafin (NCAR), and Wayne Shiver (UCAR). EPRI began negotiations with UCAR for the lease of a dedicated supercomputer at NCAR. Although the original plan had been to retain the XMP-48, other options involving CRAY Research and IBM were now being considered.

In January 1991, a provisional MECCA technical committee was formed, which consisted of Chuck Hakkarinen, Jean-Yves Caneill (EdF), Ulrich Cubasch (Max-Planck), Larry Gates (LLNL), Gerald Meehl (NCAR), Shaw Nishinomiya (CRIEPI), Dave Schimel (UCAR), Maurizio Sciortino (ENEA), and Albert Semtner, Jnr (Naval Postgraduate School). The technical committee began scoping out a plan for the selection of the numerical simulations and the analysis plan. Gates led in the development of the analysis plan, which included the concept of an analysis team which would serve all of the experimenters in providing common comparisons of their climate simulations. In March 1991, MECCA published the plan and issued a solicitation for proposals to the community. At this time, the MECCA Technical Committee was officially established.

In May 1991, the MECCA Policy Committee approved the first set of proposed simulations to be conducted. Immediately after the MECCA computer, Castle, was dedicated on 16 July 1991, the first simulations were underway. These included:

- Sensitivity of CCM2 to resolution (Williamson, NCAR)
- Transient response to doubled CO_2 (Washington and Meehl, NCAR)
- Sensitivity of dynamics to CO_2 (Saltzman, Yale University)
- Effects of Amazonian and Southeast Asian deforestation on climate (Henderson-Sellers, Macquarie University).

Even as Castle became operational and simulations were begun, efforts to broaden the support for MECCA continued. In particular, certain Federal agencies were contacted to ascertain their interest in becoming partners (and co-sponsors) of MECCA. These invitations were received with interest, but did not generate any funding. For some of the same reasons, CSMP was having difficulty generating major support from agencies other than NSF.

As the first MECCA simulations were being conducted, it became obvious that to make the whole greater than the sum of the individual projects, it would be necessary to have an analysis team of scientists looking at all the results and synthesizing them in a way that addressed MECCA objectives. In February 1992, the MECCA Policy Committee nominated candidates for leading the analysis team. After interviewing these candidates, Ann Henderson-Sellers from the Climatic Impacts Centre at Macquarie University in Sydney, was appointed leader of the analysis team in July 1992. This team consisted of a group of 14 staff located at Macquarie University and several sponsors and individuals at other institutions. The goals of the MECCA Analysis Team (MAT) were to:

- ensure systematic examination of existing and future MECCA projects,
- synthesize MECCA experiments with results from other international projects, and
- relate model uncertainties to questions of importance to policy makers.

MAT has over the years become involved in many projects and developed a high international profile (Henderson-Sellers et al. 1995). Members of MAT have presented many scientific papers, delivered talks and lectures and represented (and hosted) MECCA at many conferences. A major achievement of MAT was the publication of the *Climate Change Atlas: Greenhouse Simulations from the Model Evaluation Consortium for Climate Assessment.*

1.3 Summary

MECCA was formed in 1991 after several years of effort on the part of UCAR, NCAR and EPRI to respond to the growing national and international concerns about possible climate change due to increasing emissions and atmospheric concentrations of greenhouse gases.

UCAR, NCAR and EPRI recognized that a major limitation blocking the rate of progress in climate system models, powerful tools in projecting possible future climates under various scenarios of interest to policy makers, was the lack of adequate supercomputing facilities. The development of MECCA and its support of the dedicated numerical laboratory at NCAR was an important step in meeting the scientific community's need for computer resources dedicated to climate modeling.

MECCA was successful in several regards. It attracted some of the world's leading climate modelers and assessment scientists who produced important advances in the science (see, for example, Kittel 1995). MECCA thus accelerated the rate of progress on climate modeling and assessment and gained international recognition. For example, MECCA was recognized in the 1992 IPCC assessment and contributed to the 1992 United Nations Conference on Environment and Development (UNCED) in Rio de Janeiro.

MECCA demonstrated that a numerical laboratory, dedicated to a few large climate modeling projects, could make possible important climate simulation studies that would have been difficult or impossible to carry out in the multi-user environment of most supercomputer centers. Thus it helped demonstrate the need for the Climate Simulation Laboratory at NCAR, a multi-agency facility dedicated to climate system modeling.

MECCA also stimulated productive, long-term interactions among the international scientific community, interdisciplinary scientists, industry, universities and national laboratories, leveraging the resources of all the participants. MECCA proved that an industrial consortium could support unbiased, peer-reviewed research and that academic scientists could work productively and harmoniously with industry.

References

Committee on Earth Sciences (1989) *Our Changing Planet: A U.S. Strategy for Global Change Research. A report by the CES to accompany the U.S. President's Fiscal Year 1990 budget*, pp. 138. OSTP, Washington, D.C.

Electric Power Research Institute (1990) *Proceedings of the Workshop on Greenhouse Gases and Climate Change—Research Needs and Industrial Participation*, Boulder, Colorado, 24–26 January.

Giambelluca, T. and Henderson-Sellers, A. (eds) (forthcoming) *Climate Change, People and Policy: Developing Southern Hemisphere Perspectives*. Based on a graduate summer school on Coupled Climate System Modeling held at Macquarie University, 8–19 February 1993. John Wiley & Sons, Ltd.

Hakkarinen, C.S. (1995) Preface, Results from the Model Evaluation Consortium for Climate Assessment (MECCA). *Global and Planetary Change*, **10**, p. 1.

Henderson-Sellers, A. and Hansen, A.-M. (1995) *Climate Change Atlas: Greenhouse Simulations from the Model Evaluation Consortium for Climate Assessment.* The Netherlands: Kluwer Academic Publishers.

Henderson-Sellers, A., Howe, W. and McGuffie, K. (1995) The MECCA analysis project. Special issue of *Global and Planetary Change*, **10**, 3–21.

Kalnay, E., Kanamitsu, M., Kistler, R., Collins, W., Dearen, D., Gandin, L., Iredell, M., Saha, S., White, G., Woollen, J., Zhu, Y., Chelliah, M., Ebisuzaki, W., Higgins, W., Janowiak, J., Mo, K.C., Ropelewski, C., Wang, J., Leetmaa, A., Reynolds, R., Jenne, R. and Joseph, D. (1996) The NCEP/NCAR 40-Year Reanalysis Project. *Bulletin of the American Meteorological Society*, **77**, 437–471.

Kittel, T. (1995) Results from the Model Evaluation Consortium for Climate Assessment (MECCA). Special issue of *Global and Planetary Change*, **10**, p. 238.

Mueller, P.K. and Hakkarinen, C.S. (1993). *Confidence in Greenhouse Gas Forced Climate Predictions: The Story of the Model Evaluation Consortium for Climate Assessment*, paper presented at Fifth New Zealand Coal Conference, Wellington, NZ, January.

NASA (1986) *Earth System Science Overview—A Program for Global Change*, 20546. Washington, D.C.: NASA.

Pielke, Jr. R.A. (1995) Usable information for policy: An appraisal of the US Global Change Research Program. *Policy Sciences*, **28**, 39–77.

Trenberth, K.E. (1992) *Climate System Modeling*. New York: Cambridge University Press.

UCAR (1987) *The Atmospheric Sciences: A Vision for 1989–1994*. Report of the NSF-UCAR Long-Range Planning Committee, Colorado: UCAR.

UCAR (1989) *The UCAR Climate System Modeling Initiative: Report of the First CSMI Workshop*, Boulder, CO, 25–27 January. Colorado: UCAR.

UCAR (1990) *Proposal for a Dedicated Supercomputer for Climate Modeling and Assessment*, submitted to the Electric Power Research Institute, 27 July.

Appendix 1.1

Climate Systems Modeling Initiative (CSMI) Joint CSMI Steering Committee Consensus Statement 20 December 1988

1. The Joint CSMI Steering Committee (JCSC), recognizing the scientific importance and timeliness, strongly supports the proposed UCAR Climate Systems Modeling Initiative (CSMI), involving the University community, NCAR and other national and international research institutions.
2. The JCSC supports the focused effort of the CSMI to reduce the uncertainties in predicting regional climate change on time scales of 10 to 50 years due to increasing greenhouse gases and other changes in climate forcing.
3. In addition to the practical aspects of the problem, which are driven by policy requirements, the JCSC recognizes the many exciting scientific opportunities in the CSMI.
4. The JCSC recognizes the essential participation in the initiative of a broad range of scientific disciplines, including atmospheric science, oceanography, biology, chemistry, and other related fields.
5. In order to achieve the focused goal of improving three-dimensional coupled climate systems models, the CSMI should involve the development and improvement of a hierarchy of models, including process models of components of the climate system and models of forcing and interactive processes.
6. The JCSC recommends that the international scientific community be involved in the CSMI through coordination with international programs and related research activities in other countries. In addition, we wish to seek co-sponsorship of the program with a few other countries.
7. The JCSC recommends that the CSMI be coordinated closely with the US effort in Global Change.
8. The JCSC recommends that proceeding with the first CSMI planning workshop, to be held 25–27 January 1989, to sustain the momentum gathered so far to preserve the possibility of starting funding in FY 1990 from the private sector. Rapid progress is required in the planning to take advantage of the window of opportunity that exists in national interest in the problem.

CHAPTER 2

EVOLUTION OF THE MECCA PROJECT

Wendy Howe and Ann Henderson-Sellers

2.1　Brief history of the MECCA project

Responding to US governmental climate change initiatives in 1989, Mr Richard Balzhiser, President of the Electric Power Research Institute (EPRI) in Palo Alto, California, instructed his staff to investigate the recommendations and to make them a reality (Hakkarinen 1994, pers. comm.).

In response to this directive a workshop, 'Greenhouse Gases and Climate Change: Research Needs and Industrial Participation', was organised by the EPRI Environment Division in January 1990 in Boulder, Colorado, to explore the resources needed to improve climate models. Scientists from the United States' Federal Government, national laboratories, academia and industry were invited. Some of the issues arising from this workshop included EPRI's visibility and leadership in climate change research, the desire to stay abreast of the current research by being involved in their own research, and the acknowledgment of the possibility of regulations designed to reduce greenhouse gas emissions being considered.

Following on from the January workshop, an Industry Liaison Meeting was held in Boulder, Colorado in October 1990 and was attended by representatives from EPRI, UCAR and relevant members of academia. The purpose of this meeting was to ascertain the current state of climate modelling and to identify current issues of concern. The importance of establishing and ultimately reducing the uncertainties associated with current climate model predictions was emphasised. The goals of the meeting were to influence policy decisions with 'hard science', to improve economic and business forecasts and to demonstrate industry concern about, and desire to participate in, major topical issues of the decade. The acronym MECCA (Model Evaluation Consortium for Climate Assessment) was introduced at this meeting (Hakkarinen 1994, pers. comm.).

In November 1990, following from this meeting, approaches were also made to other industry representatives enlisting their support in a collaborative effort focussed on climate model evaluation. (Anthes, in Chapter 1, describes the development of MECCA from the

Table 2.1 MECCA Consortium members (1995)

Member Organisation	Date of Joining	Funding and In-Kind Contributions (US$ million)
Electric Power Research Institute (EPRI), USA	1990	6.7
University Corporation for Atmospheric Research (UCAR) (with funding from the National Science Foundation)	1990	7.1
Electricité de France (EdF)	1990	0.8
Italian Agency for New Technologies, Energy and Environment (ENEA)	1990	0.4
Central Research Institute of the Electric Power Industry (CRIEPI), Japan	1990	1.6
N.V. KEMA, Netherlands	1992	0.2
Southern California Edison Company (SCE)	1992	0.2
National Supercomputing Center for Energy and the Environment at the University of Nevada at Las Vegas (UNLV)	1992	0.5
US Environmental Protection Agency (EPA)	1993	0.1
Total		17.6

point of view of its joint co-originator: UCAR.) The aim of this project was to establish an international collaborative research effort into climate change. The project, through the use of dedicated supercomputing facilities, focussed on assessing the reliability of global climate models.

2.1.1 Membership and funding

Many industry representatives, both in the United States and internationally, were approached to join the MECCA project. Uptake was not as enthusiastic as had originally been hoped for the reasons described in Chapter 1. The MECCA Consortium finally comprised the Electric Power Research Institute, the University Corporation for Atmospheric Research (with support from the National Science Foundation) of the USA, Japan's Central Research Institute of the Electric Power Industry, the Italian Agency for New Technologies, Energy and the Environment, Electricité de France (the French national electric utility), N.V. KEMA of The Netherlands, Southern California Edison (SCE) of the USA and the National Supercomputing Centre for Energy and the Environment (SCEE) at the University of Nevada, Las Vegas (Table 2.1).

Each member of the Consortium negotiated their terms of membership to accommodate their particular institutional goals, funds and constraints. Five of the sponsors provided cash through contracts negotiated and administered by EPRI without any stipulations, other sponsors contributed a mix of cash, other resources, personnel and supercomputing facilities, whilst other sponsors stipulated that a portion of the experiments cover one or more geographic domains pertinent to their world region. Even allowing for sponsors' specific requests, it was felt that the entire matrix of areas of interest of the Consortium members would complement and support the basic goals of the MECCA project. Also, members of the Consortium could participate in the decision-making processes of MECCA via several routes, including acting as members of the Policy and/or Technical Committees (Mueller & Hakkarinen 1993).

2.2   Scientific experiments and integration

2.2.1   Experiment selection process

MECCA research began in mid-1991 and the modelling experiments were carried out in two phases (Figure 2.1). Most Phase 1 experiments were completed within the first nine months or so, with Phase 2 experiments scheduled over the subsequent two years.

The call for participation in Phase 1 of the MECCA project was made in April 1991. Phase 1 experiments were supposed to be 'sensitivity experiments' and were described as follows:

> In this phase it is envisaged that simplified or low-resolution models (including uncoupled atmospheric and oceanic models and atmospheric models coupled to mixed-layer models of the upper ocean) will be used to examine the sensitivity of simulated climate to a variety of alternative model parameterisations, numerical solution techniques, and alternative boundary conditions. It is recommended that these sensitivity experiments be carried to statistical equilibrium so that definitive (although model-specific) conclusions regarding the reliability of the results (i.e. the uncertainty of the simulated climate) may be reached. It is also recommended that, whenever possible, this experimental ensemble include cases of increased atmospheric CO_2 or greenhouse gases (MECCA Experiment and Analysis Plan, August 1991, p. 8).

Most, but not all the Phase 1 experiments concentrated on sensitivity studies involving relatively short-time periods (decades). The objective of these experiments was to try to establish which assumptions are most important in controlling the differences in the outcome of a General Circulation Model (GCM) simulation. Some of the parameters of interest in Phase 1 experiments included atmospheric CO_2 concentration, heat transfer between the oceans and the atmosphere, air motion dynamics, vegetation changes and changes in parameterisation of tropical precipitation. Phase 1 also included two sets of longer climate simulations by Oglesby and Saltzman, and Schneider and Kinter.

The program announcement for proposals for Phase 2 experiments, which were to be submitted to Dr Chuck Hakkarinen, the MECCA Project Manager, by 13 October 1992, was designed to produce experiments which focussed on a smaller number of high-resolution transient-forcing experiments in which the simulated greenhouse gas concentration increases gradually over a long period of time (up to 100 years). However, the submitted proposals did not match the call, and the selected experiments in Phase 2

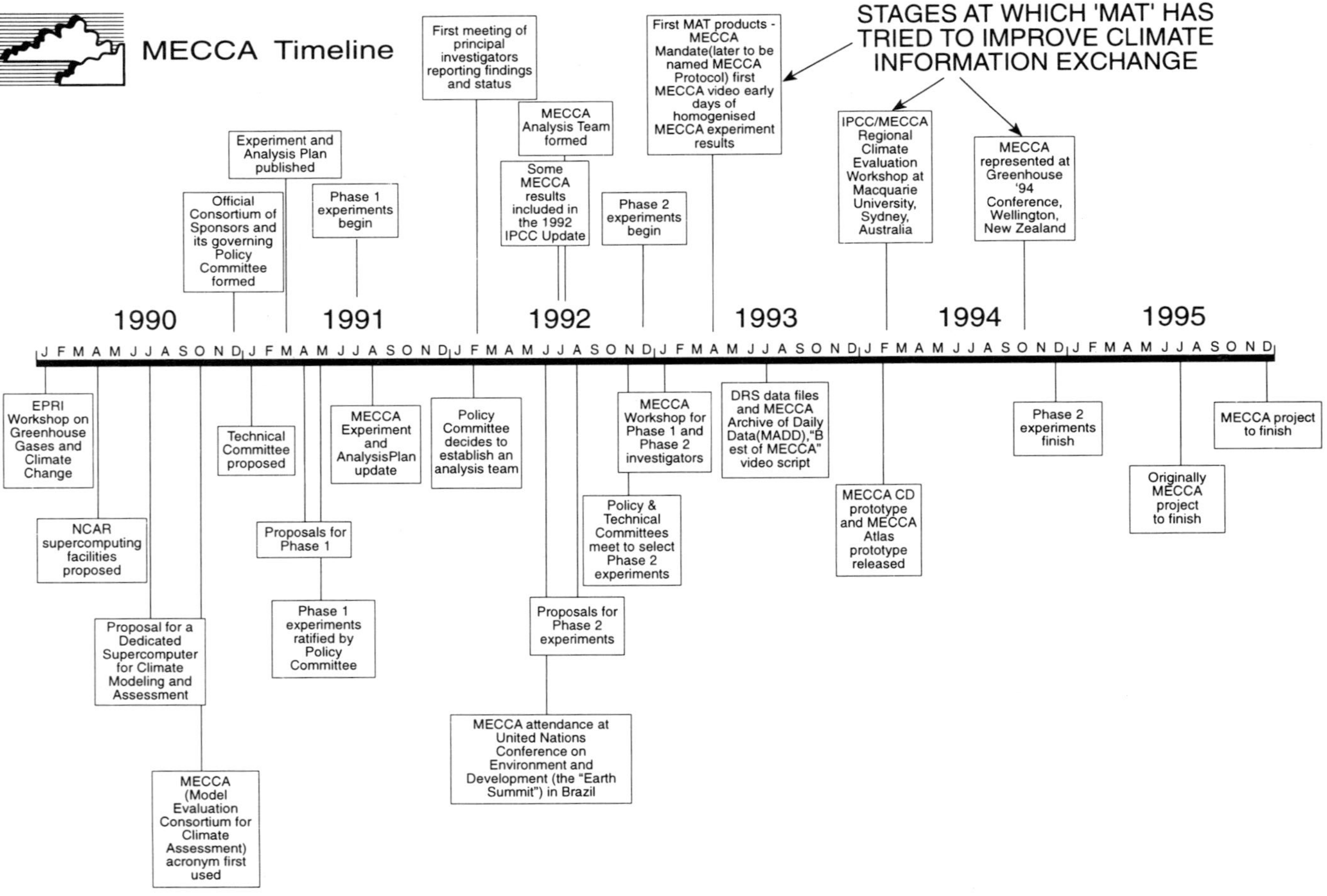

Figure 2.1   MECCA timeline

were somewhat similar to those in Phase 1 except that there were no long integrations and there was a slightly increased emphasis on analysis.

Both the Phase 1 and Phase 2 experiments underwent a review process by members of the MECCA Technical Committee and MECCA Analysis Team Advisory Committee. However, the final composition of the experimental program was decided upon by the Policy Committee. This had some very positive outcomes, particularly the inclusion in Phase 2 of the sulphate aerosol experiment by Penner, Wigley, Jaumann, Santer and Taylor.

2.2.2 Principal Investigators and their experiments

Fourteen Phase 1 experiments and fourteen Phase 2 experiments were selected for MECCA sponsorship. The Principal Investigators (PIs) and titles of the proposed experiments are listed in Table 2.2. (The proposals by Keith & Cubasch were discontinued during Phase 2.)

Table 2.2 MECCA Experiments and Principal Investigators

Phase 1

- GCM Sensitivity to Changing Atmospheric Carbon Dioxide Concentration, Dr Robert J. Oglesby, Purdue University and Dr Barry Saltzman, Yale University, USA.
- Tropical Deforestation: Local, Regional and Global Consequences, Professor Ann Henderson-Sellers, Macquarie University, Australia
- Climate Sensitivity Experiments with Improved GCMs, Dr Warren Washington and Dr Gerald Meehl, NCAR, USA
- Climate Model Sensitivity to Grid Resolution, Dr David Williamson, NCAR, USA
- Regional Climate Change Scenarios for the United States and Europe, Dr Filippo Giorgi, NCAR, USA, Dr Maria Marinucci, NCAR, USA, Dr Guido Visconti, University L'Aquila, Italy and Dr Gerardo De Canio, ENEA, Italy
- GCM Surface Sensitivity Experiments, Dr Starley Thompson, Dr David Pollard and Dr Jon Bergengren, NCAR, USA
- Climate Modelling with Emphasis on Tropical Rain, Dr T.N. Krishnamurti, Florida State University, USA
- Assessment of the Climate Effect of Other Greenhouse Gases, Dr Wei-Chyung Wang, State University of New York at Albany, USA
- Sensitivity of an Atmospheric GCM to Mass Flux Schemes, Dr Bryant McAvaney, Bureau of Meteorology Research Centre, Australia
- Sensitivity of Equilibrium Climate in a GCM to Cloud Parameterization Schemes, Dr Valentin Meleshko, Voikov Main Geophysical Observatory, Russia
- Sensitivity of Climate Simulations to Changes in Ocean Circulations, Dr William Holland, NCAR, USA, Dr Nadia Pinardi, IMGA-CNR, Italy and Dr Vincenzo Artale, ENEA, Italy
- Internal Variability in Long-Term Climate Simulations, Dr Edwin Schneider and Dr James Kinter III, University of Maryland, USA
- Sensitivity of Ocean Model Circulation to Atmospheric Forcing, Dr Robert Chervin, NCAR and Dr Bert Semtner Jnr, Naval Postgraduate School, USA
- Regional Climate Change Impacts on Northern Great Plains Ecosystems, Dr Tim Kittel, Climate System Modeling Program, NCAR, USA

Phase 2

- Model Assessment of the Enhanced Greenhouse Effect, Phase 2, Dr Wei-Chyung Wang, State University of New York at Albany, USA
- Development of Regional Climate Change Scenarios for Europe and East Asia, Dr Filippo Giorgi et al., NCAR, USA
- Greenhouse-Gas Climate Sensitivity Experiments with Improved Coupled Atmosphere and Ocean Models, Dr Warren Washington and Dr Gerald Meehl, NCAR, USA
- Development of State-of-the-Art Interactive Land Model for Greenhouse Projections, Dr Robert Dickinson, University of Arizona, USA
- Validation Processes in GCM by Ensemble Averages over Local Atmospheric States, Dr David Keith, NCAR, USA (withdrawn after approval)
- Simulated Crop Response to Climate Change in the US Great Plains using Scenarios from a Nested Regional Climate Model, Dr Linda O. Mearns, NCAR and Dr C. Rosenzweig, GISS, USA
- Assessing the Reliability of Climate Simulations: High Resolution Experiments, Dr Joseph Tribbia et al., NCAR, USA
- To Evaluate the Basic Performance of Climate Models with respect to the Transfer and Scale Interaction in the Wave-number Domain, Dr H.L. Tanaka et al., University of Tsukuba, Japan
- North Atlantic-Mediterranean Conveyer Belt Experiments, Dr William R. Holland et al., NCAR, USA
- Climate Studies of the Direct and Indirect Effects of Sulfate Aerosols, Dr Joyce Penner et al., Lawrence Livermore Research Laboratories, USA
- Applying GCM Outputs to Ecological Impact Studies, Dr Tim Kittel et al., NCAR, USA
- Interannual Variability of Arctic Sea Ice in Coupled GCMs, Dr Lawrence Mysak, Canada
- Regional Climate Evaluation Intercomparison among GCMs, Dr Ulrich Cubasch, Max Planck Institute, Germany (results unavailable)
- Applying GCM Outputs to Integrated Assessment, Dr Chuck Hakkarinen et al., EPRI, USA

2.2.3 MECCA Analysis Team

In February 1992 the MECCA Policy Committee decided that an analysis team was needed in order to achieve the MECCA goals. It was deemed to be desirable that analysis of Phase 1 and Phase 2 experiment results occurred in parallel with the modelling activities and so, in July 1992, the MECCA Analysis Team (MAT) was created by nominating as MAT leader Professor Ann Henderson-Sellers (Figure 2.2). Various members joined and left MAT as funding and research interests fluctuated (Table 2.3).

MAT was established with the primary goal of assessing the transfer of uncertainty in the scientific prediction of climatic change into policy development through impact and reduced-form models. MAT's goals were to:

1. ensure systematic examination of existing and future MECCA projects;
2. synthesise MECCA assessments with results from other international projects; and
3. relate model uncertainties to questions of importance to policymakers.

Figure 2.2 MECCA Analysis Team. Standing (L to R): Xiaohua Yuan, Craig Brescianini, Margriet Nakken, Jay Larson, Zav Kothavala, Kendal McGuffie, Olga Balachova, Christopher Gross. Sitting (L to R): Jeljer Hoekstra, Catherine Ciret, Ann Henderson-Sellers, Margaret Dudgeon, Wendy Howe. Absent: Anne-Maree Hansen, Aihong Zhong.

Table 2.3 Members of the MECCA Analysis Team

Name	Position and Activity
Ann Henderson-Sellers	Team leader: Project and research co-ordinator
Wendy Howe	Project manager: Communication and documentation
Olga Balachova	PhD student: Atmospheric branch of the hydrological cycle
Craig Brescianini	Computer scientist: MADD Archive, DRS formatting
Catherine Ciret	PhD student: Land ecology and GCMs
Margaret Dudgeon	Administrative officer: CIC
Tim Durbidge	Computer scientist: Data homogenisation
Christopher Gross	Undergraduate: MADD Archive; Data homogenisation
Ann-Maree Hansen	Undergraduate: Visualisation of GCM results
Neil Holbrook	Research scientist: ENSO
Geoff Horne	Masters student: Visualisation of GCM output
Zavareh Kothavala	PhD student: Surface hydrology fields in GCMs
Christopher Landsea	Research scientist: Regional simulation evaluation, tropical climates
Jay Larson	Research scientist: Cloud parameterisation, Strange attractors
Shuhua Li	PhD student: Tropical cyclone intensity from GCMs
Sascha Schubert	Research scientist: Downscaling
Heather Tonkin	Masters student: Extreme tropical weather events
Christian Werner	PhD student: Radiation in GCMs
Xiaohua Yuan	PhD student: Regional climate evaluation and crop impact models
Huqiang Zhang	PhD student: Tropical deforestation using GCMs
Aihong Zhong	Masters student: Cryospheric components in GCM simulations

<u>MECCA Consortium members</u>

16 August 1993–August 1994	Mr Jeljer Hoekstra, Researcher on Climate Change, N.V. KEMA, Environmental Research Department, Arnhem, The Netherlands
July 1992–December 1995	Dr Tim Scheitlin, UCAR, Colorado, USA

<u>MAT Visitors</u>

23 June–mid-August 1993	Dr Linda O. Mearns, Scientist II, National Center for Atmospheric Research, Boulder, Colorado
early August–September 1993	Dr Igor I. Mokhov, Senior Scientist, Institute of Atmospheric Physics, Russian Academy of Sciences, Moscow
22 August–1 September 1993 and 6 March–14 March 1994	Dr Robert J. Oglesby, Assistant Professor of Atmospheric Sciences, Department of Earth and Atmospheric Sciences, Purdue University, West Lafayette, Indiana
Feb 1994–June 1994	Margriet Nakken, Researcher from Wageningen Agricultural University, Wageningen, The Netherlands

<u>MAT Consultant</u>

July 1992–December 1995	Dr Kendal McGuffie, Department of Applied Physics, University of Technology, Sydney

---

Note: The funding base and approximate time contribution during the 3.5 years of MAT activities is given in Table 18.3.

An important part of MAT's work has been to interact effectively with the MECCA Phase 1 and Phase 2 PIs. This has been a challenging task. Various methods of communication were used including personal contact through workshops, email and facsimile or via the MECCA (EPRI) Project Manager. An example of an early stage of this communication exchange between MAT and the MECCA experimenters is the questionnaire circulated in August 1992.

The stated aims of MECCA are to provide a detailed quantification of the uncertainties in climate models' projections of greenhouse gas-induced climate changes. The responsibility of MAT was to provide a synthesis of Phase 1 experiments and results and the first questionnaire was designed as one means of achieving this goal.

Circulated to all Phase 1 team members on 7 August 1992 with a reply by deadline of 31 August 1992 the questionnaire was designed to:

1. obtain up-to-date information about Phase 1 projects;
2. obtain information on Phase 1 analysis;
3. gain comparable information from Phase 1 projects;
4. obtain direct assessments of uncertainty from Phase 1 projects;
5. attempt to relate Phase 1 project results to wider community studies;
6. obtain input to assessments of uncertainty related to policy;

7. identify and locate Phase 1 project results for the Analysis Team;
8. raise awareness among the Phase 1 investigators of reduced form modelling and to obtain their input; and
9. solicit suggestions of ways in which the Phase 1 projects could be synthesised.

By 31 October 1992 (two months after the original requested reply date) only eleven of the fourteen Phase 1 team members had responded. None of the responses answered all the questions fully and only four groups' answers could be classed as fairly complete. One PI was hostile and most others could best be described as uninterested in contributing to this aspect of the analysis process. The responses obtained from the questionnaire did not altogether fulfil expectations. Some details about some of the Phase 1 projects were obtained but, on the whole, the information supplied was scant, often terse and generally received only after repeated requests. Despite such rather negative aspects, the responses contained some interesting views on MECCA and on the Phase 1 experiments.

As the MECCA project was nearing completion, and in an effort to benefit from the wide range of experiences that participation in MECCA had offered, a second questionnaire was sent to all MECCA PIs. It was also felt that as Phase 2 experiments and results had not been included in any previous analysis that this questionnaire would represent a way of including and evaluating information from all of the MECCA experiments.

This second questionnaire was designed to:

1. obtain up-to-date information about Phase 1 and Phase 2 projects;
2. obtain information on Phase 1 and Phase 2 analysis processes;
3. gain comparative information from Phase 1 and Phase 2 projects;
4. obtain direct assessments of model evaluation methods from Phase 1 and Phase 2 projects;
5. raise awareness among all the MECCA investigators of their valuable contribution to the overall MECCA Project;
6. solicit the opinions of participants concerning their participation in the MECCA Project; and
7. solicit the opinions of participants concerning their participation in any future MECCA-type project.

It was interesting to note the responses to this last questionnaire. Dr Chuck Hakkarinen, the EPRI Project Manager for the MECCA Project, to whom the draft questionnaire was sent as a matter of courtesy, made some suggestions concerning the general wording of the questionnaire, which were accepted and acted upon. However, he did not view the questionnaire with enthusiasm, commenting that, in his opinion: 'Many of the investigators may be very reluctant to answer — in 25 words of less — what the results of their work were, or how closely the work conducted matched the work originally proposed'. He went on to say, 'I personally do not think it is wise to invite everybody who has been involved in the project into a public debate and retrospection now.' (The tension between MAT and the EPRI-based MECCA Project Manager is discussed in Chapter 18.)

Despite the concerns expressed, the responses received were positive and useful, although one respondent declined to answer the questionnaire, stating: 'Its been two years since we had EPRI money, and 18 months since the MECCA computer time ran out'. More positive comments included: 'MECCA accelerated our pace of science and got many computer simulations accomplished that were impossible in a batch, very-many-user environment'; 'Our MECCA Phase 1 and Phase 2 experiments produced different benefits . . . Phase 2, a very interesting intercomparison of two apparently identical GCM simulations which offers real scientific outcomes in the form of a methodology for testing the control climates of GCMs prior to their use in climate change experiments'; as well as, 'We got more computer time to do more experiments that would have taken much longer to do elsewhere; MAT was a pioneering concept, and future such efforts will use it as a model'; and, 'The interface with the impacts community via MAT was very enlightening'.

2.3 Results from the MECCA Analysis Team

The results of the MECCA project can be considered as falling into three groups:

1. input to international science;
2. individual scientific achievements of MECCA investigators; and
3. a series of 'products'.

All of these achievements, described in detail throughout this book, developed as analysis demonstration tools and for public and policy education. The MAT products include:

- the MECCA Archive of Daily Data (MADD) (Table 2.4);
- the MECCA Protocol (Appendix 2.1);
- MECCA video(s);
- the MECCA Atlases: *Climate Change Atlas: Greenhouse Simulations from the Model Evaluation Consortium for Climate Assessment*, and the 'Regional Atlas: Greenhouse Simulations of Western USA' (published on the EPRI Web site);
- the MECCA prototype compact disc containing data from Greenhouse Model Simulations (duplicated on a WWW site); and
- the MECCA book, *Assessing Climate Change: Results from the Model Evaluation Consortium for Climate Assessment.*

2.3.1 The MECCA Archive of Daily Data (MADD)

In order to be able to conduct quantitative intercomparisons of the results from the MECCA experiments there was a need for a standard set of variables generated from the GCM simulations. Because of the very different type of experiments selected for Phase 1 (and the varying types of individual archiving strategies and degrees of cooperation with the PIs), it was found that of the fourteen experiments, only six were sufficiently similar for intercomparison purposes. Of these six experiments, only eight variables were able to

Table 2.4 MECCA Archive of Daily Data (MADD)

Variables		Climate Models	
1.	Surface Temperature (K)	1.	BMRC (McAvaney)
2.	Total Precipitation (mm/day)	2.	CCM0 (Washington and Meehl)
3.	Sea-Level Pressure (hPa)	3.	CCM1 (Oglesby and Saltzman)
4.	Snow Depth (mm of H20 equivalent)	4.	CCM1-OZ (Henderson-Sellers et al.)
5.	Evaporation (W/m2)	5.	CCM1W (Wang)
6.	Lowest-Level Air Temperature (K)	6.	GENESIS (Thompson et al.)
7.	Total Cloud Amount (fraction)			
8.	Sea-ice Extent/Landmask (flag)			

Note: Phase 1 Archive consists of 8 variables from 6 climate models.

be retrieved and converted into Data Retrieval and Storage System (DRS) files (Table 2.4). The establishment and maintenance of MADD from January 1993, with limited personnel and limited assistance, has been a long and time consuming process. However, when comparing MADD with other international archives, for example the IPCC regional evaluation archive, it is seen to represent a unique and highly worthwhile achievement (Table 2.5).

2.3.2 The MECCA Protocol

In an effort to increase the understanding of both modelling communities and the decision makers of the limitations associated with the use of climate model output, MAT formulated, in May 1993, a protocol designed to assist in clarifying the respective expectations of each group. The aim of the MECCA Protocol was to encourage closer co-operation, mutual understanding and education between climate modellers and those researchers using the output of climate models for the improvement of interdisciplinary research (Figure 2.3).

The MECCA Protocol is implemented through a documented set of standard procedures. These documents include explanatory notes, a set of questions prepared for climate modellers, another set of questions prepared for impacts modellers and a detailed list of available MECCA experiment results. These are presented in detail in Appendix 2.1.

2.3.3 The MECCA video

In May 1993 MAT produced a video that was distributed to all the MECCA Phase 1 and Phase 2 PIs. The video illustrated the comparison of GCM and observed data by means of animation. Accompanying the video was a document outlining the aim of the video and the information contained in the animations. A questionnaire comprising twelve questions was also distributed requesting feedback. Ten questionnaires were returned with an average of eight responses for each question. The feedback from the participants was discussed within technical, analytical and communicative categories which arose from the overall opinion of the video.

Table 2.5 Comparison of MADD and the IPCC Archive

MECCA	IPCC
6 GCMs/modelling centres (BMRC, CCM0, CCM1, CCM1W, CCM1-OZ, GENESIS)	4 GCMs/modelling centres (DKRZ, NCAR, GFDL, HADLEY)
Phase 1: mixed layer oceans Phase 2: some model versions include 3-D oceans	3-D Oceans
Phase 1: all assume instantaneous increases in CO_2 Phase 2: some experiments include transient increases in CO_2	Some archived experiments include transient increases in CO_2
Archives: 1. Daily data 2. Monthly averages	Archives: 1. Seasonal means 2. Variance of seasonal means 3. Daily variance averaged over 10 seasons
Retains the original spatial resolution used in the GCM. R15 (48×40) R21 (64×56)	Aggregated or interpolates all models to the same spatial resolution. T21 (64×32) $2^\circ \times 2^\circ$ (180×90)
8 variables	22 variables
Consistent for all models: Surface Temperature (K) Total Precipitation (mm/day) Sea-level Pressure (hPa) Snow Depth (mm of H_2O) Evaporation (W/m^2) Lowest-level air temp. (K) Cloud (fraction) Sea-ice Extent/Landmask (flag)	Only 4 consistent across all models: Surface Temperature (K) Wind (m/s) Cloud (fraction) Soil Moisture (cm)
Data archived as DRS files	Data archived as ASCII files
Results are not across-model homogeneous	Results are not across-model homogeneous

The technical section examined responses that were relevant to methods, details and interpretations of the visual representation in this media. This included colour schemes, graphic architecture, speed of animation, layout, audio, and information content. The majority of comments concerned speed, colours and labelling. Many respondents suggested modification of these attributes. In relation to speed the suggestions were to slow the frame rate and freeze the frames occasionally to give the viewer time to absorb and understand the information before progressing. Recommendations in respect to colour included choosing more directly related colour schemes (e.g. red to indicate hot, blue for cold) and checking combinations for 'busy-ness'. It was suggested that the time indicator be modified so the transition between months

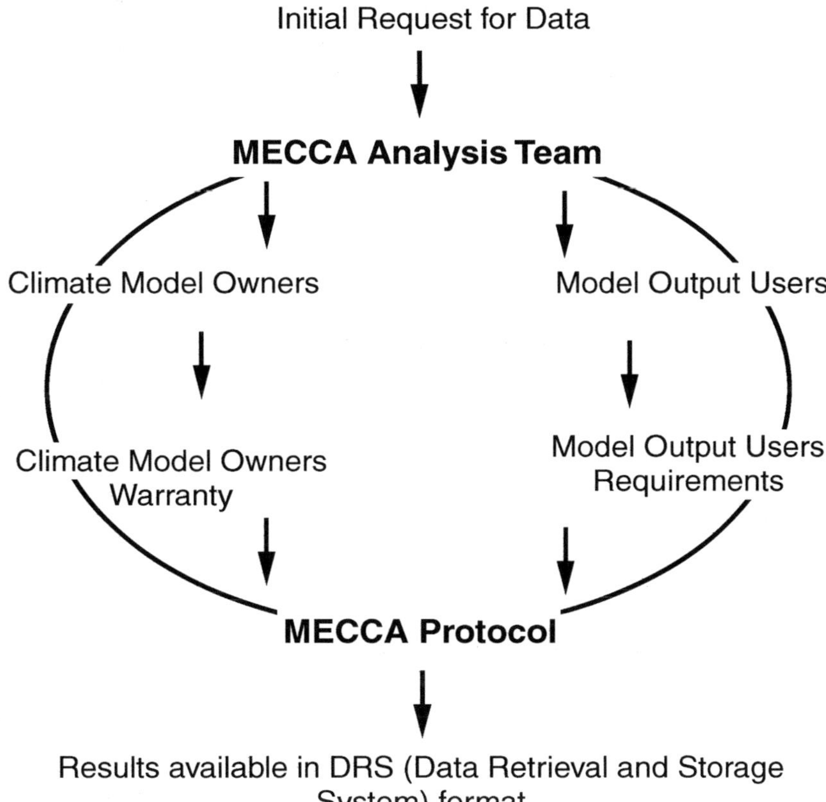

Figure 2.3 Flow diagram of the MECCA Protocol

is smoother, non-distracting and sits passively within the frame to be referred to when required. All participants agreed that in most instances an audio explanation would enhance and be useful in the video. Preferences for the layout of graphics within the frame, for example single maps or comparisons, were roughly equally divided and dependent upon viewer requirements and the message being conveyed. Technically, the video according to the participant feedback was satisfactory, although most claimed little or no expertise in this area.

Regarding the viewers' responses for analysis purposes, answers from the ten viewers in this context were highly subjective and appeared to depend upon their particular field of interest. Although the video was not distributed to encourage or highlight analysis of the particular fields displayed (outgoing longwave radiation, clouds and planetary albedo) *per se*, many responses were based on the information content of the video.

An important general response was that viewers would have liked to view other fields, as well as, or instead of, those displayed. Had this been possible (for example, by means of a Web site) then the changes in opinions from a technical and communicative aspect would be interesting. Some parties agreed that animation (usually interactive) would be the best way to analyse results (using a variety of fields). The general feeling was that video was a poor vehicle for analysis because it was very restricted in terms of interactivity; it has pre-determined speed, layout and colour and it was not real-time animation. The degree to which the video was able to convey information to the viewer was interesting. Only a few modellers assessed the usefulness of these examples in the context of their own data. Overall, as there was no direct question concerning the analytical uses of video and given that the video was not designed for such a function, it was interesting to see how many of the viewers responded from an analytical point of view.

At best the feedback was positive regarding the media and less so regarding the content of this particular product. Addressing the key question raised here could, however, aid development of effective videos for climate evaluation and assessment which will deliver the desired information efficiently to the target audience.

2.3.4 The *Climate Change Atlas: Greenhouse Simulations from the Model Evaluation Consortium for Climate Assessment*

The Climate Change Atlas: Greenhouse Simulations from the Model Evaluation Consortium for Climate Assessment is a means of visually representing, comparing and communicating the voluminous numerical results derived from a small subset of MECCA GCM simulations (Colour Plate 18.2). It was produced as a guide for scientists and policymakers on the quantification, consensus and (un)certainty derivable from these models. The displays were designed to be readily comprehensible to a broad community who are motivated by the need to provide and utilise climatic change assessments and, importantly, to understand their reliability in the development of policy initiatives.

The Atlas contains results from six Phase 1 MECCA experiments that employed a variety of enhanced greenhouse simulations. The six models are BMRC, CCM0, CCM1, CCM1-OZ, CCM1W and GENESIS. In all cases the models comprise an atmospheric GCM coupled to a simplified upper ocean sub-model. The enhanced greenhouse gas results are derived from experiments in which an instantaneous increase of trace gases was imposed and the model climate allowed to come to equilibrium.

The Atlas was not intended to cover all aspects of the MECCA experiments nor was it intended to provide a complete record of the four selected variables: surface temperature, precipitation, snow-cover and sea-ice extent. Rather, it was designed specifically to offer a means of visual comparison of selected climatic characteristics. Written by Ann Henderson-Sellers and Anne-Marie Hansen, the *Climate Change Atlas: Greenhouse Simulations from the Model Evaluation Consortium for Climate Assessment* was published in 1995 by Kluwer Academic Publishers of The Netherlands, with financial support from N.V. KEMA.

2.3.5 The MECCA prototype compact disc containing data from greenhouse model simulations

MAT worked on a number of different ways of presenting information from the MECCA experiments to the wider community. One of the avenues investigated was via a compact disc (CD) containing preliminary information on some of the MECCA Phase 1 experiments. The resulting compact disc contained approximately 500 Mb of general circulation model results, text, a colour atlas, as well as movie and static colour displays (Figure 2.4). On this CD are daily data sets of total precipitation and surface temperature from five of these GCMs for the last five years of their simulations of the present-day climate and of enhanced greenhouse conditions. The GCM results used in this CD were CCM0, CCM1, CCM1-OZ, CCM1W and GENESIS (Colour Plate 2.1)

The development of MADD facilitated the production of the CD. MADD binary data files of daily data from the MECCA experiments constructed using the Data Retrieval and Storage System (DRS) were included on the CD as one means of disseminating MECCA results. To access these files it is necessary to transfer them to a UNIX-based computer system which has the DRS libraries installed. A sample dataset, dictionary file and FORTRAN program are included in the DRS FORTRAN code folder on the CD.

Created in November 1993 and released in February 1994, the MECCA CD represented a first attempt to disseminate information about climate model simulations to a wider community and was assessed by climate modellers, impact modellers, policymakers and other interested organisations and concerned individuals. The contents have recently been made available on the EPRI web site.

2.3.6 The *Regional Atlas: Greenhouse Simulations of Western USA*

The *Regional Atlas: Greenhouse Simulations of Western USA* was the second in a series designed to display a small subset of results from MECCA climate model simulations. Created in August 1995 the *Regional Atlas* was commissioned by one of the MECCA sponsors, Southern California Edison, which has published it on the EPRI web site (http://www.epri.com/MELCA/pics/us-atlas/index.html). (EPRI is an electric industry research institute in which SCE, a regulated electric utility, is currently a member.)

On the recommendation of the MECCA Policy Committee the *Regional Atlas* focuses on the Western United States and incorporates results from two additional climate models plus observational data. The eight models are BMRC, CCM0, CCM1, CCM1-OZ, CCM1W, GENESIS, 2 versions of LLNL and MM4 (regional model). The four variables selected to represent the regional climate and climate changes under greenhouse conditions include surface temperature (that is, the land surface, ocean surface or the surface of snow or ice covers) and precipitation. Also included in the Atlas are snow-cover depth and extent, and evaporation. These two additional variables were selected because they permit insight into potential changes in water resources in the Western United States under greenhouse warming.

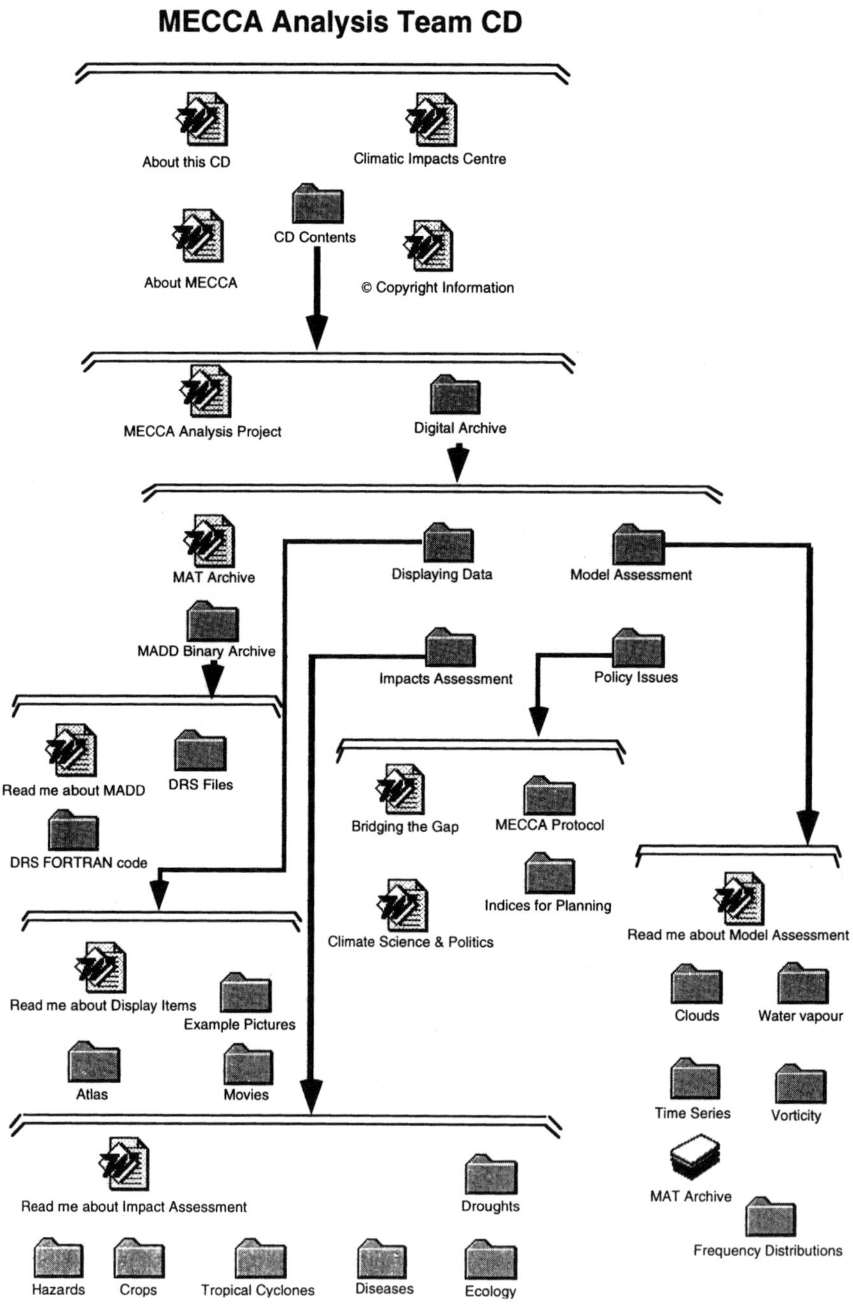

MECCA Analysis Team CD

Figure 2.4 Flow diagram of the MECCA CD

2.3.7 The MECCA book

As the MECCA project was nearing completion, and in an effort to commemorate the wide range of experiences that participation in MECCA offered, a book was proposed. The original proposal suggested contributions from all MECCA PIs but this was not considered a practical approach and a series of alternatives was debated. The design and contents of this book were approved as a 'final report' on MECCA at the Policy Committee meeting in May 1995. The response from and co-operation of those authors contributing to this book has been overwhelmingly positive: only two people declined to participate and only one author withdrew after agreeing to be a contributor.

2.4 Summary

This chapter has provided a summary of the integrative activities of the MECCA Analysis Team. We believe that all the MAT products have contributed to understanding and reducing uncertainties associated with climate change, and that, perhaps, the MECCA Protocol is the most successful and probably the most important contribution made by MECCA.

Acknowledgments

This chapter is CIC contribution number 95/37.

References

Henderson-Sellers, A. and Hansen, A-M. (1995) *Climate Change Atlas: Greenhouse Simulations from the Model Evaluation Consortium for Climate Assessment.* Dordrecht: Kluwer Academic Publishers.

Henderson-Sellers, A., Howe, W. and McGuffie, K. (1995) The MECCA Analysis Project. *Global and Planetary Change,* **10**, 3–21.

MECCA Experiment and Analysis Plan, RP3267-2 Report, August 1991, Version 3.1

Mueller, P.K. and Hakkarinen, C. (1993) Confidence in Greenhouse Gas Forced Predictions: The Story of the Model Evaluation Consortium for Climate Assessment, paper presented at the Annual Meeting of the New Zealand Coal Research Association, October.

Appendix 2.1

The MECCA Protocol

The use of climate model output for impacts assessment and decision and policy making has, in the past, proved rather difficult. There has been some apprehension on the part of climate modellers to make their computational results available to impact modellers. This is largely because climate modellers fear the incorrect use and interpretation of their results, through inadequate knowledge of climate modelling and the meaning of climate model output. Many impact modellers have had difficulty extracting the desired information from the climate model output, and have complained that the climate modellers do not generate the many essential fields required by their impact models. Policy and decision makers have also expressed dissatisfaction with the output of climate model data whilst extrapolating the information needed for their purposes. Therefore, in an effort to increase the understanding of both modelling communities and the decision making communities of the limitations concerned with the use of climate model output, the MECCA Analysis Team has formulated a policy designed to assist in clarifying the respective expectations of each group. The aim of this 'information exchange protocol' is to encourage closer cooperation, mutual understanding and education between climate modellers and those researchers using the output of climate models for the improvement of interdisciplinary research.

MECCA Analysis Team Protocol for Use of Climate Model Results

The protocol is implemented through a documented set of standard procedures. These documents include an explanatory list of procedures, a set of questions prepared for climate modellers, another set of questions prepared for impacts modellers and a detailed list of available MECCA experiment results. This set of procedures implements the MECCA Analysis Team Protocol for Use of Climate Model Results. It is not anticipated that this policy will resolve all difficulties associated with the use of climate model output, but it is meant as a sincere attempt to improve the communication necessary for such a resolution. We ask that the questionnaires in the document be answered in the spirit in which they were written, one of desire to improve cooperation in interdisciplinary research. Your completed questionnaires can be sent to Ms W. Howe, email: wendy@mqmet.cic.mq.edu.au; fax: +61 2 850 8428.

Procedure for the Processing of Requests to use MECCA Climate Model Results

The procedure in response to requests for use of climate model results will be:

1. After receiving the initial request, MAT will forward a brief summary of the climate model results currently available to the model output user.
2. If any of these results are of interest, potential users reply to MAT requesting more details about a specified model simulation. These have been previously supplied by the climate model owners in the form of a 'Climate Model Results Warranty'.
3. MAT will then forward a 'Model Output Users Requirements' questionnaire to the model output user which should be completed and returned together with the requested warranty(s) to MAT.
4. The completed Model Output Users Requirements questionnaire will then be forwarded to the relevant climate model owner group for approval to use their results for the specified study, if so desired.
5. The combined Climate Model Results Warranty and the Model Output Users Requirements documents in combination form the 'MECCA Analysis Team Protocol for Use of Climate Model Results'.
6. Only after the MAT Protocol is confirmed can the results be made available in Data Retrieval and Storage System (DRS) format at the National Center for Atmospheric Research (NCAR) either for ftp or for direct processing.
7. Participation in the MAT Protocol carries these obligations:
   (a) verification of the model description in the Climate Model Results Warranty;

(b) adherence to the use of model results output detailed in the Model Output Users Requirements questionnaire

(c) the supply of a report of the final results by the model output users to MAT.

MAT hopes that by formalising the requests for climate model results and model output requirements in this way, the MAT Protocol will form the basis for a lasting and satisfactory partnership between climate modellers and climate modeller output users in the impacts and policy communities. Please contact us again if you have any questions. Thank you, Wendy Howe, MECCA Analysis Team Project Manager.

MECCA Climate Model Results 'Warranty'

The assumption in this draft is that the model is global in domain and based on an AGCM; later modifications to the protocol will encompass regional and ocean models.

Questions to be answered by the climate modeller

  I Background
  1. Respondee's name and contacts
  2. MECCA project title and leader(s)
  3. Model name and brief (3 sentences) description

 II Model structure
  1. Spatial resolution (lat/long and wave number, if appropriate)?
  2. Vertical resolution (no. of layers and how distributed)?
  3. Time-step (all if e.g. radiation differs from dynamics)?
  4. Time interval used to store results (sampled, averaged or accumulated)?
  5. What is specified (e.g. SSTs; clouds; ozone; sea-ice) and how?
  6. Value of solar constant?
  7. Seasonal cycle activated?
  8. Diurnal cycle activated?
  9. Please supply a global map of land/ocean divisions and give the orographic heights of the land points. (Many impacts modellers would also be interested in the vegetation and soils information, if any. If this can be supplied easily we would be pleased to receive it too).

III Control (present-day) conditions
  1. Length of integration available?
  2. Time and space averaging of output?
  3. Global mean surface temperature (=288 K?)
  4. Globally averaged precipitation (=1050 mm)
  5. Total cloud amount (=55%?)
  6. Globally averaged planetary albedo (=0.30?)
  7. Emitted longwave radiation at the top-of-the-atmosphere (N.B. (II6/4.0) $\times$ (1.0–III6)=III7)

IV Evaluation (present-day)
  1. What evaluations of your models performance have been conducted?
       (i)   Please give specific references
       (ii)   Were they with this version and this integration?
       (iii) Which observations did you use? (i.e. what was the vintage source and coverage for observations?) Why did you choose these?

 V Perturbed conditions
  What exactly was the nature of the perturbation in your model run (e.g. $2\times CO_2$)?
  1. Length of integration available?
  2. Time and space averaging of output?

3. Global mean surface temperature?
4. Globally averaged precipitation?
5. Total cloud amount?
6. Globally averaged planetary albedo?
7. Emitted longwave radiation at the top-of-the-atmosphere?

VI Do you have any other comments concerning these model results, or their use, that you wish to advise us about? In particular, could you please expand your comments to encompass any perspectives you have on the effect of any errors in the control run on the models ability to 'realistically' respond to a perturbation e.g. changes in external forcings, (i.e. list any important caveats about your model). It is hoped that by being as specific as possible in the answers to this questionnaire, climate modellers will be saved the inconvenience of repeated requests for details regarding their model output. MECCA Model Output 'Users' Requirements

Questions to be read and answered by the potential climate model output 'user'

I Background
1. Respondee's name and contact address, e-mail address and fax number.
2. Your project title and leader(s).
3. Impacts model name and brief description. Please give references, if available (approx. 3 sentences).
4. What variables do you need from the climate model?

II Proposed Climate Model
1. Which climate model results do you wish to use? (see Appendix 3).
2. Why did you choose it?
3. Have you received and read this climate modelling group's answers listed in the 'Model Results Warranty'? If not, you may be unable to answer the following questions. Please ask MAT for a copy of the Model Results Warranty.
4. How long a run do you require for your impacts model use? Climatologies from observations typically use 30 years.

III Space and Time Domains:
1. Does your use of the impact model require:
    (i) Global results?
    (ii) Results for one region? If yes, where is it? (Please give latitude and longitude boundaries)
    (iii) Results for many regions? If yes, where are they? (Please give latitude and longitude boundaries)
2. What temporal resolution of climate input does your impact model require?
3. It is generally considered important to check how realistically the climate model characterises the geography appropriate for the impact model.
    (i) How many climate model grid 'points' lie in your impact model domain?
    (ii) How close are the model 'points' to important locations in your impact model?
    (iii) Is the model orography appropriate for your impact model?
    (iv) Are the model continental surface features (e.g. vegetation, soils, ocean, rivers, cities) appropriate for your impact model domain?
    (v) Do you need any additional information about the climate model you have selected? If the answers to any of these questions are 'No' or 'Not Applicable' could you please explain why.

IV Requirements
1. Does your impact model require actual GCM data, or would a climate change scenario representing conditions in some region of a GCM be sufficient? (You may be able to answer this better after reviewing question V).

V Evaluation (of present-day conditions)
1. It is generally recognised that the climate model should capture the climate of the impact model domain realistically. Such, impact-specific, evaluation is rarely undertaken by model developers themselves. This question necessitates comparison of control climate simulations with observations. Here we invite you to look at the observational data appropriate to your impact domain BEFORE accessing the climate

model output. We assume you will evaluate only the variables you require though in some cases a fuller evaluation may be preferred.

   (i)   What is the source of your observational data?
   (ii) What is the record length?
   (iii) How does it encompass your impact model domain?

To make comparisons between your selected observations and the model results try to use the same spatial and temporal domains and sampling/averaging strategies. By doing this you are able to gain some measure of confidence in the climate model results. As part of this validation process, we encourage you to compare the means, variances and extremes for the variables that you need for your impact model.

   2. Model results should be evaluated in relation to the standard errors e.g. +/-2 SE (standard errors) of observations. You should consider testing the sensitivity of your impacts model to differences etween simulated and real climate. If you have done this, are the results published and where? If not, it is strongly recommended that such sensitivity analysis be undertaken.

VI   Perturbation

   1. Is the climate model prescribed perturbation what you desire to employ in your impact model or is it a surrogate (e.g. instantaneous doubling of CO_2 instead of gradual increase in all greenhouse gases). Yes / No

   2. If a surrogate, does the climate model perturbation realistically reflect your expectations for the climate change you are interested in studying?

VII   Application-specific

Does the climate model provide characteristics that are important for your impact model use? (e.g. to assess the impact of deforestation, it is presumably necessary to incorporate vegetation)? Yes / No

VIII   Feedbacks

Would the results of your impact model provide important feed back into the climate system i.e. in future could the impacts model become an interactive part of the climate model? Yes / No
If Yes, briefly how?

Please send your completed questionnaire, and any clarifications you may require, to Ms W. Howe, MECCA Analysis Team Project Manager, email: wendy@mqmet.cic.mq.edu.au, or fax: +61 2 850 8428. It is hoped that by being as specific as possible in the answers to this questionnaire, climate model output users will be saved the inconvenience of repeated requests for details regarding the use of climate model output.

Details of currently available results from MECCA climate model simulations. Only about one third of all MECCA model experiments have addressed the question of climate change in increased greenhouse gas conditions explicitly. Of these from Phase 1: Meleshko's results are stored as monthly means in an unknown format; Giorgi's are for restricted regions and the format is non-standard; while Phase 2 experiments are current.

There remains from Phase 1 the following:
1. Henderson-Sellers et al.
2. Thompson et al.
3. Wang
4. Washington and Meehl
5. McAvaney
6. Oglesby and Saltzman

These are listed in the order in which results can be processed to be made available to MECCA climate model output users (i.e. impact modellers and policy advisers).

The following summary details are important:

Henderson-Sellers et al.
Experiment title: Tropical Deforestation: Local, Regional and Global Consequences
Model: CCM1-OZ

Model characteristics: CCM1-OZ + BATS + mlo + q flux; seasonal cycle; diurnal cycle
Resolution: R15/L12
Experiment: Instantaneous doubling of CO_2, after equilibration
Length of runs: $1 \times CO_2 = 25$ yrs; $2 \times CO_2 = 10$ yrs
Data saved: Daily and monthly
Sensitivity (global): Delta T = 2.5 K; Delta P = 5.6%
Other comments: New model version; little 'verification' available

Thompson et al.
Experiment title: GCM Surface Sensitivity Experiments
Model: GENESIS
Model characteristics: CCM1 with many mods + LSX + mlo + q flux; seasonal cycle; diurnal
    cycle
Resolution: R15/L12; surface fields at 2 deg x 2 deg
Experiment: Instantaneous doubling of CO_2, after equilibration
Length of runs: $1 \times CO_2 = 5$ yrs; $2 \times CO_2 = 5$ yrs
Data saved: Monthly only
Sensitivity (global): Delta T = 2.27 K; Delta P = 3.31%
Other comments: Short runs; higher resolution surface information; no daily data

Wang
Experiment title: Assessment of the Climate Effect of Other Greenhouse Gases
Model: CCM1W
Model characteristics: CCM1 + mlo + q flux; seasonal cycle; no diurnal cycle; bucket land
    surface (??)
Resolution: R15/L12
Experiment: Instantaneous increase of all greenhouse gases, includes equilibration
Length of runs: $1 \times CO_2 = 100$ yrs (but only first 16.5 yrs are at NCAR); $2 \times CO_2 = 100$ yrs
Data saved: Twice daily samples (NB: no diurnal cycle)
Sensitivity (global): Delta T = 3.9 K; Delta P = 6.9%
Other comments: Long runs; Most of control run not at NCAR

Washington and Meehl
Experiment title: Climate Sensitivity Experiments with Improved GCMs
Model: CCM0
Model characteristics: CCM0 + mlo; no q flux; seasonal cycle; no diurnal cycle;
    bucket land surface
Resolution: R15/L9
Experiment: Instantaneous doubling of CO_2, includes equilibration; with and without cirrus
    feedback exps.
Length of runs: $1 \times CO_2 = 30$ yrs; $2 \times CO_2 = 30$ yrs
Data saved: Twice daily samples – for with cirrus; error in archiving without cirrus feedback
    (NB: no diurnal cycle)
Sensitivity (global)(for with cirrus feedback): Delta T (DJF) = 5.12 K; Delta T (JJA) = 4.52K;
    Delta P (DJF) = 6.5%; Delta P (JJA) = 3.0%
Other comments: Model well described in literature; includes equilibration – authors use
    only last three years; without cirrus feedback had large sensitivity archived only one
    time sample from the first day of each month; systematic anomalies in climate is to be
    anticipated as no q flux correction.

McAvaney

Experiment title: Sensitivity of an Atmospheric GCM to Mass Flux Schemes

Model: BMRC

Model characteristics: BMRC + mlo + q flux; seasonal cycle; diurnal cycle; bucket land
   surface
Resolution: R21/L9
Experiment: Instantaneous doubling of CO_2, includes equilibration
Length of runs: $1 \times CO_2$ = 15 yrs; $2 \times CO_2$ = 15 yrs (but as latter includes equilibration
   probably not much is useful). Further simulation was conducted in Australia but results
   are not available to MAT.
Data saved: Daily and monthly
Sensitivity (global): Delta T (DJF) = 2.09 K; Delta T (JJA) − 2.12 K; Delta P (DJF) = 3.17%; Delta P (JJA) = 1.59%
Other comments: Higher resolution; non-CCM1 format tapes

Oglesby and Saltzman
Experiment title: GCM Sensitivity to Changing Atmospheric Carbon Dioxide Concentration
Model: CCM1
Model characteristics: CCM1 + mlo + sometimes q flux; seasonal cycle; no diurnal cycle;
   bucket surface
Resolution: R15/L12
Experiment: Multiples of CO_2
Length of runs: Not known
Data saved: Twice daily samples (NB: no diurnal cycle)
Sensitivity (global): Not known
Other comments: Results not released to MAT

Key:
BATS and LSX = both are land surface parameter schemes
mlo = mixed layer ocean
qflux = flux corrections to bring SSTs in the model into agreement with present day values
R15 = approx. 4.5 x 7.5 degrees
R21 = approx. 3 x 3 degrees
L9 = 9 levels in vertical layer
L12 = 12 levels in vertical layer

More information is available from MAT about all these GCM experiments.

Impact modellers need to consider:

- whether 24 hour 'twilight' is satisfactory
- how many years information are needed;
- the required spatial resolution; and
- whether a bucket surface scheme is satisfactory

MAT can supply some additional information on some of these issues.

PART 2

CLIMATE AND THE ATMOSPHERE

CHAPTER 3

COMMENTARY

Kendal McGuffie

3.1 Atmospheric processes

The atmosphere is the fastest responding and, perhaps, the most complex of all the components of the climate system which also includes the hydrosphere, the cryosphere, the biosphere and the land surface. In the widespread efforts to understand the impact of continuing human activities on the components of the climate system, the atmosphere has received the most attention, partly because of the apparent importance of atmospheric processes to human society through 'the weather' and partly because of the evident complexity of the atmosphere and its rapid responses.

The occurrence of water in all three phases (vapour, liquid and ice), and the very large energy exchanges which accompany these phase changes give the earth's climate its unique character. Radiative, dynamic and chemical processes are fundamental to the circulation of the atmosphere and the nature of the climate. All these processes are strongly affected by water fluxes, through the formation of clouds (moisture processes), the flux of energy from the ocean and land surfaces in both latent and sensible forms, and the occurrence of atmospheric aerosols (droplets and particulates) from natural and anthropogenic sources. The radiative properties of the clouds are altered by the presence of aerosols. Additional condensation nuclei result in higher albedo clouds (because there are more droplets), but when the aerosol has a significant component which is dark in colour, then this can act to decrease the albedo of the clouds.

Turbulent processes, which govern the transport of energy and moisture around the planet must also be incorporated successfully in climate models if their projections are to be believed. These turbulent processes generally occur at scales too small to be resolved by the current generation of models, yet the successful depiction of large scale processes depends quite crucially on the way in which small scale processes are parameterised.

The evaluation process, which must be undertaken in relation to climate model projections, takes many forms. Three techniques which are common practice in climate assessment today can be illustrated quite readily with reference to the experiments performed under the auspices of MECCA and reported in this section. This chapter reviews these techniques.

3.2   Climate model evaluation

The climate modelling community has initiated a series of model intercomparisons and evaluations over the last decade to which the MECCA project's uncertainty assessments contribute. Evaluation can produce a range of outcomes which have been grouped as:

(a) model predictions that are unreasonable;

(b) model predictions so reasonable as to be already known;

(c) model predictions previously not considered but readily understood and accepted once examined; or

(d) model predictions that whilst being reasonable identify novel outcomes that challenge current theories (Skiles 1995).

Normal practice in model development would screen out model versions producing unreasonable results and there is little benefit in intercomparison of results which are totally reasonable and well known. Thus the intercomparisons and group evaluations tend to try to focus on results in outcomes (c) and (d): new predictions that are consistent with theory and those which challenge existing ideas.

The process of comparison of model predictions and group evaluation is complex, as it has to encompass models and modelling groups from around the world and has to be organised so that comparable results are being compared. To facilitate the process of model evaluation and intercomparison, the World Climate Research Programme's Working Group on Numerical Experimentation (WGNE) categorised intercomparisons into three levels (Figure 3.1).

Level 1, the simplest, uses only available model results and a common diagnostic set. The Intergovernmental Panel on Climate Change (IPCC) assessments are Level 1 intercomparisons. Level 2 requires that simulations are made according to pre-specified, identical conditions; that common diagnostics are employed; and that there is a common data set against which all the predictions are evaluated. Level 3, the 'best' intercomparison process, requires, in addition to the requirements of the lower two levels, that all the models employ the same resolution and that the intercomparison includes the use of common subroutines or code modules.

Until the 1990s, intercomparisons were conducted at Level 1. Most of the MECCA intercomparisons are also Level 1. Within the MECCA project, scientists have tried to match the patterns of change predicted by models to the changes which have been observed over the last hundred years or so. They have compared the performance of models with each other, searching for weaknesses and common factors which have affected the performance of models. MECCA scientists have also sought to perform detailed 'validations' of the physics of particular models and to make deductions about the uncertainties associated with the resolution of these models. While none of these procedures can make absolute statements about the validity of the climate model results, they each provide a means of evaluating uncertainty in climate model projections.

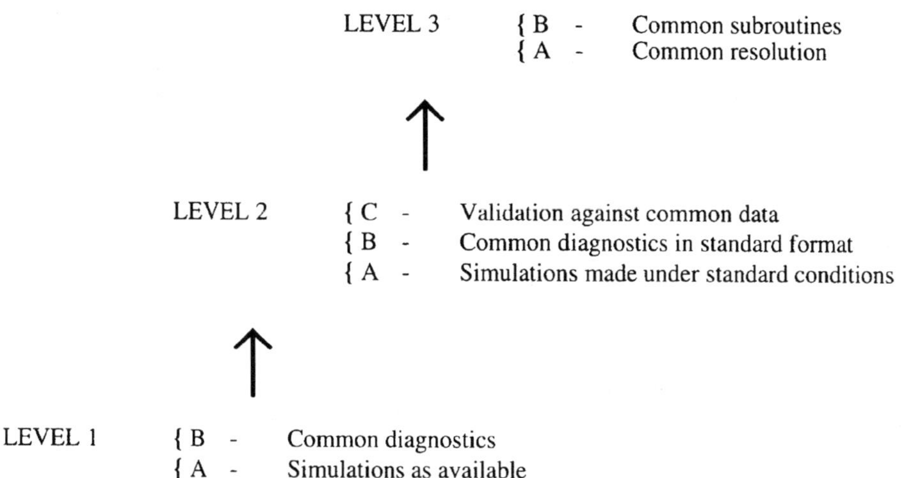

Figure 3.1 Levels of model intercomparison as defined by the World Climate Research Programme's Working Group on Numerical Experimentation in support of the World Climate Research Programme after Gates (1992). (The attributes of lower intercomparison levels are assumed to be included in higher levels when appropriate.)

3.2.1 Correlation

It is possible to construct model experiments which tell us how the climate comes to be like it is today. Historical records offer us a chance to assess how models perform and by comparing model results to observed climate changes it is possible to obtain a deeper understanding of the important features of the climate system. In particular, as illustrated by Penner, Wigley, Jaumann, Santer and Taylor in Chapter 5, the magnitude of the forcing due to aerosols can be estimated by comparing model and observed data for well-observed periods of history this century. Although the direct effect of aerosols is reasonably easy to incorporate into climate models, there is little information available on how to include the indirect effects, which are coupled closely to cloud formation and character. Penner et al. have avoided this difficulty by examining the performance of a suite of model simulations and deriving a total aerosol forcing.

The uncertainty assessment technique employed by Penner et al. also has significant difficulties. Since not all aspects of the climate system are yet included in climate models, there is uncertainty as to how well the model results should correlate with the observed changes. We do not know the relative magnitudes of the direct and indirect forcing by aerosols; neither do we know the size of the influence of the soot component. The aerosols are very short lived so the uncertainty in the prescribed geographic distribution is important and the exact magnitude of volcanic forcing is difficult to determine.

Despite these problems, this type of analysis perhaps offers some of the best evidence of the validity of climate models (cf. Wigley 1995). The technique also reveals additional problems: as we begin to understand better the role of aerosols, we must revise our ideas about the role of the oceans and about other components of the climate system.

3.2.2 Intercomparison

The most commonly used technique for evaluation of model predictions is the comparison of the results for one or many model integrations with the results of observational programmes. In some cases, very good observational data are available with which to evaluate model performance. Temperature records, derived using a relatively consistent set of objective monitoring tools, are available over much of the surface of the earth.

The cloud cover of the earth plays an important role in the determination of the amount and distribution of the solar energy which is absorbed by the climate system. Cloud-radiation interactions are known to be an important source of uncertainty in climate simulations (Houghton et al. 1990, 1992; Cess et al. 1993, 1995). Despite having been recorded for over 100 years at regular intervals and locations, cloud amount variations remain difficult to validate (Henderson-Sellers & McGuffie 1988). Before the advent of satellites in the 1960s, cloudiness statistics were based entirely on relatively subjective assessments of sky conditions (McGuffie & Henderson-Sellers 1990). Even the advent of objective satellite techniques has not brought a complete solution to the problems associated with cloudiness observation. Satellite observations are hampered by difficulties in detecting low clouds with thermal signatures which are similar to the surface and are also fraught with difficulties in the polar regions, where the visible brightness of the surface is comparable to the clouds and the thermal structure of the atmosphere makes straightforward identification of clouds difficult (Rossow & Garder 1993).

Table 3.1 shows the mean cloud amounts predicted by a range of MECCA models. Observational datasets show a range of between 0.50 and 0.60 for the global mean cloud cover. The models cover a slightly larger range and some models significantly underestimate the cloud amount. The models also vary in their representation of the seasonal cycle of cloud amount, represented in Colour Plate 3.1, in the form of Hovmöller plots for each of the models together with the results from the International Satellite Cloud Climatology Project (ISCCP) (e.g. Rossow & Schiffer 1991).

3.2.3 Evaluation

Perhaps the most rigorous technique for the evaluation of models is the detailed analysis of a physical process, rather than the analysis of the distribution of a variable. The MECCA project of Tanaka and Hasegawa, presented in Chapter 4, used this approach to examine how model resolution affects the spectral distribution of kinetic energy within a climate model. The effect of changing resolution has only recently come to the fore in atmospheric modelling because the computer resources for running high-resolution models have increased. In addition, climate models have been written and re-written without a 'hard-coded' resolution,

Table 3.1 Annually averaged cloud amounts predicted by a selection of MECCA models for present day (control) and greenhouse warmed conditions (enhanced) together with annually averaged observed data from five different observational datasets

	Northern Hemisphere		Southern Hemisphere		Global	
	Control	Enhanced	Control	Enhanced	Control	Enhanced
Model						
CCM0	0.60	0.59	0.64	0.64	0.62	0.62
CCM1-Oz	0.45	0.44	0.46	0.45	0.45	0.45
CCM0	0.48	0.46	0.53	0.50	0.51	0.48
CCM1W	0.44	0.43	0.48	0.48	0.46	0.45
MGO (variable optics)	0.52	0.52	0.51	0.49	0.50	0.50
MGO (fixed optics)	0.52	0.52	0.50	0.50	0.51	0.51
BMRC	0.55	0.55	0.55	0.54	0.55	0.55
Observations						
Berlyand and Strokina (1980)	0.59		0.62		0.60	
Warren et al. (1988)	0.59		0.64		0.61	
ISCCP (1985–92)	0.59		0.66		0.63	
METEOR	0.52		0.62		0.57	
Nimbus-7	0.49		0.52		0.50	

so that experiments such as these have become possible. Tanaka and Hasegawa have shown how the inadequacies of one aspect of a numerical scheme (resolution) can be compensated for by inadequacies in another (enhanced diffusion). This is a result which is shared by many other international model intercomparisons of the 1990s.

3.3   Summary

Quantitative evaluation of climate model projections is a challenging task. In the past, the most common technique for evaluation of model simulations was comparison of the results from one model with one observational dataset, often of uncertain validity. Since the development of intercomparison projects (e.g. Gates 1992; Randall et al. 1994; Henderson-Sellers et al. 1995.) the practices and techniques have changed. The quality of observational data has increased and the quality of correlative information associated with these data has also increased. Increasingly sophisticated techniques and protocols are applied to models and observations, and our understanding of the models and the important processes within the models continues to develop.

Among the more important current intercomparisons are the Atmospheric Model Intercomparison Project (AMIP), (established in 1989 and now moving into its second stage), the Intercomparison of Radiation Codes for Climate Models (ICRCCM) and the Feedback Analysis of GCMs and Intercomparison with Observations (FANGIO). AMIP focuses on structured (Level 2 in Figure 3.1) intercomparisons of the atmospheric component of global climate models. Participating models use observed ocean surface temperatures and sea-ice extents as well as agreed values of the solar constant (1365 W m^{-2}) and an atmospheric concentration of CO_2 (345 ppmv) as input to a fixed length simulation from 1 January 1979 to 31 December 1988.

ICRCCM has a straightforward mandate: to intercompare results from participating radiation codes in the long and short wavelength regions of the spectrum for the cases of clear and cloudy skies. Around 30–40 modelling groups participated in the first phase, representing both climate model codes and radiative transfer algorithms employed in retrieval of fluxes from satellite measurements. These intercomparisons were compared with the most detailed radiative transfer calculations available, termed 'line-by-line' calculations. In the second phase, these line-by-line calculations are augmented by observational data from satellite and surface-based field programs including the Earth Radiation Budget Experiment (ERBE), ISCCP and the Surface Radiation Budget Climatology Programme.

The FANGIO project seeks to improve understanding of the feedback processes in climate models involving cloud and radiation calculations. FANGIO investigators have calculated the value of a climate model feedback parameter for their models in the cases of clear skies, cloudy skies and for the global response overall. The range in the clear-sky values is very small, underlining the main conclusion of this intercomparison: that the threefold variation among Atmospheric General Circulation Models' (AGCMs) sensitivity to a prescribed climate change is due almost entirely to cloud feedback processes.

The three techniques which have been discussed here with reference to MECCA experiments provide different and complementary perspectives on climate model uncertainty assessment. By comparing different experiments with the same model with observational data, Penner et al. have been able to derive estimates for the total forcing due to atmospheric aerosols. The analysis of cloudiness information performed by MAT may provide information on common areas of difficulty associated with cloud radiation feedbacks. Analysis of physical characteristics of the atmospheric circulation and their relationship to the computational aspects of the model also have important implications for the uncertainties of climate model simulations. We need to understand how the forcings, the parameterization schemes and the computational constraints impact on the final model predictions if we are to improve confidence in model predictions. The most important outcomes of the international intercomparisons of climate model performance described here in this book and elsewhere are: (1) the identification of group outliers; (2) the estimation of the range of confidence (or uncertainty) inherent in predictions of any one of the 'reasonable' models; and (3) the development and dissemination of data sets for model evaluation and of tests for climate model results. MECCA results are in agreement with other evaluations from other international intercomparisons in concluding that:

- no one model performs well in all the evaluations employed;
- no one test evaluates all aspects of the participating models; and
- the group mean (after excluding unreasonable results/outliers) outperforms any one individual model.

There are dangers associated with large international comparisons. The most omnipresent is the trend towards group-median values. However, if the intercomparisons are of Level 2 or 3 the existence of observational data should balance this central tendency. To date, considerable benefits have been derived from well structured climate model intercomparisons, including those sponsored by MECCA.

Acknowledgments

I wish to acknowledge the support of the University of Technology in Sydney and the Australian Research Council Small Grants Scheme in the preparation of this chapter.

References

Cess, R.D., Zhang, M.H., Minnis, P., Corsetti, L., Dutton, E.G., Forgan, B.W., Garber, D.P., Gates, W.L., Hack, J.J., Harrison, E.F., Jing, X., Kiehl, J.T., Long, C.N., Morcrette, J.-J., Potter, G.L., Ramanathan, V., Subasalir, B., Whitlock, C.H., Young, D.F. and Zhou, Y. (1995) Absorption of solar radiation by clouds: observations versus models. *Science*, **267**, 496–499.

Cess, R.D., Zhang, M.H., Potter, G.L., Barker, H.W., Colman, R.A., Dazlich, D.A., Del Genio, A.D., Esch, M., Fraser, J.R., Galin, V., Gates, W.L., Hack, J.J., Ingram, W.J., Kiehl, J.T., Lacis, A.A., Le Treut, H., Li, Z-X., Liang, X.-Z., Mahfouf, J.-F., McAvaney, B.J., Meleshko, V.P., Morcrette, J-J., Randall, D.A., Roeckner, E., Royer, J.-F., Sokolov, A.P., Sporyshev, P.V., Taylor, K.E., Wang, W-C. and Wetherald, R.T. (1993) Uncertainties in carbon dioxide radiative forcing in atmospheric general circulation models. *Science*, **262**, 1252–1255.

Gates, W.L. (1992) The atmospheric model intercomparison project. *Bulletin of the American Meteorological Society*, **73**, 1962–1970.

Henderson-Sellers, A. and McGuffie, K. (1988) A quantitative evaluation of observations of oceanic cloudiness. *Atmospheric Research*, **21**, 241–260.

Henderson-Sellers, A., Pitman, A.J., Love, P.K., Irannejad, P. and Chen, T. (1995) The Project for Intercomparison of Land-surface Parameterization Schemes (PILPS): Phases 2 & 3. *Bulletin of the American Meteorological Society*, **76**, 489–503.

Houghton, J.T., Jenkins, G.J. and Ephraums, J.J. (eds) (1990) *Climate Change: The IPCC Scientific Assessment*. Cambridge: Cambridge University Press.

Houghton, J.T., Callander, B.A. and Varney, S.K. (eds) (1992) *Climate Change 1992: The supplementary report to the IPCC Scientific Assessment*. Cambridge: Cambridge University Press.

McGuffie, K. and Henderson-Sellers, A. (1990) Basis for integration of conventional observations of cloud into global nephanalyses. *Journal of Atmospheric Chemistry*, **11**, 1–25.

Randall, D.A., Cess, R.D., Blanchet, J.-P., Chalita, S., Colman, R., Dazlich, D.A., DelGenio, A.D., Keup, E., Lacis, A., Letrent, H., Liang, X.Z., McAvaney, B., Mahfouf, J.F., Mekshko, V.P., Morcrette, J.J., Norris, P.M., Potter, G.L., Rikus, L., Roekner, E., Royer, J.F., Schlese, U., Sheinin, D.A., Sokolov, A.P., Taylor, K.E., Wetherald, R.T. et al. (1994). Analysis of snow feedback in 14 general circulation models, *Journal of Geophysical Research*, **99**, 20757–20771.

Rossow, W.B. and Garder, L.C. (1993) Validation of ISCCP cloud detections. *Journal of Climate*, **6**, 2370–2393.

Rossow, W.B. and Schiffer, R.A. (1991) ISCCP cloud data products. *Bulletin of the American Meteorological Society*, **72**, 2–20.

Skiles, J.W. (1995) Modeling climate change in the absence of climate change data—editorial comment. *Climatic Change*, **30**, 1–6.

Wigley, T.M.L. (1995) A successful prediction? *Nature*, **376**, 463–464.

CHAPTER 4

COMPARATIVE ENERGETICS ANALYSIS OF CLIMATE MODELS WITH HIGH AND LOW RESOLUTIONS

Hiroshi L. Tanaka and Akira Hasegawa

4.1 Introduction

Global-scale atmospheric motion is excited by solar radiation. The driving force of the radiation has a global scale. Solar energy into the atmospheric system cascades down to smaller scale eddies of high–low pressure systems and, ultimately, to the molecular dissipation processes. Analysis of energy transfer among different scales of motions offers a fundamental understanding of the general circulation of the atmosphere. The operational weather prediction models and the state-of-the-art climate models must be formulated to correctly simulate the energy cascade in the wavenumber spectral domain based on the knowledge of the observational analysis.

The effect of doubled CO_2 and increasing anthropogenic gases will enhance the absorption of solar energy. The anomalous radiative driving force due to the anthropogenic gases would be initiated at the global scale and is expected to cascade down to smaller scale motions, such as high- and low-pressure systems, through the nonlinear wave–wave interactions. The model performance for simulating a new atmospheric equilibrium in response to the anomalous global heating is closely related to the accuracy of energy redistribution over the wide range of the wavenumber spectrum.

Present climate models have relatively coarse horizontal and vertical resolutions due to the inevitable restrictions in computing capability. In the coarse general circulation models, for example R15 (rhomboidally truncated at the total wavenumber 15), the spectral truncation at the medium wavenumber imposes unrealistic restrictions in spectral energy transfer. The nonlinear energy cascade within the model atmosphere must terminate at the imposed truncation wavenumber. In reality, of course, there is no artificial spectral boundary, and the energy input at the large scale is transferred to the dissipation range without interruption. Hence, the examination of the influence of the artificial spectral boundary at the truncation wavenumber upon the climatic equilibrium state is an important research subject in the climate modeling effort. The evaluation of the energy transfer between the large-scale and small-scale motions is the first step in assessing the basic performance of existing climate models.

The diagnosis of spectral energetics has long been performed by a zonal harmonic expansion since the original work by Saltzman (1957, 1970). Kung and Tanaka (1983, 1984) reported the first comprehensive analysis of global energy transformations in the zonal wavenumber domain, using the most extensive worldwide observation of the First GARP (Global Atmospheric Research Program) Global Experiment (FGGE). According to their observational analysis, nonlinear wave–wave interactions of kinetic energy are basically negative for zonal wavenumbers less than 15, indicating that the energy is generated here and transformed from this range to other scales of motions. For wavenumbers larger than 15, the nonlinear wave–wave interactions indicate positive values because the waves receive energy from the large-scale motions. The sum of the nonlinear energy interactions over the entire spectral domain is zero as required from the energy conservation law.

In contrast to the real atmosphere, the nonlinear energy transfer of a coarse resolution climate model is forced to terminate at the truncation wavenumber. Hence, the positive and negative values of the nonlinear energy interactions must be redistributed over the resolvable wavenumbers. This appears to be a serious constraint in the nonlinear energy transfer. The artificial spectral boundary is anticipated to cause a biased response to the doubled CO_2 and anthropogenic activities. Yet, it is not clear how this spectral boundary modifies the spectral energy transfer and how the model's equilibrium is maintained without the accurate energy cascade toward the dissipation range.

The complicated nonlinear wave–wave interactions with subgrid-scale motions have been parameterised by a simple diffusion or a biharmonic diffusion as seen in the NCAR Community Climate Model (CCM). The performance of the diffusion parameterisation has been questioned in its accuracy for the global atmosphere on a rotating spherical earth (e.g. Matsuno 1980). A detailed analysis is necessary to compare the spectral energy transfers in coarse- and fine-resolution models in order to assess the basic performance of the existing climate models.

In order to overcome the limitations of computer memory required for extremely high resolution climate models, an alternative approach has been developed in the field of limited area climate models. For the purpose of conducting regional climate simulations, the methods of nesting a limited area model (LAM) into a large area model (e.g. GCM) have been studied (Giorgi & Mearns 1991; Kida et al. 1991). The nesting procedure can be implemented in a one-way or a two-way interactive mode. In one-way nesting, information from the coarser resolution model is used at the horizontal boundary to drive the higher-resolution regional submodel, but information from the higher-resolution subregion does not feed back into the lower-resolution domain. In two-way nesting, the exchange of information between the lower-resolution and higher-resolution model components occurs interactively and in both directions. The nested LAM-GCM models are mostly based on the one-way nesting procedure. That is, most of nested LAM-GCM models assume that the up-scale signal transfer of the higher-resolution submodel to the lower-resolution domain can be ignored. In the framework of climate prediction for a range of decades of the future, this assumption of ignoring the two-way signal transfer could be a great danger in achieving a meaningful prediction on small scales. This cannot be justified unless the magnitude of the scale interactions (i.e. wave–wave interactions) is quantified to be small enough. The magnitude of the scale interactions, however, has not been analysed quantitatively with sufficient accuracy in the previous studies. Answering this question may be one of the most urgent requirements in endeavouring to develop a LAM nested in a GCM.

The objective of the present study is to evaluate the basic performance of present climate models with respect to the energy transfer and scale interactions in the wavenumber domain. We have conducted a sequence of diagnostic analyses of climate model simulations with different model resolutions.

First, spectral energy levels of the model atmosphere are compared with observations. The model bias in the energy level will be analysed. Second, the magnitude and direction of the nonlinear wave–wave interactions are computed for model atmospheres with different horizontal resolutions. The nonlinear energy cascade in the high- and low-resolution models are compared with that of the operational analysis with very high resolution models. Third, the variabilities in both the energy level and its interactions are examined for the high wavenumber range near the truncation of the low-resolution model. The characteristics of the variability for high- and low-resolution models are separately analysed. Finally, the effect of the artificial truncation in the wavenumber domain to the energy redistribution of the model atmosphere is discussed. The analyses of energy transfer spectra would provide some useful information to the abovementioned studies of nesting a LAM into a GCM.

4.2   Global gridded data

It is desirable that the models are integrated under the same conditions. The NCAR Community Climate Model version 2 (CCM2) is integrated under the same conditions except for the parameters in the cloud parameterisation under the MECCA project (Hack et al. 1993; Williamson 1993). The CCM2 simulations are appropriate for the main purpose of this study. As the climate models with different horizontal spectral resolutions, the CCM2 are analysed for four types of models using horizontal resolutions of rhomboidal-15 (R15), and triangular–42, –63 and –106 (T42, T63, T106) resolutions, respectively.

Moreover, the same analyses are carried out for the observed atmosphere. As the observed atmosphere, the European Centre for Medium Range Weather Forecasts (ECMWF) analyses data (TOGA Basic level III) for five years during 1986 through to 1990 are used. The discrepancies among the ECMWF, National Meteorological Center (NMC) and Japanese Meteorological Agency (JMA) analyses in the wavenumber domain are reported by Ogasawara (1995) and Tanaka and Kimura (1996), which conclude that the spectral patterns and global energy balances with these operational analyses are sufficiently close to each other.

The present spectral energetics diagnosis requires a complete set of atmospheric state variables of zonal wind speed u (m/s), meridional wind speed v (m/s), vertical p-velocity ω (Pa/s), temperature T (K), geopotential height Z (m), relative humidity RH (%) and surface pressure p_s (Pa) for instantaneous time intervals (daily values for CCM2). These variables should be given at the longitude–latitude grids over the globe at the vertical pressure levels for the troposphere and stratosphere. The vertical levels prepared by the existing diagnostic code are 1000, 850, 700, 500, 400, 300, 250, 200, 150, 100, 70 and 50 hPa.

The computation of the nonlinear wave–wave interactions requires a considerable amount of CPU time because the nonlinear interactions involve double summations over the whole wavenumbers. Therefore, we analysed for a three-month period in summer and winter of the climate model simulations.

4.3   Description of the spectral energetics

The computational analysis scheme of the standard spectral energetics in the zonal wavenumber domain is based on Saltzman (1957, 1970). Here, the meteorological variables are expanded in zonal harmonics, and the Fourier coefficients are substituted in the governing equations of primitive equations. The primitive equations may be transformed from the space domain to the wavenumber domain expressed by the Fourier expansion coefficients.

Any real, single-valued function $f(\lambda)$, which is piecewise differentiable in the interval $(0, 2\pi)$, may be written in terms of a Fourier representation,

$$f(\lambda, \phi, p, t) = \sum_{n=-\infty}^{\infty} F(n, \phi, p, t)e^{in\lambda} \tag{1}$$

where the complex coefficients, $F(n)$, are given by

$$F(n, \phi, p, t) = \frac{1}{2\pi} \int_0^{2\pi} f(\lambda, \phi, p, t)e^{-in\lambda}d\lambda \tag{2}$$

We consider the Fourier representation of meteorological variables specified along a given latitudinal circle. Thus, λ is taken as longitude, n is the wavenumber around the latitude circle. The functions $f(\lambda)$ and $F(n)$ to be considered here are listed as follows:

$$f(\lambda) \quad u \ v \ \omega \ Z \ T \ h \ x \ y \ : \ \text{space domain}$$

$$F(n) \quad U \ V \ \Omega \ A \ B \ H \ X \ Y \ : \ \text{wavenumber domain}$$

Here, h, x, y designate diabatic heating rate, external mechanical force with respect to the east–west direction, and that of the north–south direction, respectively. Since atmospheric kinetic energy and available potential energy are proportional to the wind and temperature variances, the zonal mean energy can be decomposed in contributions from every zonal wavenumber by means of the Parseval's theorem.

$$\frac{1}{2\pi} \int_0^{2\pi} \frac{1}{2}(u^2 + v^2)d\lambda = \sum_{n=-\infty}^{\infty} (|U(n)|^2 + |V(n)|^2) \tag{3}$$

$$\frac{1}{2\pi} \int_0^{2\pi} \frac{1}{2}C_p\gamma T''^2 d\lambda = \sum_{n=-\infty}^{\infty} C_p\gamma |B(n)|^2 \tag{4}$$

Energy equations are then constructed for every zonal wavenumber by differentiating (3) and (4) with respect to time (t), which describes how the temporal variation of wave energy occurs.

We can form a set of equations for the rates of change of the globally integrated kinetic energy and available potential energy in the zonally averaged component (wavenumber zero) and in each of the eddy components corresponding to nonzero wavenumbers. These equations can be written as follows:

$$\frac{dK(0)}{dt} = \sum_{n=1}^{\infty} M(n) + C(0) - D(0)$$

$$\frac{dK(n)}{dt} = -M(n) + L(n) + C(n) - D(n) \quad (n = 1, 2, 3, \dots)$$

$$\frac{dP(0)}{dt} = \sum_{n=1}^{\infty} R(n) - C(0) + G(0)$$

$$\frac{dP(n)}{dt} = -R(n) + S(n) - C(n) + G(n) \quad (n = 1, 2, 3, \dots) \tag{5}$$

Definitions of integrals appearing in (5) are written as follows, where all quantities are measured per unit mass:

total kinetic energy in eddies of wavenumber n,

$$K(n) = \langle(|U(n)|^2 + |V(n)|^2)\rangle$$

total kinetic energy in zonally averaged motions,

$$K(0) = \frac{1}{2}\langle[u]^2 + [v]^2\rangle,$$

total available potential energy in eddies of wavenumber n,

$$P(n) = \langle C_p\gamma |B(n)|^2\rangle,$$

where

$$\gamma = \left(\tilde{T} - \frac{c_p p}{R} \frac{d\tilde{T}}{dp} \right)^{-1},$$

total available potential energy in zonally averaged temperature distribution,

$$P(0) = \frac{1}{2} \langle C_p \gamma \{ [T]''^2 \} \rangle,$$

where the symbol $\langle \rangle$ designates a mass integral over the entire mass of the atmosphere.

This kinetic energy equation represents that the energy variation is caused by its zonal-wave interactions $M_{(n)}$, wave–wave interactions $L_{(n)}$, baroclinic conversion $C_{(n)}$, and energy dissipation $D_{(n)}$. Available potential energy equation is also represented by its zonal-wave interactions $R_{(n)}$, wave–wave interactions $S_{(n)}$, baroclinic conversion $C_{(n)}$, and diabatic heat sources and sinks $G_{(n)}$. Evaluating those energetic terms from the gridded data, the energy spectra and energy transformations can be examined over the wavenumber domain.

The energy transformations are defined as follows:

baroclinic conversion from available potential energy of wavenumber n to kinetic energy of wavenumber n,

$$C(n) = -\left\langle \frac{R}{p} [B(n)\Omega(-n) + B(-n)\Omega(n)] \right\rangle,$$

baroclinic conversion from zonal available potential energy to zonal kinetic energy,

$$C(0) = -\left\langle \frac{R}{p} \{ [T]''[\omega]'' \} \right\rangle,$$

zonal-wave interaction of kinetic energy from eddies of wavenumber n to the zonally averaged flow,

$$M(n) = M_1(n) + M_2(n) + M_3(n),$$

where

$$M_1(n) = \left\langle [U(n)V(-n) + U(-n)V(n)] \frac{\cos\phi}{a} \frac{d}{d\phi} \left(\frac{[u]}{\cos\phi} \right) \right\rangle$$

$$M_2(n) = \left\langle [U(n)\Omega(-n) + U(-n)\Omega(n)] \frac{d[u]}{dp} \right\rangle$$

$$M_3(n) = \left\langle 2[V(n)V(-n)] \frac{\partial[v]}{a\partial\phi} + [V(n)\Omega(-n) + V(-n)\Omega(n)] \frac{d[v]}{dp} \right.$$
$$\left. -2[U(n)U(-n)][v] \frac{\tan\phi}{a} \right\rangle,$$

zonal-wave interaction of available potential energy from eddies of wavenumber n to the zonally averaged field,

$$R(n) = R_1(n) + R_2(n),$$

where

$$R_1(n) = \left\langle C_p \gamma [B(-n)V(n) + B(n)V(-n)] \frac{\partial [T]}{a \partial \phi} \right\rangle,$$

$$R_2(n) = \left\langle C_p \gamma \left(\frac{p}{p_s}\right)^k \left\{ [B(-n)\Omega(n) + B(n)\Omega(-n)]'' \frac{d[O]''}{dp} \right\} \right\rangle,$$

wave–wave interactions of kinetic energy for wavenumber n from eddies of all wavenumbers,

$$L(n) = L_1(n) + L_2(n), \quad \left(\sum_{n=1}^{\infty} L(n) = 0\right),$$

where

$$L_1(n) = -\left\langle \sum_{\substack{m=-\infty \\ \neq 0}}^{\infty} \frac{in}{a \cos \phi}(U(m)[U(-n)U(n-m) - U(n)U(-n-m)] \right.$$

$$+ V(m)[V(-n)U(n-m) - V(n)U(-n-m)])$$

$$+ \frac{1}{a \cos \phi}\left(U(-n)\frac{d}{d\phi}[U(m)V(n-m)\cos\phi]\right.$$

$$+U(n)\frac{d}{d\phi}[U(m)V(-n-m)\cos\phi]$$

$$\left.+V(-n)\frac{d}{d\phi}[V(m)V(n-m)\cos\phi] + V(n)\frac{d}{d\phi}[V(m)V(-n-m)\cos\phi]\right)$$

$$-\frac{\tan\phi}{a}(V(m)[U(-n)U(n-m) + U(n)U(-n-m)] - U(m)[V(-n)U(n-m)$$

$$\left.+V(n)U(-n-m)])\right\rangle,$$

$$L_2(n) = -\left\langle \sum_{\substack{m=-\infty \\ \neq 0}}^{\infty} \left(U(-n)\frac{d}{dp}[U(m)\Omega(n-m)] + U(n)\frac{d}{dp}[U(m)\Omega(-n-m)]\right.\right.$$

$$\left.\left.+V(-n)\frac{d}{dp}[V(m)\Omega(n-m)] + V(n)\frac{d}{dp}[V(m)\Omega(-n-m)]\right)\right\rangle,$$

wave–wave interactions of available potential energy for wavenumber n from eddies of all wavenumbers,

$$S(n) = S_1(n) + S_2(n), \quad \left(\sum_{n=1}^{\infty} S(n) = 0 \right),$$

where

$$S_1(n) = \left\langle C_p \gamma \sum_{\substack{m=-\infty \\ \neq 0}}^{\infty} \left(\frac{in}{a \cos \phi} B(m)[B(-n)U(n-m) - B(n)U(-n-m)] \right. \right.$$

$$\left. \left. + \frac{B(m)}{a} \left[V(n-m) \frac{d}{d\phi} B(-n) + V(-n-m) \frac{d}{d\phi} B(n) \right] \right) \right\rangle,$$

$$S_2(n) = -\left\langle C_p \gamma \sum_{\substack{m=-\infty \\ \neq 0}}^{\infty} \left(B(m) \left[\Omega(n-m) \left(\frac{dB(-n)}{dp} - \frac{RB(-n)}{c_p p} \right) \right. \right. \right.$$

$$\left. + \Omega(-n-m) \left(\frac{dB(n)}{dp} + \frac{RB(n)}{c_p p} \right) \right]$$

$$\left. \left. - \frac{d}{dp} (B(m)[B(-n)\Omega(n-m) + B(n)\Omega(-n-m)]) \right) \right\rangle,$$

rate of viscous dissipation of the kinetic energy for wavenumber n,

$$D(n) = -\langle [U(n)X(-n) + U(-n)X(n)] + [V(n)Y(-n) + V(-n)Y(n)] \rangle,$$

rate of viscous dissipation of the zonal kinetic energy,

$$D(0) = -\langle ([u][x] + [v][y]) \rangle,$$

rate of generation of available potential energy of wavenumber n due to diabatic heating,

$$G(n) = \langle \gamma [B(n)H(-n) + B(-n)H(n)] \rangle,$$

rate of generation of zonal available potential energy due to zonally averaged heating,

$$G(0) = \langle \gamma \{[T]''[h]''\} \rangle.$$

By evaluating these energetics terms, Saltzman showed, first, that the atmospheric kinetic energy is generated at the synoptic scale of zonal wavenumbers 6–10 by the synoptic disturbances and that the energy is transferred to both planetary waves and short waves. The energy transfer toward the short wavelengths corresponds to the downscale energy cascade to the dissipation range, whereas the transfer toward the large-scale (up-scale cascade) feeds the energy of the planetary waves and zonal jet stream in the atmosphere. If we sum the equations over all wavenumbers, excluding wavenumber zero, noting that $\sum_{n=1}^{\infty} L(n) = 0$; $\sum_{n=1}^{\infty} S(n) = 0$, we obtain a set of equations similar to that derived by Lorenz (1955) and reviewed by Oort (1964) under the name of 'space domain equations'.

4.4   Meridional distribution of kinetic energy

Comprehensive description of the spectral energetics analysis of this study is summarised in Hasegawa (1995). Some selected results of the analysis, especially for kinetic energy balance, are documented in the following figures.

Figure 4.1 illustrates the meridional distributions of zonal and eddy kinetic energy for the CCM2 and ECMWF datasets during (a) northern winter and (b) summer. Local discrepancies in K_M, i.e. $\bar{u}$ fields, between the ECMWF analysis and CCM2 simulations are evident. Zonal kinetic energy K_M in the CCM2 is significantly overestimated in the SH middle latitudes. In other words, the jet is too strong in the model. The discrepancies of K_M are much more significant in the winter SH than in the summer SH. The black dot denotes an observed mean level by the ECMWF analysis, and the standard deviation for the five-year analysis period is marked by a deviation bar around the black dot. In the winter SH, the CCM2 simulations, except for T42, represent two-maxima of K_M in the middle latitudes (Figure 4.1 (b)). These features of the simulated atmosphere are related to the double-jet structure produced by the subtropical jet and polar frontal jet. The double-jet structure is completely missing in T42. In the SH middle latitude, zonal-wave interactions of K in the CCM2 is stronger than in the observations, that is, the CCM2 transfers much momentum from eddies to the zonal mean field and intensifies the westerly wind. The minimum value of M in the SH middle latitudes is not well simulated in the CCM2-T42.

The eddy kinetic energy K_E simulated with the CCM2 is larger than that in the ECMWF analysis in the tropics. This bias results from both large M and C_E in simulations near the equator. The CCM2 circulation, moreover, transfers a large amount of kinetic energy from zonal mean into eddies ($K_M \rightarrow K_E$) in the winter hemisphere at low latitudes.

4.5   Vertical distribution of kinetic energy

Figure 4.2 illustrates the vertical distributions of zonal and eddy kinetic energy for the CCM2 and ECMWF datasets during (a) northern winter and (b) summer. The vertical structures of K_M indicate interesting difference in the stratosphere between northern winter (DJF) and summer (JJA). For DJF the mean level of the tropopause is seen at about 200 hPa level, and the energy level decreases in the stratosphere. For JJA, in contrast, the energy level is approximately constant in the stratosphere. The difference seems to come from the different intensity of the polar night jet and the polar frontal jet. The polar night jet in the middle atmosphere extends downward and is connected with a clear polar frontal jet in the winter SH. Although the same polar night jet exists in the winter NH, it does not extend to low levels and the polar frontal jet is not detected in the winter NH in the mean.

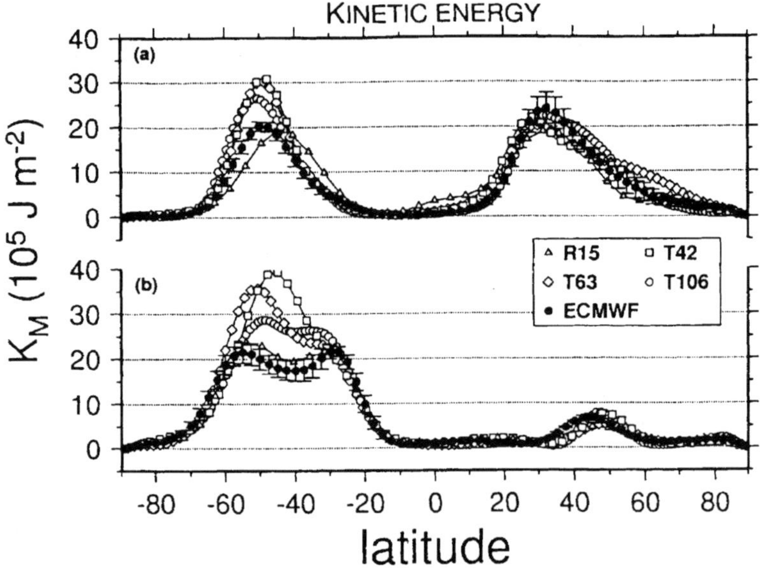

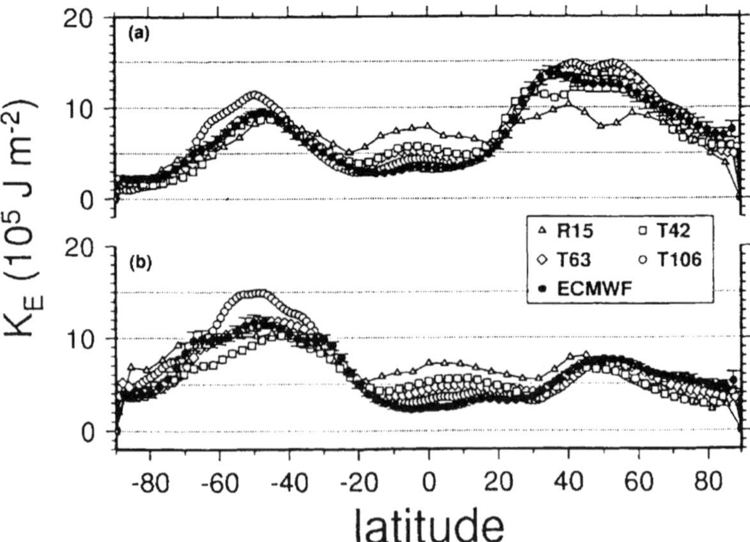

Figure 4.1 Meridional distribution of zonal kinetic energy K_M and eddy kinetic energy K_E for CCM2 and ECMWF datasets during (a) northern winter and (b) summer.

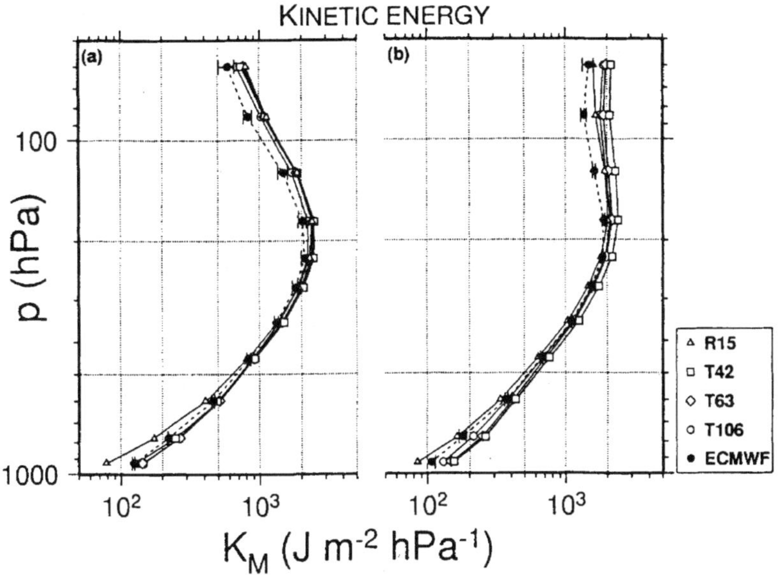

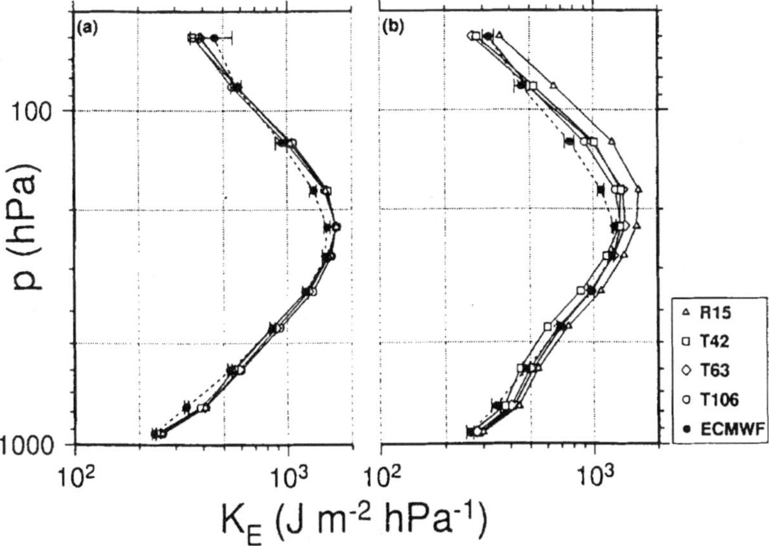

Figure 4.2 Vertical distribution of zonal kinetic energy K_M and eddy kinetic energy K_E for CCM2 and ECMWF datasets during (a) northern winter and (b) summer.

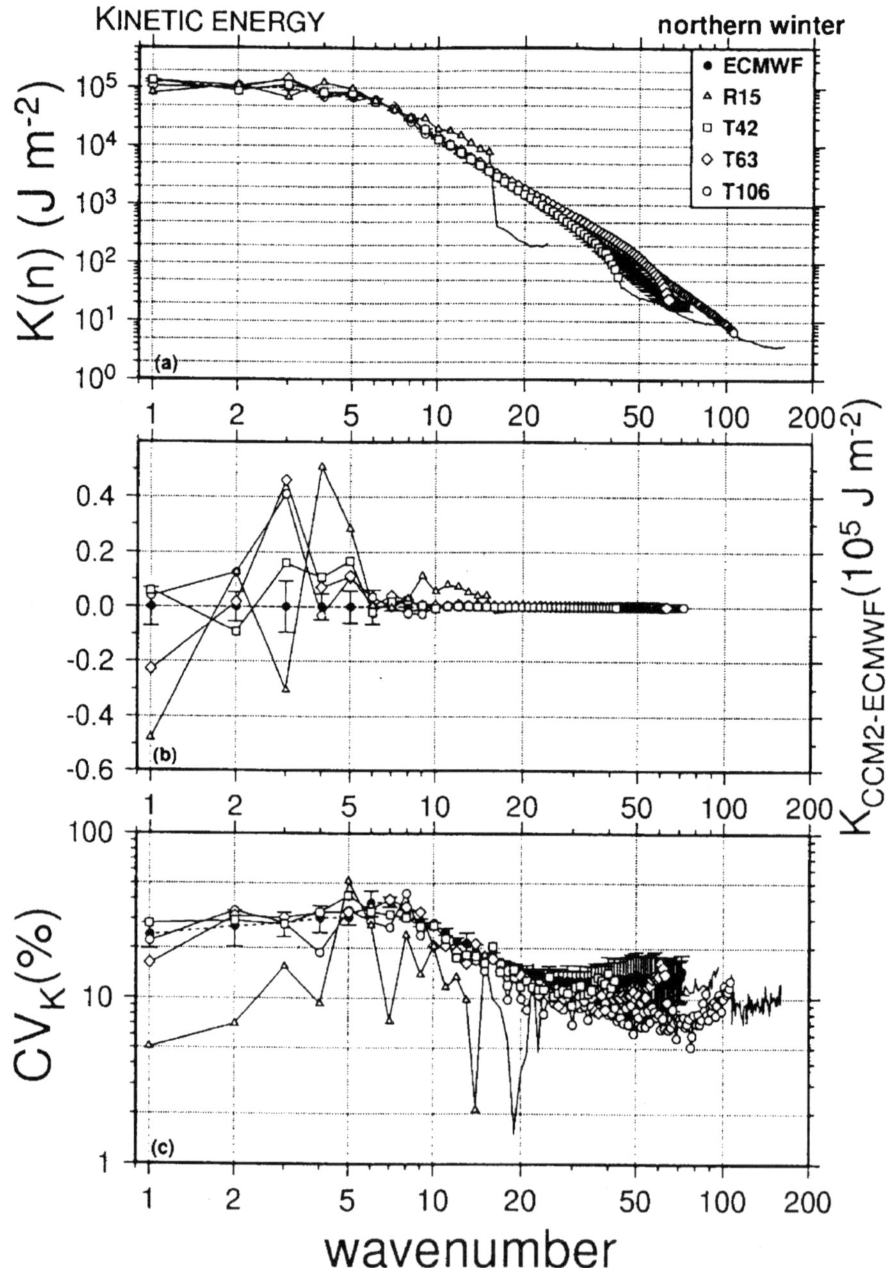

Figure 4.3 Spectral distribution of kinetic energy K_n for CCM2 and ECMWF datasets during northern winter and summer; (a) spectral distribution, (b) deviation from ECMWF analysis, (c) coefficient of variation.

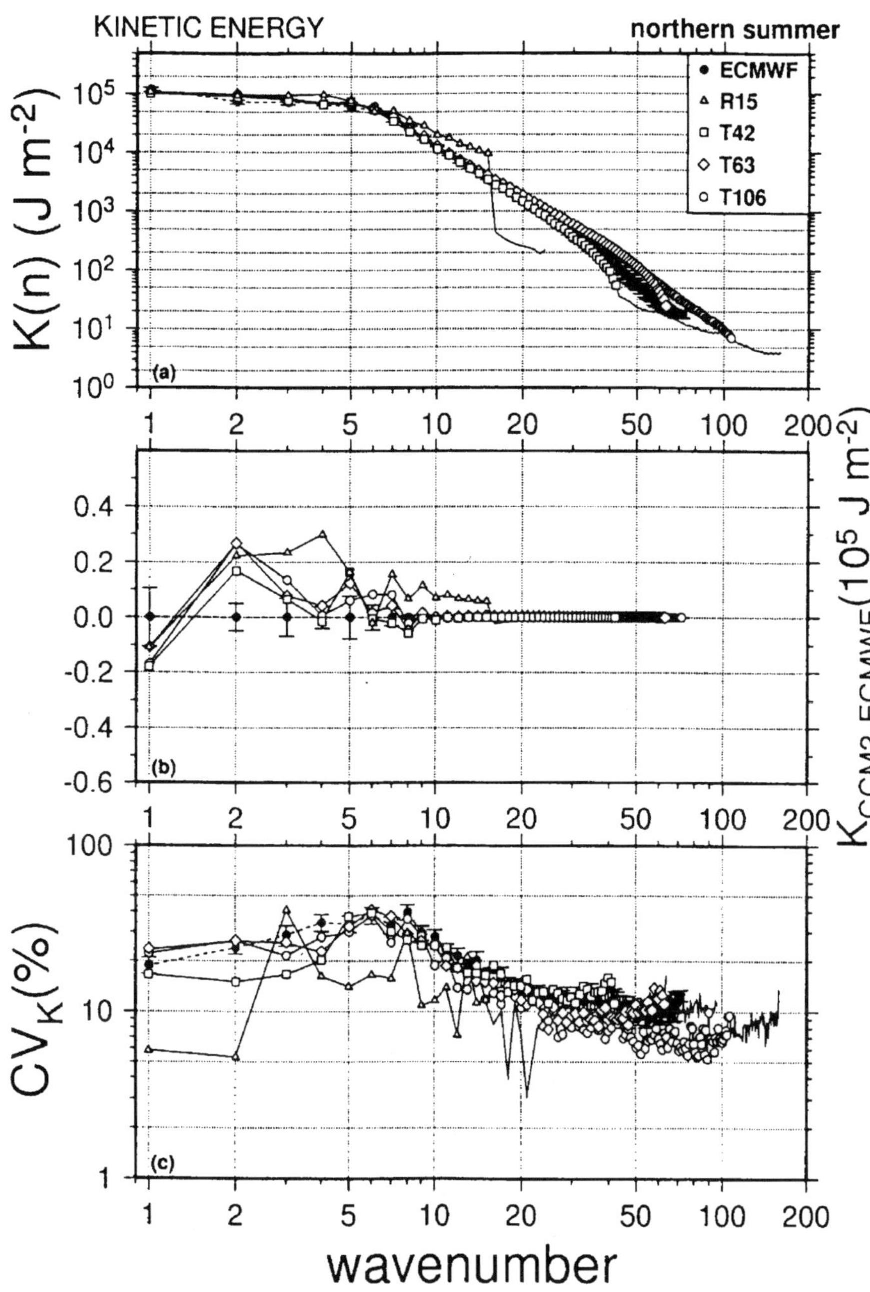

Figure 4.3 *(continued)*

Minor discrepancies are seen in the stratosphere in K_M fields between the ECMWF analysis and CCM2 simulations. The jet is generally stronger in CCM2 above the tropopause. In the troposphere, the agreement between the model and observations is rather good.

The vertical profile of K_E indicates a peak at about 250 hPa level. The peak coincides approximately with the tropopause level. The contrast between DJF and JJA is slight, showing an insignificant seasonality. The CCM2 simulations show almost a perfect fit for the northern winter, but the difference is large for the northern summer. The CCM2, especially R15, indicates excessive eddies above the tropopause level.

4.6   Spectral distribution of kinetic energy

Spectral distributions of kinetic energy K_n for the northern winter and summer are presented for CCM2 simulations and the ECMWF analysis in Figure 4.3. The upper panel of the figure illustrates the energy spectrum. The black dots are for ECMWF analysis while white symbols are for CCM2 simulations. The symbols are plotted up to the model's truncation wavenumber. No symbols are plotted beyond that wavenumber although the analysis wavenumber extends a little more depending on the gridded size of the globally interpolated data. The energy level drops by more than an order of magnitude for R15 at the zonal wavenumber 15, as is expected. The spectral slope is flat for wavenumbers 1 to about 5 and it becomes red beyond the wavenumber 5, showing a -3 power law of the 2–D or geostrophic turbulence. The CCM2 spectrum coincides with that of ECMWF except for R15 which indicates higher biases in the energy level near the truncation wavenumber.

The difference between the CCM2 simulations and ECMWF observations are plotted in the middle panel of the figure with a linear scale on the ordinate. The standard deviation bars are marked for the black symbols of ECMWF. Since the standard deviations evaluated from the time series are only about 0.1 (10^5 Jm^{-2}), almost all CCM2 simulations lie outside the standard deviation bars. Namely, the differences in the energy levels between the models and observations are significant for all simulations.

The lower panel of the figure illustrates a spectrum of the coefficient of variation, defined as the standard deviation in the time series divided by its time average. The largest value of the coefficient of variation is about 40% at the synoptic scale. The values are reduced to about 30% for the planetary waves and to 10% for short waves. It is interesting to note that the time variation for R15 is reduced consistently for almost all waves. This may be a good example to remember that the characteristics in time variation could be biased from reality for a highly truncated model, even though the climatology can be tuned at the right order.

The comparison of the results for the northern winter and summer shows that the spectral characteristics are similar to each other. The results for R15 are clearly biased from the observations. The discrepancies tend to diminish as the resolution increases.

4.7   Spectral distribution of energy transformations

Figure 4.4 illustrates the spectral distribution of baroclinic conversion $C_{(n)}$ for CCM2 and ECMWF datasets during the northern winter and summer. The baroclinic conversion

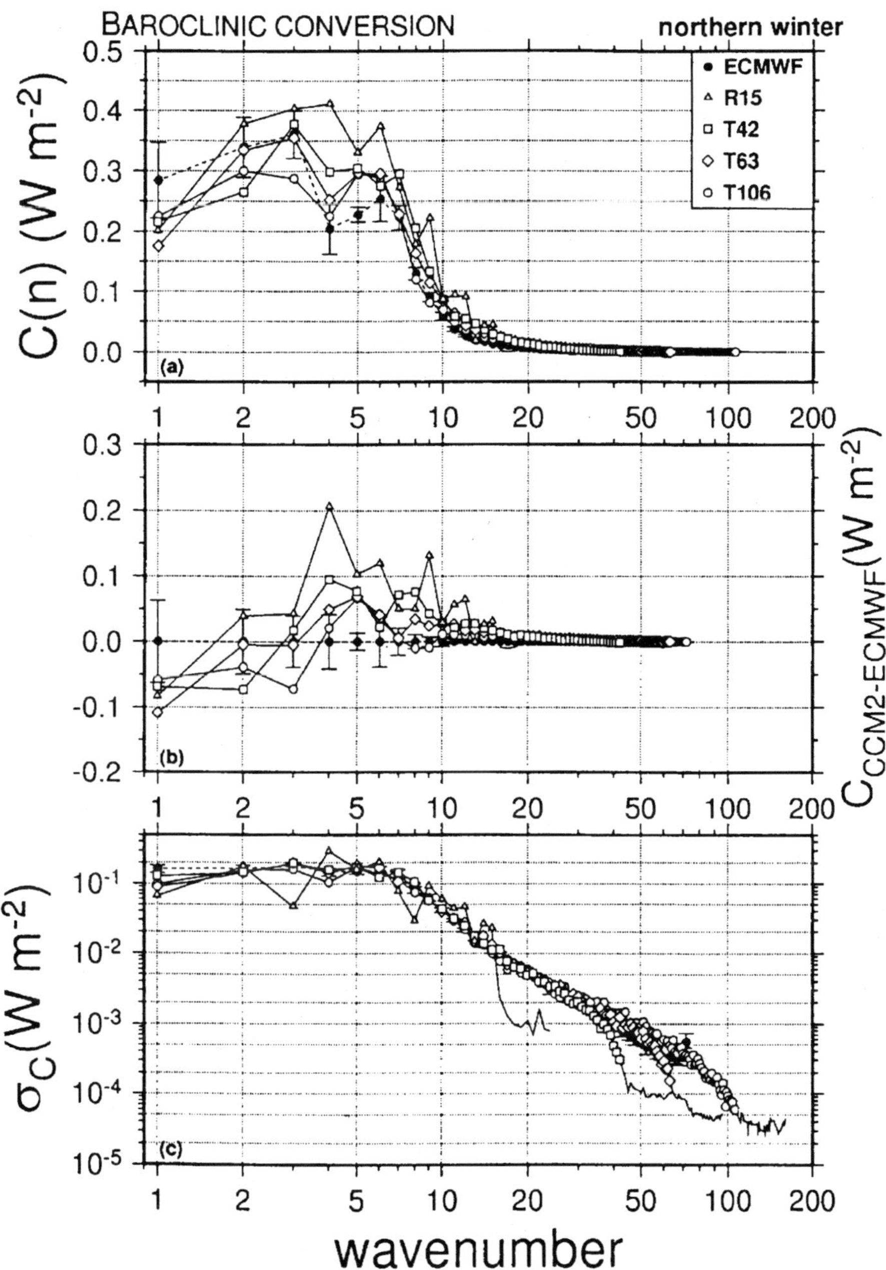

Figure 4.4 Spectral distribution of baroclinic conversion C_n for CCM2 and ECMWF datasets during northern winter and summer; (a) spectral distribution, (b) deviation from ECMWF analysis, (c) standard deviation of the time variation.

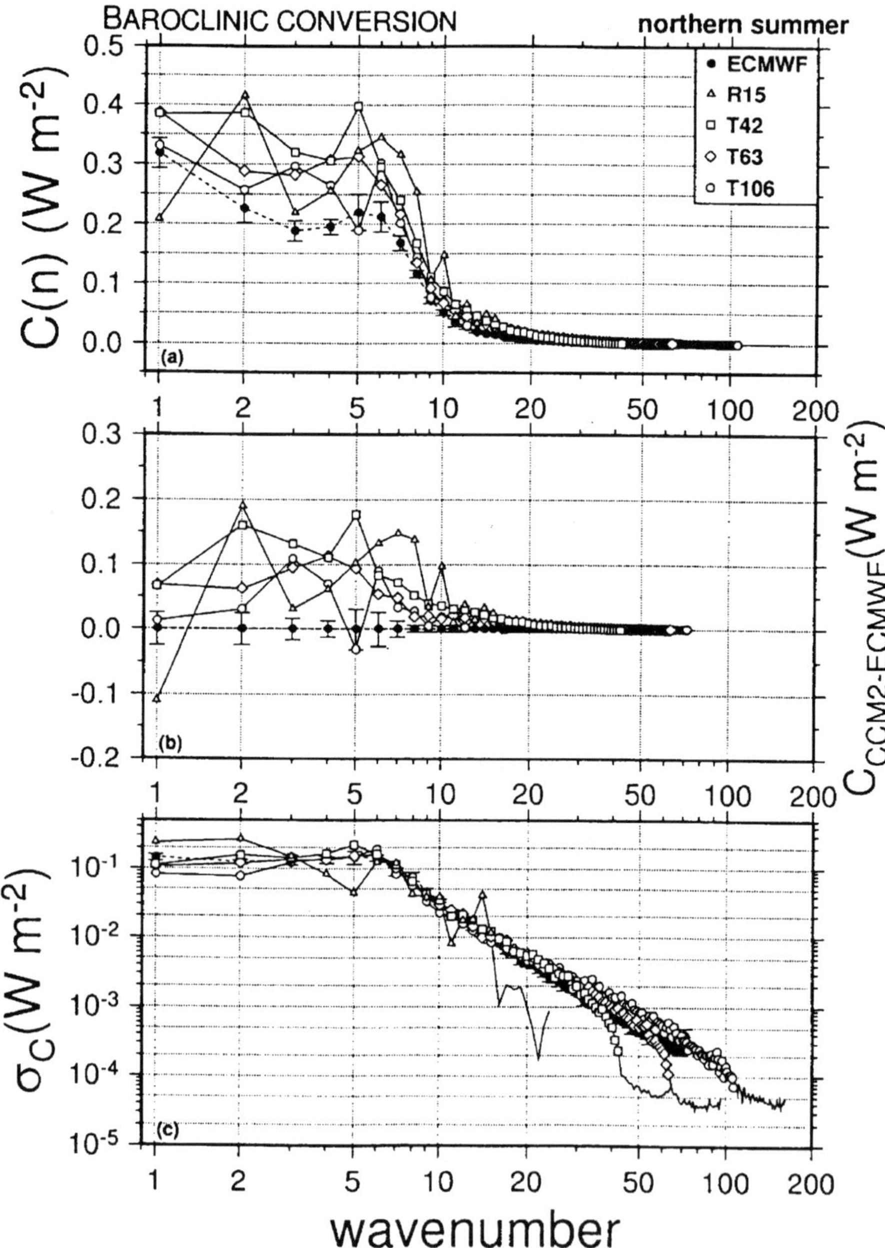

Figure 4.4 *(continued)*

from $P_{(n)}$ to $K_{(n)}$ is large at planetary waves and another peak is seen in the synoptic waves. According to the middle panel of the figure, this term is overestimated by CCM2 for almost all waves except for wavenumber 1 in the northern winter. The lower panel shows the spectrum of the standard deviation in the time series. It is noteworthy that the standard deviation is flat at the wavenumbers 1 to 6 and it decreases at the short wavelengths, showing a red spectrum with a slope of -3. The spectral characteristics observed in this energy transformation is quite similar to those in the energy spectrum itself in Figure 4.3.

Figure 4.5 illustrates the spectral distribution of zonal-wave interactions $M_{(n)}$ for CCM2 and ECMWF datasets during the northern winter and summer. The zonal-wave interactions are large for synoptic waves, which accelerate zonal motions. The model biases are greater than the standard deviation bars of ECMWF. As illustrated in Figure 4.4, the standard deviation of the time variation indicates a red spectrum beyond the synoptic wavenumbers.

Figure 4.6 illustrates the spectral distribution of wave–wave interactions $L(n)$ for CCM2 and ECMWF datasets during the northern winter and summer. This term is important for assessment of the magnitude of the scale interactions between the large and small. The term is negative at synoptic wavelengths, indicating an energy supply at this scale through baroclinic conversion. The values are positive both for planetary waves and short waves, showing that the energy is transferred from the source range to both planetary waves and short waves. There is an exception in which the value is negative at the wavenumber 2 of the northern winter. There is another type of energy source at the wavenumber 2 in the northern winter, which may be a topographic excitation of wavenumber 2.

The result of differences between CCM2 and ECMWF in the middle panel shows that the bias in R15 is unacceptable. In theory, the wave–wave interactions must be zero when summed over all the waves from one to infinity. The observed wave–wave interactions indicate negative values for synoptic waves up to the wavenumber of about 15. These negative values are compensated partly by the small positive values at large wavenumbers. This process is referred to as downscale energy transfer from synoptic waves to short waves. However, when the wavenumber is truncated at 15 in the model atmosphere as in R15, the important downscale energy cascade is forbidden. This is a strong restriction and a distortion from the reality. As a result, the wave–wave interactions for R15 indicate negative values up to wavenumber 10 and they are compensated by the unrealistic positive values at wavenumbers 10–15. The energy redistribution is evidently distorted by the artificial spectral boundary at R15.

The distortion in the wave–wave interactions become even clearer when the interaction terms are integrated with respect to the wavenumbers. The integral function is regarded as an energy flux in the wavenumber domain. Since the summation of the interactions for all waves becomes zero as required from the energy conservation law, the energy flux will converge to zero as the wavenumber approaches infinity. This physical constraint is attained at the finite wavenumbers when a model atmosphere is truncated at a finite wavenumber. For example, the energy flux must converge to zero at wavenumber 15 for the case of R15.

Figure 4.7 illustrates the energy flux over the wavenumber domain evaluated for CCM2 and ECMWF datasets during the northern winter and summer. The results for ECMWF

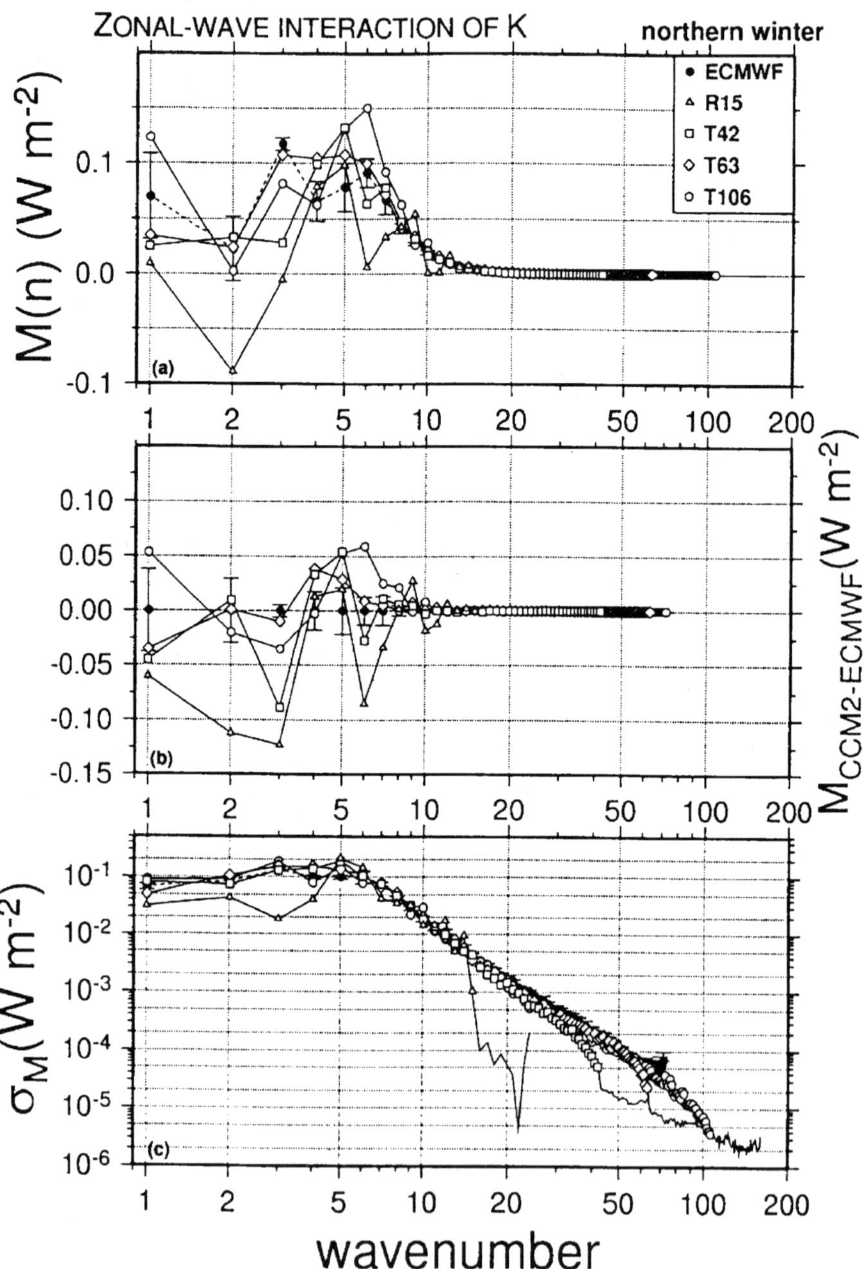

Figure 4.5 Spectral distribution of zonal-wave interaction of kinetic energy M_n for CCM2 and ECMWF datasets during northern winter and summer; (a) spectral distribution, (b) deviation from ECMWF analysis, (c) standard deviation of the time variation.

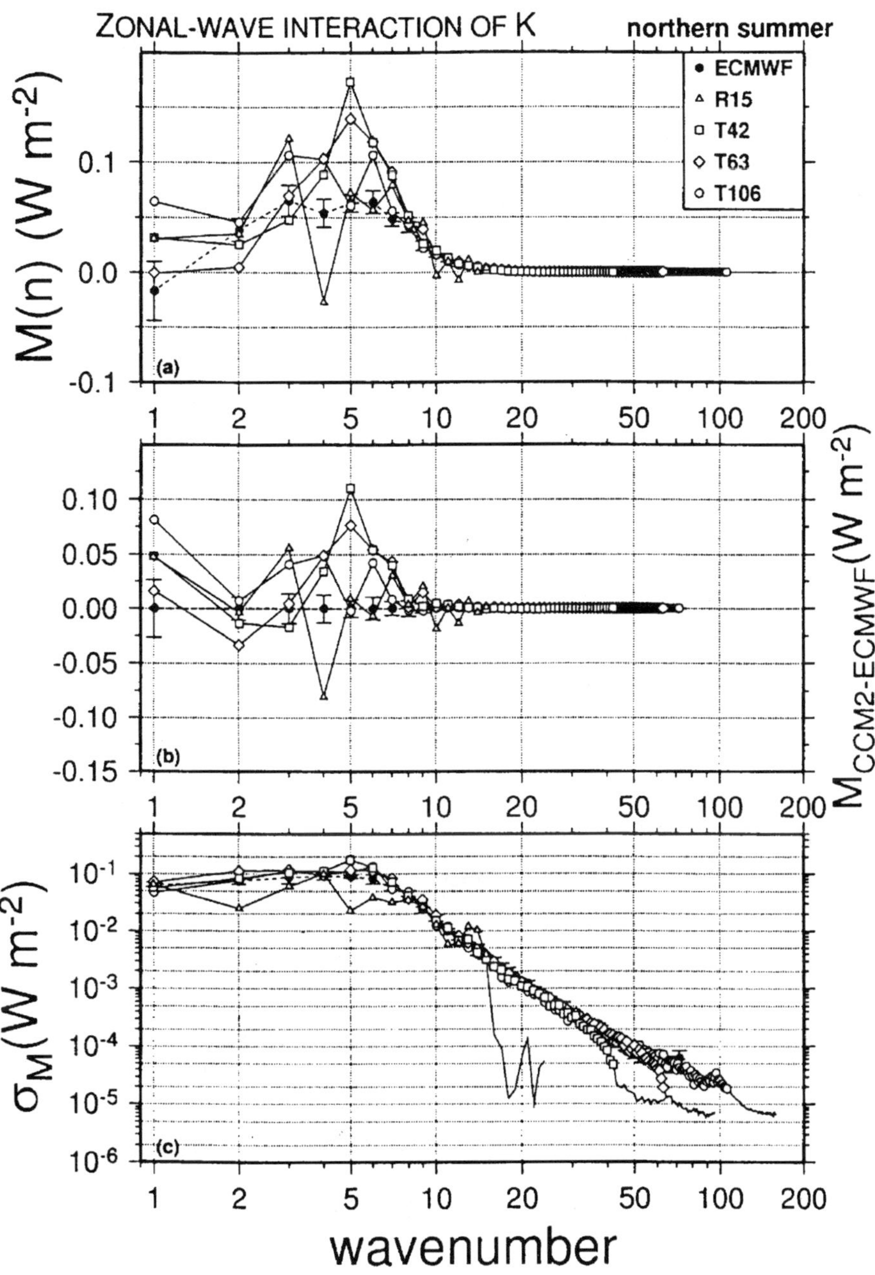

Figure 4.5 *(Continued)*

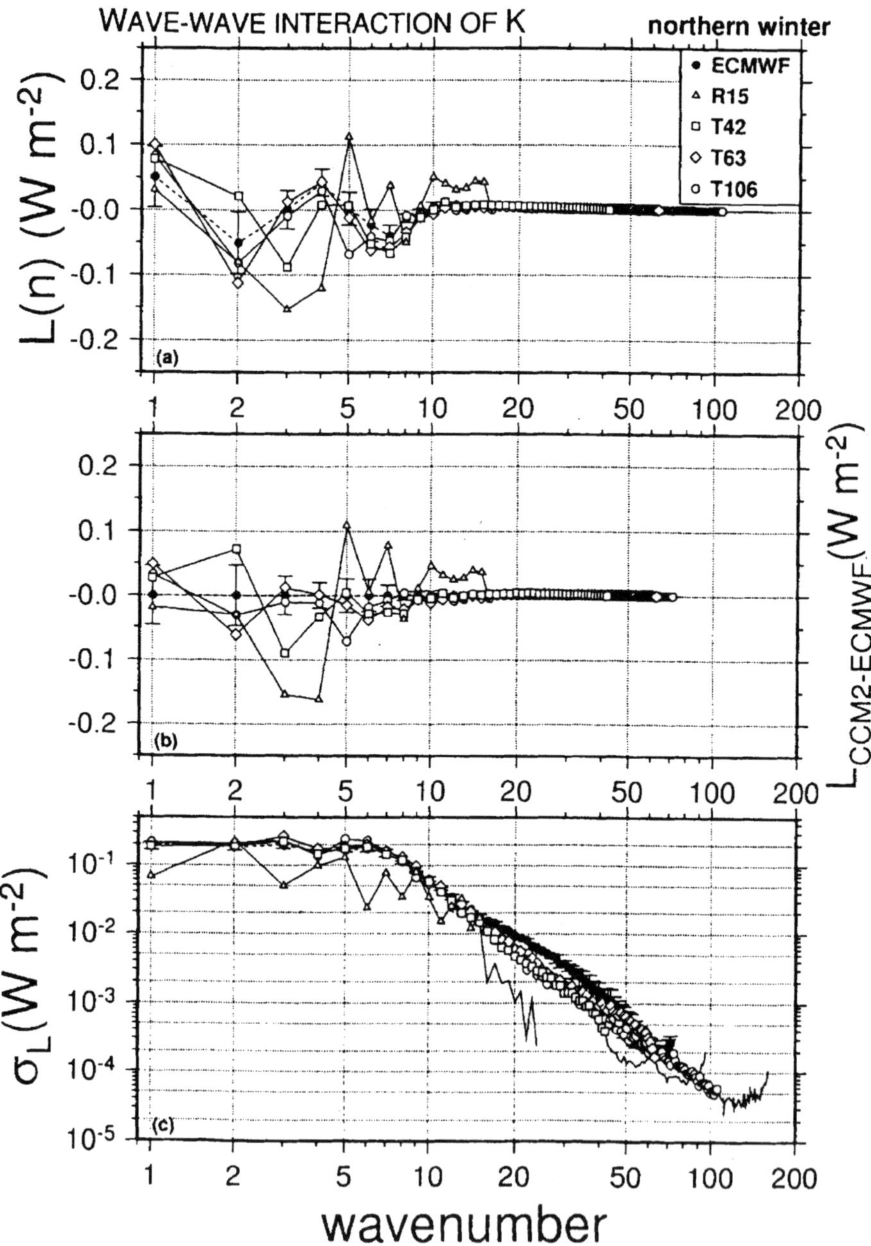

Figure 4.6 Spectral distribution of wave–wave interaction of kinetic energy L_n for CCM2 and ECMWF datasets during northern winter and summer; (a) spectral distribution, (b) deviation from ECMWF analysis, (c) standard deviation of the time variation.

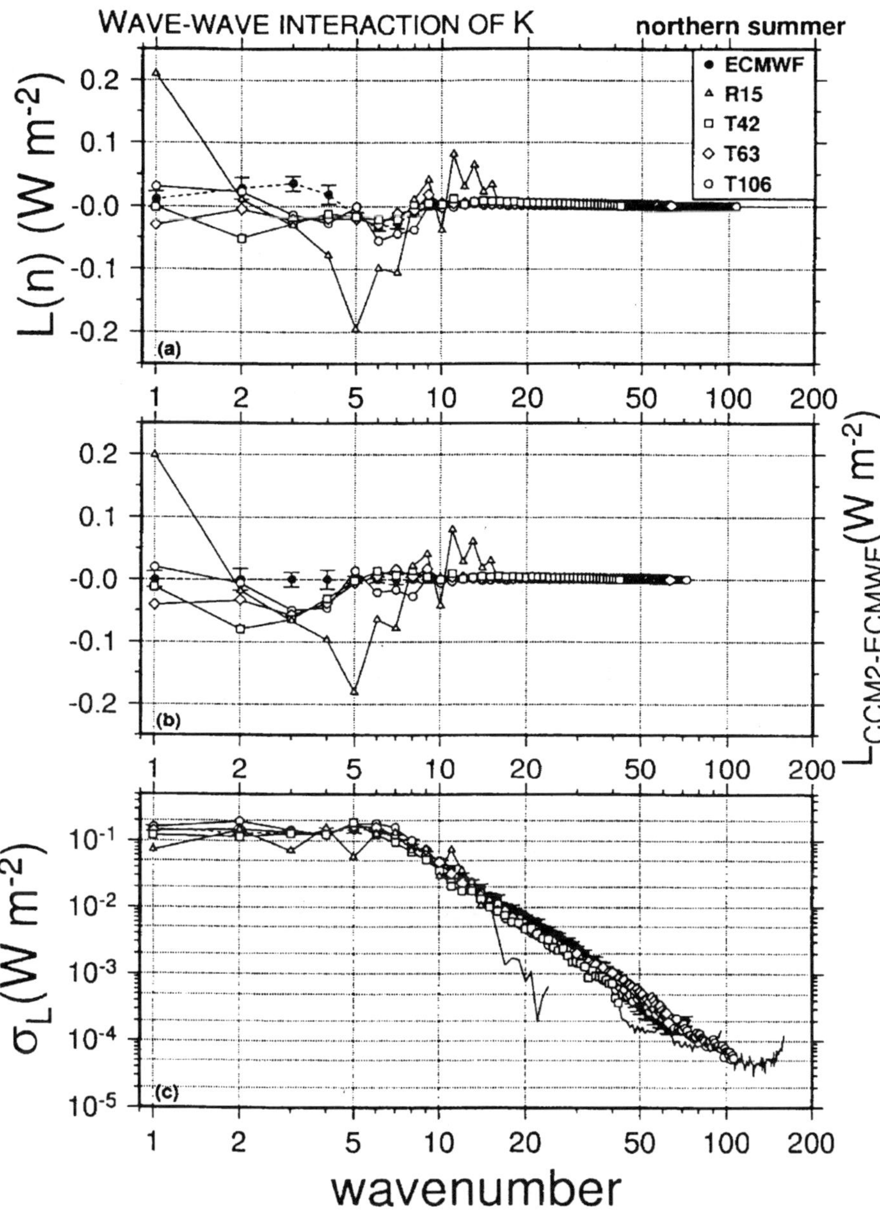

Figure 4.6 *(Continued)*

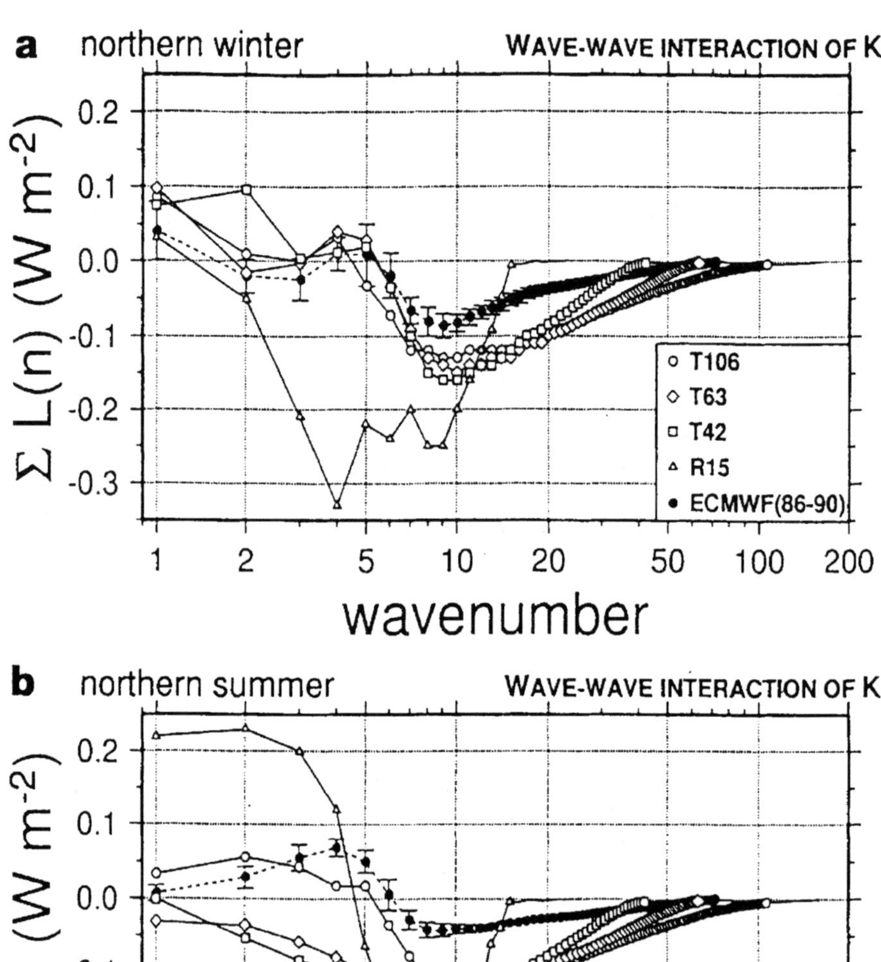

Figure 4.7 Spectral distribution of kinetic energy flux in the wavenumber domain $\sum L_n$ for CCM2 and ECMWF datasets during (a) northern winter and (b) summer.

indicate a zero crossing at the wavenumber 6 and negative values for larger wavenumbers. This implies that the energy is transferred toward larger wavenumbers (smaller scale motions). This is the flux associated with the downscale energy cascade, and the energy is transferred to smaller scale motions. When a large vortex breaks down to a number of smaller vortex and the smaller vortex breaks down to a further smaller vortex, the energy cascades down to smaller scale motions. The largest flux is seen at wavenumber 10, and the magnitude is about 0.1 W m^{-2} for the northern winter result. In the planetary waves, the values are, in general, positive as typically seen in the northern summer. In this case the energy flux directs towards the small wavenumbers (larger scale motions). This is the flux associated with the upscale energy cascade, and the energy is transferred to larger scale motions. The flux divergence represents an energy source, whereas the flux convergence represents an energy sink. Evidently, there is a large source at the synoptic scale near the zero crossing of the line where the flux strongly diverges.

Now, the results for CCM2 qualitatively indicate similar distributions of the flux as ECMWF, but the magnitudes are twice as larger for T42, T63 and T106. The result for T106 converges to zero at wavenumber 106. Similarly, the results for T63, T42 and R15 converges to zero at wavenumbers 63, 42 and 15. The energy redistribution must occur within these spectral ranges. The distortion in R15 is terrible. At the wavenumber 15 the other resolution models indicate a downscale energy flux of about 0.1 W m^{-2}. However, this flux is forced to zero in R15. The upscale energy cascade observed for the northern summer is missing in T42 and T63. Only T106 reproduces the upscale energy cascade.

4.8   Energy flow diagram

The result of the spectral energetics is summarised in the form of Saltzman's box diagram (refer to Kung 1988). Here, the kinetic energy $K_{(n)}$ and available potential energy $P_{(n)}$ are listed in the energy boxes and the energy transformations of $R_{(n)}$, $S_{(n)}$, $C_{(n)}$, $L_{(n)}$, and $M_{(n)}$ are indicated by arrows. The energy generation $G_{(n)}$ and dissipation $D_{(n)}$ are evaluated as the residual balances of the energy equations. Figure 4.8 is for the result of T106 of CCM2 during the northern winter and summer. For convenience and for simplicity, the energy boxes are combined for wavenumbers 1–3, 4–7, 8–15, 16–42, 43–63, 64–72, 73–106 and 107–160. If we sum all waves for 1–160, the resulting 4 energy box diagram is reduced to a Lorenz energy box diagram (see Lorenz 1955).

The main entrance of the atmospheric energy into the Saltzman's box diagram is at the differential heating $G_{(o)}$, where the warmer tropics are heated and the colder polar regions are cooled. Zonal available potential energy $P_{(o)}$ is excited by the differential heating with increased meridional temperature gradients. The increased meridional temperature gradients will induce baroclinic instability where warm air rises and is transported to high latitudes and cold air sinks and is transported to lower latitudes. This process is indicated by the zonal-wave interactions $R_{(n)}$ and the baroclinic conversion $C_{(n)}$. With the typical energy flow of the baroclinic instability, $K_{(n)}$ is excited mostly at the synoptic scale.

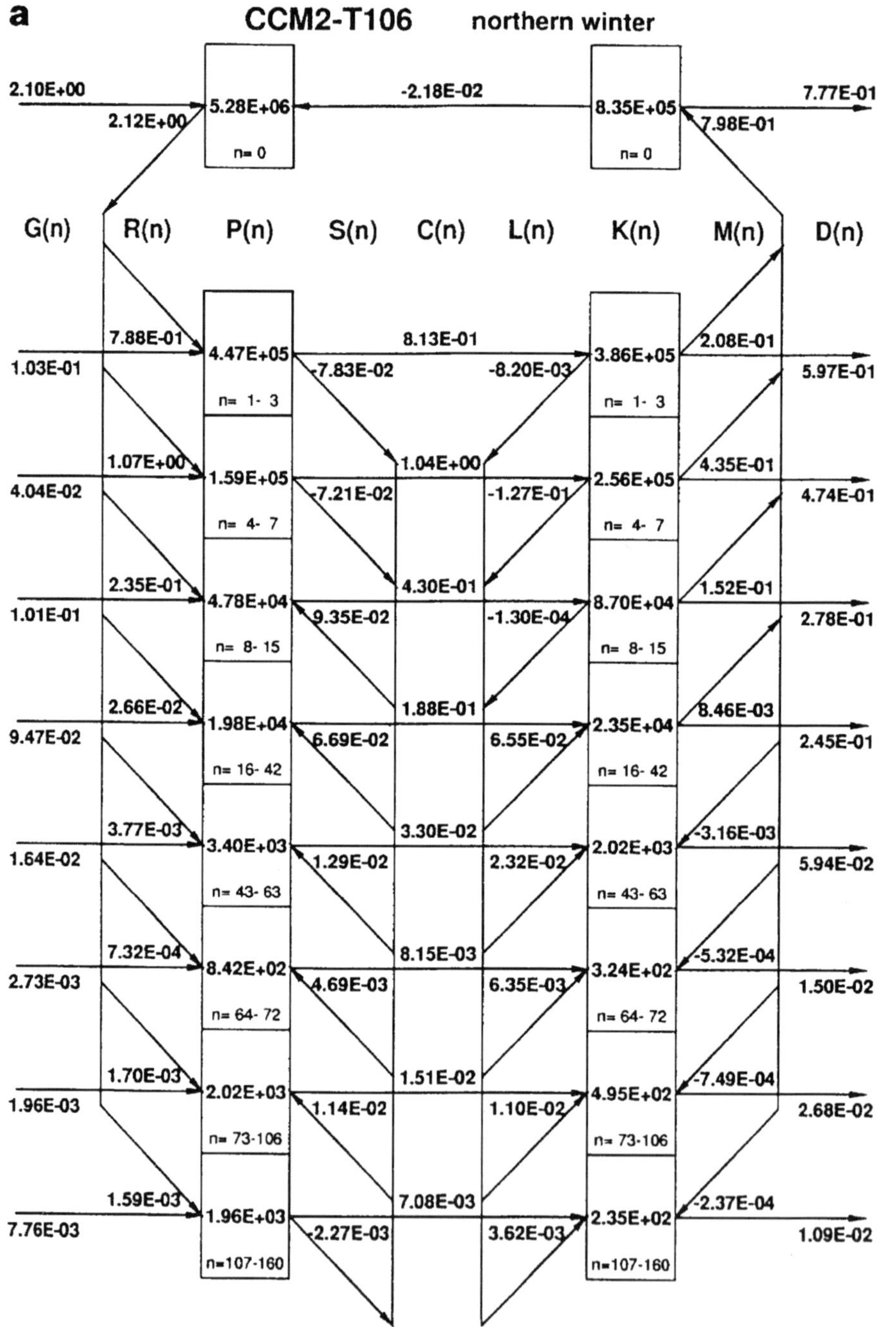

Figure 4.8 Energy flow diagram in the wavenumber domain for the CCM2-T106 simulation during (a) northern winter and (b) northern summer.

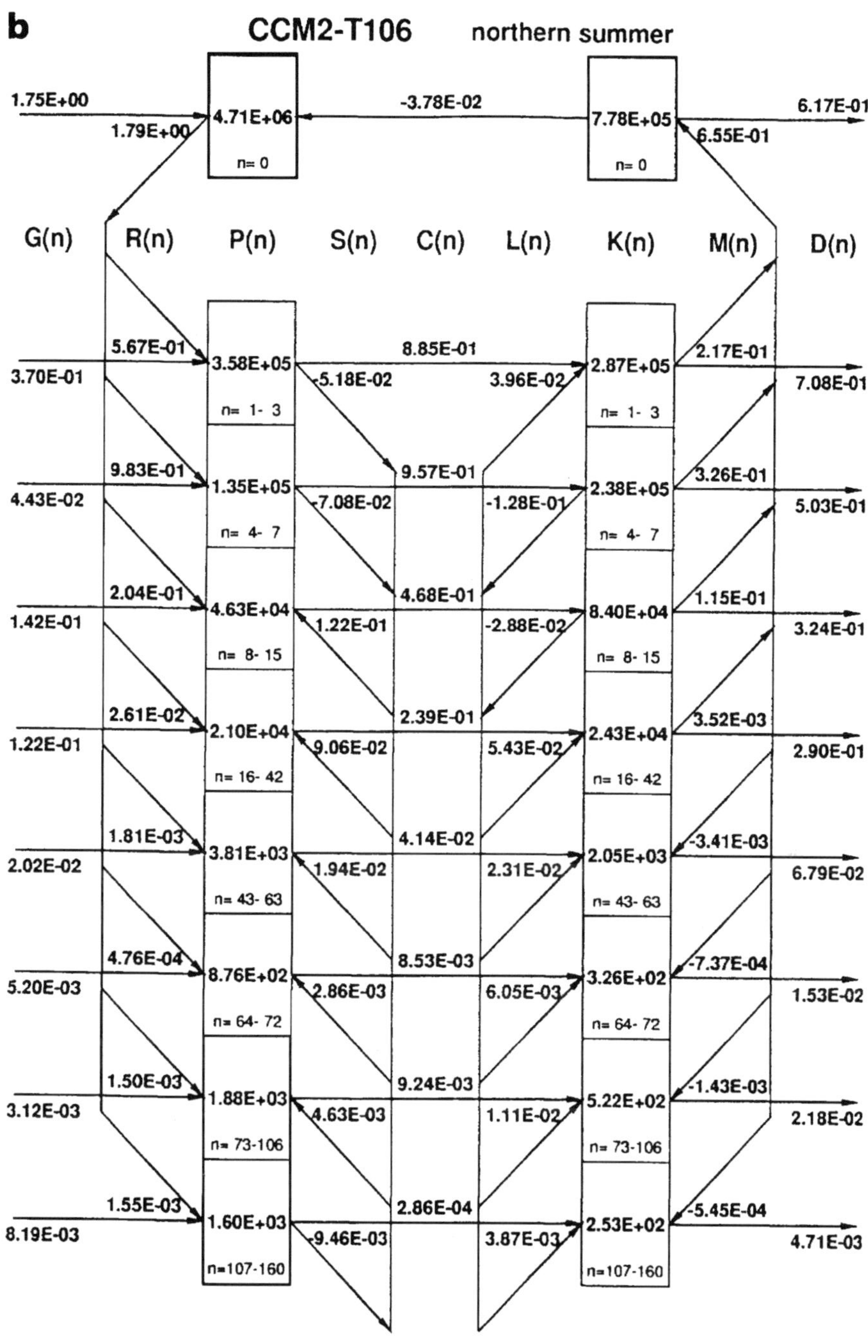

Figure 4.8 *(Continued)*

Another excitation of $K_{(n)}$ is due to the topography as a mechanical forcing in the planetary waves. Accumulated kinetic energy in synoptic and planetary waves is redistributed by the nonlinear wave–wave interactions $L_{(n)}$ to smaller scale motions by means of the downscale energy cascade. A part of the energy cascades upscale to planetary waves as seen for the northern summer. The synoptic and planetary waves accelerate the zonal jet by the barotropic conversion $M_{(n)}$ when the baroclinic waves decay. Hence, the zonal-wave interactions $M_{(n)}$ are from waves to zonal. It is found in this study that the barotropic conversion $M_{(n)}$ is negative for wavenumbers larger than 42. In other words, the short waves beyond the wavenumber 42 are not responsible for the maintenance of the zonal jet. This finding may be important for the model construction, in a sense that T42 is the minimum requirement for the sufficient two-way interactions between eddies and zonal motions. The treatment beyond these wavenumbers may be parameterised by a traditional diffusion process.

4.9   Summary

The objective of the present study is to assess the model's uncertainty associated with the energy transfer in the wavenumber domain. The energy spectrum and energy flows in the wavenumber domain are investigated with high- and low-resolution models over the truncation range of R15, T42, T63 and T106. For validation, we analysed ECMWF global analysis for five years from 1986 to 1990. The effect of the artificial truncation in the wavenumber domain to the energy redistribution of the model atmosphere is discussed in this study.

It is emphasised that the energy spectrum and energy flows, as for both direction and amount, in the CCM2-R15 simulation is much different from those in the higher resolution models. Evidently, it is difficult to employ the R15 model for the studies of climate forecasting. According to the result from T106 simulation, the wave–wave interactions of kinetic energy show an important downscale energy transfer from synoptic scales (wavenumbers 4–15) to short waves. The magnitude of the energy flux in the wavenumber domain toward the short waves are of the order of 0.1 W m^{-2} at the wavenumber 15. This energy transfer is forced as zero in R15 since the energy redistribution by the wave–wave interaction must be accomplished within the resolvable spectral range. Probably, the missing 0.1 W m^{-2} of downscale energy flux in R15 is parameterized by an unrealistically increased diffusion process in order just to draw the necessary amount of energy.

In conjunction with the climate prediction by means of a LAM nested within the GCM such as R15, we can raise a warning of a missing two-way interaction under the LAM-GCM. Suppose we attempt to save the computer resources of T106 climate prediction with a substitution of a nesting of LAM-GCM. Here, the GCM is R15 and the LAM has a resolution comparable to T106. The question is can we obtain a similar climate prediction by these two approaches: one is by T106 (expensive) and the other by LAM-GCM (cheap)? We are almost sure that the answer is no. The important finding in this study is that about 0.1 W m^{-2} of energy transfer has resulted from the complicated two-way interaction of the large- and small-scale motions at the wavenumber 15 in the case of the T106 simulation. This amount of interaction between the large- and small-scale motions is missing in LAM-R15, and the missing interactions are treated by a prescribed form of a simple diffusion process in R15. The missing energy flux of 0.1 W m^{-2} gives a measure of a residence time of 100 days by which the whole eddy kinetic energy can be replaced. Processing this amount of uncertainty within R15, can we expect a similar climate prediction of the decade ahead as in T106 using the cheap LAM-GCM? It seems that the climate prediction by LAM-R15 may allow a comparable regional resolution as in T106, but is at most a prediction by R15. What we observe with the combination of R15 and LAM is simply the prediction by R15 interpreted with the resolution of the LAM, which should be fundamentally different from the prediction by T106. The utility of the LAM-GCM depends on the application of this nesting approach. When the nesting approach is used for a short-term weather prediction, the method should work perfectly as seen in daily operational forecasting. Even for a climate study, the method of nesting will be a valuable one when it is used for a process study, for example, sea-ice dynamics and feedback process study by Lynch et al. (1995). Here, the LAM is used to help to understand the reaction of the meso-scale phenomena in the Arctic, such as arctic surface, sea-ice, lead and polynia, against a forcing by large-scale motions. The regional climate prediction, however, depends on an equilibrium of the complicated two-way interaction between the large and small. Therefore, the application to a regional climate prediction needs a further scientific basis.

The improvements from R15 to T42 resolution models are significant. The characteristics of the T63 model are similar to those in the T106 simulation. The zonal-wave and wave–wave interactions of kinetic and available potential energies in the T106 simulation, however, tend to transfer more energies into higher wavenumbers than those in the T63 atmosphere. The T42 simulation indicates a notable defect in reproducing the double-jet structure in SH winter. This discrepancy has resulted from the erroneous zonal-wave interactions of kinetic energy, $M_{(n)}$. The T106 simulation also has some defects. For example, the generation of eddy kinetic energy simulated with the T106 model is obviously different from those of the observed. In general, the T63 simulation is comparable to the T106 atmosphere, and in practice, T42 may be sufficient. It is found in this study of the high-resolution model analysis that the barotropic conversion $M_{(n)}$ is negative for wavenumbers larger than 42. In other words, the short waves beyond the wavenumber 42 are not responsible for the maintenance of the zonal jet. This finding may be important for the model construction, in a sense that T42 is the minimum requirement for the sufficient two-way interactions between eddies and zonal motions. The treatment beyond these wavenumbers may be parameterized by a traditional diffusion process.

Acknowledgments

This study was jointly supported by EPRI and CRIEPI under the MECCA project. The authors appreciate Drs H. Hiromaru, J. Tsutsui, H. Amano, S. Taguchi, A. Kasahara, D.L. Williamson and J.J. Hack, and C. Hakkarinen for the dedicated support.

References

Giorgi, F. and Mearns, L.O. (1991) Approaches to the simulation of regional climate change: A review. *Review of Geophysics*, **29**, 191–216.

Hack, J.J., Boville, B.A., Briegleb, B.P., Kiehl, J.T., Rasch, P.J. and Williamson, D.L. (1993) Description of the NCAR community climate model (CCM2). Technical Report NCAR/TN-328+STR, NCAR.

Hasegawa, A. (1995) Comparative energetics of the climate models with high and low resolutions. *Masters Thesis*, Graduate Program in Geoscience, University of Tsukuba.

Kida, H., Koide, T. , Sasaki, H. and Chiba, M. (1991) A new approach for coupling a limited area model to a GCM for regional climate simulations. *Journal of the Meteorological Society of Japan*, **69**, 723–728.

Kung, E.C. and Tanaka, H.L. (1983) Energetics analysis of the global circulation during the special observation period of FGGE. *Journal of Atmospheric Science*, **40**, 2575–2592.

Kung, E.C. and Tanaka, H.L. (1984) Spectral characteristics and meridional variations of energy transformations during the first and second special observation periods of FGGE. *Journal of Atmospheric Science*, **41**, 1836–1849.

Kung, E.C. (1988) Spectral energetics of the general circulation and time spectra of transient waves during the FGGE year. *Journal of Climate*, **1**, 5–19.

Lynch, A.H., Chapman, W.L., Walsh, J.E. and Weller, G. (1995) Development of a regional climate model of the western Arctic. *Journal of Climate*, **8**, 1555–1570.

Lorenz, E.N. (1955) Available potential energy and the maintenance of the general circulation. *Tellus*, **7**, 157–167.

Matsuno, T. (1980) Lagrangian motion of air parcels in the stratosphere in the presence of planetary waves. *Pure Application in Geophysics*, **118**, 189–216.

Ogasawara, N. (1995) Comparative study of the spectral energetics of the general circulation with JMA, NMC and ECMWF global analyses. *Graduate Thesis*, University of Tsukuba.

Oort, A.H. (1964) On estimates of the atmospheric energy cycle. *Monthly Weather Review*, **92**, 483–493.

Saltzman, B. (1957) Equations governing the energetics of the large scales of atmospheric turbulence in the domain of wavenumber. *Journal of Meteorology*, **14**, 513–523.

Saltzman, B. (1970) Large-scale atmospheric energetics in the wavenumber domain. *Review in Geophysical Space Physics*, **8**, 289–302.

Tanaka, H.L. and Kimura, K. (1996) Normal mode energetics analysis and the intercomparison for the recent ECMWF, NMC and JMA global analyses. *J. Meteor. Soc. Japan*, **74**, 525–538.

Williamson, G.S. (1993) CCM2 datasets and circulation statistics. Technical Report NCAR/TN-391+STR, NCAR.

CHAPTER 5

ANTHROPOGENIC AEROSOLS AND CLIMATE CHANGE: A METHOD FOR CALIBRATING FORCING

Joyce E. Penner, Tom M.L. Wigley, Peter Jaumann, Ben D. Santer and Karl E. Taylor

5.1   Introduction

Anthropogenic emissions of sulfur dioxide from industrial activity, fossil fuel combustion and biomass burning are now known to be large enough, (relative to natural sources), to perturb the chemistry of vast regions of the troposphere. Sulfur dioxide undergoes photochemical transformations in the troposphere in both the gas phase and aqueous phase which lead to the formation of sulfuric acid, H_2SO_4, which, when formed via the gas phase, is a condensable gas that may either adhere to pre-existing particles or form new particles in the atmosphere. The resulting increased particle concentration in the atmosphere may decrease the globally-averaged solar radiation absorbed by the earth–atmosphere system by as much as 1 W m^{-2} (Charlson et al. 1992), although estimates of the anticipated global average climate forcing, (through direct reflection of solar radiation), by anthropogenic sulfate aerosols have ranged as low as –0.3 W m^{-2} (Kiehl & Briegleb 1993). The cloudy-sky (indirect) forcing by anthropogenic aerosols, wherein aerosol cloud condensation nuclei concentrations are increased, thereby increasing cloud droplet concentrations and cloud albedo and possibly influencing cloud persistence, may also be significant. Estimates of this forcing range from 0 to ~–1.5 W m^{-2}, but are highly uncertain.

Here, we outline a method whereby simulations of the climate response to anthropogenic sulfate aerosols together with an analysis of the similarity between the predicted patterns of response and the historical patterns of temperature change, may be used to deduce a range of forcings that is consistent with the data. Our preliminary analysis, presented below, is consistent with the conclusion that the effective total forcing by northern hemisphere anthropogenic aerosols since the middle of the 19th century must be at least half the magnitude of today's effective greenhouse forcing. The analysis is also consistent with the conclusion that the aerosol forcing could be nearly as large as the greenhouse forcing.

Because the lifetime of aerosols is short relative to the major greenhouse gases and because anthropogenic sulfur is emitted mainly in the northern hemisphere, the indirect and direct forcing is regionally distinct and primarily associated with the industrial and population centers in the eastern US, Europe and eastern Asia. This non-uniform distribution, in conjunction with the greenhouse forcing, leads to a differential spatial forcing with net heating in some areas and net cooling in others. For this study, we examined the combined effects of increases in CO_2 and anthropogenic sulfate aerosol in a coupled climate/chemistry model. We extended the GCM experiments of Taylor and Penner (1994), which were made possible through a grant from MECCA, to further examine the climate sensitivity parameter and its dependence on the regionally distinct forcing by sulfate aerosols. Thus, we performed four calculations: (1) a pre-industrial case with no anthropogenic aerosol and 275 ppm CO_2; (2) no anthropogenic aerosol and near present-day atmospheric CO_2 concentration of 345 ppm; (3) present-day (1980) anthropogenic sulfur emissions and 275 ppm CO_2; and (4) present-day anthropogenic sulfur emissions and 345 ppm CO_2. Unlike our previous results, we now find that the climate sensitivity parameter is approximately constant and independent of the type of forcing. As in our previous study, the differential spatial forcing with areas of net cooling and areas of net warming leads to a complex spatial response pattern with areas of cooling in parts of the northern hemisphere and areas of warming elsewhere.

The magnitude of forcing by anthropogenic aerosols is still not known with accuracy, especially if consideration of the indirect forcing by aerosols is taken into account (Penner et al. 1994a; Shine et al. 1995). For example, the indirect forcing by anthropogenic sulfate aerosols has variously been estimated as ranging from –0.5 to –1.5 W m^{-2} (Jones et al. 1994; Chuang et al. 1994; Boucher & Lohmann, 1995)—IPCC gives the uncertainty range as 0 to –1.5 W m^{-2} (Shine et al. 1995). For the direct forcing IPCC gives –0.4 W m^{-2} as its central estimate with a range from –0.25 to –0.9 W m^{-2}. If the climate sensitivity were known with accuracy and the natural variability on time scales of roughly 50 years and longer were small, then one might hope to determine the magnitude of aerosol forcing by fitting the predicted global average change for various levels of aerosol forcing to the observed record of change. Previous attempts to estimate aerosol forcing empirically have used this approach (Schlesinger & Ramankutty 1992). However, this method is unable to derive a single value for the aerosol forcing because the climate sensitivity is not known with accuracy.

In this chapter, we explore whether the predicted *pattern* of temperature change for different levels of sulfate aerosol forcing can be used together with the observed pattern of temperature change to estimate the magnitude of the aerosol forcing. In our analysis, we essentially assume that forcing by other anthropogenic aerosols is either small relative to that of anthropogenic sulfate or has a pattern of temperature response that is roughly similar to that from sulfate forcing (e.g. forcing mainly associated with northern hemisphere industrial regions). The analysis also depends on the assumption that the pattern of natural variability of the climate system is different from that due to anthropogenic factors. To establish our procedure, we first examine the issue of whether the regional response to changes in CO_2 and anthropogenic sulfate aerosol is essentially linear, that is, are the regional patterns of temperature change predicted for the combined CO_2 plus sulfate aerosol perturbation essentially equal to the sum of the temperature changes predicted for CO_2-only forcing plus anthropogenic sulfate aerosol-only forcing? We find that the response pattern is (to first-order) linear, so that we can explore uncertainties in the magnitude of the aerosol forcing by linearly combining results, rather than performing a whole suite of climate model experiments with different aerosol forcings. We then carry out some preliminary analyses to show how pattern correlations between modeled and observed temperature changes, and the trends in these correlations, vary as the relative forcings of aerosols and CO_2 are changed. Further work is needed to determine whether an optimum aerosol to CO_2 (or greenhouse gas) forcing ratio can be identified which maximizes the pattern correlations and their trends.

5.2   Model description

The Lawrence Livermore National Laboratory (LLNL) tropospheric chemistry model, called GRANTOUR (cf. Walton et al. 1988), was used in conjunction with the Livermore version of the NCAR CCM1 model. The latter GCM was coupled to a simple 50-meter mixed-layer ocean model with interactive sea ice and specified meridional oceanic heat flux (Covey & Thompson 1989). For the chemical simulations, the anthropogenic sulfur inventories developed by Spiro et al. (1992) for SO_2 from industrial, fossil fuel and biomass burning were used except over North America where the inventory developed by Benkowitz (1982) for fossil fuel and industrial SO_2 emissions was used. These inventories represent emissions for $\sim$1980. They are smaller today in both the US and Europe but have increased over Asia (Penner et al. 1994b). The natural sulfur emission inventories from Spiro et al. (1992) were also used in our simulations except that the oceanic emissions of DMS were doubled to better reflect the mid-range of estimates (Penner 1994). This also provided a more reasonable simulation of observed sulfate concentrations in remote regions (Penner et al. 1994b).

The gas phase reactions of DMS and SO_2 with OH for the model of the sulfur cycle presented below, assumed a background concentration of OH that was specified according to the latitudinally- and seasonally-varying calculated concentrations from the LLNL two-dimensional model (see Penner et al. 1991). (H_2S emissions are small relative to other sources of sulfur and were assumed to undergo the same reactions as DMS with identical reaction rate coefficients.) The reaction rate coefficients were consistent with those recommended by DeMore et al. (1992) and Atkinson et al. (1992) (see Table 5.1). They were varied monthly according to the average temperature and pressure at each location in the model. Aqueous reaction of SO_2 to form $SO_4^=$ in clouds was treated in a simplified manner. This process was assumed to have an average e-folding lifetime of 30 hours at $40°N$ at the surface in summer and was scaled, proportional to the square of the locally specified concentration of OH at other locations. This formulation is a simplified attempt to account for the observed seasonal variations in sulfate concentrations and wet deposition in North America, by assuming that the rate of formation of $SO_4^=$ in clouds depends

Table 5.1 Reactions and rate coefficients used in the model calculations (cm^3 s^{-1})

Reaction	Rate coefficient at ground level and 298°
DMS + OH $\rightarrow$ SO_2 (multi-step)	6.13×10^{-12}
SO_2 + OH $\rightarrow$ $SO_4^=$ (multi-step)	8.89×10^{-13}
Aqueous reactions converting SO_2 to $SO_{43=}$ OH concentration	Parameterized based on the latitudinally and monthly varying

Table 5.2 Deposition velocities and precipitation scavenging coefficients used in the model

Species	Deposition velocity (cm s^{-1})	Large-scale precipitation scavenging coefficient (cm hr^{-1})	Convective precipitation scavenging coefficient (cm hr^{-1})
SO_2	0.8	0.8	1.5
$SO_4^=$	0.1	5.0	1.5

both on the frequency of interception of a cloud by an air parcel (average cloud-free periods vary from 10 to 80 hours (Lelieveld & Crutzen 1990, 1991)) and on the local concentration of H_2O_2 (which was assumed to be proportional to the square of the specified concentration of OH). The formulation provides a reasonable account of the observed variations in both sulfate concentrations and deposition in precipitation in the eastern US (Penner et al. 1994b). In Europe, observed seasonal wet deposition of sulfate shows less variation between summer and winter than the modeled deposition patterns, in part due to the increased winter emissions expected in that region which were not represented in the model.

In addition to the photochemical reactions in the model, each species may experience deposition if parcels are in the lowest 100 mb. The deposition velocities are set by average observed values. In addition, trace species may be scavenged by precipitation with scavenging rates proportional to the amount of precipitation in the model and scavenging coefficients set to reproduce measured washout ratios. The values used in the present simulations are shown in Table 5.2. Other details regarding the model formulation can be found in Walton et al. (1988) and Penner et al. (1991, 1994b).

Model calculations of the tropospheric sulfate aerosol abundance compare well with data when simulations use the wind and precipitation fields from a version of the NCAR CCM1 (GCM) that used fixed sea surface temperatures (Penner et al. 1994b). When the GCM is coupled to the mixed-layer ocean, however, sufficiently different fields are developed that surface concentrations and column abundances of sulfate increase by about a factor of two in the source area over Europe. As noted in Taylor and Penner (1994) this model error leads to an overestimate in global-mean sulfate forcing in the coupled model of approximately 0.1 W m^{-2}.

5.3   Model predictions of forcing and temperature response

In order to assess anticipated patterns of temperature response to anthropogenic emissions of sulfur dioxide and carbon dioxide, we performed four calculations. The first, or pre-industrial simulation, assumes a CO_2 concentration of 275 ppm and no anthropogenic sulfate aerosols. We refer to this low-sulfur, low-CO_2 simulation as the control (CTRL) simulation. The second simulation (low sulfur, high CO_2; C) assumed a near present-day CO_2 abundance of 345 ppm. The third simulation (high sulfur, low CO_2; S) used pre-industrial CO_2 concentrations with present-day sulfur emissions, and the fourth (high sulfur, high CO_2; SC) assumed present-day sulfur emissions and 345 ppm CO_2. Averages over years 10 to 20 of these simulations formed the basis of our previously published results (Taylor & Penner 1994).

Here, we extend these simulations for another 10, 30, 20, and 10 years for CTRL, C, S, and SC, respectively. Figure 5.1 shows the global and annual average surface air temperature predicted after the first 10 years of simulation. As noted there, the average temperature in the C simulation, which had decreased by about 0.2°C during years 11 to 20, decreases another 0.4°C in years 20 to 30, but appears to have settled after that time. The average temperature in the S simulation had decreased 0.5°C in years 11 to 18, but had then increased again by 0.2°C in years 18 to 20. In years 20 to 40, the simulated global average temperature first decreases, then increases by $\sim$0.2°C. Both SC and CTRL exhibit decadal scale variations of order 0.2°C to 0.3°C over years 11 to 30.

The variability in the global average surface air temperature associated with these simulations was unexpected. Indeed, the C simulation, after 30 years of integration, only exhibits fluctuations on the order of $\pm$0.1°C, while the other simulations, which all include sulfate forcing, have much larger temperature variations. Apparently, very long simulation times are needed in order to obtain stable response patterns.

The temperature response to forcing may be analyzed in terms of climate sensitivity. This parameter is useful because it allows one to compare the responses of different climate models to different forcings on a common basis. The climate model sensitivity parameter measures the global average temperature change relative to the total global forcing. Table 5.3 gives the forcing and response results for the CO_2-only and for the sulfate aerosol-only calculations. These results are based on an average of the last 20 years from each of the simulations shown in Figure 5.1. The difference in temperature is the difference between the C case and CTRL (pre-industrial) case for forcing by CO_2 and the difference between the S case and CTRL (pre-industrial) case for forcing by $SO_4^=$ only. Our previous results, which were based on the second 10 years of simulation in each of the four cases, yielded markedly different climate sensitivity parameters for the CO_2 and sulfate aerosol forcing cases. Our present results yield much closer climatic sensitivity values. This result agrees with more recent simulations (Cox et al. 1995; Ramaswamy & Chen 1995; Roeckner et al. 1995). Our conclusion is still preliminary because, given the length of simulations reported here (Figure 5.1), we cannot rule out climate variability in the model on time scales longer than a few decades. We note that as a result of these extended runs, the climate model sensitivity to doubled CO_2 is now estimated as 4.8°C.

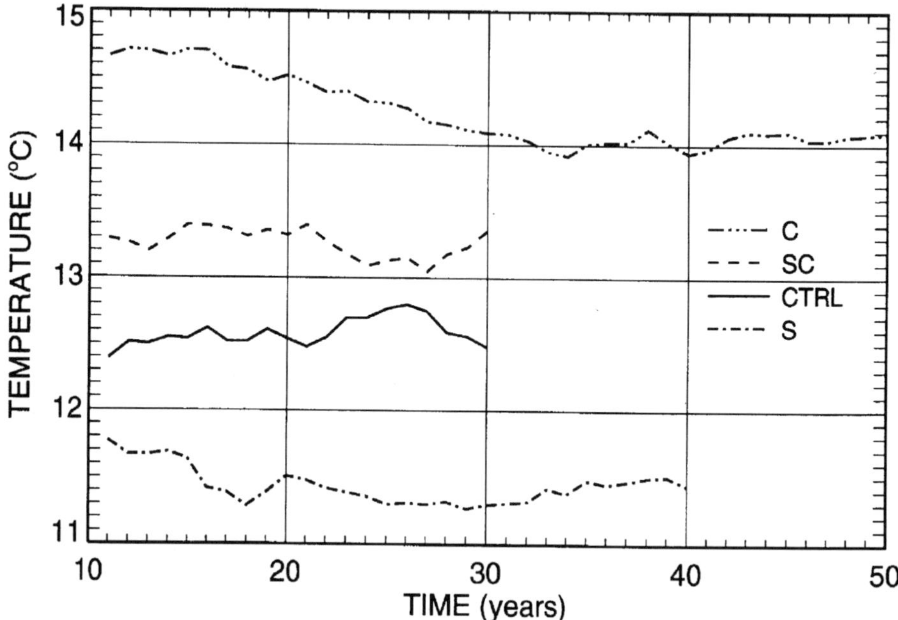

Figure 5.1 Global and annual average surface air temperature predicted after the first 10 years of simulation for the low sulfur, low CO_2 simulation (CTRL), the high sulfur, high CO_2 simulation (SC), the low sulfur, high CO_2 simulation (CTRL), and the high sulfur, low CO_2 simulation (S).

Table 5.3 Forcing, climate response, and climate sensitivity for CO_2 and sulfur simulations

	CO_2 only	$SO_4^=$ only
Forcing, ΔQ (W m^{-2})	1.26	–0.95
Response, ΔT (K)	1.5 ± 0.2	-1.2 ± 0.2
Sensitivity, $\lambda = \Delta T / \Delta Q$ (K W^{-1} m^2)	1.2 ± 0.2	1.2 ± 0.2

The main features of our previously published climate response results remain valid, namely, that the pattern of response exhibits a regional character that is distinct from the forcing pattern. This is demonstrated in Table 5.4 which presents the calculated forcing in each hemisphere together with the change in temperature in each hemisphere from the CO_2 only, sulfate only, and combined cases. In the CO_2 only case, the forcing is nearly equal in each hemisphere, and the response in each hemisphere is within 0.2°C of the global average response. For the $SO_4^=$ only case, the forcing is more than a factor of five larger in the northern hemisphere, but the response in each hemisphere is again within 0.2°C of the global average response. Thus, although the forcing may be concentrated in

Table 5.4 Hemispheric forcing and climate response

	Forcing (W m^{-2})		ΔT (°C)	
	NH	SH	NH	SH
CO_2 only	1.3	1.2	1.3	1.6
$SO_4^=$ only	−1.7	−0.3	−1.4	−1.0
$SO_4^= + CO_2$	−0.4	0.9	0.3	1.0

Table 5.5 Seasonal forcing and response

	Forcing (W m^{-2})		ΔT (°C)	
	DJF	JJA	DJF	JJA
CO_2 only	1.2	1.3	1.4	1.5
$SO_4^=$ only	−0.5	−1.5	−1.2	−1.2
$SO_4^= + CO_2$	0.7	−0.2	0.6	0.7

one hemisphere, significant surface temperature changes occur in both hemispheres. In the combined forcing case, there is net cooling in some regions of the northern hemisphere, but the hemispheric average temperature change in the northern hemisphere is positive. In this case, the magnitude of the difference between the northern hemisphere and southern hemisphere response is larger than it is in the other two experiments.

Finally, one may also note that in the simulations that include sulfate, the global average forcing varies significantly with season. However, the global average temperature change is about the same in all seasons. Table 5.5 shows the forcing and response by season. In northern hemisphere winter (DJF) the forcing by sulfate aerosols is only one-third that for the northern hemisphere summer (JJA), but the calculated global average temperature change is similar in both seasons. Thus, the model's response is strongly moderated by seasonal ocean heat storage.

5.4   A method for calibrating anthropogenic aerosol forcing

Detection of an aerosol and greenhouse forcing signature in the climate record is an important area of active research. Initial results show that if sulfate aerosols are included in the model, the predicted pattern of temperature response is more similar to the patterns recorded in the historical surface air temperature record than if sulfate aerosols are not included (Mitchell et al. 1995; Santer et al. 1995, 1996). However the magnitude of forcing by anthropogenic aerosols is still not well known, especially if consideration of the indirect forcing by aerosols is taken into account (Penner et al. 1994a; Shine et al. 1995). The direct forcing by anthropogenic sulfate aerosols is estimated by IPCC to lie between –0.25 and –0.9 W m^{-2} since 1850 (Shine et al. 1995). This should be offset somewhat by the effects of carbonaceous aerosols ('soot') associated with fossil fuel combustion (Penner et al. 1993; Haywood & Shine 1995; Penner 1995; Cooke & Wilson 1996; Liousse et al. 1996). IPCC 'best guess' value for the direct forcing by sulfate and soot aerosols is –0.3 W m^{-2}. The indirect forcing by anthropogenic sulfate aerosols has been estimated in a number of studies (e.g. Jones et al. 1994; Chuang et al. 1994; Boucher & Lohmann 1995) with widely varying results. The estimated range from IPCC (1995) is 0.0 to –1.5 W m^{-2}, with a highly uncertain 'best guess' of –0.8 W m^{-2}. A further complication is that present climate simulations do not account for the possibility of important changes in forcing due to smoke associated with biomass burning (Penner et al. 1992; Liousse et al. 1996) and/or dust aerosols associated with changes in land use (Tegen & Fung 1995), or due to ammonium and nitrate aerosols associated with fertilizer and land use changes (in the case of ammonium) or with fossil fuel use (in the case of nitrate aerosols) (Penner et al. 1994a). An inability to characterize aerosol forcing accurately as a function of space and time is, therefore, an important impediment in climate detection studies.

One possible approach to improving this situation is to develop a set of simulations in which greenhouse forcing remains constant at, for example, today's level of forcing, while aerosol forcing is varied in accord with a reasonable range of estimates. The simulation with aerosol forcing that was best able to explain the historical temperature record may then be identified as that which has the 'best' forcing. In principle, the evaluation of such experiments could be carried out at both the global-mean and regional levels. We note, however, that it is not possible to estimate aerosol forcing with any accuracy by comparison of the predicted *global-mean* temperature change with the historical record of global-mean change. This is because the predicted global-mean temperature change is sensitive to the model's climate sensitivity parameter, which varies widely among different models (IPCC 1990). In addition, uncertainties in other anthropogenic and natural forcings as a function of time and the unknown contribution to past warming due to natural variability (Wigley & Raper 1991) make empirical estimates of aerosol forcing from global-mean data poorly constrained. An alternative approach is to look at the *patterns* of change associated with different aerosol forcing amounts, and see whether the fit between such model-derived patterns and observations is optimized for some particular aerosol magnitude, extending the work of Santer et al. (1995, 1996). While natural variability may still play a role in the uncertainty associated with deriving an aerosol forcing from a pattern analysis, it is less likely to do so.

Santer et al. (1995) used a standard centered pattern correlation to determine the similarity between the predicted spatial patterns of near-surface temperature change for the simulations described here and the historical record of temperature change. The 50-year trends in the pattern correlation statistic, $R(t)$, for the SC simulation for JJA and SON were significantly different from zero when judged relative to similar 50-year trends developed by correlation of the predicted patterns with temperature records from two separate long-term unforced climate simulations.

The centered pattern correlation statistic correlates patterns in which the global-mean change has been subtracted. Thus it is a measure of the similarity in the *regional* features of temperature change only and is not sensitive to the global-mean temperature change. Here, we suggest that this same statistic may be used together with model simulation patterns that result from a variety of forcing levels to determine the forcing that gives the largest (and most significant) 50-year trends in $R(t)$. The use of 50-year trends is justified based on the significance in this trend-length found by Santer et al. (1995), the fact that the forcing change over a 50-year period is relatively large, and the fact that the historical data cover a reasonable portion of the globe if we restrict our analysis to trends of this length. The use of the centered pattern correlation statistic is important here, because any degree of likeness due to a likeness in the global-mean temperature change is removed from the analysis. This ana-lysis procedure, therefore, is independent of the climate sensitivity of the particular model being used as long as the predicted patterns are also independent of the climate sensitivity.

The preliminary analysis presented below relies on the assumption that the predicted pattern of temperature change when sulfate and greenhouse forcings are both included is approximately a simple linear sum of the component patterns (as is the global annual average temperature change when the climate sensitivity parameter is constant, cf. Table 5.3). If true, then instead of producing a series of simulations with varied sulfate forcing, it should be possible to use weighted linear combinations of the response patterns for CO_2-only forcing and sulfate-only forcing (in combination with the pattern correlation analysis with the historical record) to gauge the relative magnitudes of the positive and negative forcings. We have previously demonstrated that the pattern correlation coefficient between forcing and response for our simulations is never larger than 0.5 (Santer et al. 1995).

First, we test the linearity hypothesis using the model results from the extended Taylor and Penner (1994) simulations. Colour Plate 5.1 shows the annually averaged surface temperature response to forcing by sulfate only (panel a), CO_2 only (panel b), the combined forcing by CO_2 plus sulfate aerosols (panel c), and the result of adding the fields from the sulfate-only and CO_2-only fields (panel d). Comparing panel (a) with panel (b), there is a high degree of similarity in the regions where maximum warming and maximum cooling occur. Thus the strong cooling in the North Atlantic east of Greenland in the simulation with sulfate-only forcing is reflected by the high warming in this region for CO_2-only forcing. The land mass over Europe and western Asia is strongly cooled in the sulfate forcing experiment and somewhat less strongly warmed in the CO_2 forcing experiment. The region off the coast of Antarctica also responds strongly in both simulations. In the combined experiment (panel c), these areas of response are again evident. The strong sulfate forcing in the northern hemisphere leads to a net cooling dominating the northern hemisphere response regions and warming dominating the southern hemisphere regions.

The added fields (panel (a) plus panel (b)) are shown in panel (d). If the response patterns combine linearly, we would expect this panel to look very similar to panel (c), and there are notable similarities. These again emphasize the regions where maximum response was evident in panels (a) and (b). However, the global-mean temperature in panel (d) is less than that of panel (c) (0.3°C versus 0.6°C).

Table 5.6 Between experiment pattern correlation results and standard deviation statistics

Season	r(SC, S)	r(SC, C)	r(SC, S + C)	r(S, C)	S.D.(S)/S.D.(SC)	S.D.(C)/S.D.(SC)
DJF	0.48 (0.46)	−0.18 (0.01)	0.46 (0.58)	−0.73	1.94	1.65
MAM	0.49 (−0.26)	0.29 (0.79)	0.68 (0.72)	−0.31	0.98	0.87
JJA	0.58 (−0.26)	0.32 (0.79)	0.77 (0.74)	−0.30	0.81	0.66
SON	0.59 (0.39)	0.20 (0.52)	0.75 (0.79)	−0.40	0.77	1.00
ANN	0.60 (0.10)	0.01 (0.63)	0.77 (0.80)	−0.61	1.38	0.97

Note: Unbracketed results are for comparisons limited to the observed data mask for 1954 (see Santer et al. 1996). Bracketed correlation results are for the full global data field. The r(SC, S + C) results were determined using the actual patterns; identical results may be obtained from the r(SC, S) and r(SC,C) values using a relationship equivalent to equation (A1).

For our analysis, a more relevant test of linearity may be obtained by comparing the addition of panels (a) and (b) after the global-mean temperature change is subtracted with the predicted response in panel (c) after subtraction of the global mean. Colour Plate 5.2 shows the pattern of response for the annual average surface temperature change corresponding to the four panels of Colour Plate 5.1 after subtraction of the global average temperature change (thus, the global average temperature for the fields in Colour Plate 5.2 is zero). The degree of similarity between panel (c) and panel (d) is even more noteworthy. In fact, the pattern correlation coefficient between the fields in Colour Plate 5.2(c) and those of (d) is 0.80, indicating that the assumption of a linear response is not unreasonable. The pattern correlation coefficients for DJF, MAM, JJA and SON are 0.58, 0.72, 0.74 and 0.79, respectively (see Table 5.6). These correlations are for the full spatial field. Table 5.6 shows that (except for DJF) very similar correlations also occur when the field is restricted to the observational data mask (i.e. the region for which observed data coverage is relatively stable over the last 50 years). We may tentatively conclude, then, that the equilibrium pattern of model temperature response to a combination of greenhouse forcing and any set of sulfate-like forcings can be formed from a linear combination of the patterns obtained with sulfate-only forcing and CO_2-only forcing.

The most straightforward way to form new response patterns would be to form linear combinations of the sulfate-only and CO_2-only simulations, i.e. combinations of the form $\alpha S + C$. However, we could also use linear combinations of the form $SC + \alpha S$ or $SC + \beta C$ to represent response patterns for different relative S and C forcing weights. Because the SC pattern is a reasonable 'first guess' for the combined greenhouse gas-plus-aerosol response pattern and therefore, the perturbation to this pattern needed to explore different forcings is smaller than that in the case $\alpha S + C$, we believe this latter approach is the most appropriate strategy. We note that as $\alpha \rightarrow \infty$ or as $\beta \rightarrow \infty$ the pattern correlations formed by comparing $SC + \alpha S$ or $SC + \beta C$ with the observed patterns approach the 'pure' case (i.e. those obtained for observations versus pure S or pure C). On the other hand, as either α or β becomes large and negative, we anticipate possible departures between the linear approximation to the model-predicted response and that from an actual climate model simulation. In the Appendix we demonstrate that the correlation coefficient $r(SC - C, S)$ is $\cong 0.78$ and 0.86 while that for $r(SC - S, C) \cong 0.67$ and 0.77 for JJA and the annual average, respectively. Thus, the use of values of α or β as small as -1 should still provide reasonable approximations to actual climate model simulations. For consistency between the patterns formed by $SC + \alpha S$ and those formed by $SC + \beta C$, we require that the pattern for $\beta = -1$ be the same as the pattern for $\alpha = \infty$ and that the pattern for $\beta = \infty$ be the same as that for $\alpha = -1$. This consideration leads to the requirement that $\beta = -\alpha/(1 + \alpha)$ or, equivalently, $\alpha = -\beta/(1 + \beta)$.

How should we relate the response patterns $SC + \alpha S$ or $SC + \beta C$ to the amplitude of the forcing? Clearly, the pure S or pure C pattern (with global mean subtracted) cannot be related to a unique forcing amplitude because the pattern of response for pure S or pure C forcing is assumed to be independent of forcing. (This is the pattern-equivalent statement to the statement that the climate sensitivity, or $\Delta T/\Delta Q$, is independent of the type of forcing). All we are able to do is to determine the relative importance or ratio of forcing for S and C for any given linear combination of the $SC + \alpha S$ or $SC + \beta C$ patterns. From the definition of α, we have:

$$\frac{\Delta Q_S^*}{\Delta Q_C^*} = [(1 + \alpha)](S/C)_{\text{TP}}$$

or, equivalently,

$$\Delta Q_S^* = [\Delta Q_C^*(1 + \alpha)](S/C)_{\text{TP}}$$

where $(S/C)_{\text{TP}}$ is the ratio of forcings in the Taylor-Penner experiment (viz. $(-0.9)/1.3$), and $\frac{\Delta Q_S^*}{\Delta Q_C^*}$ is the ratio of forcings for the response pattern formed from $SC + \alpha S$. (If we began with $SC + \beta C$, the corresponding result would simply replace $1 + \alpha$ by $1/(1 + \beta)$.)

Let us now suppose that ΔQ_S^* is the total forcing due to factors with sulfate-like response patterns and ΔQ_C^* is the forcing for factors with CO_2-like response. Thus, we group direct and indirect forcing by sulfate aerosols in ΔQ_S^* as well as aerosol forcings associated with industrial sources of nitrates, ammonium and carbonaceous aerosols from fossil fuel sources:

$$\Delta Q_S^* = \Delta Q_{SO_4} + \Delta Q_{NO_3} + \Delta Q_{NH_4} + \Delta Q_{\text{CARB}} + \Delta Q_{\text{indirect}}.$$

The main CO_2-like anthropogenic forcings are those from CO_2, CH_4, N_2O, and halocarbons. Here, we are explicitly neglecting forcing by biomass aerosols, anthropogenic dust, and tropospheric O_3 because the spatial and temporal patterns of these forcings are not likely to be similar to either CO_2 or anthropogenic sulfate aerosols. The CO_2-like forcings include, in addition to the anthropogenic forcings noted above, that from solar variations over the last 100 years. Hence,

$$\Delta Q_C^* = \Delta Q_{CO_2} + \Delta Q_{CH_4} + \Delta Q_{N_2O} + \Delta Q_{\text{halo}} + \Delta Q_{\text{solar}}.$$

Table 5.7 Relationships between α, β and ΔQ_S^*

α	β	$\Delta Q_S^*/\Delta Q_C^*$	ΔQ_S^* (W m^{-2})
−1	∞	0.0	0.0
−0.8	4.0	−0.14	−0.35
−0.6	1.5	−0.28	−0.71
−0.4	0.67	−0.41	−1.03
−0.2	0.25	−0.55	−1.39
0	0	−0.69	−1.74
0.25	−0.2	−0.86	−2.17
0.67	−0.4	−1.15	−2.90
1.5	−0.6	−1.73	−4.36
4.0	−0.8	−3.45	−8.69
∞	−1.0	∞	1

[1]Undefined.

[2] ΔQ_S^*: The latter are based on an estimated value of $\Delta Q_C^* = 2.52$ W m^{-2} (see text).

Using the IPCC recommended values we find:

$$\Delta Q_C^* = 1.52 + 0.44 + 0.15 + 0.11 + 0.3 = 2.52 \text{ W m}^{-2}$$

where we have lumped together the 100-year solar forcing with those estimated from the pre-industrial and present-day estimates for the other forcings. Then, as an example, if α (or β) were zero we would find

$$\Delta Q_S^* = 2.52(-0.9/1.3) = -1.74 \text{ W m}^{-2}$$

This is almost double the current IPCC best-guess value for the net forcing by anthropogenic sulfate aerosols and soot. Table 5.7 shows the appropriate correspondence between α, β and ΔQ_S^* for different levels of relative forcings. From here it can be seen that a substantial negative value of α or positive value of β ($\alpha < -0.4$; or $\beta > 0.7$) is required to match or be smaller in magnitude than the IPCC best-guess value for sulfate and carbonaceous aerosol forcing (which is −1.1 W m^{-2}).

We now turn to the question of whether trend estimates for the pattern correlation between patterns for $SC + \beta C$ or $SC + \alpha S$ and the historical data can be used to give a 'best' guess for the optimal $\Delta Q_S^*/\Delta Q_C^*$. Appendix A provides a discussion and justification for the use of $SC + \beta C$ as the most reasonable single choice of pattern approximations.

If we apply this strategy, how do the pattern correlations between modeled and observed data change as we increase or decrease the C loading relative to that in SC? We could answer this question by 'brute force' simply by producing composite patterns as described above and then carrying out the same pattern correlation analyses as performed with SC in Santer et al. (1995). There is, however, a more elegant method (see Wigley et al. 1995). By using an expression similar to equation (A1) in Appendix 5.1 it is possible to derive an analytic relationship for $r(SC + \beta C, D)$, where D represents the observed data field, in terms of the $r(SC, D)$ and $r(C, D)$ correlations, which are already known (as $R(t)$ time series) from the Santer et al. work. (Similarly, $r(SC + \alpha S, D)$ can be expressed in terms of $r(SC, D)$ and $r(S, D)$.)

A comprehensive analysis is given in Wigley et al. (1995). Here we show some illustrative results.

Pattern correlations, $R(t)$, for a variety of different C and S forcing levels for JJA are shown in Figure 5.2. Here, $R(t)$ is the test statistic used by Santer et al. (1995) and defined by $r(SC + \beta C, D)$. The following β values used are: $\beta = -0.8$ (equivalent to $\alpha = 4.0$ if $SC + \alpha S$ were used); $\beta = -0.4$ (i.e. $\alpha = 0.67$); $\beta = -0.25$ ($\alpha = 0.33$); $\beta = 0.0$ ($\alpha = 0.0$); $\beta = 0.33$ ($\alpha = -0.25$); $\beta = 1.5$ ($\alpha = -0.5$); and $\beta = 5.0$ ($\alpha = -1.25$). The model patterns ($SC + \beta C$ or $SC + \alpha S$) are constants, but the observed temperature change pattern is a function of time. Hence, the correlations are also time-dependent. When $\alpha = \beta = 0$, the correlation reduces to $r(SC, D)$, which is the case considered by Santer et al. When β becomes large, $r(SC + \beta C, D)$ tends to $r(C, D)$, the 'pure C' case. For $\beta = -1$, $r(SC + \beta C, D)$ is approximately the 'pure S' case. These cases were considered in Santer et al. Our interest here, however, is in the intermediate cases.

Figure 5.2 shows the JJA $R(t)$ time series and Figure 5.3 shows the running 50-year trends in $R(t)$. As shown in the Appendix, the results for small $|\beta| \leq (-0.25$ and $0.33)$ are similar to the $\beta = 0$ (pure SC) case due to the relatively small effect of small additions of C to the pattern for SC. For this range of β values, the $(SC + \beta C, SC)$ pattern correlation is always above 0.96 (see Figure 5.5a). Trends for this β range are uniformly positive and high, showing (as in Santer et al. 1995) an increasing trend in the similarity between model results and observations. For larger $|\beta|$ the results are still positive, but slightly degraded. One can infer from this that the range of β values that best fit the data is, roughly, –0.3 to +0.3. From Table 5.7, this implies a global-mean aerosol-like forcing in the range –1.2 to –2.4 W m^{-2}. The lower bound is close to the best-guess IPCC value (for sulfate direct plus indirect plus soot), while the upper bound is significantly larger than the IPCC estimate.

Results for other seasons (a more comprehensive treatment will be published elsewhere) support the conclusion that the aerosol forcing might be significantly larger than the IPCC estimate. Indeed, for the SON case the optimum $R(t)$ results are obtained for the 'pure S' limit. This result could be anticipated from the results given in Santer et al. (1995) which show noticeably greater trends in $R(t)$ for S compared with those for SC, and noticeably weaker (negative!) trends for C compared with SC. Given the near-linearity of the compositing method, it is *a priori* unlikely that any weighting could improve on the 'S' end-member case. Detailed calculations support this conclusion.

5.5 Summary

The above analysis shows that the trend in the historical patterns of temperature change is consistent with a substantial northern hemisphere anthropogenic aerosol forcing.

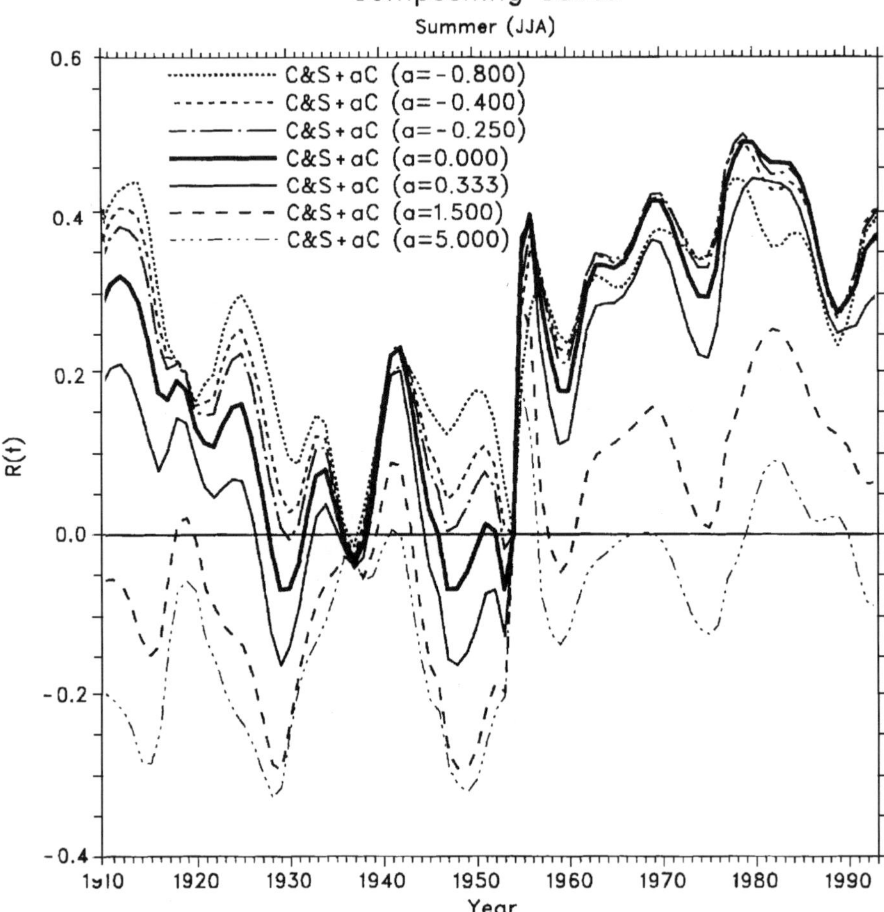

Figure 5.2 Pattern correlation statistic $R(t)$ for JJA between the observed data and linear combinations of $SC + \beta C$ for β = –0.8, –0.4, –0.25, 0.0, 0.333, 1.5 and 5.0 (β is labelled a).

Unfortunately, the significance of the lower and upper bounds for forcing derived above (i.e. –1.2 W m^{-2} to –2.4 W m^{-2}) cannot be adequately judged because we have no way of discerning whether differences in the trend of $R(t)$ of the order of 0.1 (as shown in Figure 5.3 for the cases outside the range $|\beta| < 0.3$) are significant. In part, our uncertainty is the result of imperfect linearity. However, tests using alternate approaches (e.g. $SC + \alpha S$ or simply $\alpha S + \beta C$) lead to very similar ranges for the implied aerosol forcing. A further complication results from the fact that the aerosol forcing deduced in this way is only properly termed an 'effective sulfate forcing' over the period of trend analysis. The climate system has in fact been subjected to variations in the relative magnitude of greenhouse and aerosol forcing over the last 50 years. The forcing deduced in this way is therefore only an approximate

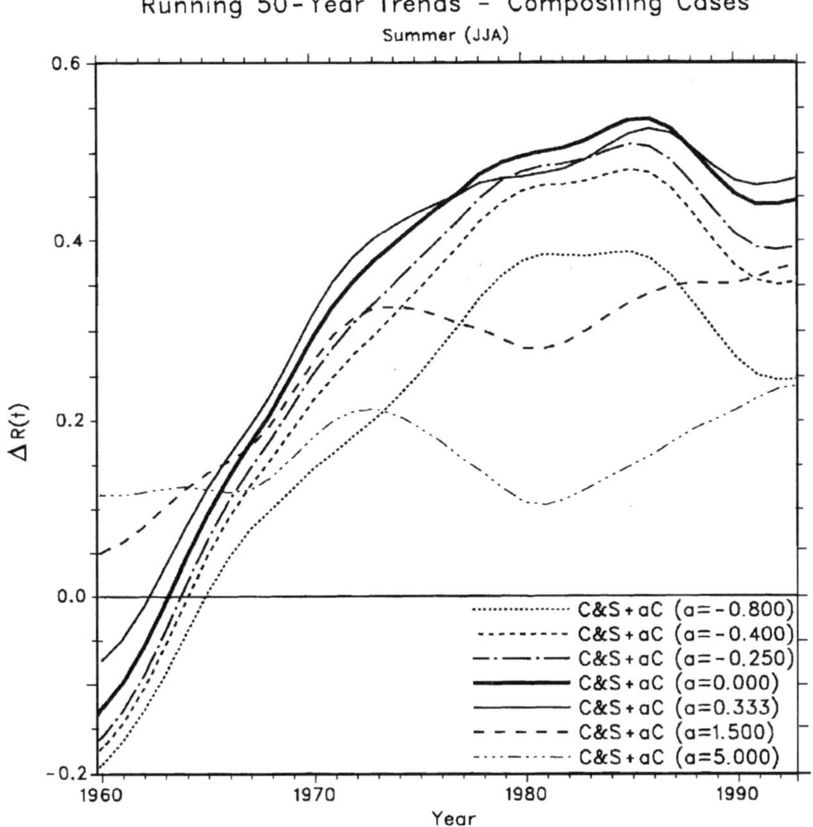

Figure 5.3 Fifty-year running trend in the JJA pattern correlation statistic $R(t)$ for the observed data and linear combinations of $SC + \beta C$ for β = –0.8, –0.4, –0.25, 0.0, 0.333, 1.5 and 5.0 (β is labelled a).

indicator for today's atmosphere. In order to infer the actual anthropogenic aerosol forcing for today, the analysis needs to account for the fact that the relative magnitudes of the forcing by anthropogenic aerosols and greenhouse gases have changed with time.

We have demonstrated that linear combinations of the patterns of response for different forcing experiments may be used to approximate the actual climate response. A further application of the assumption that response patterns may be added linearly may be to explore future scenarios for climate change. If the predicted patterns of climate response may be added linearly for variations in the location of maximum forcing, it may be possible to combine response patterns for varying assumptions regarding future emissions of anthropogenic aerosols from different regions. However, further work is needed to evaluate this assumption.

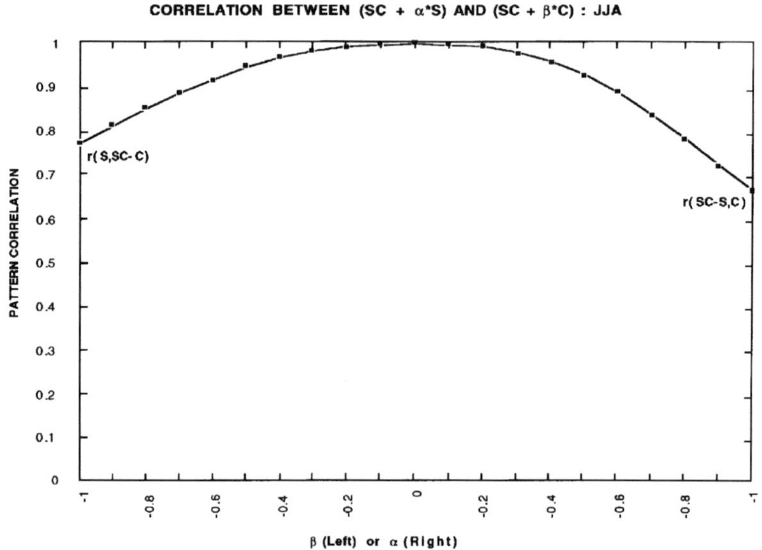

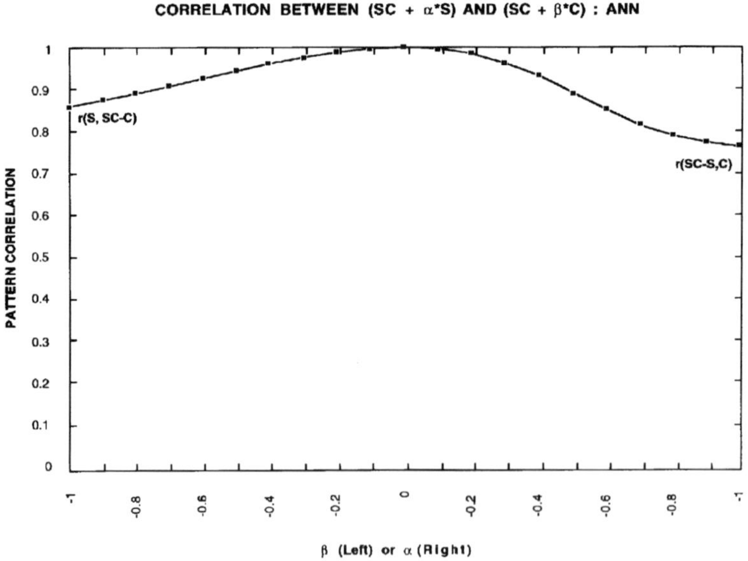

Figure 5.4 Pattern correlation between C and $SC + \alpha S$ (e.g. r$(C, SC + \alpha S)$) for $\alpha < 0$ (right half) and between S and $SC + \beta C$ for $\beta < 0$ (left half) for JJA (panel a) and the annual average (panel b).

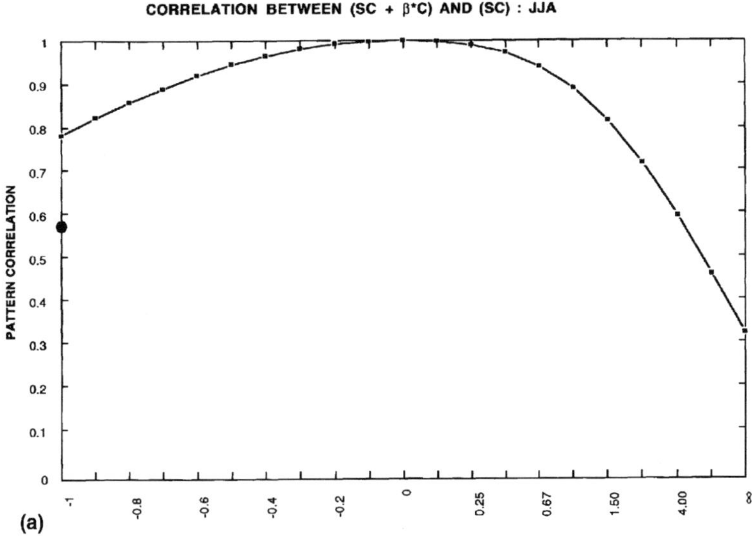

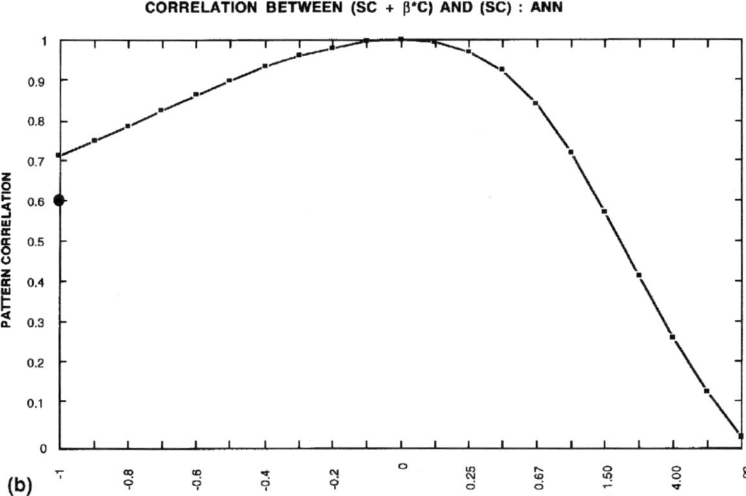

Figure 5.5 Pattern correlation between SC and $SC + \beta C$ for $-1 < \beta < \infty$ for JJA (panel a) and the annual average (panel b). (Note: scale on the left half of the figure is linear, on the right half it is not).

As a final note, in comparing the predicted and observed patterns of temperature change, we find that the pattern correlation statistic, even though it is highly statistically significant (see Santer et al. 1995), is quite low even towards the end of the record. The largest values for the correlation coefficient between the various response patterns and the observational record are only $\cong 0.5$ for the JJA season during the decade of the 1980s. Future climate model simulations, which include all anthropogenic aerosol types (e.g. biomass aerosols) as well as estimates for changes in tropospheric ozone may yield higher correlations. We hope to explore these possibilities in the future.

Acknowledgments

This work was performed under the auspices of the US Department of Energy under Contract W-7405-Eng-48. JEP is grateful for support from the NASA Aerosol Program and the DOE Climate Research Program. Work at NCAR was supported in part by the NOAA Office of Global Programs, Climate Change Data and Detection Program. US Department of Energy grant DE-FG02-86ER6037; NOAA grant NA96AANAG03.

References

Atkinson, R., Baulch, D.L., Cox, R.A., Hampson Jr., R.F., Kerr, J.A. and Troe, J. (1992) Evaluated kinetic and photochemical data for atmospheric chemistry: Supplement IV. *Atmospheric Environment*, **26A**, 1187–1230.

Benkovitz, C. (1982) Compilation of an inventory of anthropogenic emissions in the United States and Canada. *Atmospheric Environment*, **16**, 1551–1563.

Boucher, O. and Lohmann, U. (1995) The sulfate-CCN-cloud albedo effect: a sensitivity study with two general circulation models. *Tellus*, **47**, 281–300.

Charlson, R.J., Schwartz, S.E., Hales, J.M., Cess, R.D., Coakley Jr., J.A., Hansen, J.E. and Hofmann, D.J. (1992) Climate forcing by anthropogenic aerosols. *Science*, **255**, 423–430.

Chuang, C.C., Penner, J.E., Taylor, K.E. and Walton, J.J. (1994) Climate effects of anthropogenic sulfate: Simulations from a coupled chemistry/climate model. *Proceedings of the Conference on Atmospheric Chemistry*, American Meteorological Society, 23–28 January, Nashville, TN, pp. 170–174.

Cooke, W.F. and Wilson, J.J.N. (1996) A global black carbon aerosol model. *Journal of Geophysical Research*, **101**, 19395–19410.

Cox, S.J., Wang, W.-C. and Schwartz, S.E. (1995) Climate response to radiative forcings by aerosols and greenhouse gases. *Geophysical Research Letters*, **22**, 2509–2512.

Covey, C. and Thompson, S.L. (1989) *Paleogeography Paleoclimatology Paleoecology*, **75**, 331–341.

DeMore, W.B., Sander, S.P., Golden, D.M., Hampson, R.F., Kurylo, M.J., Howard, C.J., Ravishankara, A.R., Kolb, C.E. and Molina, M.J. (1992) Chemical kinetics and photochemical data for use in stratospheric modeling. Jet Propulsion Laboratory, California Institute of Technology Publication, pp. 20–92.

Haywood, J.M. and Shine, K.P. (1995) The effect of anthropogenic sulfate and soot aerosol on the clear sky planetary radiation budget. *Geophysical Research Letters*, **22**, 603–606.

Jones, A., Roberts, D.L. and Slingo, A. (1994) A climate model study of the indirect radiative forcing by anthropogenic sulphate aerosols. *Nature*, **370**, 450–463.

Kiehl, J.T. and Briegleb, B.P. (1993) The relative role of sulfate aerosols and greenhouse gases in climate forcing. *Science*, **260**, 311–314.

Lelieveld, J. and Crutzen, P.J. (1990) Influences of cloud photochemical processes on tropospheric ozone. *Nature*, **343**, 227–233.

Lelieveld, J. and Crutzen, P.J. (1991) The role of clouds in tropospheric photochemistry. *Journal of Atmospheric Chemistry*, **12**, 229–267.

Liousse, C., Penner, J.E., Chuang, C., Walton, J.J., Eddleman, H. and Cachier, H. (1996) A three-dimensional model study of carbonaceous aerosols. *Journal of Geophysical Research*, **101**, 19411–19432.

Mitchell, J.F.B., Johns, T.C., Gregory, J.M. and Tett, S.F.B. (1995) Climate response to increasing levels of greenhouse gases and sulphate aerosols. *Nature*, **376**, 501–504.

Penner, J.E., Atherton, C.S., Dignon, J., Ghan, S.J., Walton, J.J. and Hameed, S. (1991) Tropospheric nitrogen: A three-dimensional study of sources, distribution, and deposition. *Journal of Geophysical Research*, **96**, 959–990.

Penner, J.E., Dickinson, R. and O'Neill, C. (1992) Effects of aerosol from biomass burning on the global radiation budget. *Science*, **256**, 1432–1434.

Penner, J.E., Eddleman, H. and Novakov, T. (1993) Towards the development of a global inventory of black carbon emissions. *Atmospheric Environment*, **27A**, 1277–1295.

Penner, J.E., Charlson, R.J., Hales, J.M., Laulainen, N., Leifer, R., Novakov, T., Ogren, J., Radke, L.F., Schwartz, S.E. and Travis, L. (1994a) Quantifying and minimizing uncertainty of climate forcing by anthropogenic aerosols. *Bulletin of the American Meteorological Society*, **75**, 375–400.

Penner, J.E. (1994a) Atmospheric chemistry and air quality. In *Changes in Land Use and Land Cover: A Global Perspective*, edited by W.B. Meyer and B.L. Turner II, pp. 175–209. Cambridge: Cambridge University Press.

Penner, J.E., Atherton, C.A. and Graedel, T.E. (1994b) Global emissions and models of photochemically active compounds. In *Global Atmospheric-Biospheric Chemistry*, edited by R. Prinn, pp. 223–248. New York: Plenum Publishing.

Penner, J.E. (1995) Carbonaceous aerosols influencing atmospheric radiation: black and organic carbon. In *Aerosol Forcing of Climate*, edited by R.J. Charlson and J. Heintzenberg, pp. 91–108. Chichester: John Wiley and Sons.

Ramaswamy, V. and Chen, C.-T. (forthcoming) Climate forcing-response relationship for greenhouse and solar radiative perturbations. *Nature*.

Roeckner, E., Siebert, T. and Feichter, J. (1995) Climatic response to anthropogenic sulfate forcing simulated with a general circulation model. In *Aerosol Forcing of Climate*, edited by R.J. Charlson and J. Heintzenberg, pp. 349–362. Chichester: John Wiley and Sons.

Santer, B.D., Taylor, K.E., Wigley, T.M.L., Penner, J.E., Cubasch, U. and Jones, P.D. (1995) Towards the detection and attribution of an anthropogenic effect on climate. *Climate Dynamics*, **12**, 77–100.

Santer, B.D., Taylor, K.E., Wigley, T.M.L., Johns, T.C., Jones, P.D., Karoly, D.J., Mitchell, J.F.B., Oort, A.H., Penner, J.E., Ramaswamy, V., Schwarzkopf, M.D., Stouffer, R.J. and Tett, S. (1966) A search for human influence on the thermal structure of the atmosphere. *Nature*, **382**, 39–46.

Schlesinger, M.E. and Ramankutty, N. (1992) Implications for global warming of intercycle solar irradiance variations. *Nature*, **360**, 330–333.

Shine, K.P., Fouquart, Y., Ramaswamy, V., Solomon, S. and Srinivasan, J. (1995) Radiative forcing. In *Climate Change 1994: Radiative Forcing of Climate Change and an Evaluation of the IPCC IS92 Emissions Scenarios*, edited by J.T. Houghton, L.G. Meira Filho, J. Bruce, H. Lee, B.A. Callander, E. Haites, N. Harris and K. Maskell, pp. 163–203. Cambridge: Cambridge University Press.

Spiro, P.A., Jacob, D.J. and Logan, J.A. (1992) Global inventory of sulfur emissions with $1° \times 1°$ resolution. *Journal of Geophysical Research*, **97**, 6023–6036.

Taylor, K.E. and Penner, J.E. (1994) Response of the climate system to atmospheric aerosols and greenhouse gases. *Nature*, **369**, 734–737.

Tegen, I. and Fung, I. (1995) Contribution to the atmospheric mineral aerosol load from land surface modification. *Journal of Geophysical Research*, **100**, 18707–18726.

Walton, J.J., MacCracken, M.C. and Ghan, S.J. (1988) A global-scale Lagrangian trace species model of transport, transformation, and removal processes. *Journal of Geophysical Research*, **93**, 8339–8354.

Wigley, T.M.L. and Raper, S.C.B. (1991) Detection of the enhanced greenhouse effect on climate. In *Climate Change: Science, Impacts and Policy*, edited by J. Jager and H.L. Ferguson, pp. 231–242. Cambridge: Cambridge University Press.

APPENDIX 5.1

Here we address the question: which of the linear combinations $SC+\alpha S$ or $SC+\beta C$ is better? To answer this we need to present a simple algebraic result for the correlation (r) between the $SC+\alpha S$ and $SC+\beta S$ patterns which we also use in determining the $R(t)$ values for the composite patterns. We also consider how rapidly the SC pattern is changed by adding or subtracting multiples or fractions of the S or C patterns.

It is easy to show that

$$r(SC+\alpha S, SC+\beta C)=[1+Ar(SC,S)+Br(SC,C)+ABr(S,C)]/[XY]^{1/2} \tag{A1}$$

where

$$A=\alpha S.D.(S)/S.D.(SC)$$

$$B=\beta S.D.(C)/S.D.(SC)$$

$$X=1+A^2+2Ar(SC,S)$$

$$Y=1+B^2+2Br(SC,C)$$

and $S.D.$ denotes the spatial standard deviation (for the S, C and SC patterns). As noted in the text, for the $SC+\alpha S$ and $SC+\beta S$ patterns to be consistent, they must have the same relative weights on their S and C components. This requires that $\beta=-\alpha/(1+\alpha)$.

Since all of the terms in equation (A1) are known (see Table 5.6), we can now use it to determine how the two alternative S–C patterns (i.e. $SC+\alpha S$ or $SC+\beta C$) are correlated with the pure S or pure C patterns for different α's and β's (i.e. different relative S and C weights). The left half of Figure 5.4 shows the correlation between the pure S case and $SC+\beta C$ for varying $\beta<0$, while the right half shows the correlation between the pure C case and $SC+\alpha S$ for varying $\alpha<0$. In an ideal situation, $r(S,SC-C)$ and $r(C,SC-S)$, the left and right hand extremes of Figure 5.4, would be identically equal to one (as would be the case if $r(S,SC+\alpha S)$ were plotted for the left half of Figure 5.4 and $r(C,SC+\beta C)$ were plotted for the right half of the figure). Then, the two methods for producing new composite patterns would give the same results. In fact, the methods do give highly correlated results implying that either method could be used in the $R(t)$ analysis.

Figure 5.4 can be used to argue that the $SC+\beta C$ method is probably better than using $SC+\alpha S$. This conclusion arises from looking at the extreme cases $\beta=-1$ and $\alpha=-1$ (corresponding to the left and right axes of Figure 5.4). For the first case ($\beta=-1$), the correlations shown are for $r(S,SC-C)$. This tells us how well the S pattern is simulated by subtracting the C pattern from the SC pattern. The correlations are $r=0.86$ for annual data and $r=0.77$ for JJA data. Simulating the C pattern by subtracting S from SC (i.e. $\alpha=-1$, giving $r(SC-S,C)$) gives corresponding correlations of $r=0.76$ and $r=0.67$. Since these latter correlations are lower, it appears likely that adding or subtracting fractions or multiples of the C pattern from SC is preferable to producing an equivalent result by adding or subtracting S multiples from SC. This issue is discussed further below.

Equation (A1) can also be used to provide another test of the linear superposition method, and to show how quickly the composite patterns become dissimilar to the original SC pattern as C (or S) multiples are added or removed. If we set $\alpha=0$ in equation (A1) it reduces to a simple formula for $r(SC,SC+\beta C)$ which we can plot as a function of β (Figure 5.5). For the lower and upper extreme positive values of β ($\beta=0$ and $\beta=\infty$) the linear combination $SC+\beta C$ must give precisely correct results: $\beta=0$ just reproduces the SC pattern, while large β gives patterns that tend exactly to the C pattern. Given that the method is exact for these limiting cases, we may be reasonably confident that the linear combination pattern for $\beta>0$ is a good simulation of the pattern that would result from an equivalent GCM experiment.

Figure 5.5 shows that the $SC+\beta C$ pattern remains similar to the SC pattern up to appreciable values of β and only becomes noticeably different from SC when the S/C forcing ratio reduces to about one-half of its value in the SC experiment. This is an important result, since it implies that results for S and C weights in an appreciable band around the weights for the SC experiment may give results quite similar to the SC results. In other words, the proposed re-weighting method may not be as sensitive to uncertainties in the relative S and C forcings as one might have hoped.

For negative β, the $SC+\beta C$ pattern does not have the same limiting constraints that exist for positive β. For $\beta=-1$, Figure 5.5 gives the correlation between SC and SC minus C (which is the linear combination approximation to S)—a case for which we know the correct answer. For annual data, the correlation between $SC-C$ and SC is 0.78, compared with the true correlation $r(S,SC)$ of 0.58 (given in Table 5.6). For JJA data, the corresponding results are $r=0.71$ and $r=0.60$. While not perfect, these are still quite good results: however, they indicate that it may be better to use the $SC+\alpha S$ combination for negative β (i.e. positive α). In other words, the optimum strategy may be to produce patterns with higher S loadings than SC by adding fractions or multiples of S, and to produce patterns with lower S loadings than SC by adding fractions or multiples of C.

PART 3

OCEANS AND CLIMATE

CHAPTER 6

COMMENTARY

Neil Holbrook

6.1   Introduction

The climate system is influenced by its interaction with the ocean in three distinct ways: the absorption and exchange of carbon dioxide with the atmosphere; the surface fluxes of heat, freshwater and momentum; and the sequestration of heat into the deep ocean through subduction and convection (Houghton et al. 1990). Studies from the following three chapters presented in Part 3 address the sensitivity of the climate system to some of these important processes. These include sensitivities to ocean model type (either mixed-layer or fully dynamical models) and resolution to cloud-albedo and sea-ice feedbacks discussed by Washington and Meehl in Chapter 7, in ocean heat and mass transports to changes in wind stress discussed by Chervin, Craig and Semtner Jnr in Chapter 8, and in North Atlantic Deep Water formation (NADWF) to changes in fresh/saltwater fluxes from the Mediterranean Sea discussed by Hecht, Holland, Artale and Pinardi in Chapter 9. This chapter reviews some of the results obtained from these modelling studies and discusses some of their limitations.

6.2   Dynamical ocean models in climate studies

The studies described in the three chapters presented in Part 3 involved the use of fully 3-dimensional (3-D) dynamical ocean models. Until recently, computer limitations had prevented the use of fully 3-D ocean models with sub-1° spatial resolutions for global variability and change experiments. In fact, the recent MECCA Phase 1 climate change experiments (early 1990s) used mixed-layer (or 'slab') oceans in their climate models. Slab ocean models are the simplest form of ocean model that permits feedbacks between the ocean and atmosphere. Slab models provide heat capacity, can be run with a seasonal cycle and are computationally inexpensive compared with dynamical ocean general circulation models (OGCMs). However, slab ocean models do not include advection which is a major simplification of the real ocean. Consequently, important ocean processes such as poleward heat transport and upwelling are neglected (although errors in the simulation of sea surface temperature which arise from these deficiences are often adjusted for artificially using the so-called 'Q-flux' correction) and El Niño-type phenomena cannot be simulated. Neelin et al. (1992) provide a review of El Niño-Southern Oscillation (ENSO) simulations in a number of dynamical OGCMs.

6.3   How important is the El Niño-Southern Oscillation?

The ENSO phenomenon dominates the interannual climate variability across the tropical and subtropical Pacific Ocean and has profound effects on the precipitation patterns in this region (e.g. Ropelewski & Halpert 1987, 1989) (see Figure 6.1). For example, during El Niño conditions, much of eastern and north-eastern Australia and the western Pacific becomes much drier while the central and eastern tropical Pacific is much wetter than normal. During severe El Niños, such as that experienced during 1982–83 and 1991–92, eastern Australia can experience quite serious droughts (parts of central and eastern Australia experienced drought during most of the pentad 1991–95 due to a quasi-persistent El Niño during this time) while parts of central and western America may experience flooding. ENSO is not only confined to the Pacific but it is also linked to the monsoon circulation in the Indian Ocean and has a counterpart in the Atlantic Ocean. There are strong links between ENSO and tropical cyclone activity in not only the Pacific Ocean (e.g. Nicholls 1984; Revell & Goulter 1986; Dong 1988; Basher & Zheng 1995), but also in the Atlantic Ocean (e.g. Gray et al. 1992, 1993, 1994). Furthermore, it has been shown that the phenomenon has global teleconnections (Glantz et al. 1991) and plays a major role in the global climate system.

Overall, ENSO is a crucial component of the earth's climate system which can be both economically and socially devastating. Unfortunately, this phenomenon is still relatively poorly simulated in climate models although significant advances have recently been made due to the work undertaken during the Tropical Ocean Global Atmosphere (TOGA) program (1986–1995). The inclusion of more sophisticated ocean models in simulations of climate are likely to improve ENSO predictive skill. The inclusion of dynamical ocean models in the MECCA experiments described in Part 3 is a positive move towards this goal.

6.4   The importance of eddies in OGCMs

Ocean eddies are an important component of the ocean circulation as they transport significant amounts of energy and momentum (e.g. Wilkin & Morrow 1994). Ocean eddies often have spatial scales on the order of tens of kilometres. Conversely, atmospheric eddies (e.g. tropical cyclones) typically have scales on the order of a thousand kilometres. Due to the present limitations of supercomputers, OGCMs are generally run on spatial grids of $1°$ or coarser. For experiments on time scales of centuries to millenia, OGCM grids are typically much coarser. Consequently, subgrid scale eddy mixing processes in OGCMs need to be parameterised.

SCHEMATIC OF AREAS WITH A CONSISTENT ENSO PRECIPITATION SIGNAL

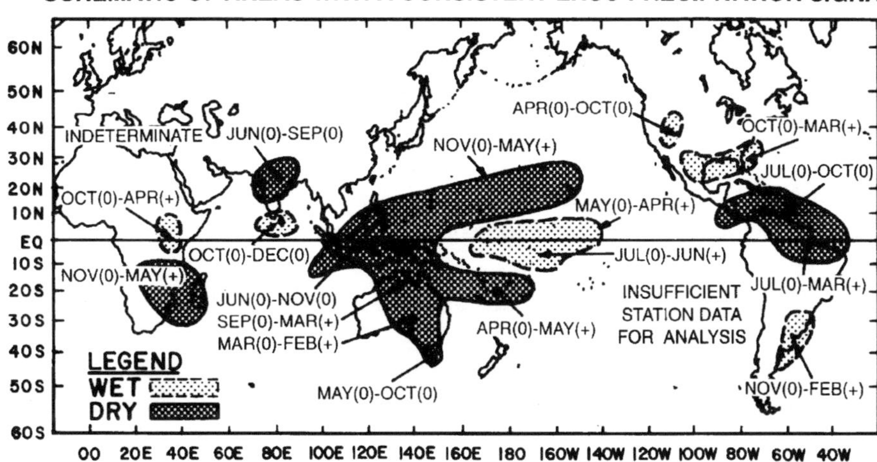

Figure 6.1 Schematic representation of the principal ENSO-related precipitation patterns. (0) indicates the year of ENSO and (+) indicates the year following ENSO (after Ropelewski & Halpert (1987). Kindly reproduced with permission from the American Meteorological Society).

Prior to the early 1990s, heat transports were only poorly represented in coarse-resolution OGCMs although significant improvements in the simulation of poleward heat transports by these OGCMs have recently been provided by the subgrid scale eddy formulation of Gent and McWilliams (1990). Further discussions of these improvements are provided by Neelin and Marotzke (1994), Danabasoglu et al. (1994) and Gent et al. (1995).

In contrast, Semtner and Chervin (1988, 1992) have shown, using a $1/2°$ resolution OGCM, that they were able to simulate a relatively realistic ocean circulation including marginally resolved eddies. Notably, the eddy circulation appears to be quite well represented near to the equator where the signal associated with El Niño is also largest. An advantage of this model is that the 'artificial' eddy subgrid scale parameterizations are less important than for those OGCMs on coarser grids. The MECCA-sponsored investigation by Chervin et al., detailed in Chapter 8, examined and evaluated ocean model sensitivity to changes in wind stress forcing using a version of this ocean model.

From their experiments, Chervin et al. showed that eddy heat transports tend to be largest during the northern hemisphere fall and winter. Furthermore, they also showed that these transports tend to be reduced during El Niño years and increased during cold eastern Pacific periods. These results have implications for the importance of resolution in OGCMs, particularly for interannual studies. Given that eddy transports vary in intensity between El Niño and La Niña periods, it is likely that interannual studies of equatorial and western boundary heat transports will be sensitive to resolution and hence eddy parameterizations in ocean models.

More recently, $1/4°$ and $1/6°$ OGCMs have been developed and experiments with these higher resolution models are currently in progress (Bert Semtner 1995, pers. comm.). These models incorporate a free surface rather than the rigid lid approximation used in the MECCA experiments. One important advantage of the free surface version of the OGCM is that more realistic topography can be incorporated at a lower cost compared with the rigid lid version. However, the major disadvantage of performing any of these high-resolution integrations is that global scale ocean experiments on the order of eddy resolution are hundreds or even thousands of times more costly to run than coarse resolution ones.

6.5   OGCM sensitivity to wind stress on seasonal and interannual time scales

Understanding the 'natural' variability of ocean heat transports and circulation on seasonal and interannual time scales is crucial to improving our understanding of climate variability on decadal or longer time scales. The redistribution of heat, through changes in oceanic advection and convection, may potentially affect not only local and regional climates but also the global energy balance. The MECCA contribution by Chervin et al. is an important evaluation of the sensitivity of ocean heat transports and circulation simulated in a marginally eddy resolving OGCM to changes in surface wind stress forcing. These sensitivity experiments describe the ocean's response to the dynamic component of the atmospheric forcing. The model thermodynamics are, however, somewhat artificial in that the heat and salt fluxes are restored at the sea surface and below 710 m. Despite this, the thermocline was free to move during the analysis part of the integration; an important diagnostic for evaluating El Niño-type variations. Hence, useful information is provided about the dynamic sensitivity of the ocean thermocline to changes in wind stress forcing during the 1980–89 decade. Some of the important results from their experiments are reviewed here.

Chervin et al. indicated that Pacific Ocean heat transport variability is extremely important to the climate in the Pacific region and to the climate of the globe as a whole. These results suggest that, since global meridional heat transports are dominated by the Pacific Ocean, changes in Pacific circulation would have a large effect on the global meridional heat transport. This highlights the importance of the Pacific Ocean in global heat transport studies.

With regard to the interannual variability, Chervin et al. showed that there was very little difference in the monthly mean heat transports when forced with daily winds compared with monthly winds. From this, it can be inferred that large scale monthly mean ocean circulation changes are relatively insensitive to forcing on the time scale of only a few days when compared with the response to forcing on monthly time scales.

Finally, Chervin et al. suggested that temperature variability sometimes plays a role in heat transport variability. Since observed ocean temperature data are much more readily available than observed ocean velocities, this result suggests that there might be some value in examining correlation relationships between ocean heat transports and ocean temperatures where observations permit. In regions where strong correlations exist between these two quantities, it might be possible to evaluate the variability of model heat transports against observed ocean temperatures.

6.6   OGCMs and coupled climate model simulations of long-term change

Coupled ocean-atmosphere experiments of long-term climate change do not currently have the luxury of using sub-1° resolution OGCMs. Given the potentially large computer costs involved with long-integration greenhouse experiments using coupled ocean-atmosphere global climate models (OAGCMs), choice of model resolution is a major consideration. Washington and Meehl in Chapter 7 address this issue for the ocean model case by examining the sensitivity of the OGCM to model resolutions of both 0.5° and 1°. Their results are compared with earlier studies which used a resolution of 5°. They found that the use of a 1° × 1°, 20-level global ocean model produces an ocean simulation that is a great improvement over the earlier 5° version while still capturing most features of the 0.5° version on time scales of a century. For example, they were able to show that the strength of the Deacon Cell is about 30 Sv for both 0.5° and 1° resolutions although it has greater vertical extent at the 0.5° resolution. They also point out that the global overturning in the Northern Hemisphere, with sinking originating near 65°N, is weaker in the 1° version (maxima of about 7.5 Sv in the 1° case against 12.5 Sv in the 0.5° version). Interestingly, the 'conveyor belt' (Broecker 1991) flow is qualitatively similar for both the 0.5° and 1° resolutions. Hence, there appears to be sufficient justification for using the coarser 1° ocean model version for these long-integration experiments.

In the enhanced greenhouse experiments, Washington and Meehl also coupled their ocean–atmosphere GCM with a simple three-level version of the thermodynamic sea-ice model by Semtner (1976) which has been extended to include the model sea-ice dynamics scheme from Flato and Hibler (1990). The inclusion of this thermodynamic-dynamic sea-ice model in the coupled climate model significantly improves the earlier climate model simulations of previous experiments which incorporated only a simple, thermodynamic, zero-layer sea-ice model (Semtner 1976). Overall, this global climate model has arguably one of the most sophisticated model configurations developed for studies of long-term change currently in use.

An interesting result described by Washington and Meehl in Chapter 7 and in their 1996 article 'El Niño-like climate change in a model with increased atmospheric CO_2 concentrations' is the tropical Pacific conditions which might arise in an enhanced greenhouse environment. They found that tropical Pacific conditions in a simulated doubled-CO_2 climate resemble 'conditions currently experienced during El Niño events, as well as the observed decadal time scale features of the 1980s and early 1990s'. This result is consistent with results from a principal components analysis of observed upper ocean temperatures in the Southwest Pacific Ocean which show fluctuations in the subtropical gyre that include both El Niño and decadal changes since the early 1970s (Holbrook & Bindoff 1997).

6.7   The deep ocean circulation: Centennial and longer time scales

Probably the most poorly understood component of the ocean circulation is that of the deep ocean thermohaline circulation. The deep ocean circulation transports heat and salt on time scales of hundreds to thousands of years and is an important long-term feedback factor in climate change. A major source region for the deep ocean circulation is the mid-latitudes of the North Atlantic. Here, it has been suggested that NADWF is primarily driven by the excess salt left behind as the result of vapour export (Broecker et al. 1985), and is reputed to continually feed the so-called 'conveyor belt' (Gordon 1986; Broecker 1991; Semtner & Chervin 1988, 1992)

Manabe and Stouffer (1988) demonstrated, using a coupled ocean-atmosphere model, the Atlantic ocean circulation could assume two stable but quite different modes: one with a strong thermohaline component akin to the conveyor and one with no thermohaline circulation. Furthermore, Aagaard and Carmack (1989) showed that the addition of relatively small amounts of freshwater to the surface can stabilise the water column enough to prevent the convection which feeds the conveyor. These studies motivated further research examining the stability of NADWF to freshwater anomalies imposed at high northern latitudes (Maier-Reimer & Mikolajewicz 1989; Marotzke 1989; Stocker & Wright 1991; Wright & Stocker 1991).

In their MECCA experiments, Hecht et al. (Chapter 9) examined the sensitivity of the overturning circulation in the North Atlantic Ocean to changes in salt flux across the Mediterranean Sea boundary. The Mediterranean Sea is a strong source of salt into the North Atlantic Ocean and contributes to the overall buoyancy-driven flows of that region. Three experiments were performed. The first of these used the simple Newtonian restoration near Gibraltar, the second included no Mediterranean waters whatsoever and the third allowed exchange of waters near Gibraltar (diagnosed model fluxes) between the North Atlantic and Mediterranean Sea models. In essence, the first experiment provided a modest decrease in NADWF over time, the second resulted in a dramatic collapse of NADWF while the third experiment maintained most of the intensity of NADWF over the model integration. These results are important and provide some valuable insight into the possible oceanic processes which maintain NADWF, a significant component of our climate system.

One caveat on the Hecht et al. simulations is the possible sensitivity of the model results to the Han (1984) surface restoring heat flux formulation. Power and Kleeman (1993) have shown that for modelling studies with mixed boundary conditions (they refer to various studies which incorporated a prescribed salt (freshwater) flux and the Haney (1971) surface restoring heat flux formulation), restoration of upper level temperatures to currently observed estimates is strictly only valid for estimates of surface heat flux when modelling the current climate since it is not clear that these values will remain the same as the climate evolves in time. Furthermore, they point out that parameters in the heat flux formulation are often based upon observations made at low- and mid-latitudes (Haney 1971; Han 1984; Oberhuber 1988), yet these values are often applied at high latitudes by ocean modellers.

Power et al. (1994) also showed that the stability of NADWF in such numerical studies may be particularly sensitive to the surface heat flux formulation. In particular, it has been demonstrated that these numerical studies are particularly sensitive to the parameterization of the 'Haney relaxation time (τ)' (Power & Kleeman 1994). Power and Kleeman indicated that for $\tau < 25$ days, a substantial collapse of the overturning ensues. For larger values of this relaxation time, there is a much more modest collapse. In the Hecht et al. study, a different heat flux formulation was used and the source of the flux perturbation is somewhat different. However, it is quite likely that there may be some analogy between, for example, the NADWF collapse demonstrated in the second experiment and those of the studies described by Power and Kleeman (1994). It should be noted, however, that at the time the MECCA experiments were commenced, little was known about the flux sensitivity which has only been illucidated in the last few years. Dr Bill Holland, the lead investigator in the MECCA experiments described in the Hecht et al. study, has pursued these NADW stability issues with Dr Frank Bryan (Holland & Bryan 1994a, 1994b) and Dr A. Capotondi (Bill Holland 1995, pers. comm.).

6.8   How successful were these MECCA ocean simulations?

The three studies presented in Part 3 provide an excellent cross-section of ocean processes and sensitivities on annual, seasonal, interannual, decadal and centennial time scales as simulated in high-resolution numerical models. Prior to the commencement of MECCA in 1991, coupled climate model simulations on century time scales using high-resolution OGCMs had not been performed. However, with the advent of improvements in supercomputer technology and the support provided by MECCA, such experiments were possible. Perhaps the projected persistence of ENSO over the Pacific Ocean as suggested in the enhanced greenhouse experiment by Washington and Meehl would not yet have been described if MECCA had not existed. Each of the simulations described in Part 3 required huge amounts of very costly computer time on state-of-the-art supercomputers. MECCA was critically important as the means for providing these facilities which greatly accelerated the progress of this research.

For the most part, the Principal Investigators closely followed their original proposals which were mainly a series of sensitivity experiments. Given the limitations of observed databases, in particular the very poor coverage of ocean data available for variability studies, the model experiments presented in Part 3, together with similar model experiments performed elsewhere, will serve to further our understanding of these physical processes. It is imperative, however, that model results continue to be evaluated against available observations where possible. Overall, it is crucial that modelling sensitivity studies such as those presented in Part 3, together with model validation, are ongoing in order that model uncertainties are reduced, confidence is increased and consequently skill in climate 'predictability' continues to improve over time. It is only when such uncertainties are reduced sufficiently that policies and active strategies will be formulated.

During certain months of the year, various ocean and atmosphere models are beginning to have some El Niño forecast skill a number of months in advance, for example, Barnston et al. (1994). However, models differ widely in their predictions and they are very often wrong. So, although the policy community and public at large are aware of the potentially damaging effects of ENSO, there remains little confidence in model predictions. For the deep ocean circulation, there is far greater uncertainty. Although the results of these MECCA experiments are very likely to be of value to the ocean modelling community as a whole, and implementation of some of these findings may well improve future simulations, it is unlikely that the uncertainties have been reduced sufficiently to have a bearing on climate-related policy decisions.

Amongst the many achievements, however, the MECCA experiments described in Part 3 have helped to identify important model resolutions and forcing functions. These findings have addressed some of the previous uncertainties associated with model sensitivities and will therefore provide a much better platform for future experiments. Furthermore, personal discussions with the Principal Investigators indicate that some of the results from their experiments have since stimulated valuable new ideas and future directions for their work. Overall, I believe that as far as the climate-related ocean modelling research work is concerned, MECCA has been very successful.

Acknowledgments

I am grateful for the helpful discussion with Dr Scott Power regarding NADWF sensitivity in OGCMs and the comments provided by Professor Ann Henderson-Sellers on previous drafts of this chapter. Finally, I am thankful for the support provided by Southern California Edison. This paper is CIC contribution number 95/33.

References

Aagaard, K. and Carmack, E.C. (1989) The role of sea ice and other freshwater in the arctic circulation. *Journal of Geophysical Research*, **94**, 14 485–14 498.

Barnston, A.G., van den Dool, H.M., Zebiak, S.E., Barnett, T.P., Ji, M., Rodenhuis, D.R., Cane, M.A., Leetmaa, A., Graham, N.E., Ropelewski, C.R., Kousky, V.E., O'Lenic, E.A. and Livezey, R.E. (1994) Long-lead seasonal forecasts—where do we stand? *Bulletin of the American Meteorological Society*, **75**, 2097–2114.

Basher, R.E. and Zheng, X. (1995) Tropical cyclones in the Southwest Pacific: Spatial patterns and relationships to Southern Oscillation and sea-surface temperature. *Journal of Climate*, **8**, 1249–1260.

Broecker, W.S. (1987) The biggest chill. *Natural History Magazine*, **97**, 74–82.

Broecker, W.S. (1991) The great ocean conveyor. *Oceanography*, **4**, 79–89.

Broecker, W.S., Peteet, D. and Rind, D. (1985) Does the ocean-atmosphere have more than one stable mode of operation? *Nature*, **315**, 21–25.

Danabasoglu, G., McWilliams, J.C. and Gent, P.R. (1994) The role of mesoscale tracer transports in the global ocean circulation. *Science*, **264**, 1123–1126.

Dong, K. (1988) El Niño and tropical cyclone frequency in the Australian region and the northwest Pacific. *Australian Meteorological Magazine*, **36**, 219–225.

Flato, G.M. and Hibler III, W.D. (1990) On a simple sea-ice dynamics model for climate studies. *Annals of Glaciology*, **14**, 72–77.

Gent, P.R. and McWilliams, J.C. (1990) Isopycnal mixing in ocean circulation models. *Journal of Physical Oceanography*, **20**, 150–155.

Gent, P.R., Willebrand, J., McDougall, T.J. and McWilliams, J.C. (1995) Parameterizing eddy-induced tracer transports in ocean circulation models. *Journal of Physical Oceanography*, **25**, 463–474.

Glantz, M.H., Katz, R.W. and Nicholls, N. (1991) *Teleconnections linking worldwide climate anomalies.* Cambridge: Cambridge University Press.

Gordon, A.L. (1986) Interocean exchange of thermocline water. *Journal of Geophysical Research*, **91**, 5037–5046.

Gray, W.M., Landsea, C.W., Mielke, P.W. and Berry, K.J. (1992) Predicting Atlantic seasonal hurricane activity 6–11 months in advance. *Weather Forecasting*, **7**, 440–455.

Gray, W.M., Landsea, C.W., Mielke, P.W. and Berry, K.J. (1993) Predicting Atlantic basin seasonal tropical cyclone activity by 1 August. *Weather Forecasting*, **8**, 73–86.

Gray, W.M., Landsea, C.W., Mielke, P.W. and Berry, K.J. (1994) Predicting Atlantic basin seasonal tropical cyclone activity by 1 June. *Weather Forecasting*, **9**, 103–115.

Han, Y.J. (1984) A numerical world ocean general circulation model part II. A baroclinic experiment. *Dynamics of Atmospheres and Oceans*, **8**, 141–172.

Haney, R.L. (1971) Surface thermal boundary conditions for ocean circulation models. *Journal of Physical Oceanography*, **4**, 241–248.

Holbrook, N.J. and Bindoff, N.L. (in press) Interannual and decadal temperature variability in the Southwest Pacific Ocean between 1955 and 1988. *Journal of Climate*.

Holland, W.R. and Bryan, F.O. (1994a) Ocean processes in climate dynamics: Global and Mediterranean examples. In *Sensitivity studies on the role of the ocean in climate change*, pp. 111–134. Dordrecht: Kluwer Academic Publishers.

Holland, W.R. and Bryan, F.O. (1994b) Ocean processes in climate dynamics: Global and Mediterranean examples. In *Modeling the wind and thermohaline circulation in the North Atlantic Ocean*, pp. 135–156. Dordrecht: Kluwer Academic Publishers.

Houghton, J.T., Jenkins, G.J. and Ephraums, J.J. (eds) (1990) *Climate Change.* New York: Cambridge University Press.

Maier-Reimer, E. and Mikolajewicz, U. (1989) Experiments with an OGCM on the cause of the younger Dryas. In *Oceanography*, pp. 87–100. Mexico: UNAM Press.

Manabe, S. and Stouffer, R.J. (1988) Two stable equilibria of a coupled ocean-atmosphere model. *Journal of Climate*, **1**, 841–866.

Marotzke, J. (1989) Ocean circulation models: combining data and dynamics. In *Instabilities and steady states of the thermohaline circulation*, pp. 501–511. Dordrecht: Kluwer Academic Publishers.

Meehl, G.A. and Washington, W.M. (1996) El Niño-like climate change in a model with increased atmospheric CO_2 concentrations. *Nature.* **382**, 56–60.

Neelin, J.D., Latif, M., Allaart, M.A.F., Cane, M.A., Cubasch, U., Gates, W.L., Gent, P.R., Ghil, M., Gordon, C., Lau, N.C., Mechoso, C.R., Meehl, G.A., Oberhuber, J.M., Philander, S.G.H., Schopf, P.S., Sperber, K.R., Sterl, A., Tokioka, T., Tribbia, J. and Zebiak, S.E. (1992) Tropical air-sea interaction in general circulation models. *Climate Dynamics*, **7**, 73–104.

Neelin, J.D. and Marotzke, J. (1994) Representing ocean eddies in climate models. *Science*, **264**, 1099–1100.

Nicholls, N. (1984) The Southern Oscillation, sea surface temperature, and interannual fluctuations in Australian tropical cyclone activity. *Journal of Climatology*, **4**, 661–670.

Oberhuber, J. (1988) An atlas based on the COADS data set, Report No. 15, The budgets of heat, buoyancy and turbulent kinetic energy at the surface of the global ocean. Max-Planck-Institut fur Meteorologie.

Power, S.B. and Kleeman, R. (1993) Multiple equilibria in a global ocean general circulation model. *Journal of Physical Oceanography*, **23**, 1670–1681.

Power, S.B. and Kleeman, R. (1994) Surface heat flux parameterization and the response of ocean general circulation models to high-latitude freshening. *Tellus*, **46A**, 86–95.

Power, S.B., Moore, A.M., Post, D.A., Smith, N.R. and Kleeman, R. (1994) Stability of North Atlantic Deep Water Formation in a Global Ocean General Circulation Model. *Journal of Physical Oceanography*, **24**, 904–916.

Revell, C.G. and Goulter, S.W. (1986) South Pacific tropical cyclones and the Southern Oscillation. *Monthly Weather Review*, **114**, 1138–1144.

Ropelewski, C.F. and Halpert, M.S. (1987) Global and regional scale precipitation patterns associated with the El Niño/Southern Oscillation. *Monthly Weather Review*, **115**, 1606–1626.

Ropelewski, C.F. and Halpert, M.S. (1989) Precipitation patterns associated with the high index phase of the Southern Oscillation. *Journal of Climate*, **2**, 268–284.

Semtner, A.J. (1976) A model for the thermodynamic growth of sea ice in numerical investigations of climate. *Journal of Physical Oceanography*, **6**, 379–389.

Semtner, A.J. and Chervin, R.M. (1988) A simulation of the global ocean circulation with resolved eddies. *Journal of Geophysical Research*, **93**, 15 502–15 522.

Semtner, A.J. and Chervin, R.M. (1992) Ocean general circulation from a global eddy-resolving model. *Journal of Geophysical Research*, **97**, 5493–5550.

Stocker, T.F. and Wright, D.G. (1991) A zonally averaged ocean model for the thermohaline circulation. Part II: Interocean circulation in the Pacific-Atlantic Basin System. *Journal of Physical Oceanography*, **21**, 1725–1739.

Wilkin, J.L. and Morrow, R.A. (1994) Eddy kinetic energy and momentum flux in the Southern Ocean: Comparison of a global eddy-resolving model with altimeter, drifter, and current-meter data. *Journal of Geophysical Research*, **99**, 7903–7916.

Wright, D.G. and Stocker, T.F. (1991) A zonally averaged ocean model for the thermohaline circulation. Part I: Model development and flow dynamics. *Journal of Physical Oceanography*, **21**, 1713–1724.

CHAPTER 7

CLIMATE MODEL SIMULATIONS OF GLOBAL WARMING

Warren M. Washington and Gerald A. Meehl

7.1   Introduction

In this chapter, we summarise the results of several experiments with versions of coupled climate models designed for global warming studies.

This ocean model addresses the problem of improved ocean boundary currents, as well as heat and salt transports. The new ocean model contributes to reduced model errors near sea-ice boundaries in the coupled model. We also investigated the effects of penetrative cumulus convection and cirrus albedo on global warming estimates; the relative roles of cloud albedo and the super greenhouse effect; improved sea-ice treatment in climate models and its effect on climate change; and new global and regional warming estimates with a new-generation coupled climate model.

The experiments described in this chapter have been published or submitted for publication elsewhere. These articles summarise research on the many uncertainties in the global-warming area and cover such subjects as the development of a new ocean model for coupling to the atmosphere and sea-ice components. For a more complete discussion and extensive analysis of each experiment please refer to the relevant reference.

7.2   A world ocean model for global change experiments

Previous coupled model greenhouse studies have typically used an ocean model with resolution of the order of 5° (e.g. Gates et al. 1985; Schlesinger et al. 1985; Stouffer et al. 1989; Washington & Meehl 1989; Manabe et al. 1990; Cubasch et al. 1992; Meehl 1990, 1992). In this section, we introduce a global ocean model with considerably higher resolution, 1° in latitude and longitude with 20 levels in the vertical. We show improvements in the 1° over our earlier 5° version and document the extent to which the 1° version captures elements of the ocean circulation simulated by a version of this model with 0.5° resolution. The 1° model is the same as the 0.5° version, except for the resolution and the addition of the Arctic Ocean, and may be adequate to represent many aspects of ocean heat transport of interest for climate studies (Bryan 1991; Covey 1992).

We attempt to bring the model ocean as close as possible to the observed state of the ocean before coupling. This is accomplished by forcing the entire three-dimensional temperature and salinity structure of the ocean into agreement with observations. The ocean model is subsequently run for some period with observed surface forcing (e.g. Washington & Meehl 1989). In this way, the various inherent imbalances in the ocean model (mainly due to model inadequacies and resolution) produce only a very slow secular drift, and the state of the ocean is relatively close to the observed state of the ocean. The slow drift is much smaller than the climate-sensitivity signals caused by, say, increased greenhouse gases. There are less systematic errors in the spun-up model, but the slow secular drifts imply that such spin-up procedures are appropriate only for experiments on the order of 100 years.

In this regard, comparisons with higher-resolution versions (such as the 0.5° version in the present study) that presumably produce improved ocean simulations are necessary to indicate the types of errors that can be expected from lower-resolution ocean models (e.g. the 1.0° version here) when coupled to an atmospheric model. Compromises in resolution must be made at this stage because of computer resource restraints. High-resolution, state-of-the-art global ocean models will likely become a possibility for global, coupled-model century-time-scale integrations in the next five years. As we show in this study, however, major improvements in the ocean simulation are possible with medium-resolution versions (e.g. 1°) of current high-resolution ocean general circulation models (GCMs) over those produced by the previous generation of coarse-grid ocean GCMs.

Although resolution is not the only consideration in ocean model performance, some of the major systematic errors noted in the global coupled simulation with the 5° ocean model (e.g. Washington & Meehl 1989; Meehl 1990) were attributed in large part to coarse ocean model resolution.

For example, a major shortcoming of ocean models with coarse resolution (e.g. 5°) is the large meridional heat transport by subgrid-scale diffusion (Meehl et al. 1982). Diffusion prevents the formation of sharp gradients in temperature, salinity, and ocean-current distributions, important factors in obtaining the approximately correct sea-ice distributions and ocean-heat transports. We show that the resolution in the 1° model allows reasonable simulation of major regional features, such as the Gulf Stream, the Kuroshio Current, and the Antarctic Circumpolar Current system, albeit without well-resolved eddies. The simulated weak eddies or waves that do appear in the 1° version are important features associated with establishing these gradients. In regard to regional features, the model simulates many of the features of the 'conveyor-belt' circulation (Broecker 1991) that connects the North Atlantic deep circulation with the tropical Pacific, as demonstrated by Semtner and Chervin (1992).

Concerning whether or not the surface of the observed present ocean is ever in thermal equilibrium with changes at the air–sea interface, it is possible that atmospheric forcing is changing rapidly for anthropogenic reasons and that the deep ocean is slowly responding to it. We would also expect a greenhouse warming (warming mostly at the top of the oceans) to result in less thermodynamic coupling of the upper ocean to the middle and bottom portions of the ocean in most regions because of less convective overturning, even though the deep ocean circulation is of considerable importance in the greenhouse problem in regions of deep mixing (e.g. the North Atlantic and the Antarctic circumpolar oceans) (e.g. Stouffer et al. 1989; Washington & Meehl 1989).

The basic features of this ocean model are discussed in many publications beginning with Bryan (1969). We used continental outlines and ocean bathymetry from the 0.5° version (Semtner & Chervin 1988) as the basis for the 1° version. The technique was the same as that of Semtner and Chervin, except we took into account the Arctic Basin. We interpolated a NCAR map of the continental outline to a 1° resolution, smoothed the coastlines, eliminated many small islands and allowed for flow through a single strait in the western Pacific Ocean.

Use of the same 20 vertical levels of the 0.5° version is designed to give higher resolution within the thermocline (10 levels in the upper 710 m) and near the bottom. Following the procedure of Semtner and Chervin (1988) and to facilitate comparison to their 0.5° results, all forcing fields are annual means.

Using robust diagnostic forcing with a free thermodynamic thermocline of observed annual mean temperature and salinity fields from Levitus (1982), we compare 1° model experiments with similar experiments previously run with the 0.5° version (Semtner & Chervin 1988, 1991a, 1991b, 1992). Forcing of temperature and salinity in the top layer and below 710 m is on time scales of one month and three years, respectively. Although this technique (full forcing experiment) weakly constrains the mean temperature and salinity fields close to the observed in the average sense, it allows dynamics locally to generate ocean mesoscale waves and sharp gradients. The experiment that does not restore the simulation to observations below the thermocline will be referred to as 'top forcing'. In both experiments, the time step is 4 h for prognostic variables.

The model starts from initial conditions of no currents and isothermal conditions and runs for 14 years until an approximate steady state of the zonal profiles of currents and temperature is achieved with full forcing. From this point, the full-forcing experiment continues to 100 years with virtually no drift. The second experiment with only top forcing starts at year 14 of the full-forcing experiment and is integrated for an additional 100 years. This allows the subsurface ocean to be 'initialized' with the observed values of temperature and salinity for 14 years with the robust diagnostic technique. In this study, all results are from the top forcing experiment unless specifically noted. At the end of the integration, there is still some temperature and salinity drift in the top forcing experiment compared to the full forcing, but the drifts are small.

The five-year average zonal mean patterns for years 96–100 of the top forcing experiment of temperature and salinity are shown in Figure 7.1. Although the ocean model is run with top forcing from year 14 to 100, the basic ocean structure remains close to the observed, and the major features of the zonally averaged temperature and salinity structure are preserved.

The 1° model thermocline structure shows much more detail than the 5° model (also shown in Figure 7.1) used in our previous greenhouse studies. The sharp discontinuity at about 65°N is the observed interface between the North Atlantic and Arctic Oceans. The salinity field shows several interesting features in the zonal average, such as relatively fresh water at the surface of the Arctic Ocean, high salinity mainly contributed by the outflow of

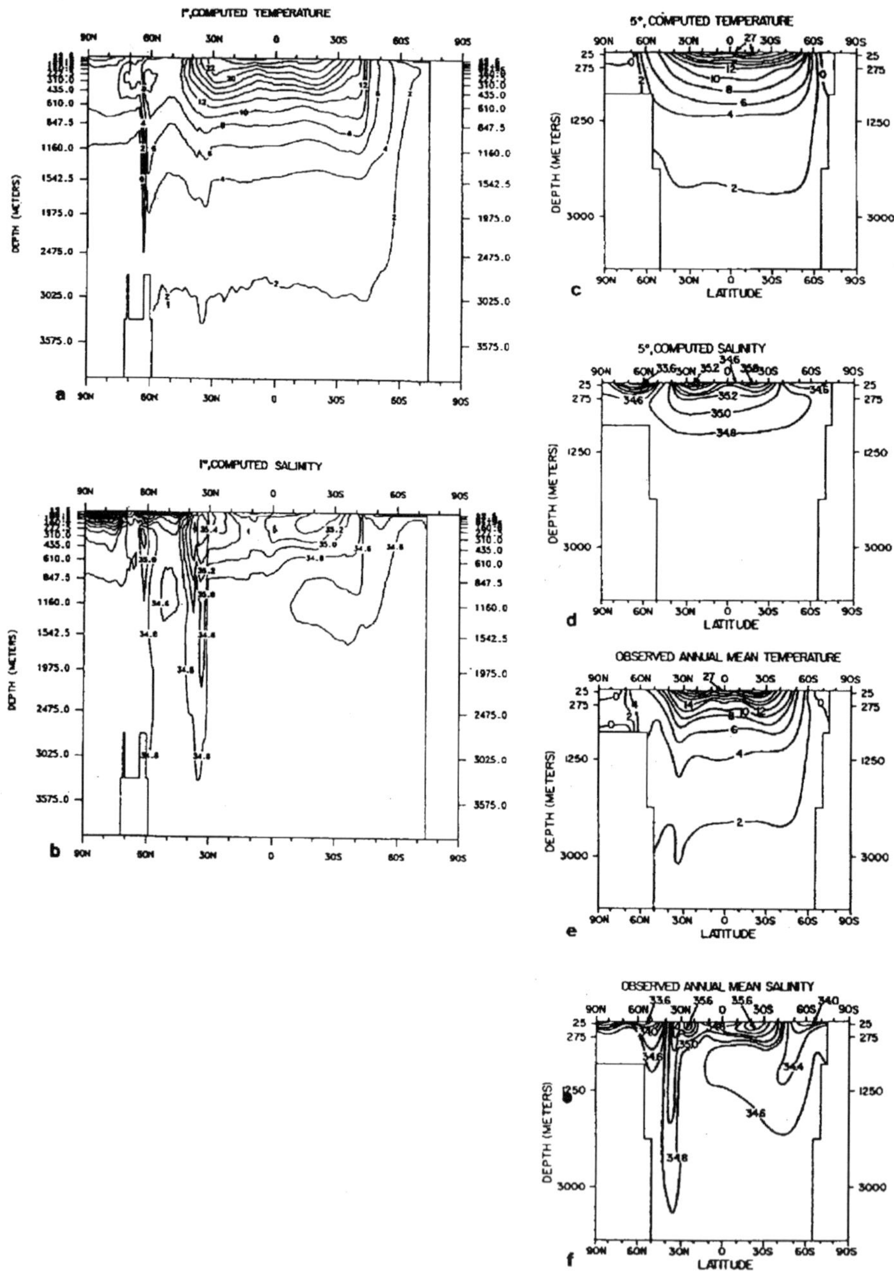

Figure 7.1 Latitude-depth distribution of zonally averaged: (a) temperature; (b) salinity from the 1° top forcing experiment; annual mean (c) temperature; (d) salinity from the 5°, 4-level model; (e) observed annual mean temperature from Levitus (1982); and (f) observed annual mean salinity from Levitus (1982).

the Mediterranean Sea near 30°N, and relatively high salinity in the subtropics near 30°S. There are low salinity values near the surface in southern high latitudes with a tongue of low-salinity water extending northward at around 1000 m (Antarctic Intermediate Water).

These surface temperatures and salinities are largely influenced by the one-month time-scale robust forcing in the surface layer, and they reflect the observed features of the Gulf Stream off the coast of North America, the Kuroshio Current off the east coast of Japan, and the Antarctic Circumpolar Current (ACC). The sharpness of the temperature gradients is much improved over the earlier studies with a 5° horizontal grid. The equatorial cold tongue off the coast of South America is simulated, as is the warm pool in the western Pacific. Somewhat cooler water, however, is being generated between 0° and 5°N north of Australia. There appears to be evidence of the Mindanao Dome (Masumoto & Yamagata 1991) east of the Philippines. The geographic distribution of salinity shows the major extrema in the regional oceans that range from a maximum of 37% in the subtropical oceans to a minimum of 30% in the Arctic Ocean. Overall, the pattern of the 1° model of temperature and salinity resembles the observed and the results of Semtner and Chervin (1988) (see Washington et al. 1994).

Meridional overturning in the ocean is important in climate models because the oceans can affect the timing and amount of greenhouse warming through heat transport and thermal inertia. From earlier studies with coupled global atmosphere and ocean models (e.g. Stouffer et al. 1989; Washington & Meehl 1989; Manabe et al. 1991), it is evident that, because the upper ocean warms and becomes less saline in the North Atlantic and the whole ocean becomes more statically stable, the meridional overturning circulation weakens. This will affect how much heat is sequestered in the intermediate and deep ocean. Also, the complex interactions between the atmosphere and ocean involving wind stress, heat flux, and changes in salinity and ocean surface temperature can cause large changes in the regional climate pattern. Figure 7.2 shows the meridional overturning stream function for the global ocean (a, b) and Atlantic Ocean (c, d) and compares the 1° results with the 0.5° results from Semtner and Chervin. The strong cells near the surface on both sides of the equator are primarily wind driven and very similar for the two resolutions. The other strong overturning cell is in the vicinity of the Antarctic Circumpolar Current, where there is upwelling near the coast of Antarctica (the Deacon Cell). The strength of this cell is about the same in both the 1° and 0.5° versions (about 30 Sv) but has greater vertical extent in the 0.5° version.

The global overturning in the Northern Hemisphere with sinking originating near 65°N is weaker in the 1° version (maxima of about 7.5 Sv in the 1° versus 12.5 Sv in the 0.5° version). The Atlantic overturning stream function is also shown in Figure 7.2 at the two resolutions. The same equatorial pattern exists in both resolutions, and the North Atlantic patterns are comparable with weak downward motion near 60–65°N, with a greater vertical extent in the 0.5° model.

Gordon (1986) has suggested a conveyor-belt circulation based upon observational studies. To investigate this circulation in the 1° model, Washington et al. (1994) show several regions of the global ocean separated into surface currents and flows at specific

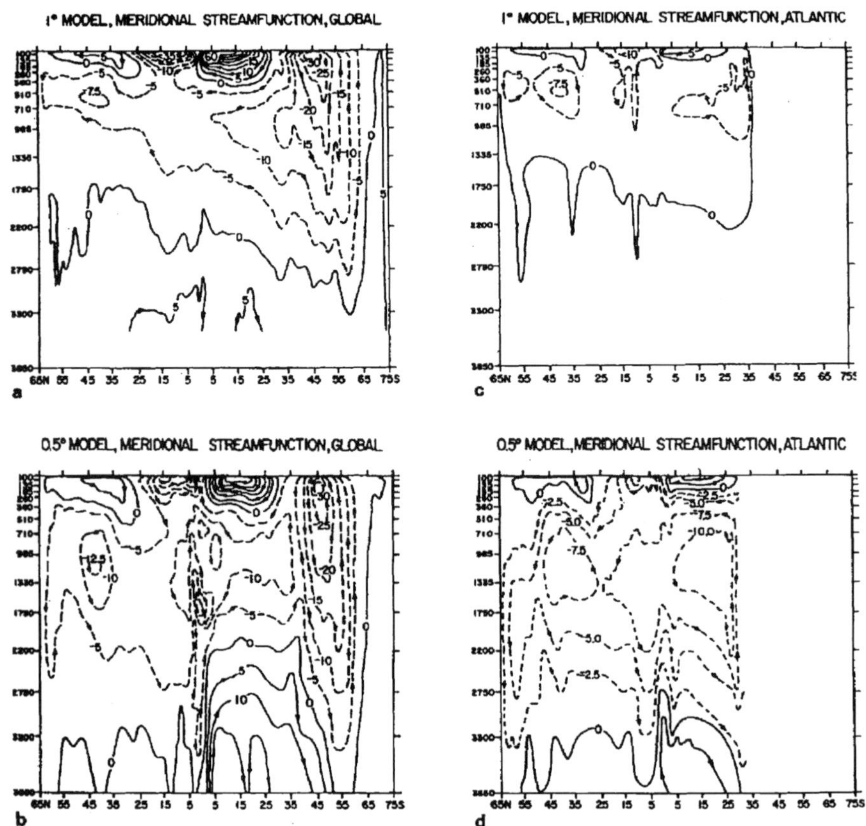

Figure 7.2 Meridional overturning stream function (Sv): (a) global domain, 1° top forcing model; (b) global domain, 0.5° model; (c) Atlantic Ocean, 1° top forcing experiment; and (d) Atlantic Ocean, 0.5° model. (1 Sv = 106 m³ s⁻¹).

depths. The return flows of the thermohaline circulation are mostly at the ocean bottom. This conveyor-belt flow is qualitatively similar for both the 0.5° and 1° versions, again keeping in mind the change in resolution of bottom topography.

In summary, the 1°, 20-level global ocean model produces an ocean simulation that is a great improvement over the earlier 5° version while capturing most features of the 0.5° version on time scales of a century with observed top forcing and small-magnitude drift after a century. Compromises in ocean model resolution must be made to conduct greenhouse-sensitivity experiments with global coupled models with today's computer capability. The fact that many observed ocean features can be simulated with a 1°, 20 levels, global model shows that the addition of the Arctic Ocean in this model does not present serious difficulties in the simulation compared to the 0.5° version without the Artic.

Washington et al. (1994) show results from two experiments with sea surface temperatures (SSTs) specified in a region of the North Atlantic near Greenland. These experiments indicate a high sensitivity to the seasonal cycle of SSTs. After 10 years, for perpetual midwinter SSTs in that region, the thermohaline circulation in the Atlantic intensifies to roughly an order of magnitude more than that with annual mean SSTs. The Gulf Stream circulation also intensifies, and the North Atlantic Current penetrates farther northward. For perpetual early winter SSTs in that same region, the thermohaline circulation increases by about a factor of three. Additionally, greater overturning in the Atlantic can intensify the vertical temperature gradient in the thermocline in both the tropical Atlantic and the Pacific via the conveyor-belt circulation noted by Semtner and Chervin (1992).

7.3   Convection and tropical cirrus-albedo effects

Washington and Meehl (1993) present the results of two climate simulations with an atmospheric model coupled to a simple mixed-layer ocean with sea-ice component to investigate the role of penetrative cumulus convection and tropical cloud-albedo feedback effects. Both simulations include a penetrative convection scheme that has the effect of pumping more moisture higher into the troposphere. One also includes a simple prescribed functional dependence of cloud albedo in areas of high SST and deep convection. Previous analysis of observations has shown that, in regions of high SST and deep convection, the upper-level cloud albedos increase as a result of the greater optical depth associated with increased moisture content. Based on these observations, we prescribe increased middle- and upper-level cloud albedos in regions of SST greater than 303 K where deep convection occurs. This crudely accounts for a type of cloud-optical property feedback but is well short of a computed cloud-optical property scheme. Since great uncertainty accompanies the formulation and tuning of such schemes, the prescribed cloud-albedo feedback is an intermediate step to examine basic feedbacks and sensitivities.

Washington and Meehl (1993) compare two model versions, one with penetrative convection and cloud-albedo feedback and one with convective adjustment and no cloud-albedo feedback, to a model from the Canadian Climate Centre (CCC) having convective adjustment and a computed cloud-optical properties feedback scheme, as well as to several other GCMs. The addition of penetrative convection increases tropospheric moisture, cloud amount, and planetary albedo and decreases net solar input at the surface. However, the competing effect of increased downward infrared flux (from increased tropospheric moisture) causes a warmer surface and increased latent heat flux. Adding the prescribed cirrus-albedo feedback decreases net solar input at the surface in the tropics, since the cloud albedos increase in regions of high SST and deep convection. Downward infrared radiation (from increased moisture) also increases, but this effect is overpowered by the reduced solar input in the tropics. Therefore, the surface is somewhat cooler in the tropics, latent heat flux decreases, and global average sensitivity to a doubling of carbon dioxide with regard to temperature and precipitation/evaporation feedback is reduced.

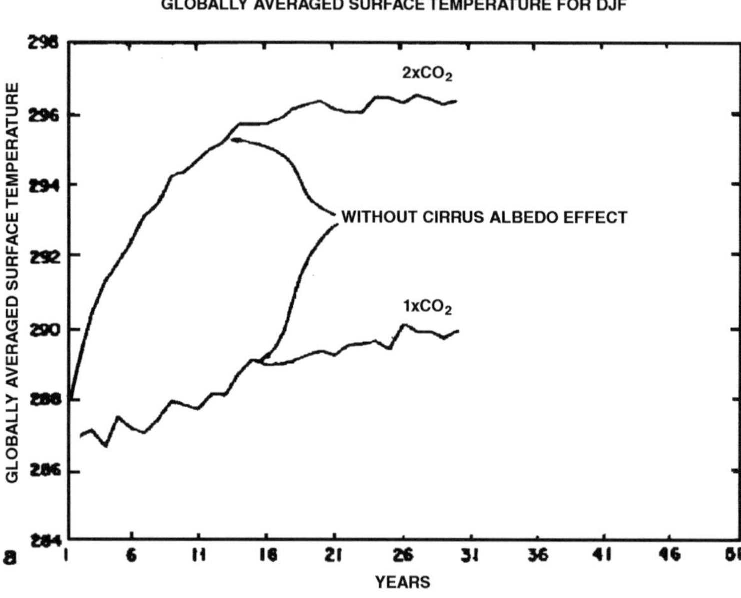

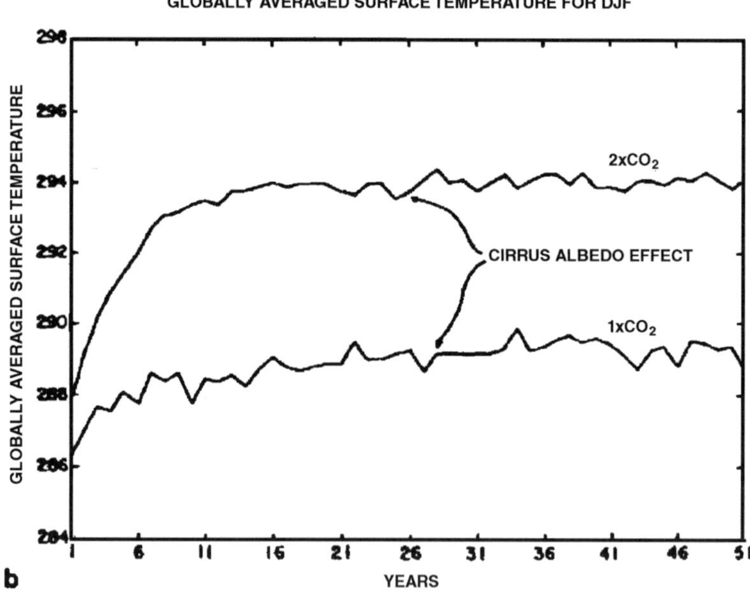

Figure 7.3 Time series plots of global DJF surface air temperature for the control $(1 \times CO_2)$ and $2 \times CO_2$ experiments: (a) without cirrus-albedo feedback and (b) with cirrus-albedo feedback.

Similar processes, evident in the CCC model with convective adjustment and a computed cloud-optical properties feedback scheme, occur over a somewhat expanded latitudinal range. The addition of penetrative convection produces global effects, as does the prescribed cirrus-albedo feedback, although the strongest local effects of the latter occur in the tropics.

The global mean December, January, and February (DJF) surface temperatures ($= 0.991$) are plotted as a function of time in Figure 7.3 for the NCAR experiments for the control ($1 \times CO_2$) and doubled CO_2 ($2 \times CO_2$). At the top are experiments without cloud-albedo feedback; at the bottom are results with cloud-albedo feedback. The starting points of all four experiments are the same, and they evolve to their respective equilibria as a consequence of the various feedbacks in the formulations. In the control experiment with cirrus-albedo feedback, the model comes to equilibrium at a cooler temperature after year 20 or so than the control experiment without cirrus albedo. The greatest effect appears to be in the doubled CO_2 experiments. By including the mass flux convective scheme without cirrus albedo, the annual global mean surface temperature warming in the $2 \times CO_2$ experiment is 6.2 K compared to 4.0 K equilibrium warming produced by the same model with convective adjustment (Figure 7.3; Meehl & Washington 1990). The model with the mass flux penetrative convective scheme with cloud-albedo feedback produces a globally averaged warming of 4.6 K. Although the cloud-albedo feedback effect is tied to warm SSTs that exist only near the equator, the effect is not local but extends throughout the globe.

The results shown here highlight uncertainties involved with different types of convective schemes and their interactions with cloud-albedo feedback in global climate sensitivity studies. Inclusion of penetrative convection greatly magnifies temperature sensitivity, and cloud-albedo feedback can moderate the sensitivity. Most noticeable in the tropics, these effects are manifested by global changes. By including both penetrative convection and cloud-albedo feedback, increases of globally averaged precipitation are somewhat less than in earlier model versions that used convective adjustment and omitted any cloud-albedo feedback.

7.4   Cloud-albedo feedback and the super greenhouse effect

Meehl and Washington (1995) further explore the roles of cloud albedo and the super greenhouse effect climate models. In studies noted above, Washington and Meehl (1993) and Boer (1993) discussed two mechanisms that influence climate sensitivity, and they are shown schematically in Figure 7.4. In the context of the Washington and Meehl (1993) experiment, the addition of the penetrative mass-flux convection scheme leads to increased moisture higher in the troposphere. This, in turn, produces increased cloudiness. At this point, two feedback loops occur as a consequence of the increased clouds and tropospheric moisture. The loop on the left side of Figure 7.4 is labelled 'cloud-albedo feedback' and is a negative feedback that decreases climate sensitivity in the following way. The increased clouds are associated with increased albedo. A consequential increase of solar radiation reflected back to space leads to decreased absorbed solar radiation at the surface. This, in turn, contributes to decreased surface temperature, decreased evaporation and precipitation, a decrease of moisture in the troposphere, a decrease of clouds, and so on.

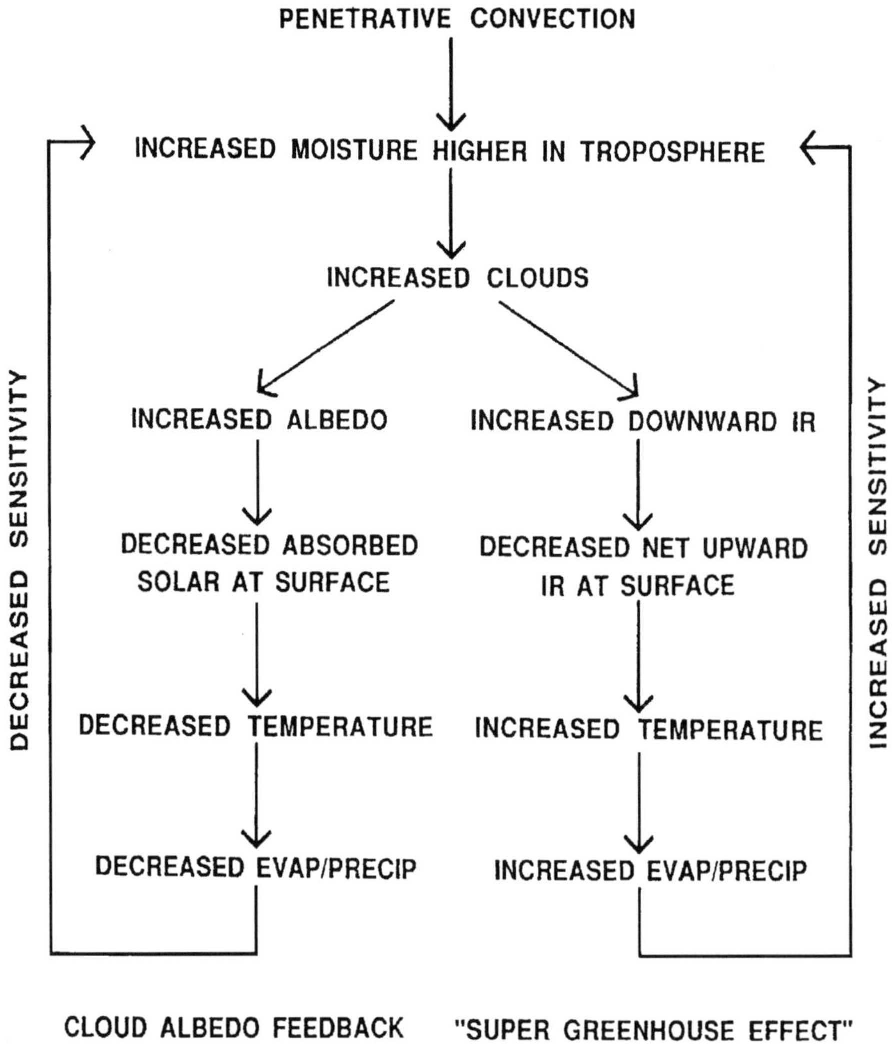

Figure 7.4 Schematic of physical interactions of cloud-albedo feedback and super greenhouse effects.

If the feedback loop on the right side of Figure 7.4 is followed, the same increase of moisture and clouds in the troposphere associated with the penetrative convection can lead to increased downward infrared radiation with a consequential decrease of net upward infrared at the surface. This contributes to increased surface temperature, increased evaporation and precipitation, and an even further increase of moisture and clouds in the atmosphere. Cloud-albedo feedback at the left of Figure 7.4 is a negative feedback contributing to decreased climate sensitivity, and the so-called 'super greenhouse effect' (e.g. Ramanathan

& Collins 1991) at the right side of Figure 7.4 is a positive feedback that should lead to enhanced climate sensitivity. The competition or, more accurately, the balance between these two feedbacks was identified by Washington and Meehl and Boer in the earlier studies as producing a more moderate sensitivity of globally averaged precipitation with a given amount of globally averaged surface air temperature warming compared to other models that did not have any cloud-albedo feedback. Washington and Meehl also noted that the inclusion of a penetrative convection scheme seemed to necessitate a comparable upgrade to some type of cloud-albedo feedback scheme to avoid having the climate system overwhelmed by the super greenhouse effect.

These two competing radiative feedbacks are examined in a sensitivity study with a global coupled ocean-atmosphere GCM (Meehl & Washington 1995. Cloud-albedo feedback is strengthened in the model by lowering the SST threshold in the specified cloud-albedo feedback scheme. This has the effect of expanding somewhat the areas eligible for coincident warm SSTs and deep convection to produce increases of upper-level cloud albedos. The competing super greenhouse effect combines with the enhanced cloud-albedo feedback to produce decreased maximum values of SST by about 1°C. There is also a cooling of the tropical troposphere with attendant global changes of atmospheric circulation reminiscent of those observed during La Niña or cold events in the Southern Oscillation. As noted by Meehl and Washington (1995), the strengthening of the cloud-albedo feedback only occurs over limited regions where there are warm tropical oceans (e.g. the western Pacific warm pool). In those regions, there is increased albedo, decreased absorbed solar radiation at the surface, lower temperatures, and decreased precipitation and evaporation. However, the weakened convection over these regions alters the large-scale circulation in the tropics such that there is increased upper-level divergence over tropical land areas. This results in increased precipitation and intensified monsoonal regimes. The increases of precipitation produce increased clouds and albedo and wetter and cooler land surfaces. These additional contributions to decreased absorbed solar at the surface combine with similar changes over the tropical oceans to produce the global cooling associated with the stronger cloud-albedo feedback. Thus, the Meehl and Washington (1995) study points to the uncertainties involved with the parameterization of cloud albedo and the major implications such parameterizations can have concerning the maximum values of SST, global climate sensitivity, and climate change.

7.5 Improved sea-ice treatment in greenhouse experiments

The treatment of sea ice in global climate models has gone through a great deal of evolution. Because of computer constraints in early efforts, it was necessary to deal with the physical aspects in a simplified manner to make simulations computationally feasible. The computational aspects are now much less severe because of improved generations of larger and faster computers. Two streams of research are leading to improved treatment of this important earth system modeling component. One involves more detailed observations, improved understanding, and improved modeling of Arctic and Antarctic sea ice over relatively small regions. The other is an increasingly more realistic treatment of sea ice in global models.

At NCAR, we have been conducting global climate coupled-model (atmosphere–ocean–sea ice) experiments without flux adjustment at the ocean surface. This climate system model uses a $1°$, 20-level version of the Semtner–Chervin (Semtner & Chervin 1988, 1992) ocean model described earlier. The sea-ice model dynamics come from Flato and Hibler (1990) and the thermodynamics from a three-level model of Semtner (1976).

In previous NCAR experiments on the effects of increased greenhouse gases on the climate system, we used a simple, thermodynamic, zero-layer sea-ice model (Semtner 1976). Based on the simple thermodynamics of heat conduction through sea ice between the atmosphere and ocean, this type of model captures the first-order effects of sea-ice formation and decay but neglects the important effects of heat capacity and transport and compaction of sea ice by ocean currents and winds. To improve this aspect of our coupled model, we add sea-ice-dynamics processes. (See Hibler 1979; Parkinson & Washington 1979; Washington & Parkinson 1986; Hibler 1992 and Hibler & Flato 1992, for general reviews of the state of sea-ice modeling.) In devising a simplified sea-ice dynamical model for coupling to a thermodynamical model, Flato and Hibler (1990) invoked a 'cavitating fluid' approximation to allow sea ice to have compressive strength and develop internal ice pressure. This characteristic would only take place in a converging wind field. For a diverging wind field, the internal ice pressure would be zero. Five major dynamic forces control the motion of sea ice–air stress at the top of sea ice, water stress below sea ice, gravitational stress from the tilt of sea surface (dynamic topography), Coriolis force, and pressure stresses within sea ice. All of these effects are included in the Flato-Hibler treatment, although simplified. (For example, the effects of shear and tensile strength are neglected and only compressive effects are included.)

David Pollard (NCAR) has programmed the Flato and Hibler code into a latitude-longitude grid structure such that the resolution can be easily changed (Pollard & Thompson 1994). For the coupled system, the ocean and sea-ice models both have $1°$ horizontal resolution.

As suggested by Flato and Hibler (1990), under converging conditions, sea ice builds up and causes internal compressive strength that further slows the ice build-up. The velocities are modified so that momentum is approximately conserved. As pointed out by Flato and Hibler (1990), similar but less conservative schemes have been proposed and tested by Nikiforov et al. (1967) and Parkinson and Washington (1979). One of the major terms in the momentum equations for sea ice is the gradient of dynamic topography. This term can be approximated from the ocean model, and it is essentially the slope of sea surface. To a good approximation, the upper-ocean geostrophic currents give the slope of the sea surface. Semtner (pers. comm.) has suggested using geostrophic velocities in the second layer of the ocean model rather than in the top layer, since the top-layer velocities may be altered significantly by Ekman-drift effects.

It is clear that the cavitating fluid approximation will not account for all the important dynamical sea-ice processes such as shear stress. This is partly taken into account by zero tangential flow in the ocean at the model coasts. It is also not clear to what time scales the approximation applies. Certainly, on daily scales or less, the balance equation may not apply. In any case, the approximations are a more physically plausible improvement over earlier simple treatments used in previous greenhouse studies where ice dynamics was completely ignored or a simple ice advection by ocean currents was included.

To make the seasonal aspects of sea ice more realistic, we use the improved sea-ice thermodynamical model with three layers devised by Semtner (1976) instead of the zero-layer model used in earlier greenhouse studies. This model has one layer of snow and divides the ice thickness into two layers. As Semtner has shown, the three-layer model has a better seasonal distribution of sea-ice thickness than the zero-layer thermodynamic sea-ice model.

The flux of salt from the freezing of sea water is taken into account, as well as the freshening of sea water when sea ice melts. With the parameterization of Parkinson (1979), the time change in salinity is computed from one day to the next such that total salt is conserved. Thermodynamic sea ice can grow and melt in several ways (e.g. Parkinson & Washington 1979; Harvey 1988). We have already discussed the melting and growth of an existing concentration or fraction of sea ice within a horizontal grid volume by the use of vertical fluxes of heat. If, for example, an ocean grid volume is cooled (either in a lead or open ocean if it is ice free) such that sea ice can form, the amount of sea-ice volume can be computed.

Thus, for a horizontal-grid volume, the ice volume generated is the product $C_i h_i$, where 'C_i' is the concentration of new sea ice and 'h_i' is thickness. Because only the product is known, either the concentration or the thickness has to be specified. In the case of cooling, we add a thin layer of new ice to the grid volume of thickness, thus changing the concentration. At the end of a time step, to obtain new sea-ice thickness and concentration, we average new and existing sea ice based on their respective concentrations.

Figure 7.5 shows March and September sea-ice distributions of ice thickness and fraction from the global coupled model for present-day climate. As shown, the sea-ice distributions show thicker ice in the central Arctic at the end of winter with more ocean water and thinner ice at the end of summer. The ice transport is largely influenced by daily variations in wind stress. It also shows cross-polar flow towards northern Greenland and southern flow along the east coast of Greenland.

The Antarctic sea-ice distribution shows a large seasonal area change. Also, as expected, there are areas of greater ice thickness and concentration in the Weddell Sea and close to Antarctica. Although the general character of the seasonal cycle of sea ice is simulated in the coupled models, errors in the distribution and thickness of ice can result from errors in the wind stress or temperature in the atmospheric model or errors in the ocean currents and SST. Sea ice remains one of the most difficult challenges in coupled modeling. The inclusion of more realistic processes, as shown here, contributes to a better understanding of the response of this important and highly sensitive component of the climate system.

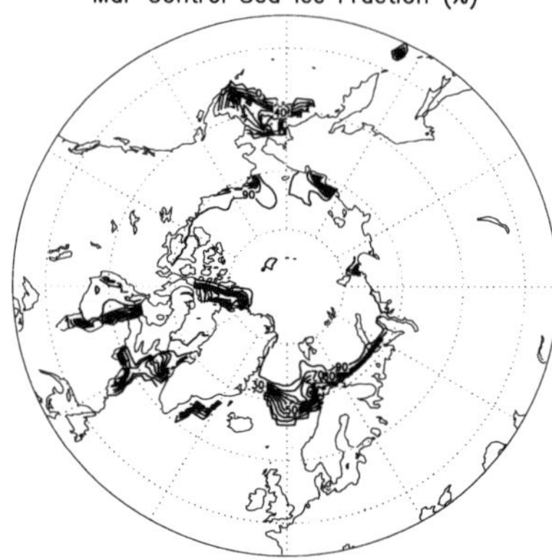
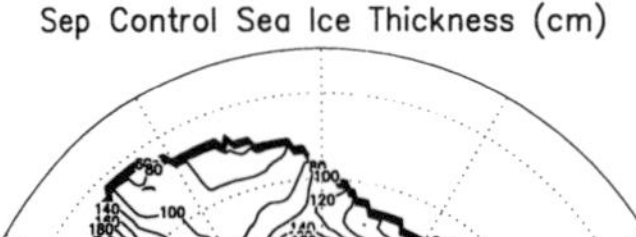
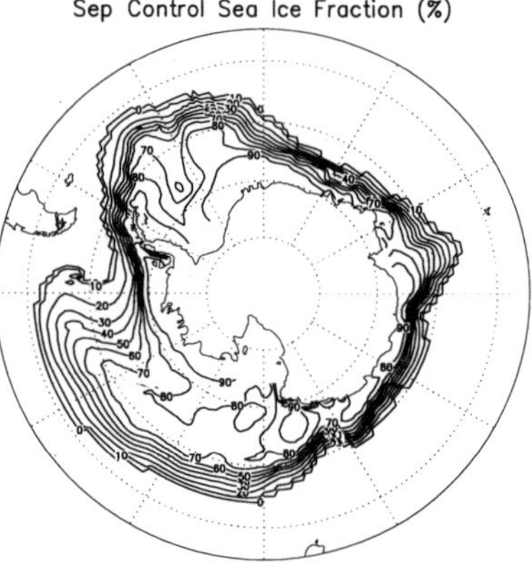
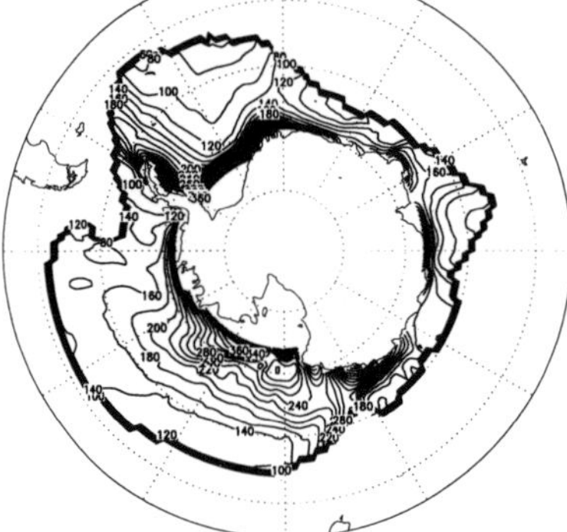

Figure 7.5 March and September sea-ice thickness and concentration (%) for the control experiment.

7.6   Early increased CO$_2$ results from the global coupled model

Washington and Meehl (1996) perform an experiment with atmospheric CO$_2$ increasing at the rate of 1% per year compounded in the global coupled ocean–atmosphere–sea-ice GCM described above. The purpose of the experiment is to assess the sensitivity of the earth's coupled climate system to this increase of CO$_2$ in terms of changes to the dynamically coupled atmospheric and oceanic circulation and alterations of sea-ice distribution. These changes were studied at the approximate time of CO$_2$ doubling near year 70 of the experiment.

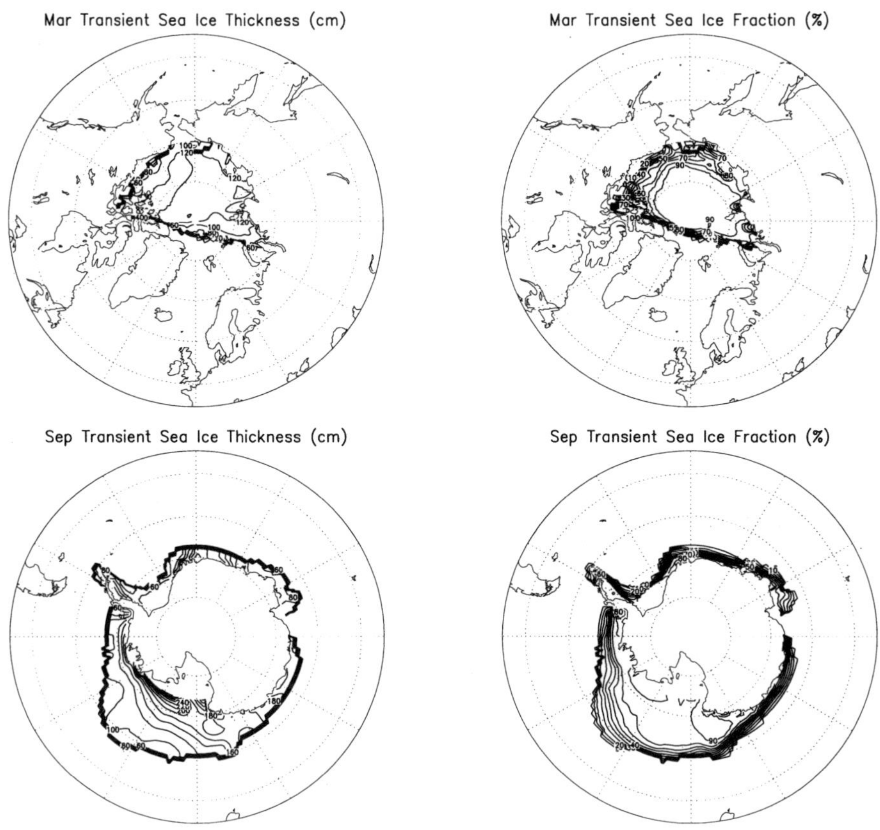

Figure 7.6 March and September sea-ice thickness and concentration (%) for transient CO_2 increase experiment.

Figure 7.6 shows the March distribution of sea ice in the transient experiment at the time of CO_2 doubling. The global warming due to increased CO_2 has caused sea ice to retreat in both the Arctic and Antarctic. This contributes to enhanced warming of surface air temperatures at high latitudes (Colour Plate 7.1). The North Atlantic Ocean warms and freshens and the thermohaline circulation in the ocean weakens. The continents warm faster than the oceans, and enhanced dry conditions over the northern continents occur during summer in the increased CO_2 environment (not shown). In the tropical Pacific, the eastern equatorial Pacific SSTs warm faster than the SSTs in the western equatorial Pacific due in part to cloud-albedo feedback and other dynamical coupled mechanisms. The result resembles conditions currently experienced during El Niño events, as well as the observed decadal time scale features of the 1980s and early 1990s. That is, in the Pacific region, the mean climate change from increased CO_2 in the model is similar to present-day El Niño conditions with enhanced moisture deficits in the Australasian region and increased precipitation in the central equatorial Pacific (Meehl & Washington 1996). Experiments with the regional effects of sulfate aerosols included show some reduction of warming over the Northern Hemisphere midlatitudes, but similar geographical responses to those in Colour Plate 7.1 are seen elsewhere (Meehl et al., 1996).

7.7   Summary

Over the last few years, we have conducted many global-change experiments with various versions of coupled atmosphere, ocean, and sea-ice models. In this chapter, we have summarised these new findings and have indicated where a complete description of the experiments can be found. Through these experiments, we have investigated the sensitivity of different components of the climate system to various parameterizations, as well as to increased carbon-dioxide concentrations. Some experiments used a version of the coupled climate system with only a simple mixed-layer ocean of thickness 50 m (Washington & Meehl 1984, 1993), and others used the same atmospheric model coupled to a global ocean GCM. The latter, our most recent global coupled model, uses a comprehensive global ocean model with a 1° resolution and 20 vertical levels. The ocean model has a resolution substantially higher than our previous version (e.g. Washington & Meehl 1989), and sea-ice treatment is more realistic with the addition of ice dynamics, a feature not usually included in sea-ice treatments of global simulations of the greenhouse effect. Experiments with this model have addressed some of the uncertainties in estimating potential global change, and MECCA has contributed to these activities by providing both increased computer capacity and international cooperation for research on global and regional climate change.

Acknowledgments

This research has been sponsored by the Department of Energy's Office of Energy Research, Office of Health and Environmental Research. It is part of the US Global Change Research Program (USGCRP) that is addressing uncertainties in understanding the global change of climate. The National Center for Atmospheric Research (NCAR) is sponsored by the National Science Foundation (NSF).

References

Boer, G.J. (1993) Climate change and the regulation of the surface moisture and energy budgets. *Climate Dynamics*, **8**, 225–239.

Broecker, W.S. (1991) The great global conveyor. *Oceanography*, **4**, 79–89.

Bryan, K. (1969) A numerical model for the study of the world ocean. *Journal of Computational Physics*, **4**, 347–376.

Bryan, K. (1991) Poleward heat transport in the ocean: A review of a hierarchy of models of increasing resolution. *Tellus*, **43**, A–B, 104–115.

Covey, C. (1992) Behavior of an ocean general circulation model at four different horizontal resolutions. In PCMDI Report No. 4, Program for Climate Model Diagnosis and Intercomparison, Lawrence Livermore National Laboratory, USA.

Cubasch, U., Hasselmann, K., Höck, H., Maier-Reimer, E., Mikolajewicz, U., Santer, B.D. and Sausen, R. (1992) Time-dependent greenhouse warming computations with a coupled ocean-atmosphere model. *Climate Dynamics*, **8**, 55–69.

Flato, G.M. and Hibler III, W.D. (1990) On a simple sea-ice dynamics model for climate studies. *Annals of Glaciology*, **14**, 72–77.

Gates, W.L., Han, Y.L. and Schlesinger, M.E. (1985) The global climate simulated by a coupled atmosphere-ocean general circulation model: Preliminary results. In *Coupled Ocean-Atmosphere Models*, edited by J.C.J. Nihoul, pp. 131–151. Elsevier Oceanography Series, 40. Amsterdam: Elsevier.

Gordon, A. (1986) Interocean exchange of thermocline water. *Journal of Geophysical Research*, **91**, 5037–5046.

Harvey, L.D.D. (1988) Development of a sea ice model for use in zonally averaged energy balance climate models. *Journal of Climate*, **1**, 1221–1238.

Hibler III, W.D. (1979) A dynamic thermodynamic sea ice model. *Journal of Physical Oceanography*, **9**, 815–846.

Hibler III, W.D. (1992) The role of sea ice dynamics in global climate change. In *Modeling the Earth System*, edited by D. Ojima, pp. 107–130. UCAR/Office for Interdisciplinary Earth Studies, Boulder, Colorado.

Hibler III, W.D. and Flato, G.M. (1992) Sea ice models. In *Climate System Modeling*, edited by K.E. Trenberth, pp. 413–436. Great Britain: Cambridge University Press.

Levitus, S. (1982) Climatological Atlas of the World Ocean, NOAA Professional Paper No. 13, Department of Commerce, Rockville, MD.

Manabe, S., Bryan, K. and Spelman, M.J. (1990) Transient response of a global ocean-atmosphere model to a doubling of atmospheric carbon dioxide. *Journal of Physical Oceanography*, **20**, 722–749.

Manabe, S., Stouffer, R.J., Spelman, M.J. and Bryan, K. (1991) Transient response of a coupled ocean-atmosphere model to gradual changes of atmospheric CO_2, Part I. Annual mean response. *Journal of Climate*, **4**, 785–818.

Masumoto, Y. and Yamagata, T. (1991) The response of the western tropical Pacific to the Asian winter monsoon: The generation of the Mindanao Dome. *Journal of Physical Oceanography*, **21**, 1386–1398.

Meehl, G.A. (1990) Development of global coupled ocean-atmosphere general circulation models. *Climate Dynamics*, **5**, 19–33.

Meehl, G.A. (1992) Global coupled models: atmosphere, ocean, sea ice. In *Climate System Modeling*, edited by K.E. Trenberth, pp. 555–581. Great Britain: Cambridge University Press.

Meehl, G.A., Washington, W.M. and Semtner, A.J. (1982) Experiments with a global ocean model driven by observed atmospheric forcing. *Journal of Physical Oceanography*, **12**, 301–312.

Meehl, G.A. and Washington, W.M. (1990) CO_2 climate sensitivity and snow-sea-ice albedo parameterization in an atmospheric GCM coupled to a mixed-layer ocean model. *Climatic Change*, **16**, 283–306.

Meehl, G.A. and Washington, W.M. (1995) Cloud albedo feedback and the super greenhouse effect in a global coupled GCM. *Climate Dynamics* **11**, 399–411.

Meehl, G.A. and Washington, W.M. (1996) El Ninõ-like climate change in a model with increased atmospheric CO_2 concentrations. *Nature*, **382**, 56–60.

Meehl, G.A., Washington, W.M., Erickson, D.J. III, B.P., Briegleb, P.J. and Jaumann, P.J. (1996) Climate change from increased CO_2 and the direct and indirect effects of sulfate aerosols. *Geophys. Res. Lett.*, **23**, 3755–3758.

Nikiforov, Y.G., Gudkovich, Z.M., Yefimov, Y.I. and Romanov, M.A. (1967) Principles of a method for calculating the ice redistribution under the influence of wind during the navigation period in Arctic seas. Trudy. *Arkticheskii i Antarkticheskii Nauchnoisslerdovatel'skii Institut*, Leningrad, 257 (Transl. AIDJEX Bulletin, 3, 40–64, 1970).

Parkinson, C.L. (1979) A simple parameterization for salt flux to upper ocean owing to freezing and melting at the surface. *Antarctic Journal*, **XIV**, 103–104.

Parkinson, C.L. and Washington, W.M. (1979) A large-scale numerical model of sea ice. *Journal of Geophysical Research*, **84**, 311–337.

Pollard, D. and Thompson, S.L. (1994) Sea-ice dynamics and CO_2 sensitivity in a global climate model. *Atmosphere-Ocean*, **32**, 449–467.

Ramanathan, V. and Collins, W. (1991) Thermodynamic regulation of ocean warming by cirrus clouds deduced from the 1987 El Niño. *Nature*, **351**, 27–32.

Schlesinger, M.E., Gates, W.L. and Han, Y.L. (1985) The role of the oceans in CO_2-induced climate change: preliminary results from the OSU coupled atmosphere-ocean general circulation model. In *Coupled Ocean-Atmosphere Models*, edited by J.C.J. Nihoul, pp. 447–478. Elsevier Oceanographic Series, 40. Amsterdam: Elsevier.

Semtner, A.J. (1976) A model for the thermodynamic growth of sea ice in numerical investigations of climate. *Journal of Physical Oceanography*, **6**, 379–389.

Semtner, A.J. and Chervin, R.M. (1988) A simulation of the global ocean circulation with resolved eddies. *Journal of Geophysical Research*, **93**, 15 502–15 522 and 15 767–15 775.

Semtner, A.J. and Chervin, R.M. (1991a) A thermohaline conveyor belt in the world ocean. US WOCE Office, Texas A&M University. WOCE Notes 2, 3, 12–15.

Semtner, A.J. and Chervin, R.M. (1991b) The thermohaline circulation of the tropical Pacific Ocean. Nova University, Dania, Florida. TOGA Notes 4, 18–24.

Semtner, A.J. and Chervin, R.M. (1992) Ocean general circulation from a global eddy-resolving model. *Journal of Geophysical Research*, **97**, 5493–5550.

Stouffer, R.J., Manabe, S. and Bryan, K. (1989) Interhemispheric asymmetry in climate response to a gradual increase of CO_2. *Nature*, **342**, 660–662.

Washington, W.M. and Meehl, G.A. (1984) Seasonal cycle experiment on the climate sensitivity due to a doubling of CO_2 with an atmospheric general circulation model coupled to a simple mixed-layer ocean model. *Journal of Geophysical Research*, **89**, 9475–9503.

Washington, W.M. and Meehl, G.A. (1989) Climate sensitivity due to increased CO_2: Experiments with a coupled atmosphere and ocean general circulation model. *Climate Dynamics*, **4**, 1–38.

Washington, W.M. and Meehl, G.A. (1993) Greenhouse sensitivity experiments with penetrative cumulus convection and tropical cirrus albedo effects. *Climate Dynamics*, **8**, 211–223.

Washington, W.M., Meehl, G.A., VerPlank, L. and Bettge, T. (1994) A world ocean model for greenhouse sensitivity studies: Resolution intercomparison and the role of diagnostic forcing. *Climate Dynamics*, **9**, 321–344.

Washington, W.M. and Meehl, G.A. (1996) High latitude climate change in a global coupled ocean–atmosphere–sea-ice model with increased atmospheric CO_2. *J. Geophys, Res.*, **101**, 12795–12801.

Washington, W.M. and Parkinson, C.L. (1986) *An Introduction to Three-Dimensional Climate Modeling*. Mill Valley, CA: University Science Books.

CHAPTER 8

MERIDIONAL HEAT TRANSPORT VARIABILITY FROM A GLOBAL EDDY-RESOLVING OCEAN MODEL

Robert M. Chervin, Anthony P. Craig and Albert J. Semtner Jnr

8.1   Introduction

The ocean heat transport is still not very well known. Estimates must be improved in order for us to understand the global climate better. There are a few direct estimates of ocean heat transport (e.g. Bryan 1962; Hall & Bryden 1982; Bryden et al. 1991), and these observational estimates are limited by the availability of data in space and time. In addition, almost no direct estimates have been computed which take into account seasonal or interannual variability.

At least two other methods have been used to predict meridional ocean heat transports. The first uses the global heat budget to estimate ocean heat transport (e.g. Michaud & Derome 1991; Trenberth & Solomon 1994). Typically, satellite radiation measurements at the top of the atmosphere are combined with estimates of the atmospheric heat transport and/or atmospheric heating, and ocean heat transport is determined as a residual. This method has become more accurate in recent years with improved estimates of the radiation fluxes at the top of the atmosphere, although there may still be some biases in the satellite data that need to be resolved. The second indirect method to compute the meridional ocean heat transport uses ocean-atmosphere surface fluxes predicted via bulk methods and ocean heat content changes as a function of time to estimate meridional ocean heat transport again as a residual (e.g. Hsiung et al. 1989). Generally, neither method has been used to predict seasonal or interannual heat transport variability, and the errors associated with these two methods are potentially significant.

In the past decade or two, ocean models and computers have evolved together such that ocean simulations on basin to global scales with eddy resolving grids are now possible. With the model data, meridional heat transport in the ocean can be analysed on annual, seasonal or interannual time scales. (See Bryan 1991 for a review.) While ocean model data can be archived with any spatial or temporal resolution within the computational and hardware limits, there are also limitations in the models, and these are usually associated with grid resolution and physical parameterisation. Modellers are continually trying to understand the models' limitations, to make the models more physically realistic, to validate the models via comparison with observations and to increase spatial and temporal resolution through computational speed-ups via hardware and software improvements. Model results will help us understand ocean heat transport better, especially when considering the amount of observational data that would have to be collected to begin to understand seasonal and interannual variability of the ocean heat transport.

Bryan (1982) reviewed the state of ocean heat transport estimates several years ago. Since that time, observational estimates have improved and models have become more realistic. As Bryan suggested at that time, the best estimates of ocean heat transport are probably going to come from a combination of modelling and observational efforts. Efforts need to continue in both areas as observations are probably the best way to determine absolute magnitudes while model results will probably help us understand the dynamics and variability better. In addition, both must be continually validated against each other.

In this chapter, we will present results from a global ocean modelling effort. Several decade long integrations were carried out to explore the effects of wind stress forcing variability on the model. The computations were carried out as part of the MECCA project. In this paper, we will focus on results related to meridional ocean heat transport. The goal is to both compare these model simulation results with observations and present results that show seasonal and interannual variability in the model. The variability results are unique because corresponding observational results are not yet available nor have they been computed from an ocean model of this scale and resolution.

The next section will discuss the model output. The following sections will present results from the experiments and will focus on the annual average meridional heat transport and seasonal and interannual variability of the heat transport in the model. In the final section, the results will be discussed and conclusions will be presented.

8.2   Model data

The global ocean model that will be used in this analysis has been discussed in detail in other publications (Semtner & Chervin 1988; Chervin & Semtner 1990; Semtner & Chervin 1992). Only a brief review will be presented here. The Parallel Ocean Climate Model (POCM) is a primitive equation model with 1/2 degree resolution globally with latitude ranges between 75°S and 65°N. The model time step for this resolution is 15 minutes. The model has 20 vertical levels, 4 in the upper 100 m, 9 in the upper 500 m, 13 in the upper 1500 m and level 20 is at 5000 m. In general, the model circulation is realistic and compares well with observations (Semtner & Chervin 1988; Semtner & Chervin 1992).

Four separate 10 year model integrations were conducted with different imposed wind stress forcing. Each experiment started from the end of a 22.5 year run with Hellerman and Rosenstein (1983) annual wind stress forcing and were integrated for an additional 10 years with either Hellerman and Rosenstein wind stresses or those derived from analyses from

ECMWF for the time span 1980–1989. The four model experiments and the wind stresses that were used are:

1. HR: Hellerman and Rosenstein monthly mean climatology
2. EC: ECMWF monthly mean climatology from years 1980–89
3. IV: ECMWF monthly mean wind stresses for years 1980–89
4. ID: ECMWF daily mean wind stresses for years 1980–89

Twice daily 1000 hPa initialised ECMWF wind speeds on a global 2 $\frac{1}{2}$ degree grid were converted to wind stresses using the Large and Pond (1982) formulation by Trenberth et al. (1989). Ten year monthly averages, monthly averages and daily averages were then computed and interpolated using a bispline to daily fields. The daily fields were then spatially interpolated to the 0.5 degree grid. There were changes in the analysis methods used at ECMWF over the 1980–1989 period. In particular, there were large changes in the analysis in September 1986, April 1987 and May 1989 (Trenberth 1992). We made no attempt to correct the raw ECMWF dataset to make the data more temporally consistent. Compared to the Hellerman and Rosenstein dataset, the ECMWF dataset has larger wind stresses in the high latitudes and less wind stress divergence in the tropics (Trenberth et al. 1990). The model forcing is updated in the model at the start of every three day period and a 'snapshot' of the model state is saved at the end of every three day period and includes the three velocity fields, temperature, salinity and streamfunction at each grid point. Monthly, annual and long term averages are then calculated from the instantaneous data. Covariances and standard deviations have also been computed.

The heat and salt flux boundary condition at the surface is set by restoring the salt and temperature field to Levitus (1982) monthly mean climatological fields with a time constant of one month. There is also a deep restoring term below 710 m in the model where the temperature and salinity fields are relaxed to Levitus annual mean conditions with a time constant of three years. In the high latitudes (75°S to 65°S, and 55°N to 65°N), the model temperature and salinity are restored to the annual mean Levitus data with a time constant of one year at all depths below the surface. This high latitude restoring attempts to simulate the horizontal fluxes from the Arctic Ocean and other marginal seas that are not included in this grid. It also supplements the convective overturning process in the model.

Each 10 year model integration took approximately 2200 CPU-processor hours on the MECCA Cray Y-MP/2 at the NCAR. The code has been optimised to run efficiently with parallel processors in multi-tasking mode, and each run took approximately five wall clock months to complete. Model output is archived on the NCAR Mass Store System and is available to interested researchers.

Colour Plate 8.1 shows the model domain and an instantaneous 'snapshot' of the simulated temperature in the top level of the model for the experiment with daily wind forcing (ID). The richness of eddy activity is obvious in the vicinity of western boundary currents, the tropics and the entire path of the ACC.

Figure 8.1 shows the standard deviation of the streamfunction for each of the ten year experiments. The largest variability in the HR experiment is in the tropics, the Kuroshio

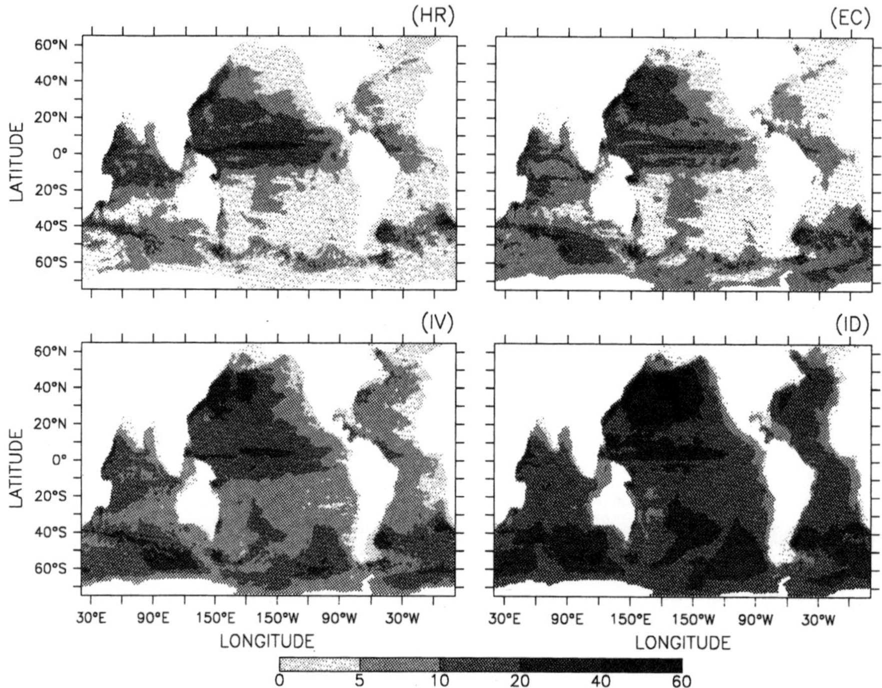

Figure 8.1 Plots of the streamfunction variability for the HR, EC, IV and ID experiments. Shading levels are 0, 5, 10, 20, 40 and 60 Sv.

region and the ACC. The EC experiment is qualitatively similar to the HR results but has greater variability in the ACC and western boundary current regions and less variability in the tropics compared to the HR results. This is directly related to differences in the wind stress forcing between the HR and EC experiments. As the forcing frequency increases in the IV and ID experiments, the amount of variability increases considerably. The ID experiment has very few regions with streamfunction standard deviations of less than 10 Sv.

In the following sections, various aspects of the model heat transport will be discussed. In the first section, the annual average heat transport will be presented.

8.3   Annual average heat transport

For each experiment, the annual average meridional heat transport at each grid point for each year was computed from the three day instantaneous output. Integrating this field zonally and with depth yields an annual average meridional heat transport in petawatts (PW). Figure 8.2 shows the five year average meridional heat transport for years six to ten for the HR and EC experiments. In that figure, the Atlantic, Indo/Pacific and global results

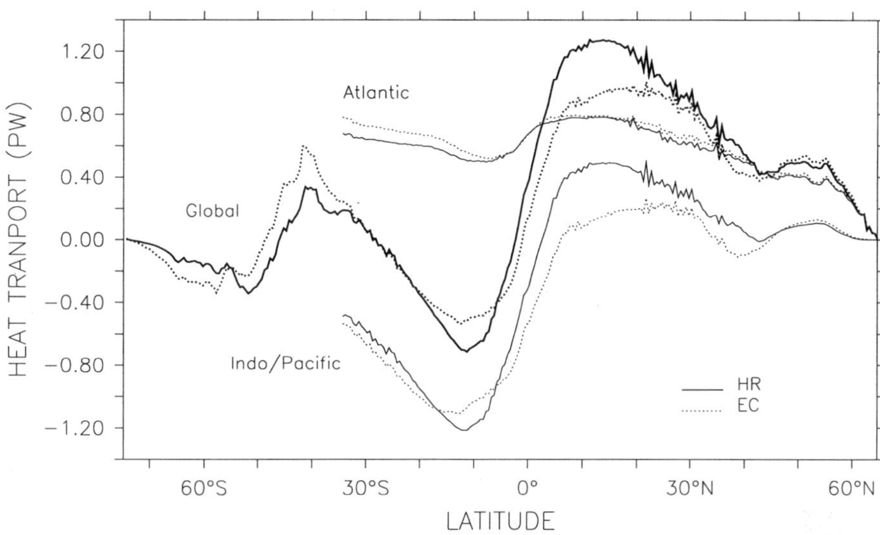

Figure 8.2 Meridional heat transport for the Atlantic, Indo/Pacific and global ocean regions for the HR and EC model experiments.

are shown. Because these two experiments are forced with climatological winds, there is little interannual variability over different years, although there is a significant seasonal signal which will be discussed in the next section. The standard deviation of the annual mean global heat transport about the five year mean is less than 0.04 PW except near 10°S and 30°N where the standard deviation approaches 0.08 PW.

Only results from years six to ten were used to compute the long-term mean heat transports in Figure 8.2 because during the first few years, there is a transient response in the model as the model equilibrates to the seasonal forcing. As described above, the initial state of the model is the end of an integration forced with Hellerman and Rosenstein annual average winds. This forcing has no seasonal variability and is substantially different from the forcing imposed in the four experiments described here. In these results, a relatively consistent seasonal cycle in the heat transport is not established until about year three. Before year three, the largest adjustments occur between 10°S and 20°N. In this region, the annual average heat transports differ from the long-term means by up to 0.6 PW in the HR experiment and 1 PW in the EC experiment. Poleward of 10°S and 20°N, the heat transport differences during the first three years are less than 0.1 PW for the HR experiment and 0.2 PW for the EC experiment compared to the long-term mean. We have also computed the multi-year average annual heat transport for the IV and ID experiments. They are nearly identical to the annual average heat transports computed for the EC experiment. This might be expected since the long-term average forcing in each of the EC, IV and ID experiments is nearly identical.

The annual average global heat transport shown in Figure 8.2 indicates that over most latitudes, the meridional heat transport is poleward. The exceptions are in the areas between the equator and 3°S where the 0 PW transport is shifted southward and between 25°S and 50°S where the heat transport is equatorward with magnitudes up to 0.6 PW for the EC experiment and 0.3 PW for the HR experiment. The greatest poleward heat transports occur in both experiments near 12°N and 12°S with magnitudes of 1.4 PW and 0.7 PW respectively in the HR experiment. The heat transports in the tropics are less in the EC experiment while the heat transports at higher latitudes are less in the HR experiment. At high southern latitudes, the southward heat transport is nearly double in the EC experiment compared to the HR experiment. Trenberth et al. (1990) compared the EC and HR wind stress fields and noted large differences in the Southern Ocean and in the North Atlantic and North Pacific. Winds in the EC climatology tend to be stronger in those regions, especially during the winter months. In the equatorial region, differences between the HR and EC wind stress climatologies are relatively small, especially in the zonal component. However, the divergent component of the wind field in the equatorial region seems to be reduced in the ECMWF climatology. The enhanced high latitude wind stresses of the EC experiment results in increased streamfunction variability in the high latitudes and decreased streamfunction variability in the tropics as shown in Figure 8.1.

In Figure 8.2, the global heat transport is separated into Atlantic and Indo/Pacific terms. In the Atlantic, the heat transport is northward at all latitudes with the EC experiment having slightly larger transports. The heat transport in the Atlantic does not vary much with latitude, ranging between 0.9 PW and 0.5 PW between the southern ocean and 55°N. The heat transport is indicative of a strong thermohaline circulation in the Atlantic where cold deep water is transported south and replaced by warm surface water. Most of the latitudinal variability of the global heat transport occurs in the Indo/Pacific basin. In that basin, the transports are poleward at all latitudes except between the equator and about 5°N where the 0 PW transport is offset to the north and near 40°N where the heat transport is close to 0 PW. The maximum poleward heat transport in the northern hemisphere Indo/Pacific is about 0.6 PW in the HR experiment and 0.3 PW in the EC experiment near 12°N. In the southern hemisphere, it is about 1.2 PW for the HR experiment and 1.0 PW for the EC experiment near 12°S.

The annual average heat transport can be compared to a number of ocean heat transport computations based on observations. Probably the most current results are those of Trenberth and Solomon (1994) for the year 1988. They find that the oceanic component transports heat poleward at all latitudes with maximum magnitudes of about 2 PW at 15°N and 15°S. In the Atlantic, they computed northward heat transports at all latitudes with a maximum transport of about 1 PW at 15°N. Trenberth and Solomon estimate an error of 0.3 PW at these latitudes. Overall, the differences between the model results and Trenberth and Solomon are relatively large. The model heat transports are too small by about a factor of 2 in many regions. In addition, the sign of the heat transport disagrees in the region between 25°S and 50°S. This may be related to model eddy resolution issues in the ACC. McCann et al. (1994) also noted this discrepancy in the HR experiment in their analysis and suggested

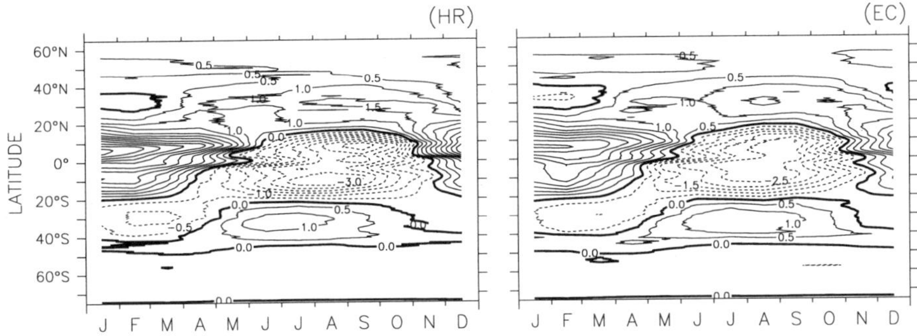

Figure 8.3 Monthly mean global meridional heat transports computed by averaging the monthly data from years 6 to 10 for the HR and EC model experiments. (Units are petawatts.)

that Southern Ocean water mass formation was a likely problem in the model which could cause the anomalous heat transport in the Southern Ocean. They also pursue other issues and we refer you to that paper for a more detailed discussion of this feature. When the 1988 annual average heat transport from the ID experiment is compared to Trenberth and Solomon (1994), these general differences still exist. Trenberth and Solomon (1994) pointed out many problems that are associated with the computation of both the atmospheric and oceanic heat transports from observations. The model probably has several deficiencies as well. These include the surface boundary condition parameterisation especially with regard to the heat and freshwater flux, the closed basin northern boundary condition, the topographical smoothing and resolution of the model and the sub grid scale parameterisation.

8.4   Seasonal variability

The previous section describes the annual average heat transport. In this section, the seasonally varying heat transport will be analysed. Figure 8.3 shows the global meridional heat transport for the HR and EC experiments as a function of latitude and month. These values were computed by averaging the monthly mean heat transports from years six to ten from the experiments. Averaging these plots in the time axis results in the annual average global heat transports presented in Figure 8.2.

There is a huge seasonal signal in the global heat transport shown in Figure 8.3. Between 20°S and 20°N, heat is transported from the summer hemisphere to the winter hemisphere with magnitudes exceeding 3.5 PW southward and 5 PW northward for the HR case. The EC case has slightly lower transports in this region. In the bands between about 20 and 40 degrees latitude, the heat transports are nearly 180 degrees out of phase with the tropical heat transports. Near 30°S where annual mean heat transports are northward (Figure 8.2), transports in excess of 1 PW are observed in both the HR and EC results in June and July (northward) and February (southward). The apparent error in the sign of the annual mean

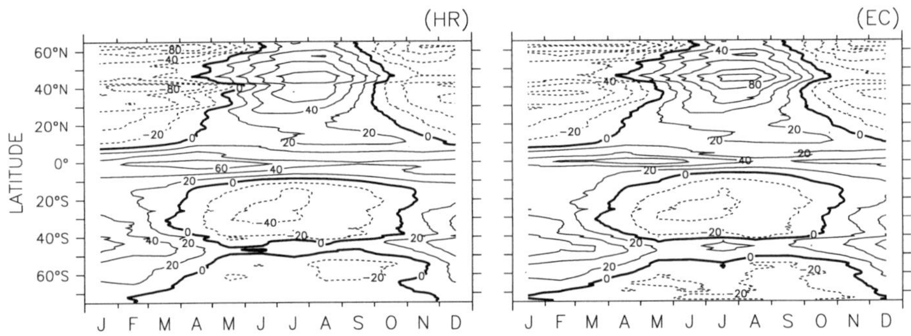

Figure 8.4 Zonal mean surface heat fluxes from the HR and EC model experiments. (Units are W m^{-2}.)

global heat transport at 30°S in Figure 8.2 may result from errors in the amplitude and phase of the seasonal cycle of the heat transport since the seasonal cycle has a large amplitude relative to the mean. To some degree, this same statement can be made about the annual average heat transport for most latitudes. In the northern hemisphere around 30°N, the seasonal cycle is also very strong, and the annual average transport is northward. In the HR experiment at 30°S, maximum northward transports exceed 1.5 PW in July, August and September and southward transports exceed 0.2 PW in January. North of 40°N and south of 50°S in both the EC and HR experiments, the heat transport is generally poleward and is less than 1.0 PW during the entire year with maximum transports during the winter season. At around 20°N heat is transported poleward during the entire year and at 45°S heat is transported equatorward during the entire year.

The heat transport in the model is forced by the seasonally varying winds and surface heat and salt fluxes. The heat fluxes are important in the model in two ways. First, they provide and remove heat in different regions so there can be a net steady state heat transport. Second, the heat and salt fluxes provide some of the push for the thermohaline circulation by forcing overturning. Since our heat and salt fluxes are somewhat artificial in that they are restoration to Levitus climatology, it is not clear how well either of these two fields in the model will compare *a priori* to the forcing of the real ocean by the atmosphere. However, we have tried to reduce these errors by optimising the restoration time scale and including high latitude and subsurface restoration to Levitus annual mean temperature and salinity. We have also computed the temperature restoration in W m^{-2} *a posteriori*, and we find reasonable agreement between the model heat flux and observations in magnitude, seasonal cycle and spatial variability. The zonal average downward heat fluxes associated with the HR and EC experiments are presented in Figure 8.4. Between 10°S and 8°N, there is positive heat flux into the ocean during the entire year of about 40 to 60 W m^{-2}. Outside this tropical band, the heat flux is positive in each hemisphere during the summer season and

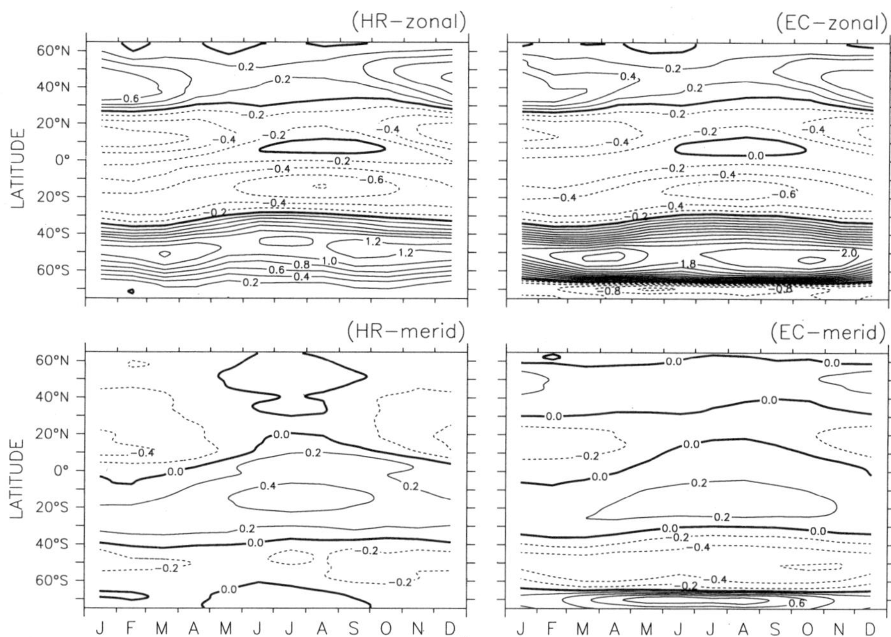

Figure 8.5 Zonal average monthly mean wind stresses from the HR and EC model experiments. (Units are dynes cm^{-2}.)

negative during the winter season, and it transitions fairly smoothly between these seasons. The maximum zonal average heat flux in the northern hemisphere is about ±100 W m^{-2} and in the southern hemisphere is about +60 and –40 W m^{-2}. The seasonal amplitude of the heat flux is slightly greater in the EC experiment compared to the HR experiment which implies that the typical departures of the model sea surface temperature from the Levitus climatology are greater in the EC experiment compared to the HR experiment. One area where there is an exception to the extra-tropical seasonal cycle of heat flux is around 45°S where the heat flux is positive throughout the year with two maximums in the EC experiment, July (20 W m^{-2}) and January (60 W m^{-2}). It is unlikely that this heat flux is realistic and it probably contributes to the erroneous annual mean northward heat transport in this region.

Differential heating across the equator between the winter and summer hemispheres could be an important driver of the meridional heat transport. However, we find that the seasonal signal of the heat transport is forced mostly by the seasonally varying zonal winds. Böning and Hermann (1994) and others have suggested that the seasonally varying zonal winds, through Ekman transport processes, dominate the dynamics that affect the seasonal signal of the meridional heat transport. We find that seasonally varying winds do correlate to a large degree with the seasonal cycle of the meridional heat transport. Figure 8.5 shows the monthly mean zonal average zonal winds from the EC and HR experiments. In a zonal average sense, the trades peak in July and August in the southern hemisphere between 5°S and 20°S and in the northern hemisphere in January and February between 10°N and 20°N. Relative to the annual mean heat transport, this should lead to enhanced northward Ekman heat transport in the tropics during the December to February season and enhanced southward Ekman heat transport during the June to October season, and this heat transport signal is noted in the model. Between 30°N and 60°N the westerlies are maximum in the northern hemisphere winter so increased southward Ekman derived heat transport would be expected and is observed in the model during that time compared to the rest of the year. In the model the seasonal signal of the meridional heat transport seems to correspond well with the variability of the zonal average zonal wind stresses. However, the magnitude and direction of the depth integrated heat transport does not correspond well with the heat transport expected from Ekman dynamics alone. Therefore we must look closer at the total heat transport to see what aspects other than Ekman dynamics may be important.

First, consider the total meridional heat transport more closely. It can be broken down into an eddy term and a product of the means,

$$\iint \overline{VT} dx dz = \iint \overline{V'T'} dx dz + \iint \overline{V}\,\overline{T} dx dz \tag{1}$$

where the overbar is the time mean and the prime is the departure from that time mean. We computed each of these terms separately and found that the seasonal cycle of the eddy term is fairly small compared to the seasonal cycle of the total heat transport. In fact the magnitude of the eddy transport is always less than 0.1 PW in the Indo/Pacific and 0.04 PW in the Atlantic except within 10 degrees of the equator where the eddy transports are equatorward with magnitudes up to 0.8 PW in the Indo/Pacific and 0.1 PW in the Atlantic. The eddy term here is defined as departures from the monthly mean. Maximum eddy heat transports in the Atlantic occur around August with another smaller peak in December. Maximum eddy heat transports in the Indo/Pacific occur in August in the HR case and in January in the EC case, although both cases have large eddy heat transports between August and January. We find that the eddy term in Equation (1) is balanced on monthly scales by mean flow, so most of the seasonal cycle of the total meridional heat transport results from the second term on the right hand side of Equation (1), the integral of the product of the mean meridional velocity and mean temperature. Because of conservation of mass and boundary constraints, the zonal and depth integral of the meridional velocity is zero in each basin, so the net meridional heat transport is really just the temperature difference of the water flowing northward versus the water flowing southward at any latitude across a basin.

Figure 8.6 is a plot of the global zonal integral of the volume and heat transport for different depth regions at 10°S as a function of month. Five year monthly mean data from the HR experiment was used in this case. Adjacent model levels were grouped together by considering the direction and seasonal signal of the transport. We retain the term 'heat transport' for individual depth ranges with non-zero volume transport. Some authors distinguish this as temperature transport. The sum in any given month of the heat

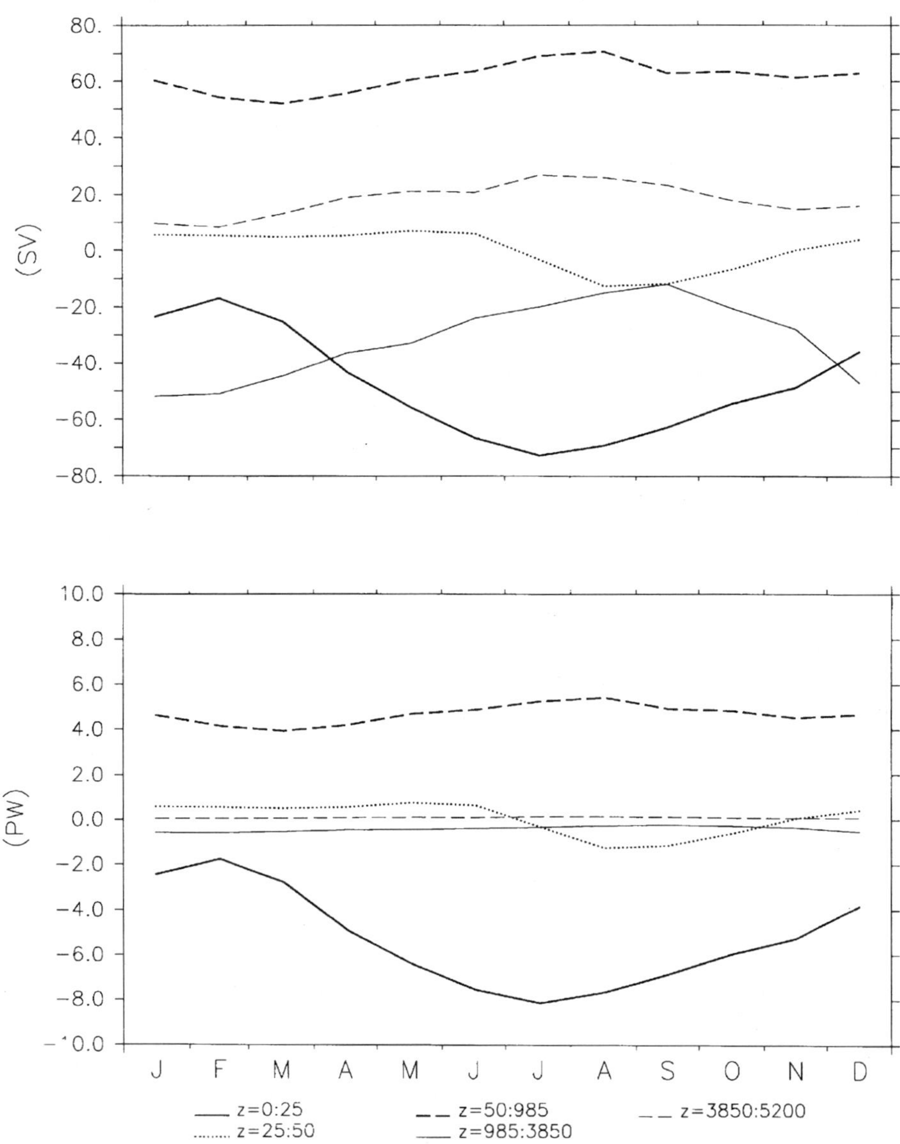

Figure 8.6 Monthly mean global volume and heat transport plotted as a function of various depth ranges from the HR experiment. Monthly data from model years 6 to 10 were averaged at 10°S in this case. (Units are Sverdrups and petawatts.)

and volume transports for all depth ranges yields the total transport for that month. The total heat transport is shown in Figure 8.3 and the total volume transport in each month is zero.

At 10°S in the HR experiment (Figure 8.6) the top layer (0–25 m) has the largest seasonal signal in both volume and heat transport. The transport is always southward with magnitudes ranging between 15 Sv and 75 Sv and 1.5 PW and 8 PW. The maximum southward transport is in July and the minimum is in February. The strength and sign of the heat transport in the upper layer seems to be directly proportional to the strength and sign of the zonal average zonal wind stress at 10°S (Figure 8.5). This is expected if Ekman dynamics dominate the upper layer transports. The second layer in the model (25–50 m) also has a fairly large seasonal cycle in both volume and heat transport. It contributes about 0.6 PW of northward heat transport at 10°S between December and June and about 1.5 PW of southward heat transport in August. In this layer, Ekman dynamics already appear to be decreasing in importance since the sign of the transports changes throughout the year which is in contrast to the sign of the zonal winds. In layers three to eleven (50–985 m) the zonal average volume and heat transport in each layer individually is northward throughout the year with magnitudes of about 5 Sv to 15 Sv and 0.5 PW to 0.8 PW. Depth integrating all the layers between 50 m and 985 m, as presented in Figure 8.6, results in northward transports of 50 Sv to 70 Sv and 4 PW to 5 PW. In these layers the seasonal cycle is relatively weak and it is out of phase by nearly six months with the upper layer volume and heat transports. Between layers 12 and 17 (985–3850 m) the zonal average heat transport is much weaker, is southward and has a total magnitude that never exceeds 1.0 PW in any month. These deepest levels contribute little to the total meridional heat transport. On the other hand, the volume transport in these layers is important. It is southward throughout the year and varies between 10 Sv and 60 Sv with a phase that is nearly six months out of phase with the volume transport in the top level. In levels 18 to 20 (3850–5200 m), the volume and heat transport reverses direction again and is northward throughout the year with magnitudes between 10 Sv and 30 Sv and heat transports of less than 0.2 PW.

Figure 8.7 is the same plot as Figure 8.6 for the EC experiments at 10°S. There are some slight differences in the volume and heat transport results, but for the most part, they can be considered identical. The results at other latitudes except within a few degrees of the equator are basically the same as those discussed for the HR and EC experiments at 10°S. Ekman dynamics are important in the upper level and dominate the seasonal signal of the heat transport. Intermediate layers have much weaker seasonal cycles of heat transport but may contribute significantly to the magnitude and direction of the depth integrated heat transport. The deepest layers contribute little to the total heat transport but they contribute substantially to the volume transport.

At the equator, the surface volume and heat transports seem to be related more to the meridional wind stress. Figure 8.8 shows the volume and heat transports as a function of depth for the HR experiment at the equator, where we again grouped similarly responding model levels together. In the top layer, the zonal average meridional heat transport is southward from January to March and northward the rest of the year. The magnitude and phase correspond with the zonal average meridional wind stress (Figure 8.5) at the equator.

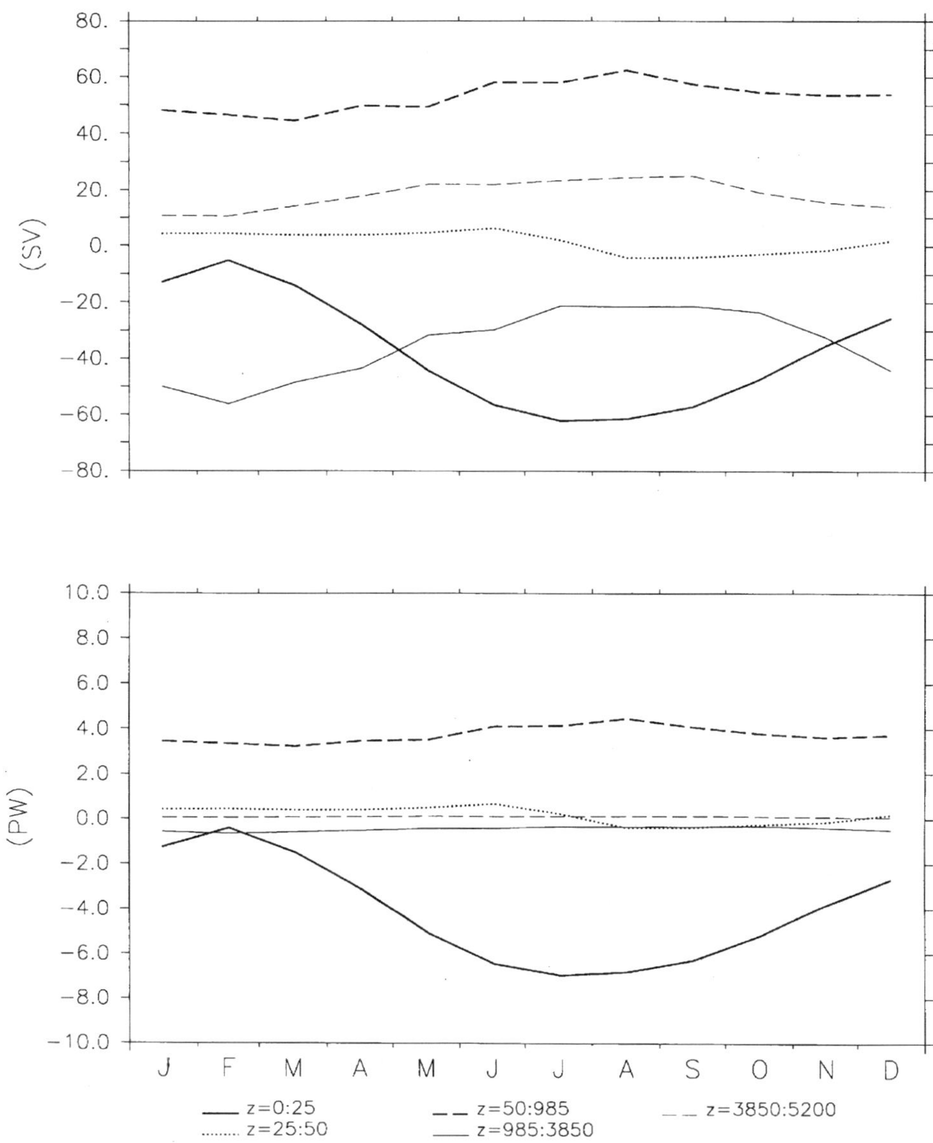

Figure 8.7 Monthly mean global volume and heat transport plotted as a function of various depth ranges from the EC experiment. Monthly data from model years 6 to 10 were averaged at 10°S in this case (Units are Sverdrups and petawatts.)

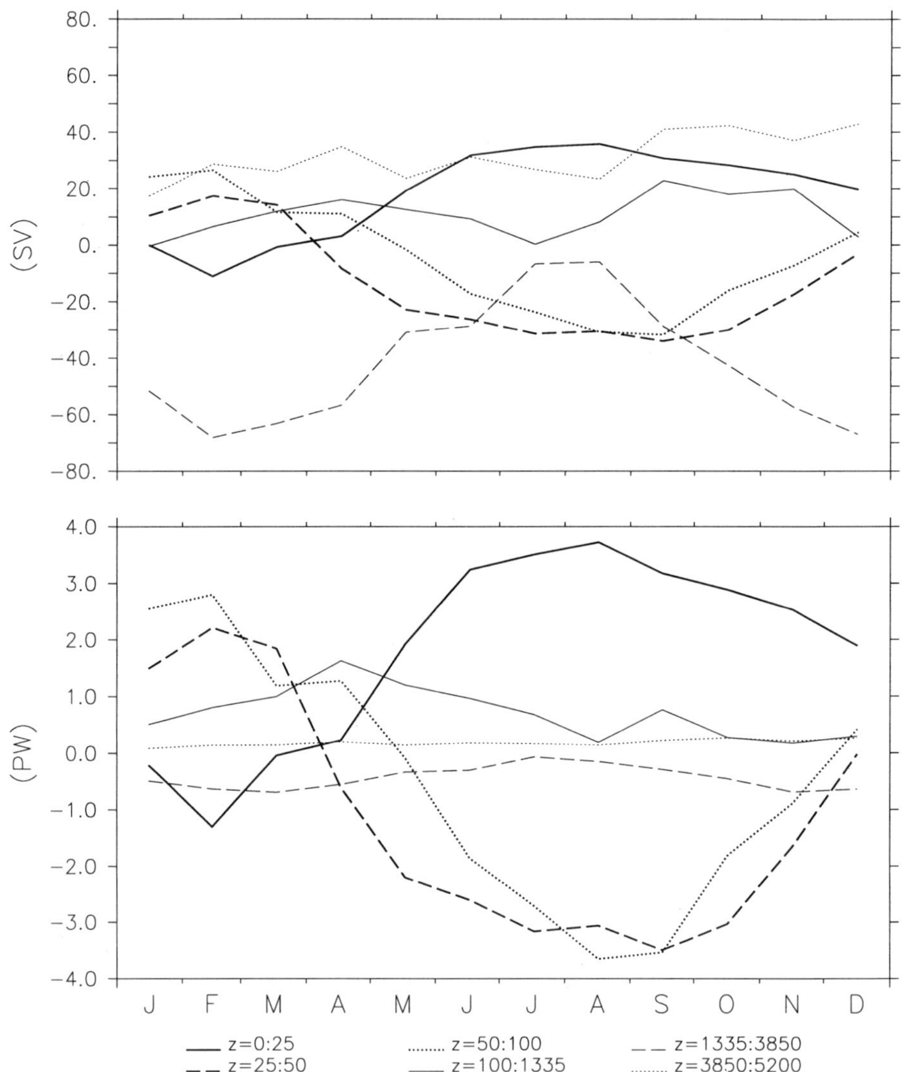

Figure 8.8 Monthly mean global volume and heat transport plotted as a function of various depth ranges from the HR experiment. Monthly data from model years 6 to 10 were averaged at the equator in this case. (Units are Sverdrups and petawatts.)

However, the seasonal signal of the surface heat transport is nearly six months out of phase with the depth integrated heat transport shown in Figure 8.3. At the equator, layers two to four (25–100 m) dominate the depth integrated heat transport seasonal cycle and these three layers overcome the relatively large and oppositely phased seasonal signal that exists in the top layer. The heat transport at the surface has about 1.5 PW of southward heat transport in February and nearly 4 PW of northward heat transport in August. Layer two (25–50 m) has about 2 PW of northward heat transport in February and 3 PW of southward transport in August.

Heat transport in layers three and four (50–100 m) is similar in phase and magnitude to layer two (25–50 m) with contributions of +3 PW and –4 PW. In layers five to twelve (100–1335 m) the depth integrated heat transport is northward during the year, layers 13 to 17 (1335–3850 m) have southward heat transport throughout the year and layers 18 to 20 (3850–5200 m) have weak northward heat transports over the entire year. It is interesting to note that in layers five to twelve (100–1335 m) the maximum volume transport occurs in September but the maximum heat transport occurs in April. Although we have been emphasising the relationship between the volume transport variability and the heat transport variability, temperature variability also plays a role in the heat transport variability in some cases. In terms of volume transport, the seasonal cycle observed in the top layer is cancelled by the volume transport in layers two to four (25–100 m) so that the top four layers have a net northward transport from December to March and southward transport the rest of the year. There is northward volume transport over most of the year in layers five to twelve (100–1335 m) with fairly large month to month variations ranging between 0 and 20 Sv. There is a large seasonal cycle of volume transport in layers 13 to 17 (1335–3850 m) with transports ranging between 0 and –70 Sv. Layers 18–20 (3850–5200 m) have 20 Sv to 40 Sv of net northward volume transport over the year. At the equator, surface layer dynamics does not dominate the seasonal signal of the heat transport. This was also noted in Brady and Gent (1994) in an equatorial Pacific model. In that case, water mass conversion both through vertical and horizontal circulation cells was more important for the seasonally varying heat transports, although the vertical overturning cells are certainly forced to some degree by Ekman upwelling at the equator.

The seasonal cycle of the heat transport in the Atlantic basin is shown in Figure 8.9 for the HR case where monthly means from the years six to ten were computed. The heat transport in the model is northward at nearly all latitudes and times of the year. The only exception is a small region in August just north of the equator. To some degree, this is slightly different to what Böning and Herrmann (1994, their Figure 8a) found in their $^1/_3$ degree Community Modeling Effort (CME) experiment, although the seasonal cycle in both cases is very similar. Böning and Herrmann found a maximum heat transport in January at 8°N of about 1.2 PW and southward heat transport between 15°N and 15°S during August and September. The maximum southward transport was about 0.7 PW at 8°N in August, so their peak to peak amplitude was about 1.9 PW, somewhat stronger than the results we find. The heat transport at 15°S in the CME results approach 0 PW because of the artificial boundary there. In the HR experiment, we find northward heat transports of between 0.3 PW and 0.7 PW at 15°S. If we subtract the heat transports at 15°S from Figure 8.9, we get a very

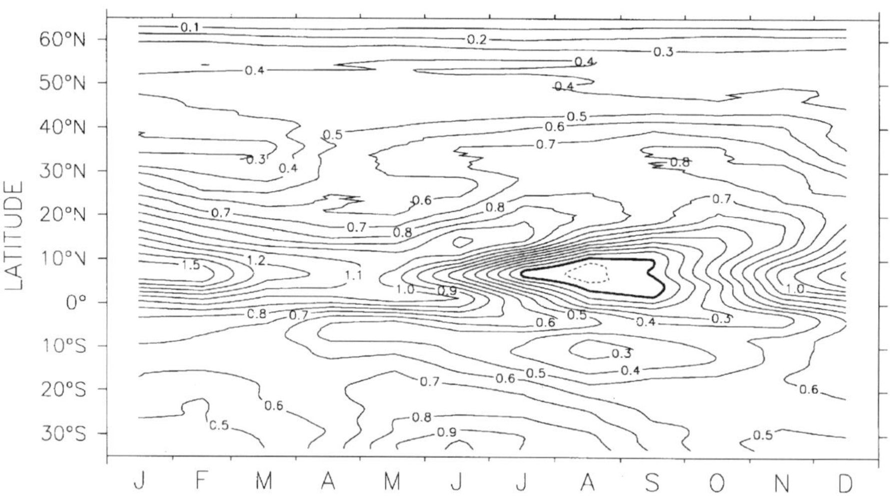

Figure 8.9 Monthly mean Atlantic meridional heat transports for model years 6 to 10 from the HR experiment. (Units are petawatts.)

similar picture to that of Böning and Herrmann. Relative to 15°S, Figure 8.9 has about 0.8 PW of northward heat transport in January at 8°N and a large latitude/time region of southward heat transport during July–September around the equator. We also find that, like Böning and Herrmann, we have southward transport in April and May south of the equator relative to 15°S. In essence, the two models agree on the seasonal cycle in the Atlantic but not on the absolute magnitudes.

In general, the seasonal cycle of the heat transport in the Atlantic is much weaker than the global ocean. The Indo/Pacific results dominate the global seasonal signal. One area where the Atlantic has a relatively large seasonal signal is between 5°N and 10°N. The peak to peak seasonal amplitude in the model is about 1.7 PW with maximum northward transport in January and minimum transport in July. In this region, the phase of the seasonal cycle matches the overall phase of the global heat transport in Figure 8.3. Over much of the rest of the Atlantic, the northward heat transport is around 0.5 PW with a peak to peak seasonal cycle of less than 0.5 PW. The Atlantic basin does not contribute much to the large seasonal cycle of the global ocean heat transport.

8.5   Interannual variability

Figure 8.10 shows the global average monthly average meridional heat transport from the IV experiment. The same results are shown for the ID experiment in Figure 8.11. The ID experiment has the same pattern although it is somewhat noisier and heat transports are generally increased by about 2% to 5% at all latitudes and months. The higher frequency wind forcing in the ID experiment does not change the monthly mean heat transports very much compared with the IV experiment. As a reminder, the IV experiment was forced with monthly mean ECMWF wind stresses interpolated to daily values and the ID experiment was forced with daily ECMWF wind stresses updated every third day. The typical seasonal cycle in Figure 8.11 looks very similar to the results from the EC experiment shown in Figure 8.3, although there is a fairly large amount of interannual variability in the results shown in Figure 8.11. The largest seasonal signal occurs between 20°S and 20°N with peak to peak amplitudes of over 10 PW in some years. Figure 8.11 (bottom) also contains a plot of the departure of the meridional heat transport for the ID experiment from the 10 year monthly average. This shows deviations from the monthly mean climatological heat transports for the ID experiment. As is clear in the top part of Figure 8.11 there is significant increase in the seasonal cycle of the heat transport between 10°S and 20°N during 1982–1983. This is a warm El Niño period in the model when the ECMWF Pacific trade winds relax, the eastern Pacific temperatures increase, the thermocline deepens and the currents in the equatorial Pacific weaken. Since the Pacific Ocean dominates the magnitude and annual cycle of the global meridional heat transport, it is not surprising that changes in Pacific circulation have such a large effect on the global meridional heat transport. El Niño leads to an enhancement of the seasonal cycle of the heat transport. Brady (1994) found that heat transport changes in an equatorial Pacific model were caused by changes in the temperature and current structure at the equator as a result of zonal wind stress changes. Much of the interannual variability in the equatorial region these results is caused by changes in the currents and temperature structure as well. There is a general increase in the northward heat transport in the tropics between mid-1986 and late-1987, a time that the model generates another El Niño in the tropical Pacific.The signal is similar to the 1982–1983 signal although it is

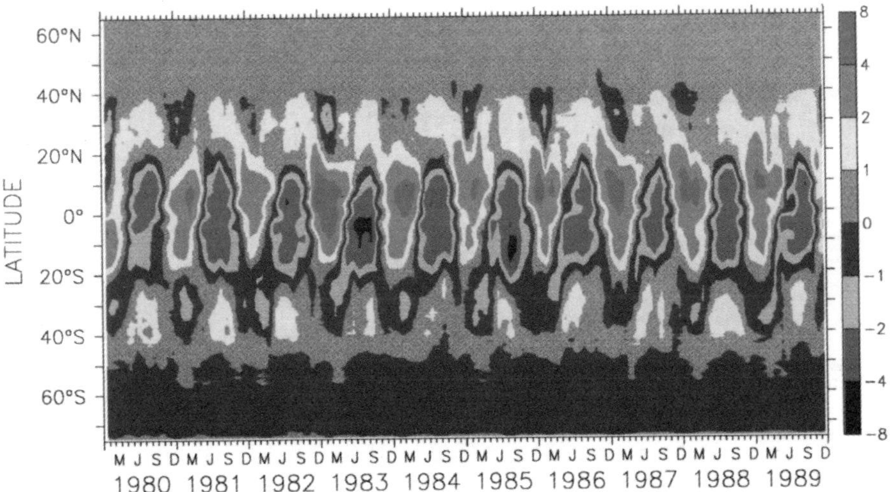

Figure 8.10 Monthly average global meridional heat transport from the IV experiment. (Units are petawatts.)

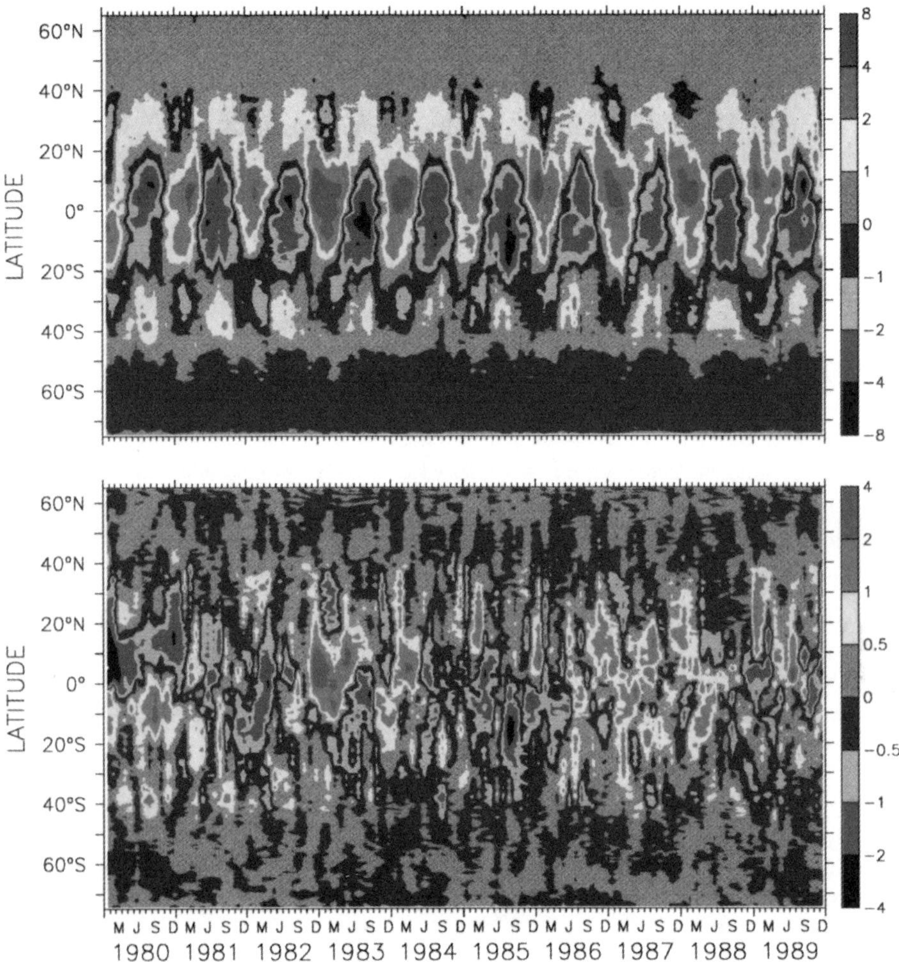

Figure 8.11 Monthly average global meridional heat transport from the ID experiment (above). Deviations of the monthly mean ID heat transports relative to the 10 year monthly mean climatology for the ID experiment (below). (Units are petawatts.)

weaker. There is a significant amount of anomalous southward heat transport between the equator and 20°S during much of late 1984, 1985 and early 1986. There is also some anomalous southward heat transport between late 1981 and early 1982 and in late 1988 through to 1989. The anomalous southward heat transport appears to be a persistent precursor to El Niño events. The anomalous heat transport in the first two years in Figure 8.11 can be attributed at least in part to the transient state of the model during this time. Outside of the tropics, the meridional heat transport is much weaker as are deviations from the average heat transport climatology. Although interannual variability accounts for a significant amount of

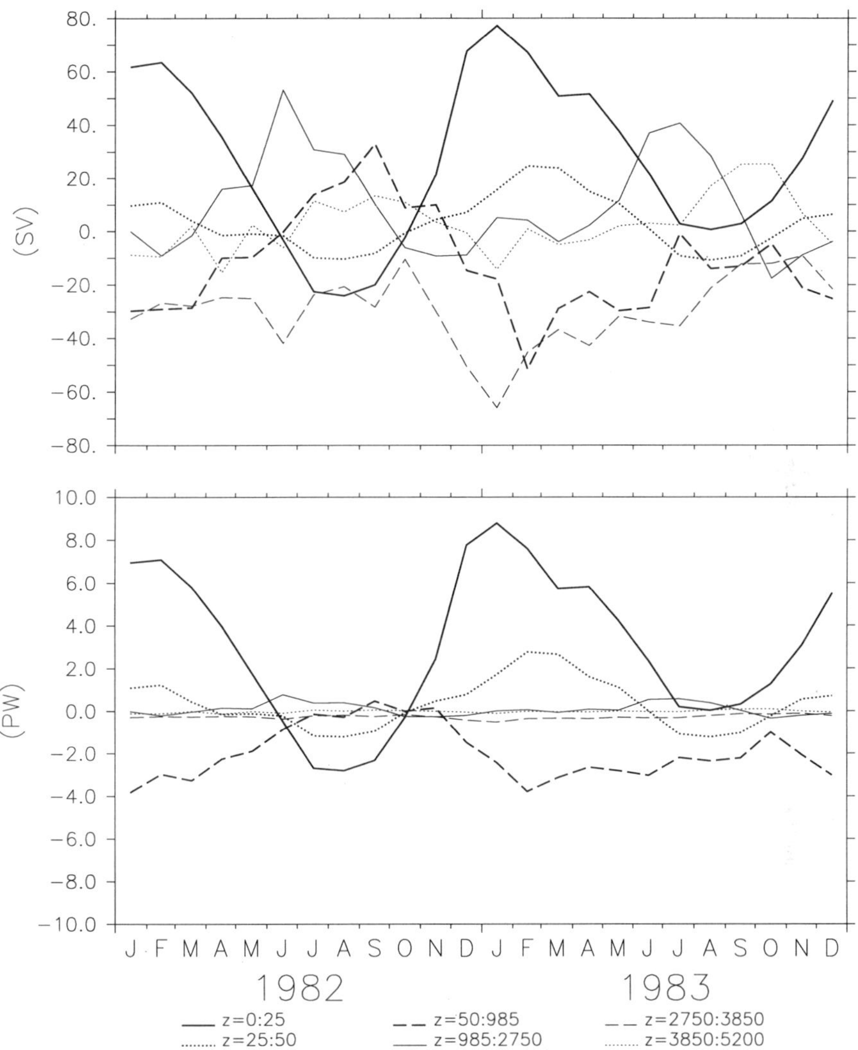

Figure 8.12 Monthly mean global volume and heat transport plotted as a function of various depth ranges for the ID experiment for years 1982 and 1983 at 5°S. (Units are Sverdrups and petawatts.)

variance in the heat transports in the ID experiment, the pattern, timing and strength of the seasonal signal of the heat transport remains fairly similar between years.

Figure 8.12 is a layer plot of the monthly mean volume and heat transports for the global ocean during the 1982–1983 period for the ID experiment at 5°N. As shown in Figure 8.11, the seasonal signal of the heat transport is enhanced during this period. During the late 1982 to early 1983 period, the heat transport in the top 985 m is enhanced such that the top 50 m has increased northward heat transport while levels three to eleven (50–985 m) have increased southward heat transport. Because of the timing, the net result is an increase in the seasonal cycle of the heat transport during this period. Figure 8.12 also shows that the excess volume transport associated with the enhanced northward heat transport in January 1983 in the top 50 m is balanced to a large degree by anomalous southward volume transport in layers three to eleven (50–985 m) and in layers 16 and 17 (2750–3850 m) during that period.

Figure 8.13 shows the same plot as Figure 8.11 but for the Atlantic basin only. As was discussed earlier, the heat transports in the Atlantic are smaller than the Indo/Pacific or global region transports. There is northward heat transport during most of the months and latitudes in the ID experiment. The largest seasonal cycle occurs between the equator and 10°N, although there is a consistent seasonal pattern at most latitudes. Figure 8.13 also shows the anomalous heat transport for the ID experiment in the Atlantic relative to the ID Atlantic monthly mean climatology. Although it is more difficult to attribute changes in the heat transport over different years to particular mechanisms, we do observe periods that contain anomalous meridional heat transport. There is an increase in the heat transport in the tropics in early 1982, late 1982, late 1983, early 1985, in the winter of 1985–1986 and during most of 1989. There is substantial anomalous southward tropical heat transport during early 1983 and most of the period from late 1986 to late 1988. The Atlantic heat transport anomalies tend to be fairly consistent over a large latitude band when they occur especially when compared to the global heat transport variability which is dominated by the Indo/Pacific.

Figure 8.14 is a plot of the monthly mean volume and heat transport at 30°N in the Atlantic for the ID experiment for various depth regions. It is similar in spirit to Figure 8.6. At 30°N in the Atlantic, there are some large interannual variations of the heat transport in the model (see Figure 8.13). In particular, there is substantial anomalous northward heat transport at 30°N in early 1982, late 1982, early 1984 and early 1989. There is substantial anomalous southward transport in late 1983, early 1985 and during the first half of 1987.

In Figure 8.14, the depth levels are systematically offset in the y axis by 10 Sv or 1 PW per group for clarity. Figure 8.14 indicates that most of the seasonal cycle of the volume transport in the top two layers (0–50 m) is offset by levels 16–20 (2750–5200 m). In addition, there is a large amount of northward transport (4–9 Sv) in levels 10–13 (510–1335 m) and southward transport (6–12 Sv) in levels 16–17 (1335–2750 m). Most of the seasonal cycle of the heat transport is contained in the top level (0–25 m) and variability on seasonal and interannual timescales is observed. The anomalous depth integrated heat transports in the Atlantic as indicated in Figure 8.13 (bottom) results from anomalous heat transports at various depth levels in the model. In general, anomalous northward transport occurs when some depth levels have either anomalous northward heat transport or lack a part of the typical seasonal southward heat transport. The opposite is true of anomalous southward heat transport. The anomalous northward heat transports in late 1982, early 1984 and early 1989 seem to be associated with in northward heat transport

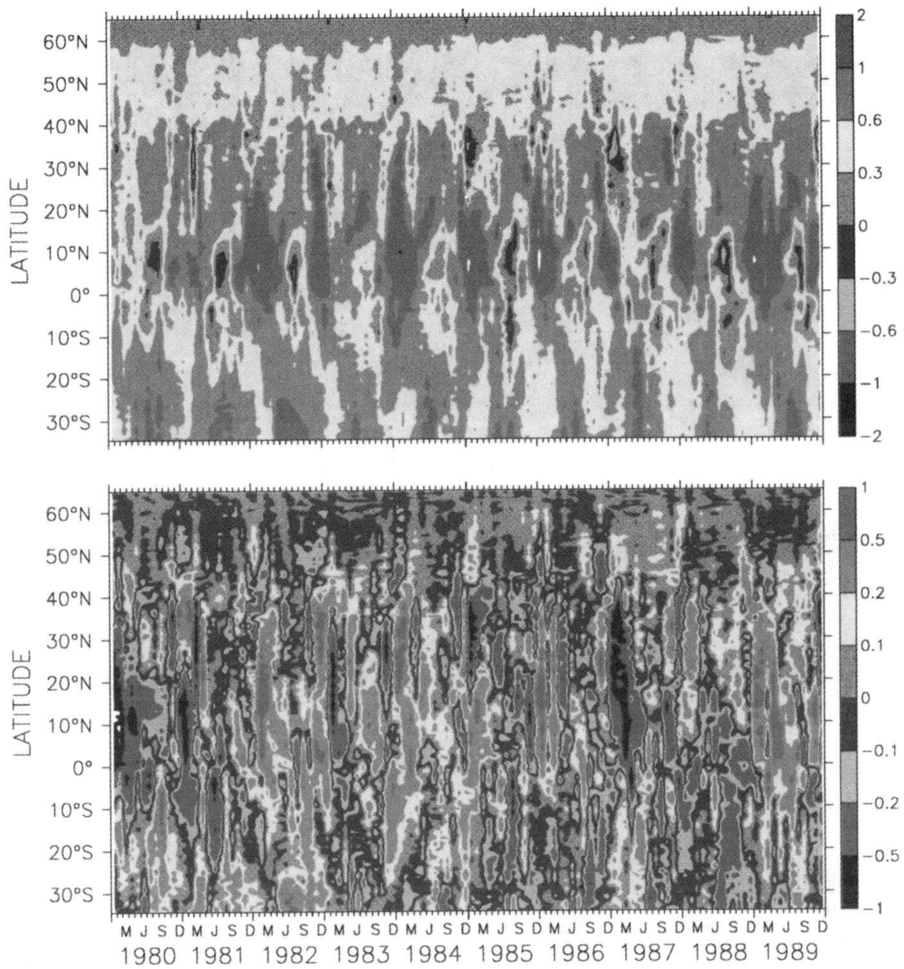

Figure 8.13 Monthly average Atlantic meridional heat transport from the ID experiment (above). Deviations of the monthly mean ID Atlantic heat transports relative to the 10-year monthly mean climatology for the ID experiment (below). (Units are petawatts.)

at the top level in the model. The anomalous northward heat transport in early 1982 seems to be associated with unseasonably large northward transports in levels three to twelve (75–1335 m). The anomalous southward heat transport in early 1985 and early 1987 seems to be linked to unseasonably strong southward transport in the top model level. In late 1983 however, the anomalous southward heat transport is present as a result of a weakening of the northward heat transport in the top level. In addition, not all strong surface events result in anomalous depth integrated heat transport. The strong southward surface heat transports in March 1981 and December 1987 are not correlated to strong anomalous depth integrated heat transports, nor are the strong northward events

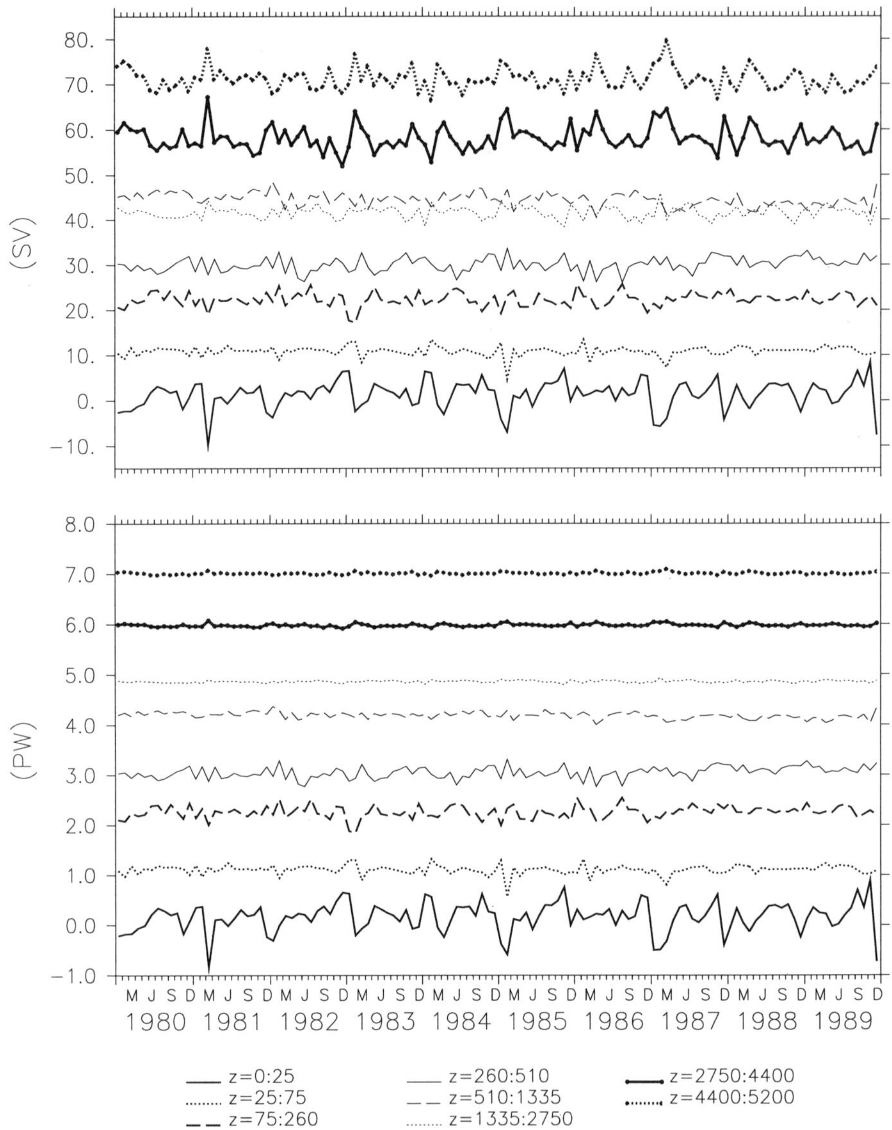

Figure 8.14 Monthly mean Atlantic volume and heat transport at 30°N plotted as a function of various depth ranges from the ID experiment. (Units are Sverdrups and petawatts.)

in November 1985 or November 1989. Figure 8.14 shows how complicated the depth integrated heat and volume balance is in the Atlantic on seasonal and interannual timescales. In general though, it is clear that the variability of the volume transport at the surface is balanced by flow below 2750 m in the model at 30°N.

8.6   Summary

We have presented results from four separate global ocean model experiments where the wind stress forcing was varied in the experiments. The HR and EC experiments were forced with monthly mean climatological winds from two different analyses for 10 years and the IV and ID experiments were forced with monthly mean and daily wind stresses from the same analysis as the EC winds. Conducting these experiments allowed us to compare heat transports in an annual, seasonal and interannual sense from several different perspectives.

In general, the annual average heat transports in the model appear to be consistently too small compared to observations and there is equatorward heat transport in a large portion of the southern ocean which is likely anomalous. There could be many reasons for these problems including errors in the temperature, salt or momentum forcing; grid resolution in the horizontal or vertical directions in the model; inaccuracies of some of the current physical parameterisations in the model; or errors in the observations. Danabasoglu et al. (1994) have shown the sensitivity of heat transports in a non-eddy resolving global ocean model to sub-grid scale parameterisation. In these POCM experiments, biharmonic sub-grid scale smoothing is used because the grid is within the eddy resolving regime. Even so, it would be interesting to incorporate and run some experiments using the isopycnal diffusion formulation outlined in Danabasoglu et al. (1994) to see whether it corrects some of the errors in the model heat transport. We hope to pursue this in the future. Another source of error in the annual average heat transport may come from the fact that the annual average is a sum of large and oppositely signed terms, so that a small error in the heat transport in any month or season could lead to a relatively large error in the annual average.

The annual average HR heat transports tend to be larger than the EC transports in the tropics and subtropics and smaller in the higher latitudes. This seems to result directly from the fact that the EC winds are stronger in higher latitudes especially in the winter but have weaker divergence in the tropics. The EC wind stresses may be overestimated by applying the Trenberth et al. (1989) algorithm to 1000 hPa winds in the high latitudes. The HR experiment has maximum heat transports in excess of 1.2 PW poleward around 12°N and 12°S. The Atlantic has annual average heat transports that are northward at all latitudes while the Indo/Pacific has poleward heat transports at most latitudes. In the southern ocean between 30°S and 50°S, the model exhibits equatorward heat transports of up to 0.6 PW and it is unclear whether this is physically reasonable, although most observational studies indicate this transport should be poleward. McCann et al. (1994) discuss this feature in detail.

The seasonal cycle of the heat transport is similar in the HR and EC experiments. There is a huge seasonal signal, particularly in the tropics. Heat is transported across the equator from the summer hemisphere to the winter hemisphere. North of 40°N and south of 50°S, there is poleward heat transport during the entire year. Off the equator, the variability of the depth integrated heat transport on seasonal timescales is driven by Ekman dynamics at the surface. The seasonal variability of the Ekman volume transport at the surface is balanced by flow in the opposite direction at depth levels below 1000 m. Model layers below the surface can have significant heat and volume transports. In fact, the intermediate layers, in sum, tend to dominate the mean annual average heat transport. In deeper layers, temperatures are low enough to contribute little to the net heat transport, although these deep layers often have significant volume transport. At the equator, the surface heat and volume transports seem to be correlated to the meridional wind stress and the seasonal signal of the upper layer does not dominate the seasonal signal of the depth averaged heat transport like it does off the equator. The depth integrated seasonal signal is dominated by the heat transport in the layers between 25 m and 100 m in the model at the equator but intermediate layers also contribute to the mean heat transport.

The model produces a fairly large amount of interannual variability in the IV and ID experiments although the largest part of the heat transport variability still seems to be in the seasonal component in these experiments. The monthly mean heat transport are very similar in the IV and ID experiments which suggests that the high frequency forcing is not needed for global scale assessments of some model fields. The amplitude of the global average seasonal heat transport is increased during El Niño years, 1982–1983 and 1986–1987, in the model in the tropical region. In the Atlantic the heat transport is generally smaller, has a weaker seasonal signal and is northward at nearly all latitudes during all months. The interannual variations are smaller in the Atlantic as well and tend to be more latitudinally homogeneous than in the global average. Monthly variations in the volume transport in the surface Atlantic is balanced by volume transport changes in the levels below 2750 m in the model. This is indicative of the thermohaline circulation in the Atlantic. It appears that interannual variations in Atlantic heat transport are forced by changes in basin scale processes such as the thermohaline circulation while interannual variations in the global heat transport which is dominated by the Indo/Pacific is forced by changes in the local circulation.

These global ocean model results suggest that there is probably a significant amount of seasonal and interannual variability in the heat transports in the ocean. Nearly all of the seasonal variability off the equator seems to be due to seasonal changes in the zonal winds through Ekman dynamics. On the equator, seasonal variability of the heat transport seems to be related to changes in the circulation. The interannual variability of heat transport in the model seems to be related to changes in the general circulation such as the thermohaline overturning, El Niño, and other large scale features. The model produces El Niño oscillations and much of the changes of the heat transport in the tropics is correlated to this feature. In the Atlantic, interannual variability seems to be dominated by basin scale processes. Although there are questions about the absolute accuracy of the heat transport in the numerical model, very few to no observational data are yet available to predict and analyse seasonal and interannual variability of the heat transport in the real ocean. As observational data become more available through satellites and other means, heat transports in the ocean will become better understood. At this point and probably for some time to come, it seems clear that a combination of modelling and observational methods is needed to understand the ocean heat transport better, particularly on seasonal and interannual timescales.

Acknowledgments

Research support was provided by the DOE CHAMMP Program on Climate Modelling, the DOE Carbon Dioxide Research Program at NCAR, and the NSF Physical Oceanography Grant No. OCE-9105162. Computing resources were provided by the Model Evaluation Consortium for Climate Assessment project and the Scientific Computing Division of the National Center for Atmospheric Research.

References

Böning, C.W. and Herrmann, P. (1993) On the annual cycle of poleward heat transport in the ocean: Results from high resolution modelling of the North and Equatorial Atlantic. *Journal of Physical Oceanography*, **24**, 91–107.

Brady, E.C. (1994) Interannual variability of meridional heat transport in a numerical model of the upper equatorial Pacific Ocean. *Journal of Physical Oceanography*, **24**, 2675–2694.

Brady, E.C. and Gent, P.R. (1994) The seasonal cycle of meridional heat transport in a numerical model of the Pacific equatorial upwelling zone. *Journal of Physical Oceanography*, **24**, 2658–2674.

Bryan, K. (1991) Poleward heat transport in the ocean: A review of a hierarchy of models of increasing resolution. *Tellus*, **43**AB, 104–115.

Bryan, K. (1982) Poleward heat transport by the ocean: Observations and models. *Annual Review of Earth Planet Science*, **10**, 15–38.

Bryan, K. (1962) Measurements of meridional heat transport by ocean currents. *Journal of Geophysical Research*, **67**, 3403–3414.

Bryden, H.L., Roemmich, D.H. and Church, J.A. (1991) Ocean heat transport across 24°N in the Pacific. *Deep-Sea Research*, **38**, 297–324.

Chervin, R.M. and Semtner, A.J. (1990) An ocean modelling system for supercomputer architectures of the 1990s. In *Proceedings of the NATO Advanced Research Workshop on Climate-Ocean Interaction*, edited by M. Schlesinger, pp. 87–95. Dordrecht: Kluwer Academic Publishers.

Danabasoglu, G., McWilliams, J.C. and Gent, P.R. (1994) The role of mesoscale tracer transports in the global ocean circulation. *Science*, **264**, 1123–1126.

Hall, M.M. and Bryden, H.L. (1982) Direct estimates and mechanisms of ocean heat transport. *Deep-Sea Research*, **29**, 339–359.

Hellerman, S. and Rosenstein, M. (1983) Normal monthly wind stress over the world ocean with error estimates. *Journal of Physical Oceanography*, **13**, 1093–1104.

Hsiung, J., Newell, R.E. and Houghtby, T. (1989) The annual cycle of ocean heat storage and oceanic meridional heat transport. *Quarterly Journal of the Royal Meteorological Society*, **115**, 1–28.

Large, W.G. and Pond, S. (1982) Open ocean momentum flux measurements in moderate to strong winds. *Journal of Physical Oceanography*, **11**, 324–336.

Levitus, S. (1982) *Climatological atlas of the world oceans*. Nat. Oceanic Atmos. Adm. Prof. Pap. 13. Washington, DC: USA Government Printing Office.

McCann, M., Semtner, A.J. and Chervin, R.M. (1994) Transports and budgets of volume, heat and salt from a global eddy-resolving ocean model. *Climate Dynamics*, **10**, 59–80.

Michaud, R. and Derome, J. (1991) The meridional heat flux in the atmosphere as derived from the ECMWF analyses for the years 1981–1986 and the oceanic heat transport inferred using the residual method. *Dynamics of Atmospheres and Oceans*, **16**, 1–3.

Semtner, A.J. and Chervin, R.M. (1992) Ocean general circulation from a global eddy-resolving model. *Journal of Geophysical Research*, **97**, 5493–5550.

Semtner, A.J. and Chervin, R.M. (1988) A simulation of the global ocean circulation with resolved eddies. *Journal of Geophysical Research*, **93**, 15 502–15 522 and 15 767–15 775.

Trenberth, K.E. (1992) Global analyses from ECMWF and atlas of 1000 to 10mb circulation statistics. NCAR Technical Note TN-373+STR, pp. 191 plus 24 fiche.

Trenberth, K.E. and Solomon, A. (1994) The global heat balance: Heat transports in the atmosphere and ocean. *Climate Dynamics*, **10**, 107–134.

Trenberth, K.E., Olson, J.G. and Large, W.G. (1989) A global ocean wind stress climatology based on ECMWF analysis. NCAR Technical Note TN-338+STR.
Trenberth, K.E., Large, W.G. and Olson, J.G. (1990) The mean annual cycle in global ocean wind stress. *Journal of Physical Oceanography*, **20**, 1742–1760.

CHAPTER 9

NORTH ATLANTIC MODEL SENSITIVITY TO MEDITERRANEAN WATERS

Matthew Hecht, William Holland, Vincenzo Artale and Nadia Pinardi

9.1 Introduction

In trying to understand the problems of human-induced climate change due to the release of greenhouse gases, it is necessary to make use of a variety of models of the climate system. In the past, only a few such models have included sophisticated ocean model components and even fewer investigations have examined the longer term evolution that may be dictated by important changes within the ocean. This is because the computational costs of running fully-coupled, global atmosphere-ocean models at reasonably fine resolution are extremely high. Nevertheless, recent work by oceanographers suggests that the ocean–atmosphere system may be in a very delicately balanced state and that major swings in the climate could be initiated by rather minor changes in today's basic state. Therefore, it is important to examine the possibility that some catastrophic change could be in store for humankind, a change larger than the few degrees of temperature change suggested by the greenhouse experiments carried out to date.

In order to make progress towards understanding such issues, we will examine one particular aspect of the ocean circulation and its stability, that of the importance of the transports of salt into the Atlantic Ocean from the Mediterranean Sea. It will be shown that the oceanic circulation is indeed sensitive to the nature of that input and that the climatic consequences of changes in salt transport can be very large on the time scale of 100 years or more. In fact, it is possible that the entire system may undergo catastrophic change and that such important climate quantities as northward heat transports in the Atlantic sector of the global ocean could be drastically modified by rather small changes in the interaction between the Mediterranean and the Atlantic.

The North Atlantic Ocean in its present state transports large amounts of heat poleward. It is becoming increasingly clear, however, that this large northward heat transport is not a dependable and time invariant component of our climate system but instead exhibits variability whose signature is found in the climate record of the northern hemisphere.

This meridional heat transport has been estimated to be 1.2 PW with an uncertainty of 0.3 PW at 25°N in the Atlantic (Hall & Bryden 1982), representing the poleward export of a sizable fraction of the total incoming solar energy which falls equatorward of 25°N. Evidence for strong and abrupt shifts in climate at high latitude in apparent response to smooth variations in orbital forcing have emerged from ice cores from glaciers on Greenland (Dansgaard 1982). The end of the last cold period has been studied in particular detail and there is evidence that the Younger Dryas period came to an end within a mere twenty years (Dansgaard 1989). Broecker et al. (1985) have proposed that the North Atlantic ocean has been responsible for this rapid switching of climate state, based on historical evidence that the response was greatest where the North Atlantic has greatest meteorological influence. More recently another collaboration of Broecker et al. (1990) have proposed that a salt oscillator is the mechanism behind these sudden shifts of climate, with freshwater anomalies at times capping off convective activity.

The very warm and saline Mediterranean waters which flow into the Atlantic at Gibraltar play a role in North Atlantic deep water formation and meridional overturning. The isopycnal layer, within which the Mediterranean outflow resides, intersects the ocean surface within the Labrador and Norwegian-Greenland seas, contributing substantially to the salinity of the waters found in these regions of deep water formation (Reid 1979). The heat content of these Mediterranean derived waters is rapidly lost by cooling at the sea surface, but the salt is retained. A vertical salinity gradient is maintained with relatively saline waters in the upper ocean and fresher water below, water that has been mixed with cold and fresh polar waters, reducing the stability of the water column.

Modelling studies also suggest that the ocean may support multiple equilibria. In this case, one would expect to find abrupt transitions from one equilibrium state to another, rather than the smoother behavior that would result from a system with only one equilibrium state. Bryan (1987) discussed these multiple equilibria within a three-dimensional primitive equation ocean model. Holland and Bryan (1993a) have shown results with two equilibrium states using the same mixed boundary conditions as we use in this paper. Two-dimensional meridional plane models have been shown to have similar behavior (Marotzke et al. 1988). In a study of the response of a meridional plane ocean model to atmospheric freshwater forcing, Zaucker et al. (1994) found that an atmospheric GCM producing an overly fresh climate at higher latitudes caused northern deep water formation to become extinguished in their ocean model. Their ocean was incapable of flushing the very large freshwater accumulation over the polar ocean.

Oceanic model response to variations in the source of Mediterranean waters has received little study. Stanev (1992) studied the influence of Mediterranean waters on Atlantic circulation at intermediate depth. His experimental design suppressed any variability in the composition of the upper 1000 m of the ocean. It is our intention to clarify the influence of Mediterranean waters on the thermohaline structure of the North Atlantic, on deep water formation and the capacity of the ocean to flush the large freshwater flux which accumulates over the polar ocean, and on the resulting conveyor belt circulation and heat transport. We perform these experiments using the Community Modeling Effort (CME) North Atlantic model (Bryan & Holland 1989; Holland & Bryan 1993a, 1993b). Our three experiments include one with a relatively ordinary prescription for the inclusion of climatological Mediterranean waters, one with no Mediterranean waters, and a third with the exchange of waters at Gibraltar diagnosed from our model of Mediterranean circulation. We emphasize that the models are completely independent except for this one-way forcing of the CME North Atlantic model by the Mediterranean model in the third experiment.

While we are modeling only a component of the climate system, and indeed only a portion of the World Ocean, our results indicate that Mediterranean waters indeed regulate the meridional overturning and heat transport of this important component of the climate system, the North Atlantic Ocean.

9.2   Models

The Community Modeling Effort (CME) North Atlantic model has been developed primarily at NCAR and at the Institut für Meereskunde in Kiel. The model is based on the primitive equation Geophysical Fluid Dynamics Laboratory (GFDL) ocean model (Bryan 1969; Cox 1984) in the configuration of MOM version 1.1. It exists in several resolutions and has been used for many experiments; see for example the papers by Böning et al. (1995), Bryan et al. (1995), and Holland and Bryan (1993a, 1993b). We use the coarsest version of the CME which has a resolution of 1 degree in latitude and 1.2 degrees in longitude with 30 levels in the vertical, reaching to a depth of 5 500 m. The domain, shown in Figure 9.1(a), is bounded at 15°S and 65°N where restoring boundary conditions are applied. These boundaries are closed to mass flux but are left open to tracer fluxes through a restoring condition. Newtonian restoring to the Levitus climatology is applied near these boundaries. The inverse restoring time constant, or restoring strength, falls linearly from $(5 \text{ days})^{-1}$ at the boundary to 0 over a distance of 6 degrees.

In the standard model configuration, Mediterranean waters are included through restoring over a region near Gibraltar of 3.6 degrees in longitude by 8 degrees in latitude. Restoring is only applied over 6 levels within this region, located at depths from 721 m to 1 875 m. This region over which restoring is applied may be seen in a plot of the Mediterranean model domain, Figure 9.1b, where the plus signs indicate the locations of columns within the CME North Atlantic domain where restoring is applied. The restoring time constant is five days.

Surface heat fluxes were included through the Han (1984) formulation, which is effectively a restoring towards an atmospheric temperature. Surface freshwater fluxes (which are actually applied through a salt flux) are parameterized through Newtonian restoring to the Levitus (1982) data set with a time constant of 5 days. The Hellerman and Rosenstein (1983) wind stress climatology is used.

Laplacian diffusivities are used. For tracers the diffusivities are 1×10^3 m²/s in the horizontal plane and 3×10^{-5} m²/s in the vertical direction. For momentum the diffusivities are 1×10^4 m²/s in the horizontal plane and 3×10^{-3} m²/s in the vertical direction.

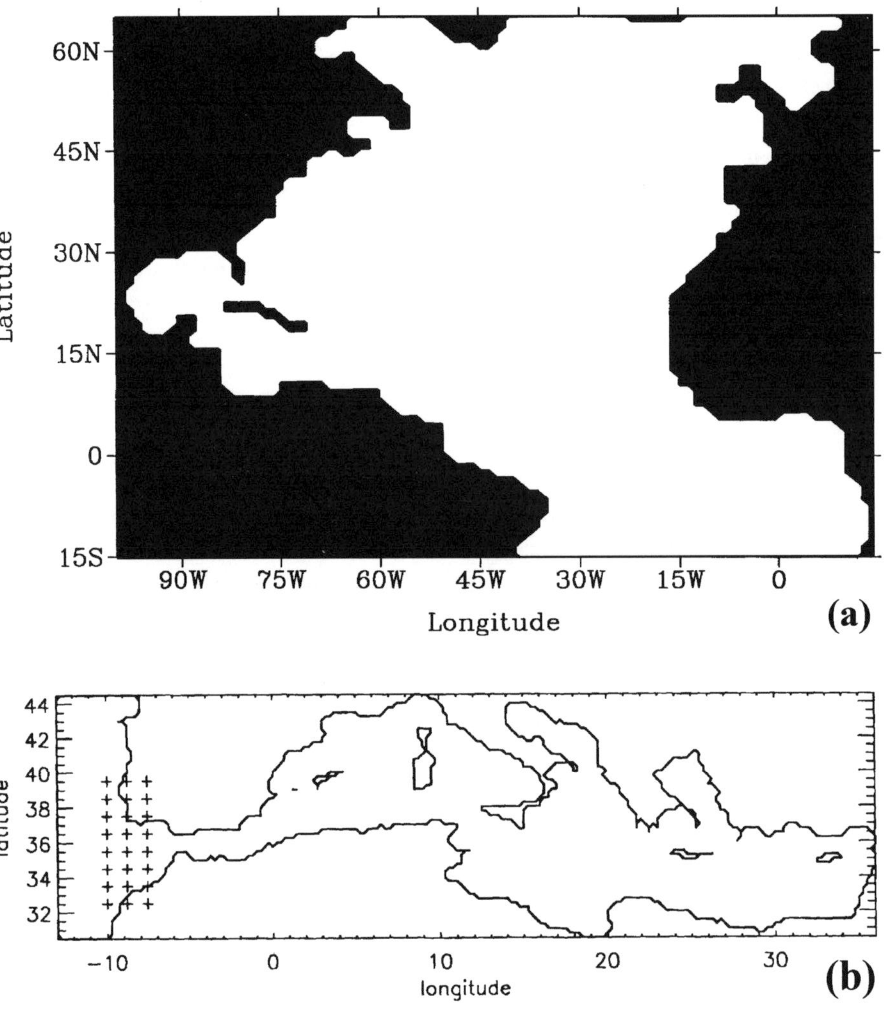

Figure 9.1 The domain of (a) the CME North Atlantic model, and (b) a Mediterranean model, from which heat and salt transports were diagnosed at 8.75 W for use in North Atlantic run JAM. At any of the 24 columns indicated by plus signs which lie within the North Atlantic ocean domain Mediterranean waters are included in run RM through restoring. The fluxes diagnosed from the Mediterranean model are applied at these same points also in run JAM.

Our primitive equation Mediterranean model, from which heat and salt transports through the Straits of Gibraltar are diagnosed, is also based on the GFDL MOM ocean model, version 1.1. It is configured at $1/4°$ horizontal resolution, contains 19 vertical levels and reaches to 3 000 m depth. In most respects the model configuration is identical to that which is documented in Roussenov et al. (1995). It differs through the use of a larger Atlantic buffer zone and simplified surface boundary conditions.

The model domains can be seen in Figure 9.1. Newtonian restoring through the entire depth of the water column is performed at the edges of the expanded Atlantic buffer zone. The region within which no restoring is applied, where the model is free to evolve, is bounded by 11°W, 32.25°N and 39.5°N. From this interior boundary toward the exterior boundary, the inverse restoring time constant, or restoring strength, varies from 0 to $(5 \text{ days})^{-1}$ with a Gaussian profile of width $3/4°$.

Surface heat and freshwater fluxes are parameterized through restoring boundary conditions. The model is restored towards the surface Levitus (1982) climatology with a 5-day restoring time constant. This corresponds to a 100 W/m^2 response to a temperature excursion of 1°C. The Levitus monthly mean values of surface temperature are used, and monthly mean values of surface salinity are interpolated from the Levitus seasonal values. Linear interpolation of the data between the two months which bound any particular time step is used to continuously supply the data to the model.

The Hellerman and Rosenstein (1983) wind stress climatology is used for the Mediterranean model runs, as it was for the CME runs.

Biharmonic diffusion is applied to tracers and momentum in the horizontal plane. The horizontal diffusivities are 8×10^{10} m^4/s for tracers and 2.4×10^{11} m^4/s for momentum. The Laplacian diffusivities which are applied in the vertical direction are 3×10^{-5} m^2/s for tracers and 1.5×10^{-4} m^2/s for momenta.

An 88 year run was performed which used the Levitus climatology as the initial state. Heat and salt transports through Gibraltar were diagnosed at monthly intervals in the final year and were used to force the CME North Atlantic model in one of three experiments, as will be explained in the next section. The results of our experiments with the Mediterranean model will be discussed in detail elsewhere.

9.3 North Atlantic experiments

The initial state of these experiments was taken from an existing 1° CME North Atlantic model run. In this existing run, Newtonian relaxation was used to accommodate heat and salt fluxes across northern and southern boundaries and at Gibraltar. These boundaries are closed to mass flux but open to tracer fluxes. Evaporation and precipitation was parameterized through Newtonian restoration of the sea surface salinity field, as discussed in the previous section. The Levitus (1982) climatology provided the observational data towards which the model was nudged. Surface heat fluxes were included through the Han (1984) formulation, as noted above.

In all three experimental cases, Han restoring was retained for the surface heat flux. This implies an assumption that the atmospheric state remains unchanged despite potentially large changes in ocean circulation. This is one of several instances in which we are making the assumption that some components of the climate system remain unchanged while we experiment with one component of that climate system. This approach is consistent with our goal to model the response of the North Atlantic component of the climate system to changes in its forcing at Gibraltar. In a model of the climate system which is intended to be as complete as is feasibly possible, many feedbacks will be included that are missing from our model simulation, but we would expect that the response found in our study would also exist within that climate model.

Whereas the surface temperature of the ocean tends towards that of the atmosphere through a natural and mutual feedback, there is no such feedback regulating the sea surface salinity. In order to allow the sea surface salinity field to evolve in response to changes in the salinity forcing at Gibraltar, we diagnose salinity fluxes from the last year of the run which provides the initial state and apply those surface fluxes to all experimental cases.

Fluxes of heat and salt are likewise diagnosed at the northern boundary of the initial state run and applied to all experimental runs. The net freshwater flux from the north across the boundary represents the large net excess of precipitation over evaporation over the polar region, most of which must be flushed into the Atlantic. The time-independence of this flux again reflects our assumption that the rest of the climate system remains in a fixed state while we experiment with the North Atlantic.

In all cases where fluxes are diagnosed, this was done at one month intervals. Linear interpolation of the fluxes between the two months which bound any particular time step is used to apply the fluxes within the model.

Restoring of both temperature and salinity at the southern boundary is retained in all three cases in order to allow for the inclusion of Antarctic bottom waters and other waters of southern origin.

The three experiments are differentiated by their treatment of Mediterranean waters. In the first case Newtonian restoration is used near Gibraltar, as was done in many other CME North Atlantic experiments. The restoring region and strength of restoring was discussed in the preceding section. This case is referred to as 'RM' for Restoring of Mediterranean waters.

In the second case no Mediterranean waters are included. This case is referred to as 'XM', for without Mediterranean waters. Not only does this provide an extreme example of the consequence of running a model with a different source of Mediterranean waters than in the RM run, but also corresponds to periods in which there has been no flow of water through the Strait of Gibraltar (during times of lowered sea levels).

In the third case the exchange of waters near Gibraltar is diagnosed from our model of the Mediterranean Sea. This case, which used diagnosed model fluxes at Gibraltar, is referred to as 'JAM', for Joint Atlantic Mediterranean experiment. We diagnosed monthly transport fields at $8.75°W$ during year 88 of the Mediterranean model run and applied those fluxes uniformly over all of the points identified with pluses in Figure 9.1b. We make use of fluxes at all 18 of the sub-surface layers of the Mediterranean model. The upper 20 layers of the CME North Atlantic model overlap with the 18 sub-surface layers of the Mediterranean model and so the diagnosed fluxes are applied to these upper 20 layers of the North Atlantic.

The annual mean of the salt sources in the RM run during years 1, 151, and 301 are shown at the six vertical levels at which restoring is applied in Figure 9.2.

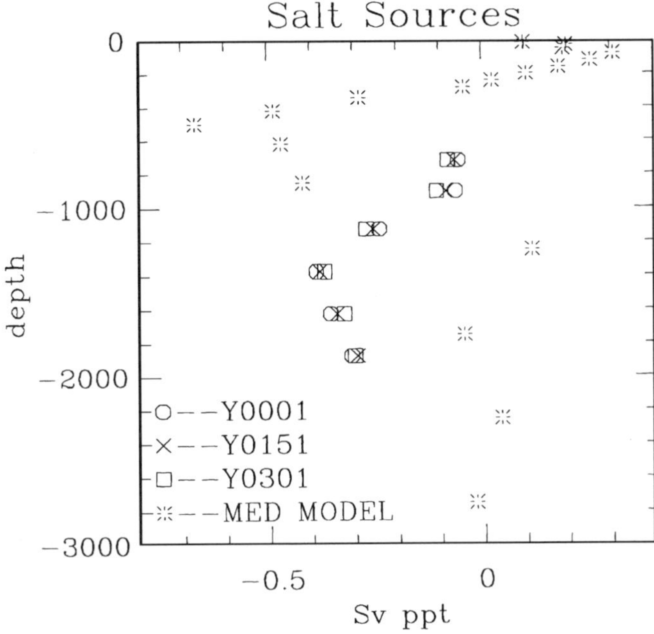

Figure 9.2 Annual mean salt sources as a function of depth for run RM during years 1, 151, and 301, and from a Mediterranean model. The salt sources from the Mediterranean model are used as the salt source for run JAM.

The annual mean salt sources at all 18 sub-surface levels of the Mediterranean model are also plotted. The total magnitude of the salt source in run RM is nearly constant, being 1.41 Sv ppt in year 1, 1.42 Sv ppt in year 151, and 1.45 Sv ppt in year 301 (one Sv equals 10^6 m^3/s). These magnitudes appear to be a bit small when compared with observational figures. If one computes an average salinity transport and standard deviation from the post-Second World War figures given by Bryden et al. (1989), one obtains a figure of 2.1 Sv ppt with a standard deviation of 0.6 Sv ppt. The magnitude of the total salinity transport from the Mediterranean model which is applied in run JAM is lower yet, at 0.92 Sv ppt. This net salt transport is very sensitively dependent on the surface boundary conditions applied over the Mediterranean domain, (as we will discuss in a forthcoming paper).

The salt source in run RM is centered around 1 500 m depth. This is where the CME North Atlantic model requires an injection of salty waters in order to produce a salinity structure resembling that of the Levitus data. In surveys the Mediterranean salt tongue is seen to be more shallow, centered around a depth of around 1 000 m (Lacombe & Tchernia 1960). On the other hand the salt source is very shallow in the JAM run; in this case the poor representation of the overflow over the Gibraltar sill is to blame. This poor representation of the overflow at a sill is a problem shared by all primitive equation models which are currently in use. As we find in this work, the unrealistically shallow source distribution has considerable implication for the resulting model circulation and thermohaline structure.

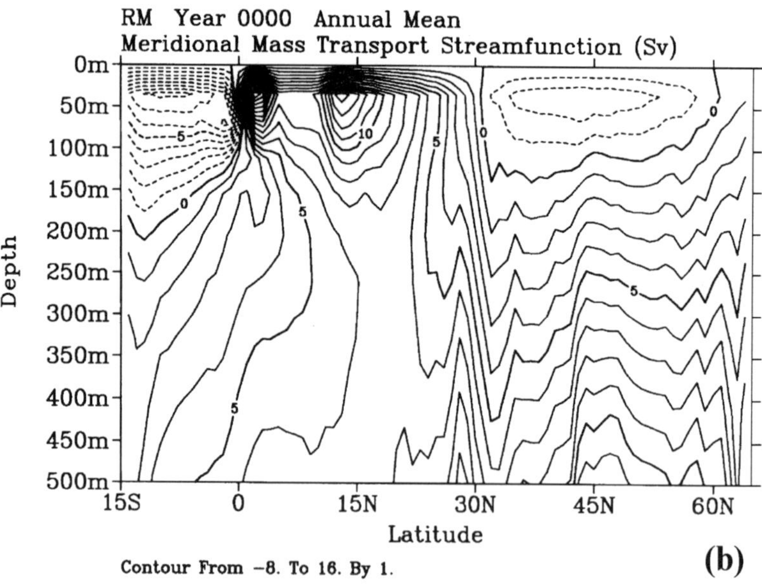

Figure 9.3 Annual average of the meridional mass transport stream function during the first year of run RM: (a) throughout entire 5 500 m depth, and (b) for the upper 500 m.

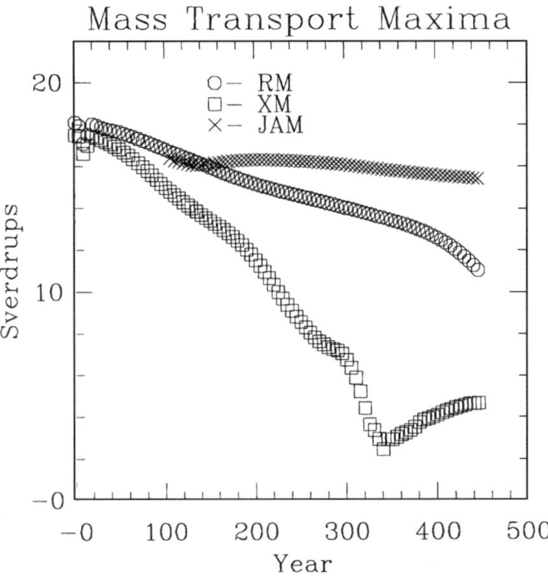

Figure 9.4 Instantaneous maxima in the deep northern overturning cell of the meridional mass transport stream function. Each point is evaluated at June 30.

9.4   Results and interpretation

The full detail of the circulation which powers the northward transport of heat and salt in the North Atlantic is contained in the time dependent velocity field. Interpretation of this circulation is greatly simplified through the consideration of a meridional mass transport stream function.

At the beginning of run RM (with restoring of Mediterranean waters at Gibraltar), a vigorous overturning cell involves all but the uppermost layers of the domain, as seen in Figure 9.3. This overturning cell, whose annual mean is between 16 and 17 Sv in our initial state, is one of the principle components of the so-called global conveyor belt circulation.

A time series of the instantaneous strength of the meridional mass transport stream function is shown in Figure 9.4. An adjustment from the boundary conditions of the initial state to those discussed in section 9.3 occurs in the first 20 years of the run. The amplitude of the transient is around 1 Sv.

A slow monotonic downward drift in the meridional mass transport stream function of run RM follows this adjustment. This downward trend intensifies at around year 400. We stopped our runs shortly after observing this behavior. We now believe that we know how to reduce or eliminate this trend through the use of surface boundary conditions which promise greater stability. Despite the existence of this trend in run RM, we are able to draw useful conclusions from a comparison of our three runs.

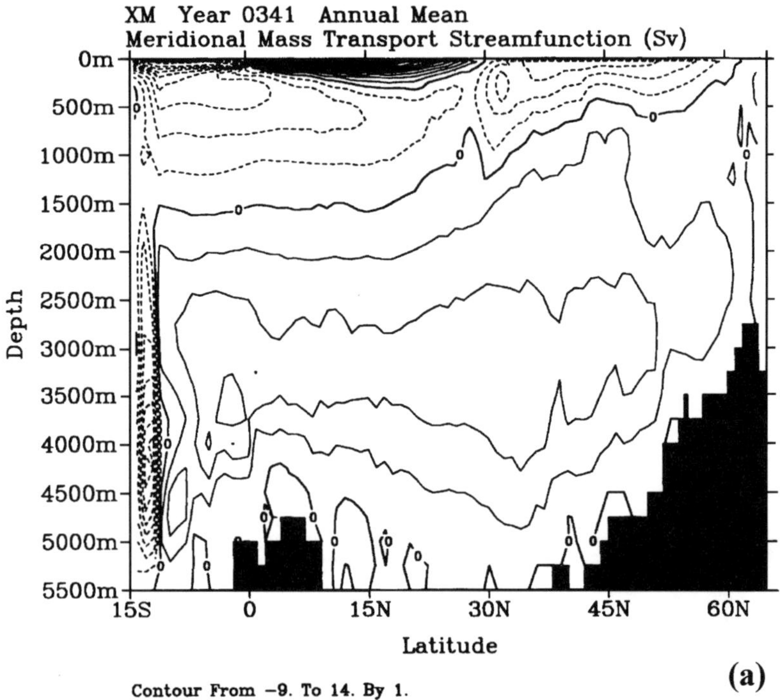

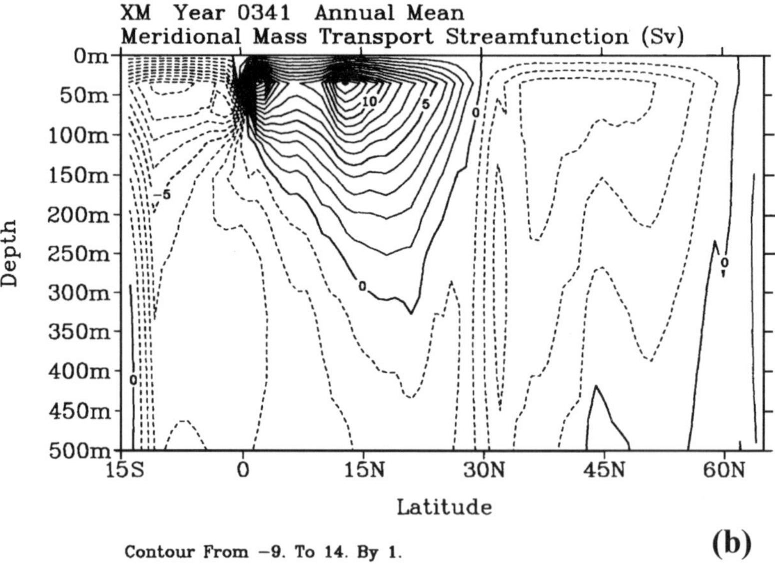

Figure 9.5 Annual average of the meridional mass transport stream function during year 341 of run XM, at which time the deep northern cell reaches a minimum: (a) throughout entire 5 500 m depth, and (b) for the upper 500 m.

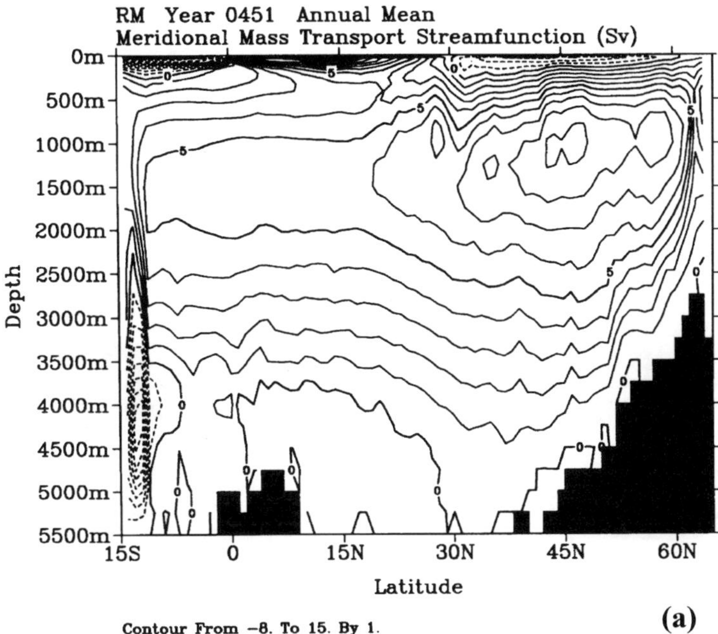

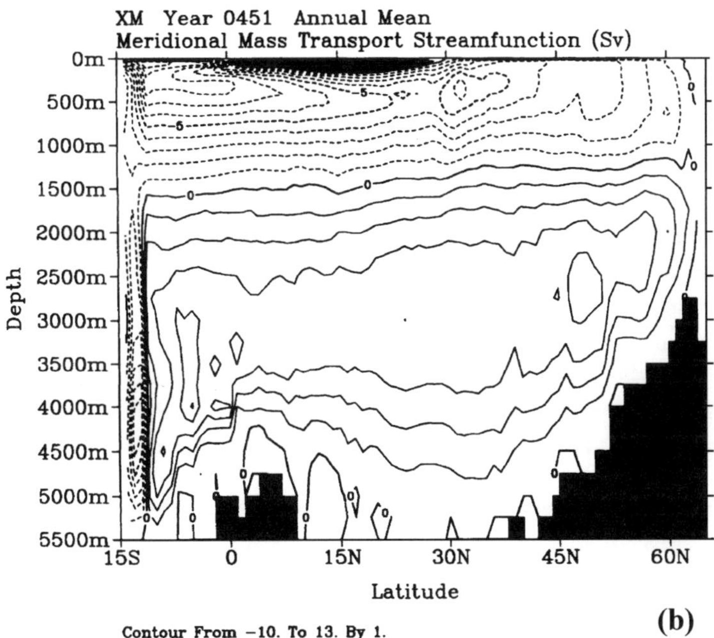

Figure 9.6 Annual average of the meridional mass transport stream function during year 451 for: (a) run RM, (b) run XM, and (c) run JAM.

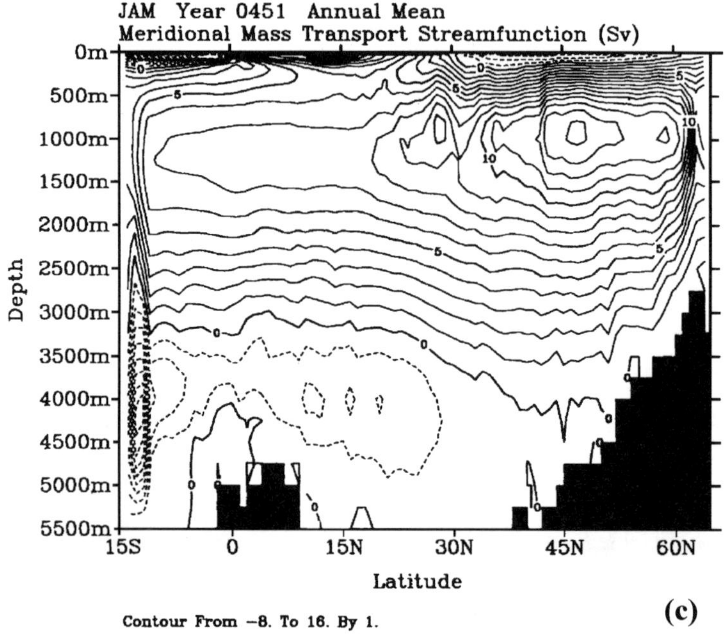

Figure 9.6 *Continued.*

The overturning in run XM (without any Mediterranean waters) collapses over a time of 340 years, as seen in Figure 9.4. In contrast, the overturning in run JAM (with Mediterranean waters diagnosed from a Mediterranean model) is seen to be quite stable. In this section we endeavor to understand the origin of these differences.

At the point of collapse the deep northern cell in run XM is nearly absent, with only a couple of sverdrups of transport evident in Figure 9.5. The shallow northern subtropical cell which was seen to be in communication with the deep northern overturning cell in Figure 9.3 is completely isolated in Figure 9.5. The equatorial upwelling which was fed in part by deep northern waters has apparently become part of a more local recirculation. The near-surface features of the meridional overturning resemble those of the initial state, reflecting the constraints posed by the surface boundary conditions. Below a few hundred meters there is little resemblance to the initial state. In addition to the collapse of the deep northern cell from some 18 Sv of overturning (at June 30) to just over 2 Sv, the several sverdrups of Antarctic bottom water which was seen in the initial state has disappeared.

At the end of the model run the meridional mass transport stream function for case RM looks qualitatively identical to the initial state, as seen in Figure 9.6(a), although the magnitude of the overturning has decreased. The overturning has rebounded somewhat from its minimum state in run XM, as seen in Figure 9.6(b), although it now involves only the deep ocean. The overturning for run JAM, as seen in Figure 9.6(c), remains vigorous.

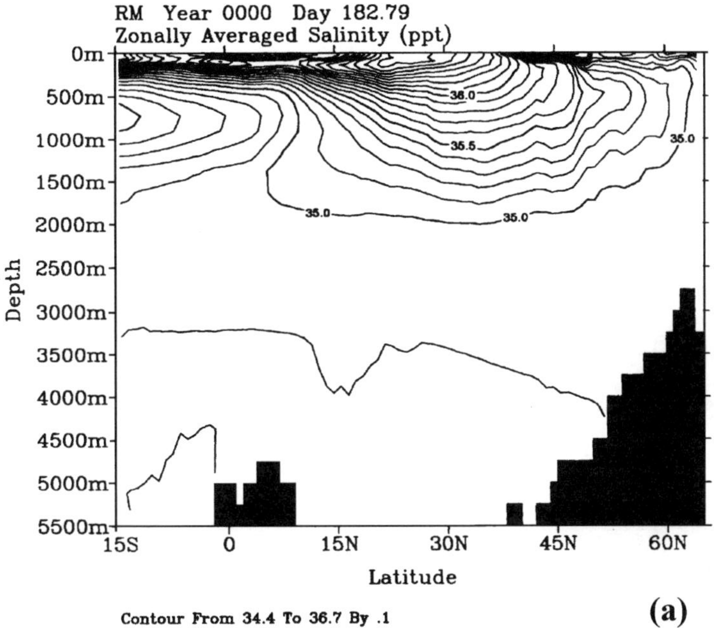

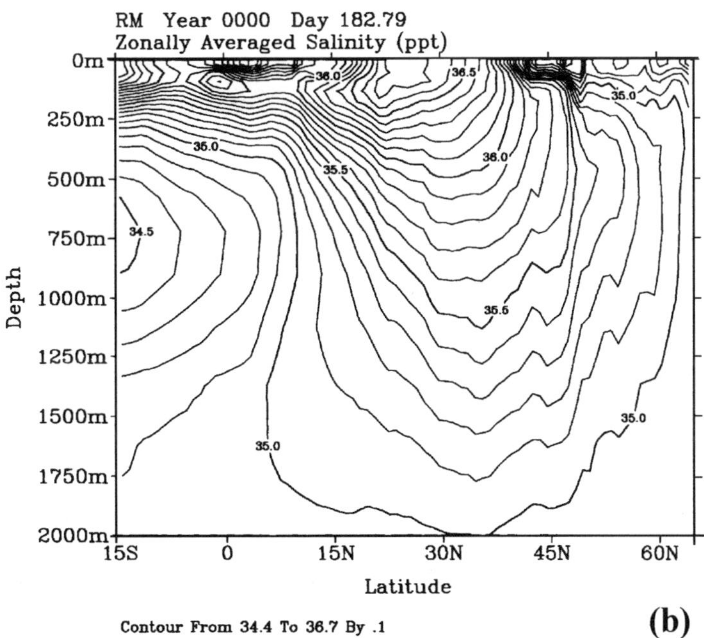

Figure 9.7 Zonally averaged salinity field at day 182 of year 0: (a) throughout entire 5 500 m depth, and (b) for the upper 2 000 m.

RM   Year 0451   Day 182.62
Zonally Averaged Salinity (ppt)
36.5
36.0
35.0
35.0
35.5
35.5
34.5
35.5
35.0
35.0
0m
250m
500m
750m
1000m
1250m
1500m
1750m
2000m
Depth
15S   0   15N   30N   45N   60N
Latitude
Contour From 33.9 To 36.6 By .1

(a)

XM   Year 0451   Day 182.62
Zonally Averaged Salinity (ppt)
34.5
32.5
35.0
34.0
34.5
34.5
34.5
34.5
0m
250m
500m
750m
1000m
1250m
1500m
1750m
2000m
Depth
15S   0   15N   30N   45N   60N
Latitude
Contour From 32. To 36.3 By .1

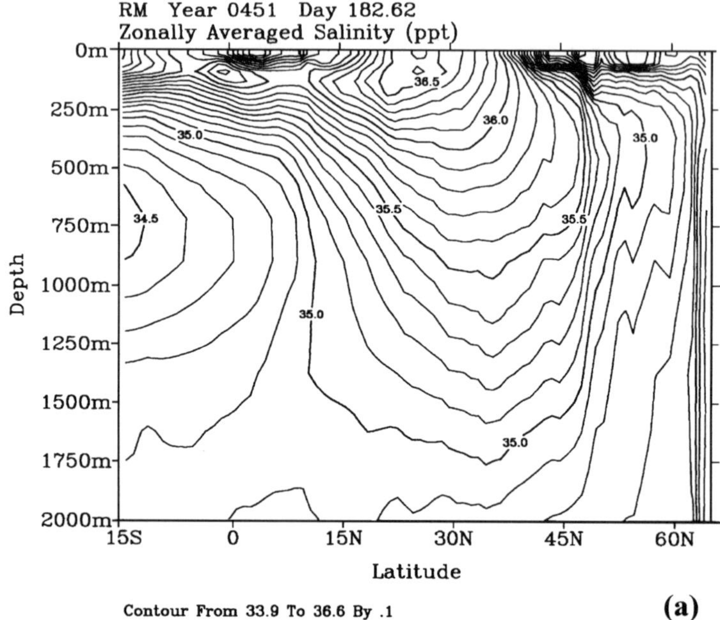

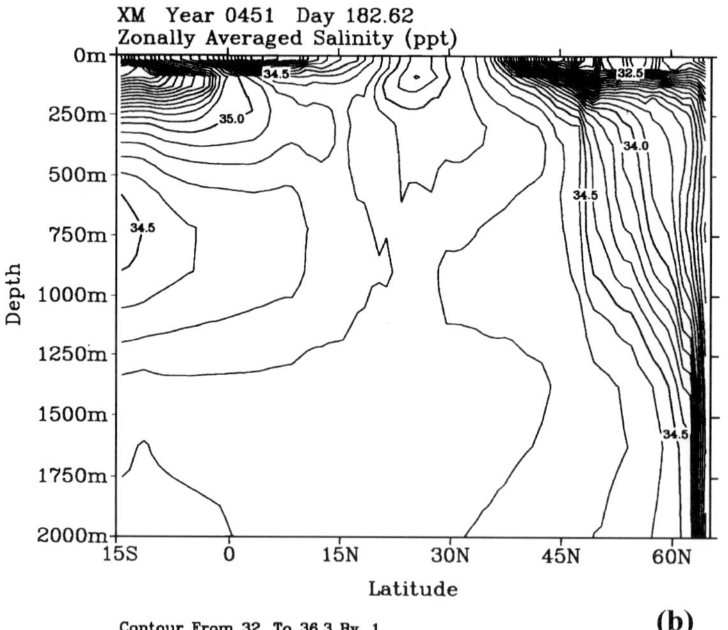

(b)

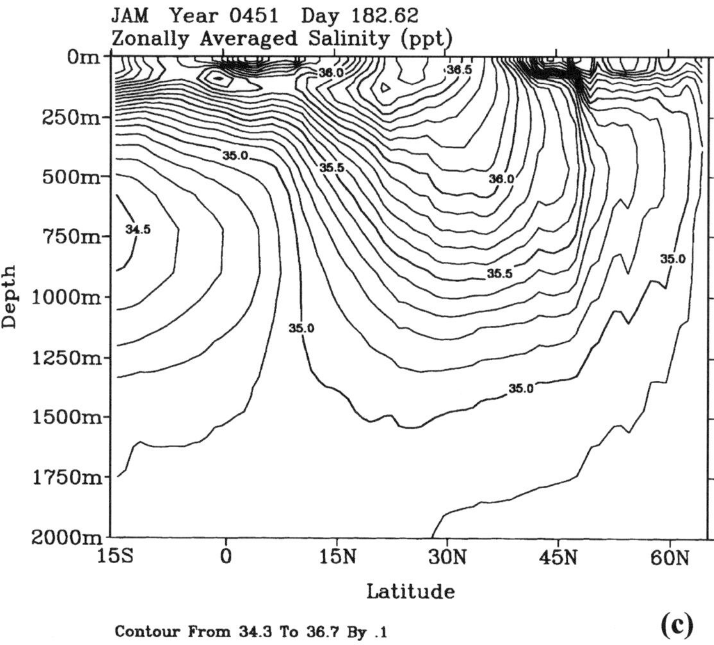

(c)

Figure 9.8 Zonally averaged salinity field at day 182 of year 451 for (a) run RM, (b) run XM, and (c) run JAM.

We next examine the salinity fields in the various model runs in order to identify differences in this component of the thermohaline structure. A Mediterranean salt tongue is clearly evident in zonally averaged plots of the salinity field. In Figure 9.7 the salinity field is intensified near the latitude of Gibraltar. The maximum in the salinity field occurs at around 35°N at intermediate depths from 1 000 m to 2 000 m, and is shifted southward of 35°N in the upper 1 000 m.

At the end of the run, the zonally averaged salinity field of run RM (Figure 9.8(a)) is little changed. The structure is almost identical to the initial state but with an 0.1 ppt shift towards fresher waters. Also evident are fresher waters along the northern boundary. This freshening remains close to the boundary throughout the upper 1 000 m but has made an incursion into the domain below that depth. The inability to completely flush the large source of freshwaters from the northern boundary of model run RM appears to be associated with the downward drift in the overturning cell which was noted above.

The Mediterranean salt tongue has completely disappeared from run XM by the final time (Figure 9.8(b)), indicating that the dissipation time for the salt tongue is less than 450 years. Tremendous freshening has occurred along the northern boundary, and contrary to case RM, the freshwater incursion is particularly evident in the upper ocean. The cold and freshwaters which flow into the northern Atlantic are mixed with saltier waters of Mediterranean

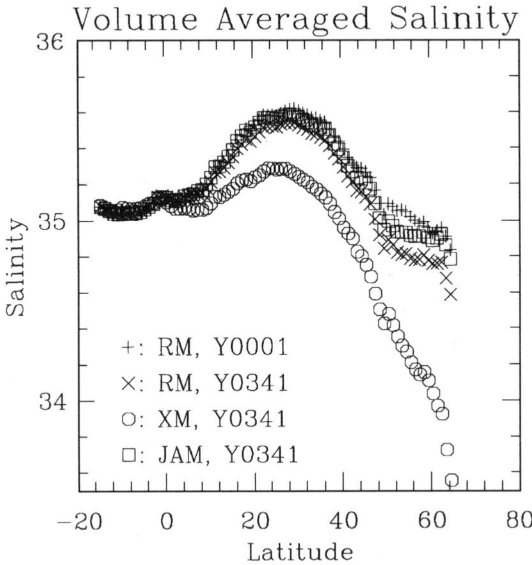

Figure 9.9 Volume averaged salinity as a function of latitude for all three cases at the intermediate time of year 341, when the overturning cell of run XM reaches a minimum.

origin in the real ocean, plunging to depths when cooled in winter. The removal of the Mediterranean salt source of some $1^1/_2$ Sv ppt in run XM has left the model incapable of flushing the larger freshwater source of some 6 Sv ppt from the northern boundary.

In the JAM run (Figure 9.8(c)), we see the same freshening of 0.1 ppt throughout the zonally integrated water column at around 35°N, but the isohalines are less peaked than in run RM. In this run, which has a shallow and somewhat anaemic Mediterranean salt source, the Mediterranean salt tongue occupies a greater breadth than it did in the initial state. The overturning cell has remained very nearly stable in intensity, but the thermohaline structure which drives the overturning has evolved to a noticeably different state. We speculate that a similarly shallow salt source of justifiably greater magnitude would cause the overturning to intensify unrealistically. This is a danger that should be noted by those attempting to model the world ocean over timescales of a century or longer.

The only freshening evident at the northern boundary of run JAM is that associated with the 0.1 ppt freshening which is apparent throughout most of the upper 2 000 m of the domain. The more vigorous overturning of run JAM has continued to flush the large freshwater source from the northern boundary.

More evidence of a small freshwater incursion at northern latitudes of run JAM is seen in Figure 9.9. While the average salinity in run JAM is consistently lower than that of the initial state throughout the northern half of the domain, it is particularly so beyond 50°N. This freshening beyond 50°N is more pronounced for run RM, due to the weakening of

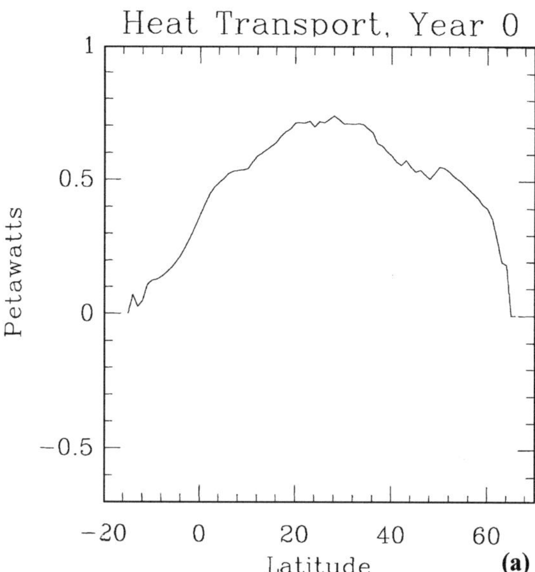

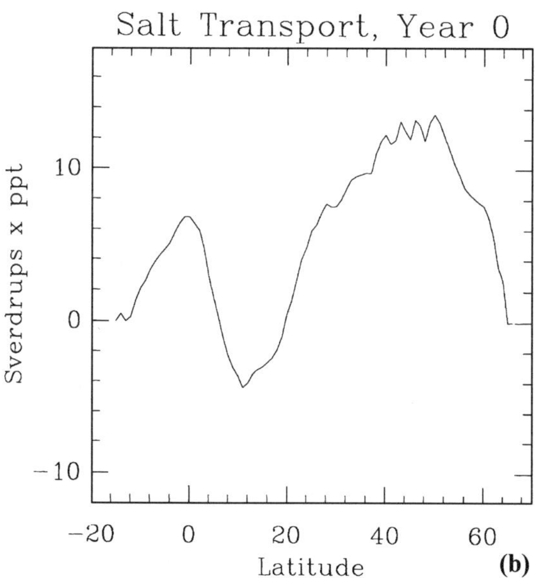

Figure 9.10 Annual mean transports over the first year of run RM: (a) heat transport, and (b) salt transport.

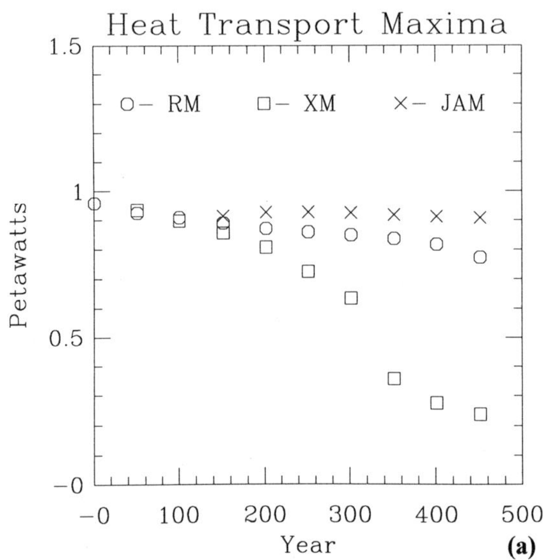

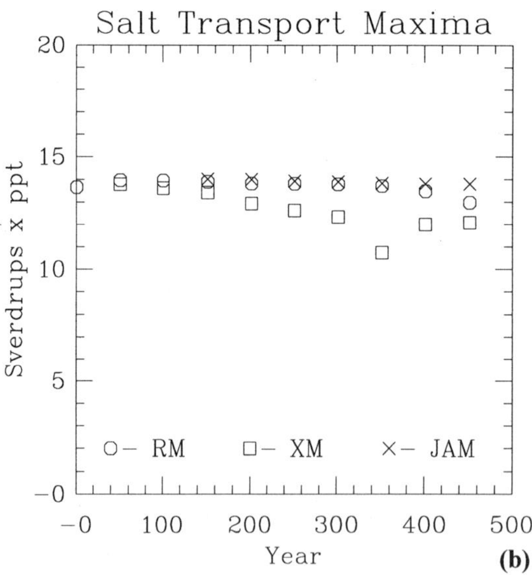

Figure 9.11 Instantaneous transport maxima recorded at June 30: (a) heat transports, and (b) salt transports. At June 30 the heat transport maxima occur at 16°N or 17°N, the salt transport maxima at 38–40°N.

the northward transport of salty surface water. Run XM is much fresher than the initial state at all but the most southern latitudes, and freshwater from the northern boundary, whose source represents the integrated net excess of polar precipitation over evaporation, has accumulated in a dramatic fashion in the northern portion of the model domain.

We also examine the meridional heat and salt transports in the model runs. Changes in the heat transport which accompany different Mediterranean source prescriptions are of particular relevance to climate modelers.

The annual mean meridional heat transport in the first year of run RM, which we refer to as year 0, is shown in Figure 9.10(a). It peaks at 0.74 PW at 28°N, somewhat weaker than the observational estimate of 1.2 PW (Hall & Bryden 1982). The salt transport (Figure 9.10(b)) peaks at a value of 13.6 Sv ppt at 50°N.

The evolution of the instantaneous transport maxima (as measured on June 30 of each year) is shown in Figure 9.11 for all cases. From a peak value of 0.96 PW, the heat transport in run RM drops off steadily for 400 years, ending at 0.77 PW. The heat transport in run XM drops rapidly, then plunges between years 300 and 350, and continuing to fall to only 0.24 PW by the end of the run. In contrast to these two cases, the June 30 heat transport of run JAM remains quite steady at around 0.91 PW.

The maximum salt transport in run RM is more constant, dropping slowly from an initial June 30 value of 13.7 Sv ppt over the first several hundred years, then falling somewhat faster over the final 50 or 100 years to a final value of 13.0 Sv ppt.

The salt transport in run XM takes a dip of several Sv ppt at year 350 (near the time of collapse) before settling back at around 12.0 Sv ppt, 1.7 Sv ppt below that of the initial state. It is notable that this difference in peak salt transport is nearly equal to the magnitude of the Mediterranean salt source which existed in the initial state and has been removed from run XM. Indeed the only other case-dependent salt source is at the southern boundary where restoring is applied, since the surface and northern boundary salt sources were diagnosed from the initial state and applied consistently to all three runs. The fact that the difference in peak salt transports between the initial state and the end of run XM is so nearly equal to the magnitude of the Mediterranean salt source, which flows in large part northward, must mean that there has been no great change in the net salt source at the southern boundary.

The salt source in run JAM is again seen to be quite steady.

At the end of the runs the annual mean heat transport of run RM has dropped from 0.74 PW at 28°N to 0.56 PW, as seen in Figure 9.12(a). The heat transport in run XM is a mere 0.08 PW at 28°N and peaks at just 0.21 PW at 58°N. The peak heat transport in run JAM remains at 0.69 PW. Similar shapes are evident in all three heat transport curves. Even that of run XM bears considerable resemblance to the other two cases and the initial state. This is reflective of the wind driven gyre structure of the North Atlantic which determines in large part the sea surface temperature gradients.

The annual mean salt transports look even more similar between the three cases owing to the case-independent salinity forcing at the surface as well as northern boundaries.

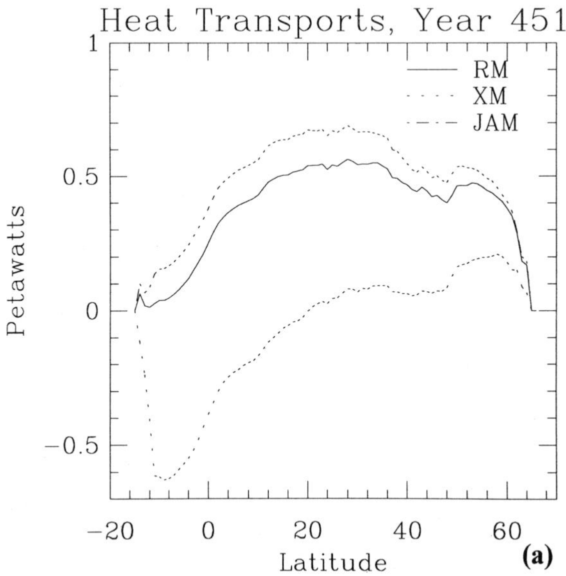

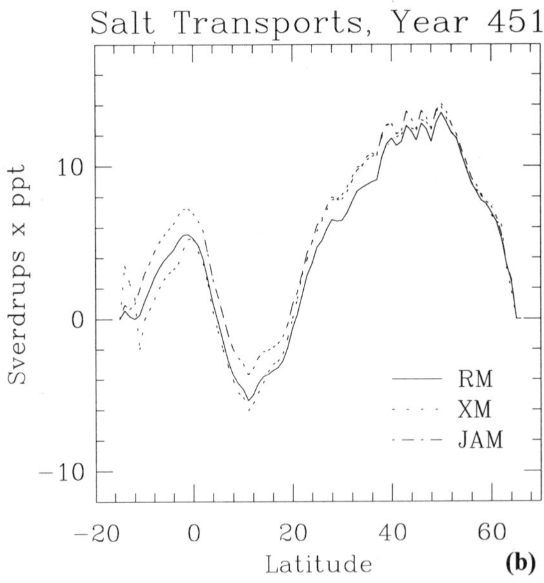

Figure 9.12 Annual mean transports from the last year of all runs: (a) heat transports, and (b) salt transports.

We again see in Figure 9.12(b) that the salt transport in case XM has fallen off from that of the other cases by 1–2 Sv ppt. The magnitude of the Mediterranean salt sources which are included in cases RM and JAM and absent from case XM falls within this range.

9.5 Summary

Our study supports the view that Mediterranean waters regulate North Atlantic overturning. The vigorous overturning which is maintained in part by saline Mediterranean waters which outcrop in the Labrador and Norwegian-Greenland seas efficiently flushes the great quantity of freshwaters which accumulate over the polar ocean, preventing a massive freshening of the northern reaches of the Atlantic.

Gross circulatory features on a time scale shorter than a century are not very sensitive to the details of the inclusion of Mediterranean waters. On these shorter time scales the structure of the Mediterranean salt tongue may be expected to change in response to the forcing at Gibraltar. On longer time scales of a century and beyond, the magnitude of the Mediterranean salt source and the depth at which it is located have considerable bearing on the resulting North Atlantic circulation.

Although the overturning remains strong in the JAM case, the run cannot be said to be better than the RM run. The relatively stable mass, heat and salt transports are accompanied by a distortion of the thermohaline structure associated with the Mediterranean salt tongue. We propose that the anomalously shallow structure of the source of Mediterranean waters in run JAM tends to invigorate the overturning, compensating for the weak magnitude of the source. Presumably the JAM Mediterranean waters more effectively maintain the vertical salinity gradient which enables deep water formation in the north due to their inclusion at relatively shallow depth. We would expect to see an unrealistically large overturning if the magnitude of the source were greater and more in accordance with observations.

It is not likely that the outflow of Mediterranean waters at Gibraltar would be represented with greater fidelity within a world ocean model than it was within our $1/4°$ Mediterranean model, whose outflow we used to force our North Atlantic model. It is unlikely that the magnitude of the salt source, which is dependent upon the fine details of surface forcing and deep water formation within the Mediterranean region, and the depth of the outflow, which is determined by poorly modeled turbulent mixing processes which describe the overflow from the sill at Gibraltar, will be determined within acceptable bounds within a World Ocean model with current modeling techniques.

While it is not at all obvious that this need be the case, it appears from various modeling results that the width of the Strait is of secondary importance to the representation of the physics of the overflow. Our experience with our Mediterranean model, which will be published in detail in the literature, are in contradiction to the prediction of strong dependence on the width of the Strait associated with Bryden and Stommel's (1984) over-mixing theory. Experience with NCAR's hierarchy of global ocean models also suggests that the width of the Strait is of secondary importance.

The results of our study emphasize the need for a better representation of overflows within ocean models, not only at Gibraltar but at other locations where important water masses must flow from one basin into another over topographic sills. Some progress has been made by Price and Baringer (1994) on this problem. The inclusion of better methods of representing overflows will significantly improve the predictive power of ocean models, and hence will significantly improve our ability to predict climate.

References

Böning, C.W., Holland, W.R., Bryan, F.O., Danabasoglu, G. and McWilliams, J.C. (1995) An overlooked problem in model simulations of the thermohaline circulation and heat transport in the Atlantic Ocean. *Journal of Climate*, **8**, 514–523.

Broecker, W.S., Peteet, D.M. and Rind, D. (1985) Does the ocean-atmosphere system have more than one stable mode of operation? *Nature*, **315**, 21–26.

Broecker, W.S., Bond, G. and Klas, M. (1990) A salt oscillator in the glacial Atlantic? *1. The Concept. Paleoceanography*, **5**, 469–477.

Bryan, F. (1987) Parameter sensitivity of primitive equation ocean general circulation models. *Journal of Physical Oceanography*, **17**, 970–985.

Bryan, F.O. and Holland, W.R. (1989) A high resolution simulation of the wind and thermohaline-driven circulation in the North Atlantic Ocean. In *Proceedings of the 'Aa Huliko'a Hawaaiian Workshop*, 17–20 January.

Bryan, F.O., Böning, C.W. and Holland, W.R. (1995) On the mid-latitude circulation in a high-resolution model of the North Atlantic. *Journal of Physical Oceanography*, **25**, 289–305.

Bryan, K. (1969) A numerical method for the study of the circulation of the world ocean. *Journal of Comparative Physiology*, **4**, 347–376.

Bryden, H.L. and Stommel, H.M. (1984) Limiting processes that determine basic features of the circulation in the Mediterranean Sea. *Oceanologica Acta*, **7**, 289–296.

Bryden, H.L., Brady, E.C. and Pillsbury, R.D. (1989) Flow through the Strait of Gibraltar. In *Seminario Sobre la Oceanografia Fisca del Estrecho de Gibraltar*, Madrid, 24–28 October 1988, edited by J.P. Almazan, H. Bryden, T. Kinder and G. Parilla, pp. 166–194. Madrid: SECEG.

Cox, M.D. (1984) A primitive equation three dimensional model of the ocean. GFDL Ocean Group Technical Report No. 1, GFDL/NOAA, Princesaton University.

Dansgaard, W., Clausen, H.B., Gundestrup, N., Hammer, C.U., Johnsen, S.F., Kristinsdottir, P.M. and Reeh, N. (1982) A new Greenland deep ice core. *Science*, **218**, 1273–1277.

Dansgaard, W., White, J.W.C. and Johnsen, S.J. (1989) The abrupt termination of the younger dryas climate event. *Nature*, **339**, 532–534.

Hall, M. and Bryden, H. (1982) Direct estimates and mechanisms of ocean heat transport. *Deep-Sea Research*, **29**, 339–359.

Han, Y.J. (1984) A numerical world ocean general circulation model Part II. A baroclinic experiment. *Dynamics of Atmospheres and Oceans*, **8**, 141–172.

Hellerman, S. and Rosenstein, M. (1983) Normal monthly wind stress over the world ocean with error estimates. *Journal of Physical Oceanography*, **17**, 158–163.

Holland, W.R. and Bryan, F.O. (1993a) Sensitivity studies on the role of the ocean in climate change. In *Ocean Processes in Climate dynamics: Global and Mediterranean Examples*, edited by P. Malanotte-Rizzoli and A.R. Robinson, NATO ASI Series, pp. 113–133. Dordrecht: Kluwer Academic Publishers.

Holland, W.R. and Bryan, F.O. (1993b) Modeling the Wind and Thermohaline Circulation in the North Atlantic. In *Ocean Processes in Climate dynamics: Global and Mediterranean Examples*, edited by P. Malanotte-Rizzoli and A.R. Robinson, NATO ASI Series, pp. 135–156. Dordrecht: Kluwer Academic Publishers.

Lacombe, H. and Tchernia, P. (1960) Quelques traits generaux de l'Hydrologie Mediterranean. *Cahiers Oceanographiques*, **8**, 527–547.

Levitus, S. (1982) *Climatological Atlas of the World Ocean*. NOAA Prof. Paper, Vol. 13. Washington, DC: US Government Printing Office.

Marotzke, J., Welander, P. and Willebrand, J. (1988) Instability and multiple steady states in a meridional-plane model of the thermohaline circulation. *Tellus*, **40A**, 162–172.

Price, J.F. and Baringer, M.O. (1994) Outflows and deep water production by marginal seas. *Progress in Oceanography*, **33**, 161–200.

Reid, J.L. (1979) On the contribution of the Mediterranean sea outflow to the Norwegian-Greenland sea. *Deep-Sea Research*, **26A**, 1 199–1 223.

Roussenov, F., Stanev, E., Artale, V. and Pinardi, N. (1995) A seasonal model of the Mediterranean sea general circulation. *Journal of Geophysical Research*, **100**, 13515–13538.

Stanev, E.V. (1992) Numerical experiment on the spreading of Mediterranean water in the North Atlantic. *Deep-Sea Research*, **39**, 1 747–1 766.

Zaucker, F., Stocker, T.F. and Broecker, W.S. (1994) Atmospheric freshwater fluxes and their effect on the global thermohaline circulation. *Journal of Geophysical Research*, **99**, 12 443–12 457.

PART 4

SENSITIVITY STUDIES

CHAPTER 10

COMMENTARY

Robert E. Dickinson, Andrea N. Hahmann and Qingqiu Shao

10.1 Introduction

The MECCA sensitivity studies report in Chapters 11, 12, and 13 address a wide range of topics under the rubric of sensitivity. One narrow definition of sensitivity is the magnitude of global average temperature response to a change in radiative forcing. Chapter 11 by McAvaney and Colman is framed in the context of this definition. More generally, sensitivity refers to the response of climate models to the change of any important model parameters. It is within this latter context that Chapters 12 and 13 are written. Chapter 13 addresses model response to the inclusion of sea-ice motion in the modeling of effects of sea ice on climate, and the response to a doubling of model stomatal resistance. Chapter 12 addresses the response of two models to tropical deforestation and discusses large differences in the answers given to this question by these two models. Although the sensitivity chapters cover a wide range of topics, they all have an underlying theme that is common to many, if not most, contemporary climate model simulations. That is, they all directly or indirectly involve the coupling between surface and atmospheric processes.

In this commentary, we discuss the significance of the results reported in these chapters, and offer some interpretation of them in terms of simple concepts and results from other related modeling studies.

In Chapter 11, McAvaney and Colman study the climate's sensitivity to convective parameterizations by using the Bureau of Meteorology Research Centre (BMRC) model. They found that the control climate is sensitive to the convection schemes, with better results given by the Tiedtke (1984) mass flux scheme than by the Kuo (1974) scheme. The two schemes show differences in intensity of convection, and therefore the cloud types and amounts. With the Tiedtke (1984) scheme, surface fluxes are larger in the subtropics due to stronger shallow convection, and smaller in western Pacific due to less penetrative deep convection. They argue that different couplings between the surface flux parameterizations and different convective schemes are sufficient to perturb the control climate. For the doubled-CO_2 climate, however, they find that the changes of surface temperature and precipitation are not sensitive to the convection schemes used, although large local variations exist. This sensitivity question should be pursued further with different models having different cloud feedbacks, since these feedbacks are directly related to convection.

By analyzing two GCM simulations of tropical deforestation, the authors of Chapter 12 addressed the important question: How good must the model control simulation be before it can be used to simulate the climate responses to deforestation (or other disturbances) in a specific region? Deforestation results from CCM1-OZ simulations (Henderson-Sellers et al. 1993; McGuffie et al. 1995) include a decrease in surface evaporation, and a larger decrease in precipitation, which are consistent with most of the previous GCM deforestation studies. In contrast, in the BMRC model, tropical deforestation leads to increases in both precipitation and surface evaporation, and a larger reduction in surface sensible heat flux.

To address these drastically different results, the authors first analyzed the control simulations. Even though overall tropical climate can be adequately simulated in both GCMs, their performance over the tropical rainforest region is different: enough precipitation is simulated in CCM1-OZ to maintain a realistic 'wet' rainforest climate, while too little precipitation in the BMRC model maintains an unrealistically 'dry' climate regime over the tropical rainforest region.

Subsequently, the authors analyzed the surface and atmospheric energy budgets, atmospheric moisture transport and near-surface atmospheric dynamics, and cloud radiative forcing in the control and deforestation simulations. For long-time averaged results, the surface energy balance and atmospheric water conservation require that

$$\Delta R_{\text{net}} = \Delta LH + \Delta SH \tag{1}$$

and

$$\Delta P = \Delta LH + \Delta M \tag{2}$$

where M is the net moisture transport, and Δ denotes the difference between control and deforestation simulations.

In the BMRC model, after deforestation, the effect of increased wind speed dominates the reduction of drag coefficient due to roughness length decrease, leading to an increased conductance for surface latent and sensible heat fluxes, and, hence, a positive ΔLH. In addition, an increased wind speed increases the moisture transported into the tropical rainforest region, leading to a positive ΔM. Through Equation (2), a large positive value of ΔP results. A negative value of ΔR_{net} results from the increase of surface albedo and the increase of cloud cover with a negative cloud radiative forcing. Through Equation (1), a large reduction in SH results.

In contrast, in CCM1-OZ, after deforestation, the reduction of drag coefficient due to roughness length decrease dominates the increase of wind speed, leading to a decrease in surface LH. In addition, the increased wind speed increases the moisture transporting out of the tropical rainforest region, leading to a large decrease in P through Equation (2). Though

ΔR_{net} is negative, it results from the increase of albedo minus the decrease of cloud cover. Because the two models provide self-consistent (e.g. satisfying Equations (1) and (2) and other physical constraints) results, it is reasonable for the authors to claim that their different control simulation performances are responsible for their drastically different deforestation results.

Following these discussions on regional climate responses to tropical deforestation, the authors go on to discuss the possible large-scale and global effect of tropical deforestation. Three possible mechanisms for these effects are briefly mentioned: Walker circulation change, meridional Hadley circulation change, and the propagation of stationary Rossby waves. Further comparison of results from the two GCMs with regards to these questions would be useful.

In Chapter 13, the authors discuss the sensitivity of the GENESIS (Global Environmental and Ecological Simulation of Interactive Systems) model to vegetation stomatal resistance change. When the vegetation stomatal resistance is increased globally by a factor of two (possibly due to the doubling of CO_2), surface transpiration decreases by as much as 20% to 50% over tropical rain forests and over the northern boreal forests in summer. This leads to a surface warming of about 2 K to 5 K. The uniform doubling of the minimum stomatal resistance also appears to shift the north–south tropical rainfall belt. On a global scale, however, these effects are smaller than those due to the doubling of CO_2.

The authors rightfully caution as to the interpretation of the above results. How much plant stomatal resistance will actually change in global ecosystems with a doubling of CO_2 is poorly known. Other responses to the increased CO_2 and the resulting climate change may have comparable or greater effects on vegetation.

10.2 Coupling of convective boundary over land to surface processes

Evapotranspiration (ET) from the land surface is most simply estimated from the Penman-Monteith (P-M) combination equation. Jarvis and McNaughton (1986) have emphasized in several past publications the potential for ultimate control of evaporation by atmospheric boundary layer (ABL) processes. They have presented theoretical models that suggest two related ideas: (1) that the ABL may act as an additional resistance coupled to aerodynamic and canopy resistances, which is large enough to weaken the dependence of ET on canopy properties as inferred from P-M with prescribed conditions for surface air, and (2) that the top of the ABL can act to contain the ABL in such a way that its properties will adjust to surface fluxes until ET depends on surface net radiation and becomes independent of canopy properties. As they recognize, these two arguments are equivalent in the limit of very large resistance to movement from the ABL to overlying air. These arguments by Jarvis and McNaughton are of considerable interest for climate sensitivity studies involving changes of surface properties as reported in Chapters 12 and 13 because they indicate that ET and its coupling to the atmosphere should be considerably less sensitive in a GCM to changes of surface properties than would be inferred from uncoupled ('stand-alone') surface models.

We argue, contrary to the Jarvis–McNaughton hypothesis, that the continental ABL generally does not act in GCMs as such a large negative feedback that climate becomes insensitive to changes in surface properties. This converse conclusion can be inferred from the wide range of recently published GCM sensitivity studies including those in the following chapters that report remarkable sensitivity of climate to changes in surface conditions.

In the following, we summarize the theoretical arguments by Jarvis and McNaughton that have suggested the possibility of negative feedback by the ABL. We contend that the ABL does not act in the manner that they imagined. Rather than have its humidity be so strongly modified by surface moisture fluxes that these fluxes become only radiatively driven, it acts to buffer its humidity changes and keep them small. Consequently, rather than strongly modify the surface fluxes, the ABL acts more to buffer changes of near-surface humidity and so to allow surface fluxes to be essentially as strongly affected by changing surface parameters as they would be in a stand-alone model.

Following the analyses of Dickinson (1983) and Jarvis and McNaughton (1986) we write the P-M equation for evaporation in energy units as

$$\frac{\lambda E}{Po} = \Gamma R_n + (1 - \Gamma)\lambda \left(q_{a_{\text{sat}}} - q_a\right)/(r_c + r_{a,w}) \tag{3}$$

where with first term R_n is the net surface radiation minus any energy flux into the soil,

$$\Gamma = 1./(1. + B_e^*) \tag{4}$$

and where B_e^* is the 'equilibrium' Bowen ratio for canopy transpiration, that is the Bowen ratio for a saturated atmosphere, and given by

$$B_e^* = B_e(r_c + r_{a,w})/r_{a,w} \tag{5}$$

where r_c is canopy (integrated stomatal) resistance, $r_{a,w}$ is the aerodynamic resistance for water flux,

$$B_e = C_p/(\lambda \partial q/\partial \Gamma) \tag{6}$$

is the equilibrium Bowen ratio for a wet surface. The qs in the second term are q_a = atmospheric specific humidity and $q_{a_{\text{sat}}}$ = saturated atmospheric specific humidity. The Γs introduced by Jarvis and McNaughton elegantly show ET as forced by an average of radiation and humidity deficit. Jarvis and McNaughton used Equation (3) and concepts of scaling from individual leaves to the whole system to argue intuitively that transport of moisture from the ABL to the overlying atmosphere could also be regarded as given by Equation (3) provided an addition resistance, say r_{ABL} was added to $r_{a,w}$ in Equation (3). Any such term would act to decrease B_e^* in Equation (5), hence increase Γ in Equation (4), and decrease the relative contribution of the second term in Equation (3) both through decreasing $1 - \Gamma$ and through increasing the $r_c + r_{a,w}$ term in the denominator. These changes all act to reduce the sensitivity of Equation (3).

The ABL could most simply be viewed as another resistance if it were a barrier to the transport of water overlaid by a well-mixed layer. However, for the daytime convective boundary, the ABL can be the opposite, mixed internally by convection but with a strongly impeded transport of water into the overlying stable atmosphere. This containment of the ABL by an overlaying stable layer might seem to match the Jarvis–McNaughton concept of an effective infinite resistance. However, examination of the processes of the convective ABL indicates that the ABL and the overlying stable atmosphere have a rather different role.

The convective ABL generally acts to entrain the dry overlying air at a sufficient rate to substantially cancel the effects of surface moisture fluxes in increasing the ABL specific humidity. With that effect and the large storage capacity of the ABL over a daytime period, it is evident that the ABL has largely the opposite effect of that argued by Jarvis and McNaughton. That is, its role is to buffer humidity change and so to hold specific humidity close to constant over the daytime period. We suggest that it is this buffering of humidity by the daytime convective ABL that explains why GCM simulations have been showing such pronounced sensitivity to changes of surface conditions.

One recent example of this buffering is the boundary layer observed during FIFE. Betts and Ball (1993) have developed area-averaged (15 km $\times$ 15 km) surface data for two of the four intensive field campaigns. (Thirty-three days from 25 June to 11 July and from 6 August to 21 August represent typical summer conditions.) These data are then averaged over the diurnal cycle. Figure 10.1 shows, for mid-day conditions, the relative change of specific humidity q in percentage over 30 minutes. The dashed lines are for standard deviation. The changes are small and q actually decreases from noon until about 2.00 p.m. when convection starts to weaken. The mean values of q, plotted in Figure 10.2, are close to a constant.

The surface Bowen ratio is also close to a constant, in mid-day (10.00 a.m. to 2.00 p.m.). Thus the decreases of q in the early afternoon must result from entrainment drying of the convective ABL. This point is illustrated by a simple mixed layer model for the ABL. For horizontal homogeneity:

$$\frac{\partial q}{\partial t} = (1 + A_q)(\overline{w'q'})_o / h \tag{7}$$

where h is the mixed layer height. The constant $A_q \equiv -(\overline{w'q'})_h / (\overline{w'q'})_o$ represents the moisture entrainment factor, where $(\overline{w'q'})_o$ and $(\overline{w'q'})_h$ are moisture fluxes at the surface and mixed layer top, respectively. The term A_q is always negative due to entrainment of dry air from above. The ABL dries out when $A_q < -1$. Betts (1992) shows that:

$$A_q = A_R(\beta_s + \varepsilon\delta)/(\beta_i + \varepsilon\delta) \tag{8}$$

where A_R is the entrainment parameter, and is about 0.4 for the FIFE cases (Betts et al. 1992). β_s and β_i are the Bowen ratios at the surface and ABL top, respectively, and $\varepsilon\delta \approx 0.07$.

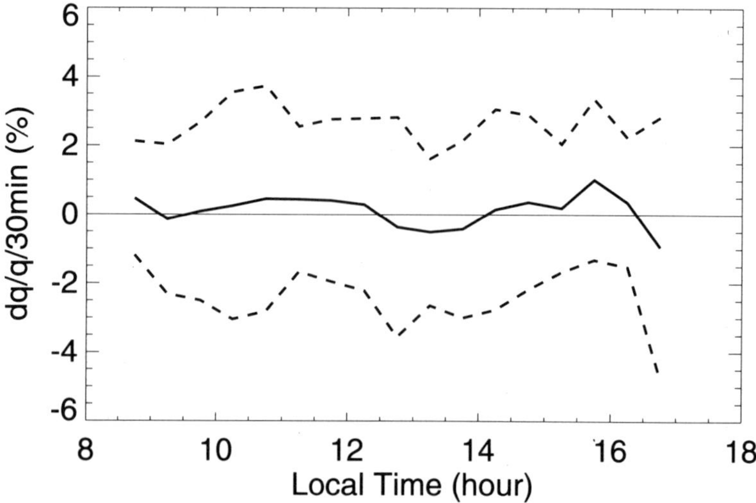

Figure 10.1 The relative change (in percentage) of specific humidity over every 30 minutes from 8:45 a.m. to 4:45 p.m. The dashed lines are for standard deviation. Data averaged over the FIFE site.

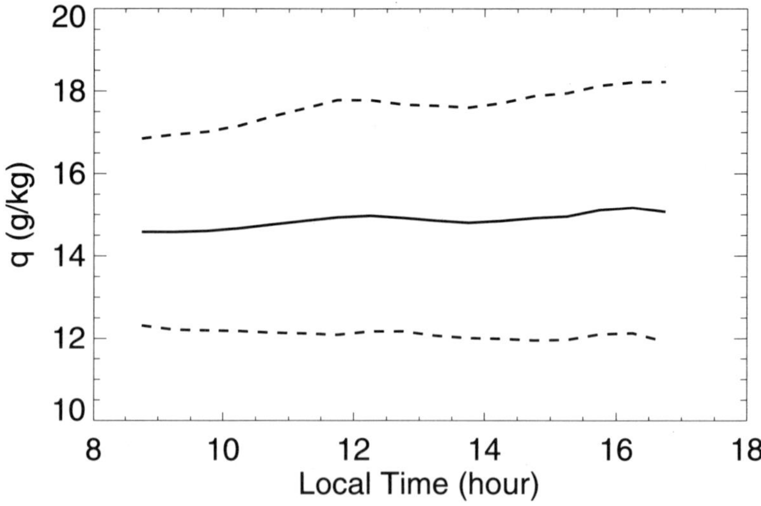

Figure 10.2 Specific humidity averaged over every 30 minutes from 8:45 a.m. to 4:45 p.m. The dashed lines are for standard deviation.

Since β_s is observed to be nearly constant, as is A_R, the only factor that can make A_q change is β_i, or the entrainment drying. For example, using from Figure 10.1 that $\frac{\partial q}{\partial t} \cong 0$, we also infer that $A_q \cong -1$ and $\beta_s \cong 0.36$ from Figure 10.3, we can get from Equation (8) that $\beta_i \approx -0.24$. For decreasing humidity $A_q < -1$ and so $\beta_i < -0.24$. Betts et al. (1992) obtained a range of values from –0.22 to –0.47 from the flight data during July and August. The variances of the data could be attributed to the effects of precipitation and synoptic-scale motions during the time.

The above analyses strongly support the argument that the ABL does not act as a barrier to the transport of moisture from surface during the daytime. Rather it is dried by entrainment at the top of the ABL to hold the specific humidity in the ABL close to constant during the daytime. Thus the deficit term in P-M equation always remains important, and the sensitivity of the climate to changes in surface conditions substantial.

10.3 The response of GCMs to Amazon deforestation

Chapter 12 emphasizes the need for proper evaluation of the GCM performance prior to the study of the response to tropical deforestation (or any other disturbance) in a specific region. With their comparison of deforestation simulations from two different models, they raise the question as to how good a model control simulation must be before it can be used to simulate the climate response to deforestation. Our experience with simulating the land surface climate and circulation over the Amazon region suggests that several factors help determine the climate of this region. Among these are the presence of an extensive tropical forest and its interaction with the atmosphere above it, and the representation of the topographic features of the terrain, especially the Andes Mountains. The first of these

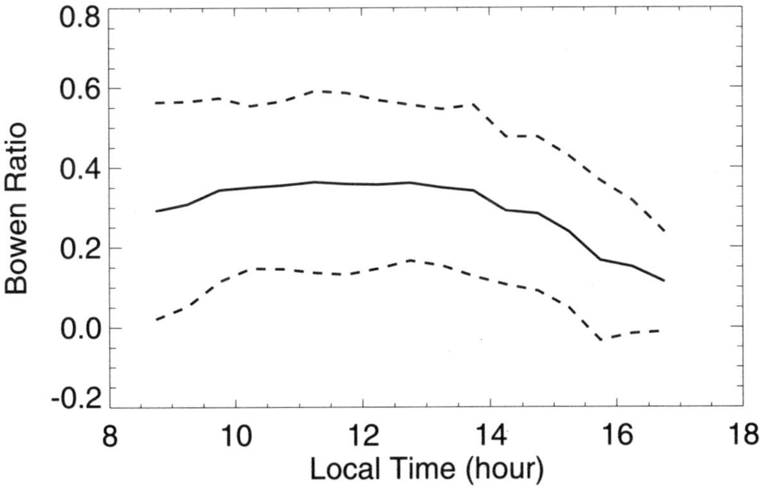

Figure 10.3 Bowen ratio averaged over every 30 minutes from 8:45 a.m. to 4:45 p.m. The dashed lines are for standard deviation.

factors is determined by the treatment of land processes, clouds, and the boundary layer in the GCM, while the second is determined by the model resolution. Multi-year integrations have recently been carried out with a revised version of the NCAR Community Climate Model (CCM2) coupled to the Biosphere–Atmosphere Transfer Scheme (BATS), referred to as RCCM2/BATS. Hahmann et al. (1995) study the sensitivity of this model to changes in the cloud optical properties. RCCM2/BATS at T42 resolution provides more realistic and highly resolved precipitation patterns than previous models used for deforestation studies.

The northern South American climate includes a large region of intense convection centered over the Amazon Basin during the austral summer. Horel et al. (1989) have used Outgoing Longwave Radiation (OLR) and European Centre for Medium Range Weather Forecast (ECMWF) initialized analysis to study the annual cycle of convective activity over the tropical Americas. The deep convection seen by the OLR data resides over the Amazon Basin region from November to March. With this area of intense convection, the upper troposphere circulation displays a closed anticyclone known as the Bolivian High and a downstream trough located over the northeast coast of South America. It is likely that the response of a GCM to deforestation is sensitive to these features.

We have used two sources of observational data for precipitation: the global climatology of Legates and Willmott (1989) and a 30-year record of daily precipitation for surface stations in the Brazilian Amazonia. Figure 10.4 displays the austral summer (December–March) total precipitation for the three model simulations and the observational estimates of Legates and Willmott. This period contains 50% to 60% of the annual precipitation in the Amazon Basin. As Hack et al. (1994) indicate, the standard CCM2 model tends to overestimate tropical precipitation. In particular, precipitation is too large over northern South America (Figure 10.1(a)). Furthermore, the region's maximum appears to be shifted eastward toward the semi-arid region of north-eastern Brazil. CCM2/BATS results show moderate improvement, and the RCCM2/BATS simulation displays much closer agreement with observed rainfall rates over this region. The simulation of the precipitation over the Amazon Basin region also depends on the model representation of the topography. Past simulations with CCM1-OZ at R15 resolution have systematically shown an eastward shift in the area of maximum precipitation. The present T42 simulation appears to be in better agreement with observations.

Three surface stations in the Amazon Basin have been selected according to the overall length of their observational record and their representativeness in terms of different annual cycles of monthly precipitation. These stations are: Boa Vista (3.45°N, 60.43°W), Benjamin Constant (4.38°S, 70.03°W), and Borba (6.75°S, 58.93°W). Figure 10.5 shows their location in the Amazon Basin. This figure also displays the annual cycle of precipitation in the three simulations for the model's grid-point closest to their respective surface station. While over southern and western Amazonia the RCCM2/BATS simulation is able to capture very well the seasonal cycle of monthly precipitation, over northern Amazonia (Boa Vista) the Northern Hemisphere summer peak in precipitation is completely missed by all model simulations. The models monthly averaged precipitation in this region exhibits a double peak during the months of May–June and October–November, probably a result of the northward

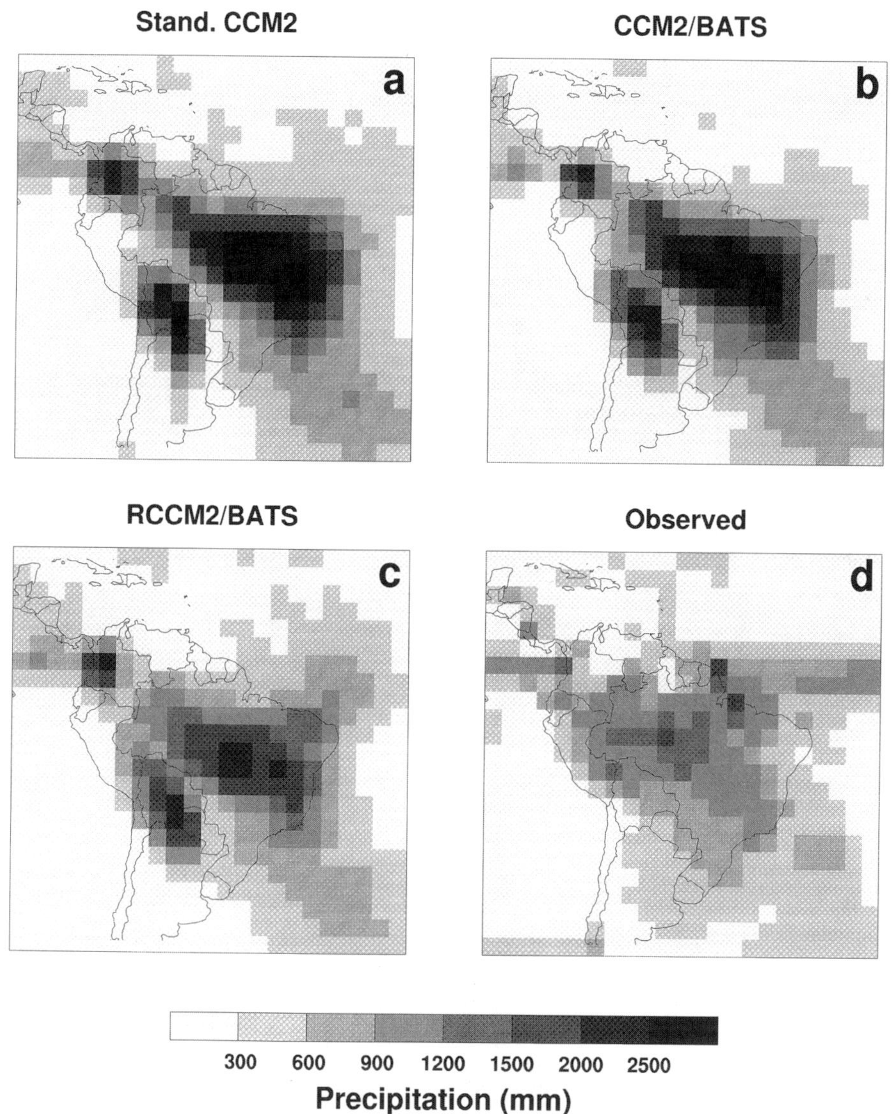

Figure 10.4 Austral summer (December–March) total precipitation over South America.

and southward migration of the model simulated ITCZ during the year. Observations of precipitation and OLR over northern South America (not shown) suggest that the displacement of the maximum convective region does not follow a simple north–south path but has a more northwest to southeast track in the transition period from the Gulf of Panama to the Amazon region (Horel et al. 1989), resulting in a single wet period in some parts of Northern Amazonia. Similar results have been presented by Nobre et al. (1991) for a modified version of the National Meteorological Center global spectral model.

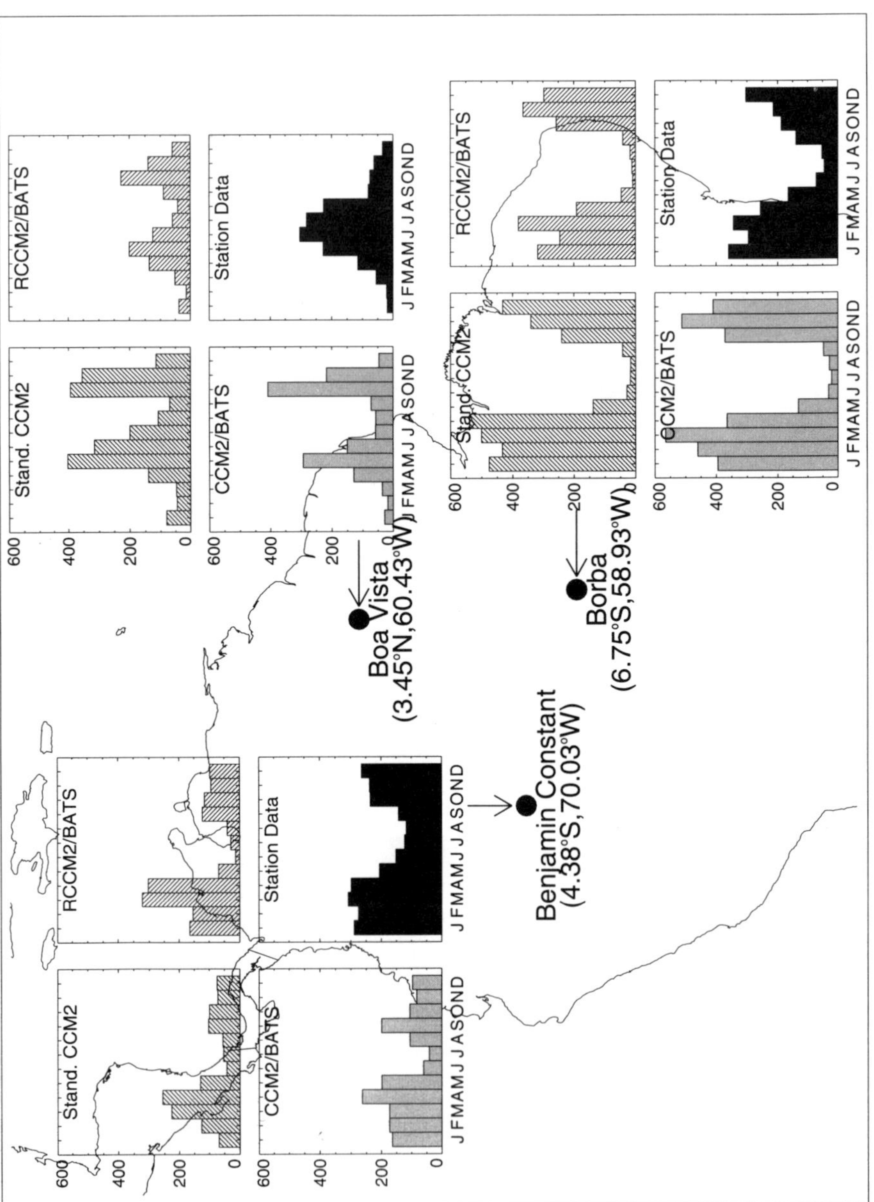

Figure 10.5 Comparison of the observed annual cycle of monthly precipitation for selected surface stations in the Amazon Basin with the results of various CCM2 simulations.

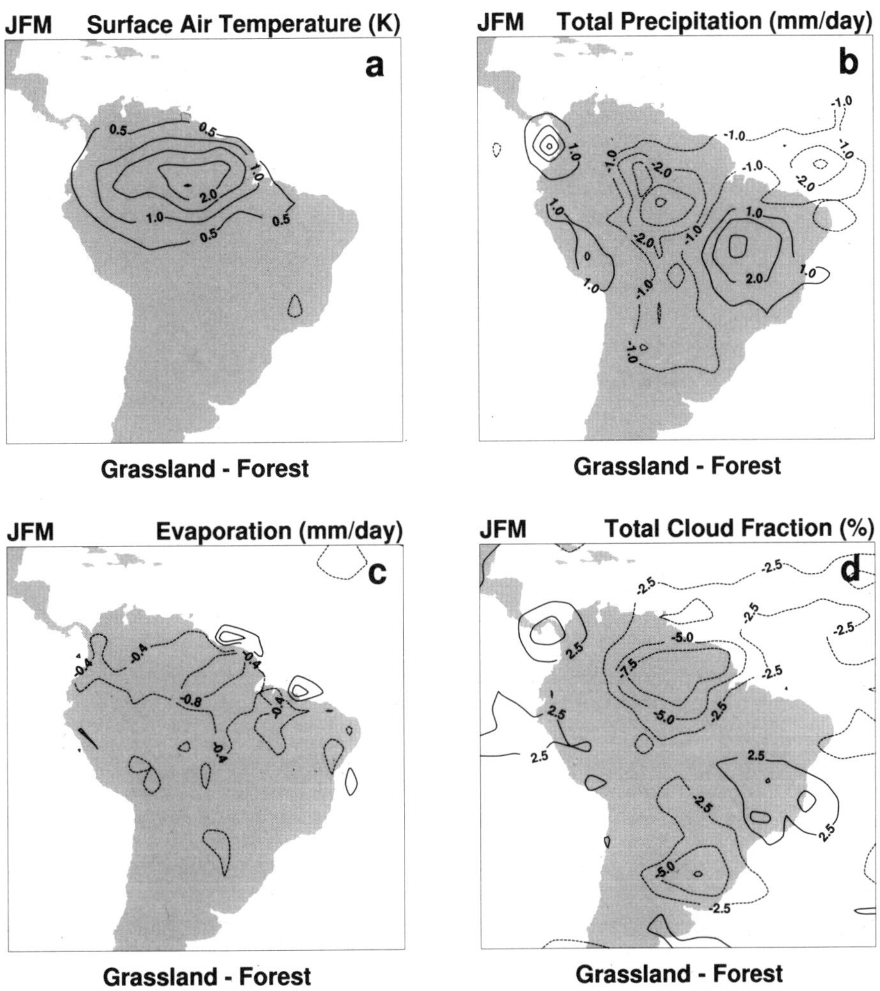

Figure 10.6 January–March changes (deforested minus control) in (a) surface air temperature (°C), (b) total precipitation (mm/day), (c) evaporation (mm/day), and (d) total cloud fraction (%) induced by the changes in surface land use.

Figure 10.6 displays the January–March changes (deforested minus control) in surface air temperature, total precipitation, evaporation, and total cloud cover induced by the changes in surface land use. The temperature differences (Figure 10.6(a)) shows distinct warming, with maximum values above 2°C, over the deforested area. Deforestation also induces changes in the precipitation distribution with large decreases over central Amazonia (up to 3 mm day^{-1}) with extensions over central Brazil and northern Argentina. Evaporation changes (Figure 10.6(c)) closely follow the deforested areas with a general reduction on the order of 0.5 mm day^{-1}. Since the reduction in precipitation is larger than that of evaporation, it follows that there must also be a reduction in moisture convergence into the Amazon region. Slight cooling is evident over southeast Brazil associated with an increase in precipitation (Figure 10.6(b)) and cloud cover fractions (Figure 10.6(d)). The precipitation reduction concentrates in the center of the deforested area with an extension into the ITCZ region over the Atlantic Ocean.

All panels in Figure 10.6 show a substantial bimodal pattern with the Amazon decrease of precipitation and temperature increase accompanied by changes in the opposite sign to the southeast of the Amazon. Similar patterns have been seen in previous studies, but not always in exactly the same locations. Evidently, how much of the region of rainfall increase overlaps the area analyzed for deforestation response strongly affects the inferred statistics. It is likely that this pattern depends on the model control climatology and possibly other model features.

References

Betts, A.K. (1992) Budget analyses of FIFE atmospheric boundary layer budget methods. *Journal of Geophysical Research*, 18 523–18 531.

Betts, A.K., Desjardins, R.L. and MacPherson, J.I. (1992) Budget analysis of the boundary layer grid flights during FIFE 1987. *Journal of Geophysical Research*, 18 533–18 546.

Betts, A.K. and Ball, J.H. (1993) FIFE-1987 compacted surface data sets. Available on floppy disk from the authors at Atmospheric Research, Pittsford, VT 05763.

Dickinson, R.E. (1983) Land surface processes and climate—Surface albedos and energy balance. In *Theory of Climate, Advances in Geophysics*, edited by B. Saltzman, **25**, 305–353. New York: Academic Press.

Jarvis, P.G. and McNaughton, K.G. (1986) Stomatal control of transpiration: Scaling up from leaf to region. *Advances in Ecological Research*, **15**, 1–49.

Hack, J.J., Boville, B.A.,Kiehl, J.T., Rasch, P.J. and Williamson, D.L. (1994) Climate statistics from the NCAR Community Climate Model (CCM2). *Journal of Geophysical Research*, **99**, 20 785–20 813.

Hahmann, A.N., Ward, D.M. and Dickinson, R.E. (1995) Surface land temperature and radiative response of the NCAR CCM2/Biosphere-Atmosphere Transfer Scheme to modifications in the optical properties of clouds. *Journal of Geophysical Research*, **100**, 23239–23252.

Henderson-Sellers, A., Dickinson, R.E., Durbidge, T.B., Kennedy, P.J., McGuffie, K. and Pitman, A.J. (1993) Tropical deforestation: Modeling local- to regional-scale climate change. *Journal of Geophysical Research*, **98**, 7 289–7 315.

Horel, J.D., Hahmann, A.N. and Geisler, J.E. (1989) An investigation of the annual cycle of convective activity over the tropical Americas. *Journal of Climate*, **2**, 1 388–1 403.

Kuo, H.L. (1974) Further studies of the parameterization of the influence of cumulus convection on large scale flow. *Journal of Atmospheric Science*, **31**, 1 232–1 240.

Legates, D.R. and Willmott, C.J. (1989) Mean seasonal and spatial variability in gage-corrected, global precipitation, *International Journal of Climatology*, **10**, 111–127.

McGuffie, K., Henderson-Sellers, A., Zhang, H., Durbidge, T.B. and Pitman, A. (1995) Global climate sensitivity to tropical deforestation. *Global and Planetary Change*, **10**, 97–128.

Nobre, C.A., Sellers, P.J. and Shukla, J. (1991) Amazonian deforestation and regional climate change. *Journal of Climate*, **4**, 957–988.

Tiedtke, M. (1984) The sensitivity of the time mean large-scale flow to cumulus convection in the ECMWF model. Report of the Workshop on Convection in Large Scale Numerical Models, ECMWF, Shinfield Park, Reading, UK. 28 November–1 December 1983.

CHAPTER 11

DEPENDENCE OF MODEL 'CLIMATE SENSITIVITY' ON CONVECTIVE PARAMETERISATION

Bryant J. McAvaney and Robert A. Colman

11.1 Introduction

The Intergovermental Panel on Climate Change (IPCC) defines 'climate sensitivity' as the increase in globally averaged surface temperature that accompanies the increase in radiative forcing due to a doubling of carbon dioxide (IPCC 1990, 1992, 1995). This one crucial property of the climate system cannot be rigorously determined from observations. Comprehensive three-dimensional climate models remain the best method of determining the climate sensitivity—typically there is a spread of between 1.5°C and 4.5°C in the results from such models. Detailed understanding of the reasons for this spread has proved elusive since the necessary systematic intercomparison experiments have not yet been conducted. The strong sensitivity to clouds, however, has long been recognised as the major candidate for differences between models (Cess et al. 1990) and it has been suggested that similar model sensitivity might occur with other model parameterizations, including that of convection.

Many aspects of a model simulation of climate are sensitive to the parameterization of convection. However, as with many aspects of physical parameterization schemes in AGCMs, there does not exist a general consensus as to which particular convection scheme is the most realistic (see, e.g. Albrecht et al. 1986). What is clear though, is that substantial changes in AGCM simulated climate may arise using different parameterizations (Tiedtke 1984). For example, significant changes have been noted between moist convective adjustment (MCA) and penetrative schemes in the representation of the present climate (e.g. Hart et al. 1990; Tiedtke 1984; Cunnington & Mitchell 1990; Slingo et al. 1994). Much of the analysis in the past, however, has been restricted to changes in parameterizations of this nature (i.e. from MCA to penetrative schemes), whereas many modellers are presently faced with choices between differing penetrative schemes (and especially the form of closure to be used). Also much less consideration has been given to the issue of changes in the climate sensitivity of a model due to changes in convective parameterization. With a doubling in atmospheric CO_2, a change in the model sensitivity with a change in convective parameterization has indeed been noted in some studies (e.g. Washington & Meehl 1984; Schlesinger & Mitchell 1987; Cunnington & Mitchell 1990). Difficulties arise, however, in the interpretation of many of these sensitivity experiments, since different experiments have generally been performed with different models (e.g. Schlesinger & Mitchell 1987) or the CO_2 climate change has involved idealised sea surface temperature changes (e.g. Cunnington & Mitchell 1990).

Our proposal to MECCA called for the Bureau of Meteorology Research Centre (BMRC) AGCM to perform two $2 \times CO_2$ equilibrium response experiments with two different convective parameterization schemes. These experiment would form an initial step towards a systematic intercomparison of the impact of convective parameterization on climate sensitivity. The proposal was part of an ambitious plan to assess the impact of various physical parameterizations and both horizontal and vertical resolution upon the climate sensitivity of the BMRC AGCM. Lack of sufficient computing resources and research staff meant that the full experimental programme was never completed. Ideally the experiments would have been extended to include other models run under controlled experimental conditions. The choice of convective parameterizations was restricted to penetrative schemes, specifically those due to Kuo (1974) and Tiedtke (1989). The shallow convection schemes (and the associated controlling parameters) were originally designed to be kept identical, but it was subsequently found necessary to adjust parameters because of strong feedback processes in the atmospheric boundary layer which produced large changes in the control climate not immediately related to the choice of penetrative convection scheme.

The mass flux convective parameterization scheme was expected to produce greater convective heating which penetrates higher into the troposphere than with the Kuo scheme. It was assumed that the subsequent modification to the diagnostic cloud field would result in a change in the strength of the cloud height and cloud amount feedbacks in the model. Colman et al. (1995) discuss in greater detail the mechanisms producing the changes in the control climate due to the change in the convective parameterization, while Colman and McAvaney (1995) discuss some of the underlying physical causes of the dependence of the BMRC AGCM climate sensitivity on the convective parameterization. McAvaney et al. (1995) presented details of a statistical evaluation of the experiments.

We chose to become involved in the MECCA programme for two basic reasons: firstly, at the time the sensitivity experiments were proposed there was insufficient computer time available on the supercomputers available to us and, secondly, the proposed sensitivity experiments seemed to 'fit' well with the objectives of MECCA.

11.2 Model description

The model used was the BMRC Atmospheric General Circulation Model (AGCM) originally described by Bourke et al. (1977) and McAvaney et al. (1978) but with a mixed-layer ocean and a simplified thermodynamic sea-ice model.

The model used the spectral formulation with a horizontal resolution truncation set at rhomboidal wave 21 and with nine unequally spaced vertical levels corresponding to sigma values of 0.991, 0.925, 0.811, 0.664, 0.500, 0.336, 0.189, 0.074, and 0.009. The model had a surface boundary layer parametrization based on the formulation of Louis (1979).

Over oceans evaporation was enhanced using the approximation of Miller et al. (1992) for the exchange coefficient. Vertical diffusion followed the stability dependant form of Louis (1979). Shallow convection followed the treatment of Tiedtke (1984) and gravity wave drag was determined using the formulation of Palmer et al. (1986). The radiation scheme used was a modified version of the Fels–Schwarzkopf scheme developed at GFDL (Fels & Schwarzkopf 1975, 1981). A conserving vertical finite difference scheme was also implemented as described in McAvaney and Colman (1993). A description of the mixed-layer ocean and thermodynamic sea-ice models is given by Colman et al. (1992).

In the then (1990) standard version of the model, convection was treated by a modified Kuo (1974) scheme, with moistening in the convective layer following the form specified by Anthes (1977). A general description of an even earlier version of the model and an analysis of the effects of revised physical parameterizations are presented by Hart et al. (1988, 1990). Further details and more recent extensions can be found in McAvaney and Colman (1993). An atlas showing the results of a non-diurnal simulation of the present climate by the AGCM with observed sea surface temperatures is given in McAvaney et al. (1991).

The penetrative convection scheme used in the first experiment (hereafter called 'K21') was a modified version of the Kuo (1974) formulation. Full details of this parameterization are given by Hart et al. (1990) and McAvaney et al. (1991), but it is briefly described here. The scheme initiates convection given a conditionally unstable stratification, a positive (vertically integrated) moisture convergence in the convective layers, and a mean relative humidity in these layers in excess of 80%. The base of the modelled convective 'cloud' is the first level above the planetary boundary layer (i.e. at or above the second sigma level), which is conditionally unstable and has a relative humidity above 90%. The highest level permitted for the convection to penetrate is the 0.189 level. The moisture convergence is partitioned into that fraction which is retained by the atmosphere and that fraction which is precipitated out, and follows the formulation of Anthes (1977). Large-scale precipitation is deemed to occur if relative humidity exceeds 100%. No evaporation of falling rain occurs.

The penetrative convection scheme used in the second experiment (hereafter called 'M21') is a version of the so called mass flux scheme formulated by Tiedtke (1989). In this formulation convection is assumed to occur in an ensemble of 'clouds' with specified bulk equations for vertical transports (in up draughts and down draughts) of mass, heat and moisture. The flux of mass into the cloud base is controlled by the large-scale moisture convergence in the layer between the surface and the cloud base. That is, a condition of moisture balance is specified in this layer whereby convergence by surface evaporation, turbulence or large-scale advection is equated to cloud base mass flux. 'Organised' entrainment is deemed to occur into the cloud sides in an amount proportional to the large scale moisture convergence. Turbulent entrainment and detrainment also occur in the lower and upper parts of the cloud ensemble respectively. The shallow convection component of the Tiedtke (1989) parameterization was not permitted to operate.

Shallow convection is parametrized following Tiedtke (1984). In this formulation a constant is added to the vertical diffusion coefficient between the second and third sigma levels (i.e. between $\delta = 0.926$ and $\delta = 0.811$) and also between the third and fourth sigma levels (i.e. between $\delta = 0.811$ and $\delta = 0.664$). Shallow convection is triggered where the lower layers are conditionally unstable with a relative humidity greater than 75%. This parametrization seeks to represent the vertical flux of heat and moisture by non precipitating convective clouds lying below a trade wind inversion (Tiedtke 1984, 1988). The diffusion constants were $10 \text{ m}^2 \text{ s}^{-1}$ and $2 \text{ m}^2 \text{ s}^{-1}$ respectively in the K21 experiment and $4 \text{ m}^2 \text{ s}^{-1}$ and $1 \text{ m}^2 \text{ s}^{-1}$ respectively in M21. These different strengths of the shallow convection were specified through a prior process of determining a 'best' control climate with each of the convective schemes, and in this sense may be thought of as a 'tuning' of each scheme to attempt to achieve its optimum performance in the model. This has the advantage of avoiding a comparison between the penetrative convection schemes in which one is producing a highly unrealistic control climate (and is thus operating in a way which would not be ordinarily used in the AGCM).

Since these experiments were completed, the 'standard' version of the BMRC AGCM now uses the full Tiedtke (1989) mass flux convection scheme at higher vertical (17 levels) and horizontal resolution (rhomboidal wave number 31).

11.3 Computer and data storage utilisation

The complete Kuo experiment (K1) had already been run on the centrally operated Cray XMP/14 (under UNICOS 4) of the Bureau of Meteorology and the daily 'history-of-state' stored and catalogued on magnetic tapes in IBM binary format via the Bureau's FACOM M380. Some pre-processing of the raw 'history-of-state' was done so that monthly, seasonal, annual and decadal averages were available. Large volumes of daily data (across several tapes) were very slow to be retrieved from the tape archive and hence minimal analysis of daily data was done. The data could only be retrieved, read and processed via the FACOM. (No suitable workstation facilities existed in BMRC at that time).

Most of the setting up, testing and debugging of the mass flux experiment was done on the Bureau's Cray. It was important that Castle, as the target MECCA machine (a Cray YMP/2), had very similar facilities.

No attempt was made to multi-process the model run and after some experimentation, it was found that the best 'throughput' on Castle for the total experiment would be achieved by first running the 'control' ($1 \times CO_2$) job and then follow with the $2 \times CO_2$ job. (The experiment K1 had already been run on the Bureau's Cray.)

11.4 Remote operation

The restriction of a relatively slow telecommunications link across the Pacific meant that all model runs submitted to Castle had to be thoroughly checked out on a much smaller Cray.

Differences in the levels of the respective operating systems and the considerable differences in basic approaches to data handling between the two institutions created many portability problems which took some time to resolve. A single on-site visit, with subsequent direct access to the NCAR computer consultants, accomplished more in a few days than had been possible over many weeks. However, reliable and automatic procedures for job submission, monitoring and data archiving on Castle were eventually achieved. Many of the procedures developed for remote running have since become 'standard' practice for BMRC AGCM runs.

An insufficient allocation of total computer time on Castle, coupled with a longer than anticipated 'shakedown' of the processing system, meant that a great deal of the experiment had to be completed on the local Cray in the Bureau of Meteorology.

11.5 The experiments

In total two full $1 \times CO_2$ ('control') and equilibrium $2 \times CO_2$ experiments were performed (K1 and M1). In each of these, the model was first run for 11 years with observed sea surface temperatures (SSTs) from Alexander and Mobley (1976). Monthly mean surface heat fluxes were then calculated for the last 10 years of this period to determine so called '*Q-fluxes*' for the slab ocean (which were thus different in each of the experiments). The depth of the ocean was specified as 50 m in all experiments. Still using observed SSTs, the model was integrated for a further seven years to iteratively determine the sea-ice bottom fluxes required to maintain an observed sea-ice area distribution with the thermodynamic sea-ice model (details of this iterative process are given by Colman et al. 1992). The slab ocean was then introduced and the model run to equilibrium both for single and doubled atmospheric CO_2 cases. The effect of the Q-fluxes was to ensure that the slab ocean temperatures remained very close to observed SSTs in each of the control experiments. Total integration for each experiment was approximately 75 years of model time. All results presented in this paper are calculated from the final 15 years of each experiment.

11.6 Results

11.6.1 Changes to control climate

In the control climate, the mass flux scheme produced a warmer, moister troposphere, with substantially more high cloud than the Kuo scheme. The precipitation distributions both agree reasonably with observations overall, although the mass flux scheme gave improvements in some regions.

The shallow convection was also varied in strength. This latter change is found to produce large increases in upwards subtropical surface fluxes. The strength and consistency of the sign of these changes are consistent with a positive feedback involving the low cloud, vertical diffusion and radiation in the model, and can produce somewhat unexpected results. In the deep tropics, both the reduction in shallow convection and the change to the mass flux convective parameterization result in decreased heat flux from the surface in the western Pacific. Both factors bring the model net heat flux into better agreement with observations. As expected, the change in penetrative convection is found to have the greatest impact in this region. The decrease in evaporation using the mass flux scheme is shown to be consistent with higher level convective entrainment, reduced low level convergence and reduced low level winds in the Tiedtke scheme compared with the Kuo scheme. Variations in surface radiation due to a decrease in the low cloud fraction reinforce the latent heat flux changes.

The impacts found indicate strong coupling between the parametrizations of surface fluxes and both types of convection, and were sufficient to perturb the implied poleward oceanic transport by around $1\,015$ W (Colman et al. 1995). This cautions users of GCMs to consider surface fluxes, and not just precipitation or atmospheric heating rates, when selecting or modifying convection schemes.

11.6.2 Changes to $2 \times CO_2$ climate

Under a doubling of atmospheric CO_2, the equilibrium responses in both experiments were extremely similar in surface and tropospheric temperatures and humidity changes. The model response is at the low end of the scale of simulated climate change, consistent with a strong negative feedback found due to clouds. This feedback is similar to that found in earlier fixed season experiments. It appears to be insensitive to differences in cloud cover simulated in the control climates of the present experiments, and some differences in cloud height and amount changes. The global average change $(2 \times CO_2 - 1 \times CO_2)$ in surface temperature for each season is shown for both experiments in Figure 11.1. The remarkable similarity between the change for each experiment and the similarity of the change in each season is very marked. In Figure 11.2 we show the percentage change in globally averaged precipitation for each season for each experiment. The differences between the experiments is more marked in this measure of climate change as is the variation in season. These differences are further accentuated when continental and regional scales are considered (Colman & McAvaney 1995).

11.6.3 Statistics

Since the differences between the climate sensitivity for each experiment were comparatively small, we chose to use the statistical package COMPARE discussed by Wigley and Santer (1990) and Santer and Wigley (1990), since this gave us a ready means of assessing the significance in changes in the mean and changes in the temporal and spatial variance. We chose to investigate both the global mean surface temperature and the precipitation and concentrated on the $2 \times CO_2$ to $1 \times CO_2$ differences for each experiment separately. For purely logistical reasons at the time (insufficient storage space on the small workstation available at the time) we did not (unfortunately) apply the technique to either, the difference between the two control $(1 \times CO_2)$ experiments, or the difference in the changes themselves. Any future experiments of this type should perform this additional cross-analysis.

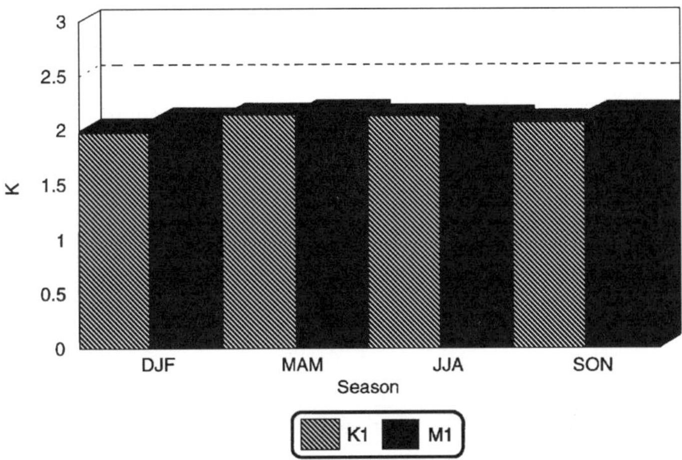

Figure 11.1 The global average change $(2 \times CO_2 - 1 \times CO_2)$ in surface temperature for each season is shown for both experiments.

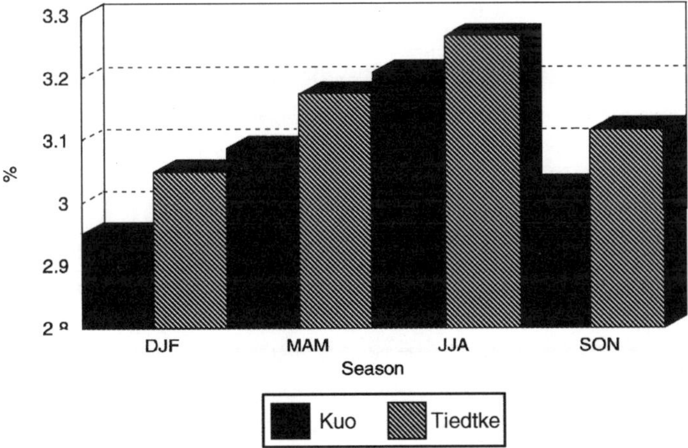

Figure 11.2 The percentage change in globally averaged precipitation for each season for each experiment.

The statistical analysis revealed that there was a high signal-to-noise ratio in the change in global surface temperature in moving from the $1 \times CO_2$ climate to the warmer $2 \times CO_2$ climate for each experiment but the difference between the experiments was small. For precipitation, the signal-to-noise ratio was much lower but the contrast between the experiments was greater.

There were significant changes in both the temporal and spatial variability of precipitation changes in both experiments with some differences apparent between the two experiments. The temporal variability of surface temperature did not change significantly in the warmer world but there were significant changes in spatial variability.

At the regional level the changes are more marked with considerable differences at times between experiments but the significance level is often much lower.

11.7 Discussion

11.7.1 Interaction with the MECCA Analysis Team

All of the other atmospheric modelling exercises in MECCA used versions of the NCAR CCM with all its attendant infrastructure and 'on-site' support. Hence it was quite easy for the expectation that 'all models do it this way' to emerge among MAT. Since there had been little, if any prior, experience in model intercomparison exercises by either the modellers in MECCA or members of MAT several iterations were necessary before a consensus arose as to the minimal requirements to be expected from model output.

As mentioned earlier, the total sensitivity experiments involved two complete $2 \times CO_2$ equilibrium runs done on different computers at different times and with data stored in two different places. Since only the 'mass flux' (M1) experiment was performed using MECCA computing facilities, and MAT was only interested in a single experiment, only the history of state for this experiment was made available outside BMRC.

Even so, since the experiment M1 was eventually completed using alternative computing resources and data had to be collected and collated at one location.

11.7.2 Data handling: Problems and solutions

In common with most climate modelling groups, the BMRC AGCM has its own internal format for its history of state and averaged files. Our group has a large investment in infrastructure supporting this data format. However, the format is extremely *unfriendly*, overly rigid and machine dependent. Because of BMRC involvement with the AMIP (Gates 1992), we chose to convert our files to the self descriptive, machine independent file structure DRS, which was well supported across many computing platforms by PCMDI at LLNL.

Ease of processing of model history tapes required that all data be stored at the one site. While storage facilities at BMRC were less than optimal, the very slow access to Castle meant that all BMRC AGCM history of state data had to be copied from the 'mass store' to tapes at NCAR and brought back to Australia. However, since some of the experiment running was completed in Australia, reverse data flow also had to occur for eventual access by MAT. Since, in BMRC, the AGCM model developers also ran the MECCA experiments, it was relatively straightforward (but not without some pain) to modify the BMRC AGCM processor to produce a (reduced) history of state in a more *friendly* self descriptive and machine independent format. A version of the BMRC processor with the DRS option installed was made available on Castle for use by MAT. The requirement that substantial resources from the BMRC would have to be directed towards effectively duplicating our modelling infrastructure at a remote site was not fully appreciated before the project began and this remains a major impediment to further work.

11.7.3   Would we do it again?

Having been involved with a number of international modelling intercomparison studies, we had thought that our experience with MECCA would be straightforward. Such was not the case. While we, as modellers and experimenters, had a clear idea of what we wanted to accomplish, we did not fully appreciate the extra resources that would be required to meet the expectations of the sponsors and MAT.

Any future attempt at a similar exercise involving more than one modelling group should take note of the following:

1. agreed protocols for model data transfer;
2. agreed standards for data storage and access; and
3. agreed standardisation of common experimental conditions.

Also reasonable 'model visualisation' tools utilising the advantages that flow from points 1 and 2 should be centrally supported and not left to individual modellers to provide.

As the international climate modelling community moves towards the CLIVAR programme and its associated numerical experimentation, it is to be hoped that the lessons from the MECCA and AMIP approaches are well learned.

We consider that successful systematic model intercomparison studies require the support of a central co-ordinating agency which is well resourced and is prepared to take on the difficult and onerous task of providing suitable software tools to support *user-friendly* storage, data analysis and visualisation packages for use by both model developers and model diagnosticians.

11.8 Summary

It has now become much clearer that changes in the formulation of the closure condition within an individual convective parameterization can have large impact upon simulated climate. From this perspective it would be most interesting to repeat our previous experiment with the mass flux scheme but change from a moisture convergence closure to a stability dependent closure scheme.

We found that the climate sensitivity of the BMRC AGCM was not very sensitive to the form of convective parameterization used. However, since the climate sensitivity is itself strongly controlled by the feedback due to clouds (and this cloud feedback is negative in the BMRC model), our experiment has served to highlight (again) the importance of clouds in controlling the response of a AGCM to a change in CO_2 radiative forcing.

Our MECCA experiment proved to be an interesting introduction into the world of privately sponsored research.

References

Albrecht, B.A., Ramanathan, V. and Boville, B.A. (1986) The effects of cumulus moisture transports on the simulation of climate with a GCM. *Journal of Atmospheric Science*, **43**, 2 443–2 462.

Alexander, R.C. and Mobley, R.L. (1976) Monthly average sea-surface temperatures and ice-pack limits on a 1=F8 global grid. *Monthly Weather Review*, **104**, 143–148.

Anthes, R.A. (1977) A cumulus parameterization scheme utilizing a one-dimensional cloud model. *Monthly Weather Review*, **105**, 270–286.

Bourke, W. (1988) Spectral methods in global climate and weather prediction models. In *The Physical Basis of Climate Modelling*, edited by M.E. Schlesinger, pp. 375–431, NATO ASI Series, D. Reidel Publishing Company.

Cess, R.D., Potter, G.L., Blanchet, J.P., Boer, G.J., Ghan, S.J., Hansen, J., Kiehl, J.T., Le Treut, H., Li, Z.-X., Liang, X.-Z., McAvaney, B.J., Meleshko, V.P., Mitchell, J.F.B., Morcrette, J.-J., Randall, D.A., Rikus, L.J., Roeckner, E., Schlese, U., Sheinin, D.A., Slingo, A., Sokolov, A.P., Taylor, K.E., Washington, W.M., Wetherald, R.T. and Yagaii, I. (1990) Intercomparison and interpretation of cloud-climate feedback processes in nineteen atmospheric general circulation models. *Journal of Geophysical Research*, **95**, 16 601–16 615.

Colman, R.A., McAvaney, B.J., Fraser, J.R. and Dahni, R.R. (1992) Mixed layer ocean and thermodynamic sea-ice models in the BMRC GCM. BMRC Research Report No. 30.

Colman, R.A., McAvaney, B.J. and Puri, K.K. (1994) Sensitivity of the BMRC climate model to changes in convective parameterization. In *BMRC Research Report* No. 46, 137–142.

Colman, R.A. and McAvaney, B.J. (1995) The sensitivity of the climate response of a general circulation model to changes in convective parametrization and horizontal resolution. *Journal of Geophysical Research*, **100**, 3155–3172.

Cunnington, W.M. and Mitchell, J.F.B. (1990) On the dependence of climate sensitivity on convective parametrization. *Climate Dynamics*, **4**, 85–91.

Fels, S.B. and Schwarzkopf, M.D. (1975) The simplified exchange approximation: A new method for radiative transfer calculations. *Journal of Atmospheric Science*, **32**, 1 475–1 488.

Fels, S.B. and Schwarzkopf, M.D. (1981) An efficient, accurate algorithm for calculating CO_2 15 = E6m band cooling rates. *Journal of Geophysical Research*, **86**, 1 205–1 232.

Hart, T.L., Bourke, W., McAvaney, B.J., Forgan, B.W. and McGregor J.L. (1988) Atmospheric general circulation simulations with the BMRC global spectral model: The impact of revised physical parameterisations. BMRC Research Report No. 12.

Hart, T.L., Bourke, W., McAvaney, B.J., Forgan, B.W. and McGregor J.L. (1990) Atmospheric general circulation simulations with the BMRC global spectral model: The impact of revised physical parameterisations. *Journal of Climate*, **3**, 436–459.

Houghton, J.T., Jenkins, G.J. and Ephraums, J.J. (1990) *Climate Change. The IPCC Scientific Assessment.* United Kingdom: Cambridge University Press.

IPCC (1990) *Climate change. The IPCC Scientific Assessment.* Edited by J.T. Houghton, G.J. Jenkins and J.J. Ephraums. United Kingdom: Cambridge University Press.

IPCC (1992) *Climate change 1992. The supplementary Report to the IPCC Scientific Assessment.* Edited by J.T. Houghton, B. Callander and S.K. Varney. United Kingdom: Cambridge University Press.

Katz, R.W. (1992) Role of statistics in the validation of general circulation models. *Climate Research*, **2**, 35–45.

Kuo, H.L. (1974) Further studies of the parameterization of the influence of cumulus convection on large scale flow. *Journal of Atmospheric Science*, **31**, 1232–1240.

Louis, J.-F. (1979) A parametric study of vertical eddy fluxes in the atmosphere. *Boundary Layer Meteorology*, **17**, 187–202.

McAvaney, B.J., Bourke, W. and Puri, K. (1978) A global spectral model for simulation of the general circulation. *Journal of Atmospheric Science*, **35**, 1 557–1 583.

McAvaney, B.J., Fraser, J.R. Hart, T.L, Rikus, L.J., Bourke, W.P., Naughton, M.J. and Mullenmeister, P. (1991) Circulation statistics from a non-diurnal seasonal simulation with the BMRC Atmospheric GCM: R21L9. BMRC Research Report No. 29.

McAvaney, B.J. and Colman, R.A. (1993) The AMIP Experiment: The BMRC AGCM Configuration. BMRC Research Report No. 38.

McAvaney, B.J., Dahni, R.R., Colman, R.A. and Fraser, J.R. (1995) The dependence of the climate sensitivity on convective parametrization: Statistical evaluation. *Global and Planetary Change*, **10**, 181–200.

Miller, M.J., Beljaars, A.C.M. and Palmer, T.N. (1992) The sensitivity of the ECMWF model to the parameterization of evaporation from the tropical ocean. *Journal of Climate*, **5**, 418–434.

Mitchell, J.F.B., Manabe, S., Meleshko, V. and Tokioka, T. (1990) Equilibrium climate change—and its implications for the future. In *Climate Change. The IPCC Scientific Assessment*, edited by J.T. Houghton, G.J. Jenkins and J.J. Ephraums, pp. 131–172. United Kingdom: Cambridge University Press.

Palmer, T.N., Shutts, G.J. and Swinbank, R. (1986) Alleviation of a systematic bias in general circulation and numerical weather prediction models through an orographic gravity wave drag parameterization. *Quarterly Journal of the Royal Meteorological Society*, **112**, 1 001–1 039.

Preisendorfer, R.W. and Barnett, T.P. (1983) Numerical model-reality intercomparison tests using small sample statistics. *Journal of Atmospheric Science*, **40**, 1 884–1 896.

Santer, B.D. and Wigley, T.M.L. (1990) Regional validation of means, variances and spatial patterns in GCM control runs. *Journal of Geophysical Research*, **95**, 829–850.

Santer, B.D., Wigley, T.M.L., Jones, P.D. and Schlesinger, M.E. (1991) Multivariate methods for the detection of greenhouse gas induced climate change. In *Greenhouse-Gas-Induced Climate Change: A Critical Appraisal of Simulations and Observations*, edited by M.E. Schlesinger, Amsterdam: Elsevier Science Publishers BV.

Slingo, J.M. (1987) The development and verification of a cloud prediction scheme for the ECMWF model. *Quarterly Journal of the Royal Meteorological Society*, **113**, 899–927.

Slingo, J.M., Blackburn, M., Betts, A., Brugge, R., Hodges, K., Hoskins, B., Miller, M., Steenman-Clark, L. and Thurburn, J. (1994) Mean climate and transience in the tropics of the UGAMP GCM: Sensitivity to convective parametrization. *Quarterly Journal of the Royal Meteorological Society*, **120**, 881–922.

Tiedtke, M. (1984) The sensitivity of the time mean large-scale flow to cumulus convection in the ECMWF model. Report of Workshop on Convection in Large Scale Numerical Models, ECMWF, Shinfield Park, Reading, United Kingdom. 28 November–1 December 1983.

Tiedtke, M. (1988) Parametrization of cumulus convection in large-scale models. In *Physically based modelling and simulation of climate and climate change—Part I*, edited by M.E. Schlesinger, pp. 375–431. Dordrecht: Kluwer Academic Publishers.

Tiedtke, M. (1989) A comprehensive mass flux scheme for cumulus parameterization in large-scale models. *Monthly Weather Review*, **117**, 1 779–1 799.

Wigley, T.M.L. and Santer, B.D. (1990) Statistical comparison of spatial fields in model validation, perturbation and predicability experiments. *Journal of Geophysical Research*, **95**, 851–865.

Zwiers, F.W. (1990) The effect of serial correlation on statistical inferences made with resampling procedures. *Journal of Climate*, **3**, 1 452–1 461.

CHAPTER 12

UNCERTAINTIES IN GCM EVALUATIONS OF TROPICAL DEFORESTATION: A COMPARISION OF TWO MODEL SIMULATIONS

Huqiang Zhang, Ann Henderson-Sellers, Bryant J. McAvaney and Andrew J. Pitman

12.1 Introduction: Simulating tropical deforestation

Tropical deforestation is of public and political concern because the continuing reduction of tropical rainforest extent (e.g. Moran et al. 1994) is believed to be likely to result in local climate changes (Salati & Vose 1984) and a huge reduction in biodiversity as the favourable climate conditions under which the complex biosphere can survive deteriorate. Global Climate Models (GCMs) are one of the tools being employed to assess the impacts of tropical deforestation on the local and regional climates and, ultimately, on biodiversity and sustainability. However, uncertainties among the GCM simulations of current climate and future climate changes have been seen in various climate modelling intercomparison projects including MECCA (Henderson-Sellers et al. 1995a) and the Atmospheric Model Intercomparison Project (AMIP) (Gates 1992). In addition, results from the Project for Intercomparison of Landsurface Parameterization Schemes (PILPS) (Henderson-Sellers et al. 1993a, 1995b) show that sensitivities of many land surface schemes, which are being coupled into current GCMs, differ even with the same atmospheric forcing.

Recent GCM simulations pertaining to tropical deforestation reflect the uncertainties which may come from the GCMs, from the land surface schemes and from the different imposed disturbances. In the published GCM simulations of tropical deforestation in the past decade, predicted changes in the local precipitation over the Amazon Basin have ranged from +33 to –640 mm yr^{-1} and changes in surface temperature from –0.1 to +3.0 K (Table 12.1). Clearly, more attention needs to be paid to try to understand the underlying reasons for the differences seen in such GCM results before efforts to try to produce more precise predictions of the impacts of tropical deforestation can be successful. The uncertainties in the GCM simulations of the impacts of tropical deforestation on the local, regional and possibly the global climate are investigated in this study by comparing two MECCA-sponsored model simulations.

Table 12.1 Comparison of previous GCM simulations of tropical deforestation over the Amazon Basin

Reference	D & H-S (1988)	L & W (1989)	Nobre et al. (1991)	L & R (1992)	D & K (1992)	Dirmeyer (1992)	H-S et al. (1993b)	P & L (1994)	Sud et al. (1996)	McG et al. (1995)	Zhang et al. (1996)	This study
	CCM0B	UKMO	NMC	UKMO	CCM1-OZ	NMC	CCM1-OZ	LMD	GLA	CCM1-OZ	CCM1-OZ	BMRC
GCM	(4.5×7.5)	(2.5×3.7)	(1.8×2.8)	(2.5×3.7)	(4.5×7.5)	(4.5×7.5)	(4.5×7.5)	(2.0×5.6)	(4.0×5.0)	(4.5×7.5)	(4.5×7.5)	(2.0×3.0)
Ocean	Fixed SST	Fixed SST	Fixed SST	Fixed SST	Slab Ocean	Fixed SST	Slab Ocean	Fixed SST	Fixed SST	Slab Ocean	Slab Ocean	Slab Ocean
Surface	BATS	Canopy	SiB	Canopy	BATS1e	SSiB	BATS1e	SECHIBA	SSiB	BATS1e	BATS1e	BEST
Integration	3/1 year	3/3 year	1/1 year	3/3 year	3/3 year	14/14 mth	6/6 year	11/11 year	3/3 year	14/6 year	25/11 year	10/6 year
Albedo Changes	0.12 to 0.19	0.136 to 0.188	0.12–0.14 to 0.16–0.24	0.136 to 0.188	0.12 to 0.19	increment = 0.03	0.12 to 0.19	0.135 to 0.216	0.097 to 0.146	0.12 to 0.19	0.12 to 0.19	0.15 to 0.205
Roughness	2.0 to 0.05	0.79 to 0.04	2.65 to 0.08	0.79 to 0.04	2.0 to 0.05	2.65 to 0.08	2.0 to 0.2	2.3 to 0.06	2.65 to 0.077	2.0 to 0.2	2.0 to 0.2	1.1 to 0.1
ΔT (K)	+3	+2.4	+2	+2	0.6 (soil)	Not given	+0.6	−0.1	+1.3	+0.3	+0.3	+0.9
ΔP (mm)	0	−490	−640	−295	−511	+33	−588	−186	−266	−437	−402	+445
ΔE (mm)	−200	−310	−500	−200	−255	−146	−232	−128	−350	−231	−222	+248
$\Delta E - \Delta P$ (mm)	−200	+180	+140	+95	+256	−179	+356	+58	−84	+205	+180	−197

Note: 1. D and H-S is Dickson and Henderson-Sellers; L & W is Lean and Warrilow; L & R is Lean and Rowntree; D & K is Dickinson and Kennedy; H-S et al. is Henderson-Sellers et al.; P & L is Polcher and Laval; McG et al. is McGuffie et al.

2. ΔT (K), ΔP (mm) and ΔE (mm) are the annual averaged changes over the Amazon Basin. Lengths of integration are represented by control/deforestation.

12.2 Model description

CCM1-OZ comprises a modified version of the NCAR CCM1 with a spatial resolution of R15 (about latitude 4.5°, longitude 7.5°) (Williamson et al. 1987) coupled to a version of the Biosphere–Atmosphere Transfer Scheme (BATS1e) (Dickinson et al. 1993) and including a mixed-layer of slab ocean model and three-layer sea-ice model (Henderson-Sellers et al. 1993b; McGuffie et al. 1995). A standard q-flux correction is employed in the model to correct for the ocean advection of energy and the prescription of a fixed mixed-layer depth of 50 m. This version of CCM1 includes the radiative scheme of Kiehl et al. (1987), the cloud radiative scheme of Slingo (1989) and the convective adjustment scheme of Manabe et al. (1965) with 100% relative humidity as the condition for condensation. The land surface scheme (BATS1e) (Dickinson et al. 1993) incorporates a single vegetation layer, a multiple-layer soil scheme, provision for snow cover on the land surface and it can treat a wide range of different surface types, soil characteristics and vegetation cover. The imposed tropical deforestation is identical to that described in Table 12.1 of Henderson-Sellers et al. (1993b). The major changes include an increase of surface albedo from 12% to 19% in the deforested landscape; surface roughness length reduced from 2.0 m to 0.2 m; and soil colour brightened by two classes and the soil texture coarsened by two classes. Results from a 25-year control experiment and an 11-year deforestation simulation by CCM1-OZ are used (Zhang et al. 1996a, 1996b).

The BMRC GCM employed in this experiment has a spectral rhomboidal truncation with 31 waves (R31): about latitude 3°, longitude 2° resolution in Gaussian grid space. The model uses the Phillips sigma vertical coordinate system with nine levels in the vertical differential scheme. The model framework comprises prognostic equations for temperature, moisture and surface pressure, the ideal gas law and the hydrostatic equations (Hart et al. 1990). In this model version, the convection is parameterized by a version of the mass flux scheme of Tiedtke (1989) and shallow convection parameterization as enhanced vertical diffusion from Tiedtke (1984). The boundary layer parameterization is based on the formulation of Louis (1979). The radiation code used in this version is a modified version of the Fels–Schwarzkopf scheme (Fels & Schwarzkopf 1975). The Bare Essentials for Surface Transfer (BEST) (Pitman et al. 1991) land surface scheme is coupled into the BMRC GCM. BEST has an explicit canopy layer, three soil layers and a prognostic snow mass (Pitman et al. 1991).

The deforestation experiment performed with the BMRC GCM 'removes' the same areas of forest in the Amazon Basin, Southeast Asia and tropical Africa by changing the fractional vegetation cover, surface roughness and soil properties. However, in this deforestation experiment, the increase of surface albedo from 15% to 20.5% is smaller than that in the CCM1-OZ experiment. Surface roughness length is decreased from 1.1 m to 0.1 m. A 10-year control experiment and a 6-year deforestation experiment have been undertaken and analysed in this study. It is hoped that by analysing the BMRC GCM simulation of tropical deforestation and comparing the results with those from the CCM1-OZ (Henderson-Sellers et al. 1993b; McGuffie et al. 1995; Zhang et al. 1996a, 1996b), a more complete understanding of the uncertainties and differences in the simulations will be extended and, hopefully, enhanced.

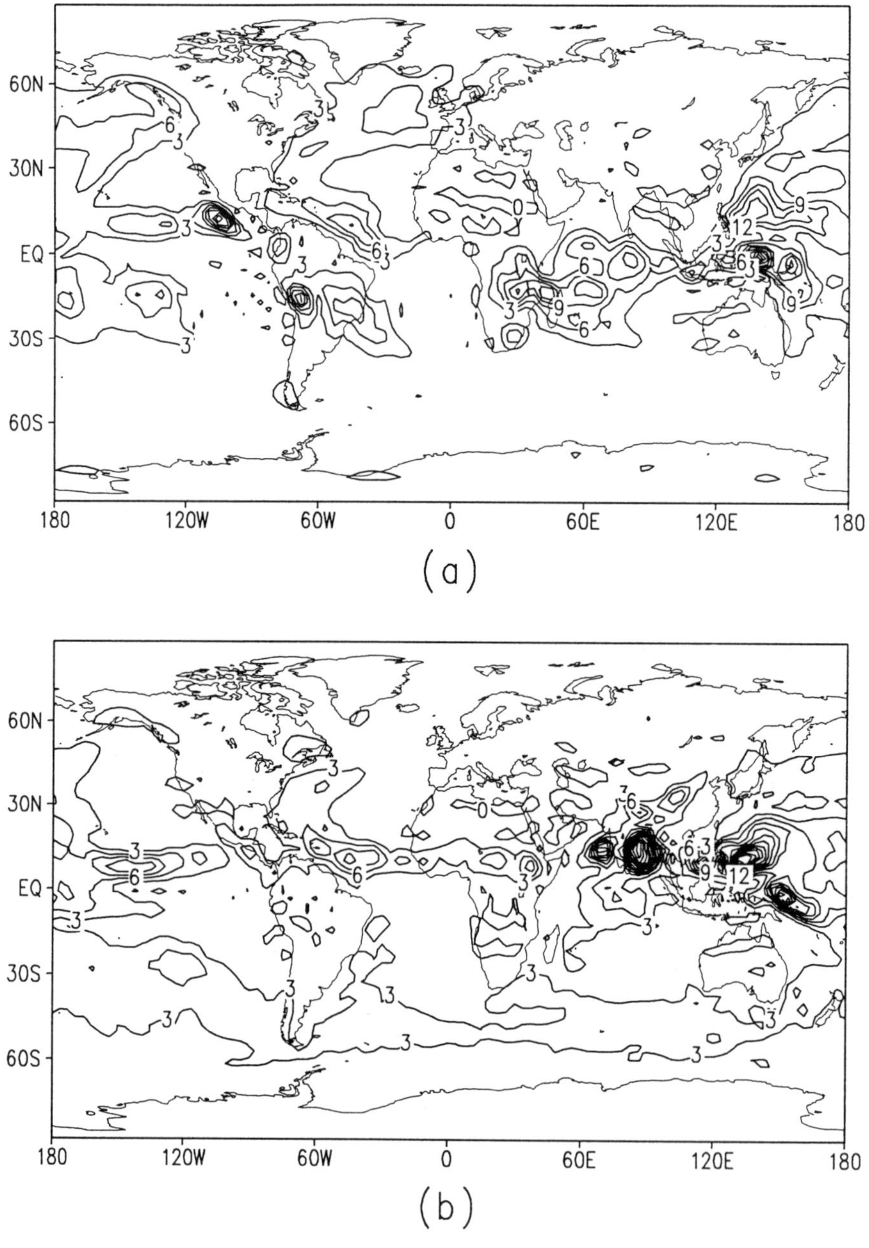

Figure 12.1 10-year averaged global climate simulated in BMRC GCM control experiment:
(a) monthly mean total precipitation in January with a contour interval of 3 mm d^{-1} and (b) as
(a) but in July.

12.3 Evaluation of control simulation

One of the prerequisites for simulating the climatic responses following deforestation is that the climate in the model's control experiment captures the observed features of current climate reasonably well, especially over the tropical rainforest regions. The control climate simulation in CCM1-OZ has been shown in Henderson-Sellers et al. (1993b) and McGuffie et al. (1995) which captures the observed climate features over the tropical rainforest regions satisfactorily. Here BMRC GCM simulated monthly mean precipitation in January and July are shown in Figure 12.1 in order to evaluate its control climate.

Precipitation over the tropics is mainly dominated by the Inter-Tropical Convergence Zone (ITCZ) which has a seasonal migration forced by the maximum incident solar radiation. The rainfall distribution over tropical regions is represented in the control experiment (Figures 12.1(a) and (b)) by the location of the high precipitation zones inside the ITCZ and South Pacific Convergence Zone (SPCZ). If attention is focused on the distribution of the precipitation over the tropical rainforest regions, then some discrepancies between the BMRC simulations and observations over the Amazon Basin and Southeast Asia are found. Here particular attention is paid to the model simulations over the Amazon Basin. As shown in Figure 12.2(a), the maximum rainfall in January (the local wet season) is simulated over the Brazilian Highlands in the south Brazil and near the Andes mountains (due to the topographic forcing), but observations (e.g. Nobre et al. 1991) show that high precipitation occurs over the central Amazon Basin.

The poor simulation of precipitation over Brazil might suggest that water recycling (e.g. Salati 1987; Zhang et al. 1996a) over the central Amazon Basin might not be simulated appropriately in the BMRC GCM. In order to investigate this hypothesis, the distribution of precipitable water (at 1200 GMT) over the Amazon Basin is shown in Figure 12.2(b). Large amounts of precipitable water are found over the central Amazon Basin rather than over southern Brazil where the maximum precipitation is located in the control experiment. As precipitation is controlled by the atmospheric dynamics and the water vapour availability over the particular region, it is reasonable to hypothesize that the dynamical processes which control the high precipitation over the Amazon Basin may not be well represented in this version of BMRC model.

12.4 Regional climatic responses to tropical deforestation

Regionally averaged surface climate changes over the Amazon Basin, Southeast Asia and tropical Africa simulated by CCM1-OZ and BMRC GCM are shown in Table 12.2. In CCM1-OZ, regional averaged precipitation is decreased over all the deforested regions: -402 mm yr^{-1} over the Amazon Basin, -250.9 mm yr^{-1} over Southeast Asia and -62.8 mm yr^{-1} over tropical Africa. Surface net radiation is reduced by between -10 W m^{-2} to -16 W m^{-2} over the deforested regions; a large reduction of surface latent heat flux is seen, especially over the Amazon Basin.

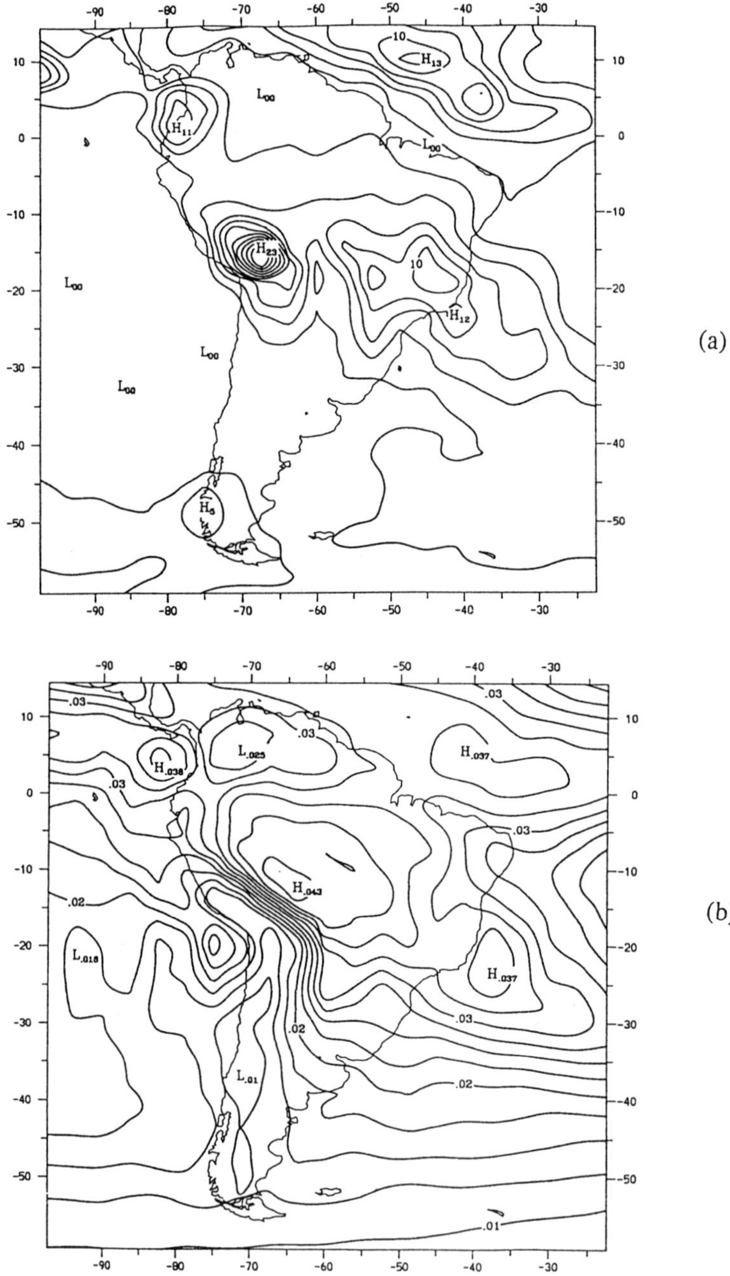

Figure 12.2 (a) BMRC GCM simulated monthly mean precipitation over South America in January with a contour interval of 2 mm d^{-1}, and (b) BMRC GCM control simulation of precipitable water in January with a contour interval of 0.0025 mm.

Table 12.2 Comparison of regional averaged annual climatic changes simulated by the CCM1-OZ and BMRC GCMs following deforestation

	Amazon Basin		Southeast Asia		Tropical Africa	
Variable (units)	CCM1-OZ	BMRC	CCM1-OZ	BMRC	CCM1-OZ	BMRC
Precipitation (mm yr^{-1})	−402.0	+445.3	−250.9	+193.5	−62.8	+109.5
Surface temperature (K)	+0.3	+0.9	−0.2	+0.8	−0.02	+1.3
Net radiation (W m^{-2})	−16.3	−9.8	−11.9	−8.4	−9.7	−4.0
Latent heat (W m^{-2})	−17.8	+21.3	−11.1	+7.3	−6.0	+6.4
Sensible heat (W m^{-2})	+1.6	−27.3	−0.9	−15.1	−3.7	−10.1

In contrast, the BMRC model predicts an increase in annually averaged precipitation over all three deforested regions; +445.3 mm yr^{-1} over the Amazon Basin, +193.5 mm yr^{-1} over Southeast Asia and +109.5 mm yr^{-1} over tropical Africa. Because of the increase in surface albedo, surface net radiation is decreased by −9.8 W m^{-2} over the Amazon Basin, −8.4 W m^{-2} over Southeast Asia and −4.0 W m^{-2} over tropical Africa. Changes in surface energy partition occur in such a way that latent heat flux from the land surface is increased, especially over the Amazon Basin by +21.3 W m^{-2}. Meanwhile sensible heat flux shows large decreases over all three deforested regions by −27.3 W m^{-2} over the Amazon Basin, −15.1 W m^{-2} over Southeast Asia and −10.1 W m^{-2} over tropical Africa. Following the changes in the surface energy distribution, the land surface in all three regions is warmed with simulated surface temperatures increasing by around +1 K.

Clearly, the two GCMs produce very different simulations of the local climatic changes following deforestation: CCM1-OZ has a drier climate with much less evaporation but no real temperature changes, while the BMRC GCM produces much more rainfall and a noticeable warmer surface climate. These very different results, an example of the uncertainty in climate model predictions, are investigated here. As in Zhang et al. (1996a), a statistical procedure is used to establish whether the BMRC GCM predicted changes are differentiable from model noise. In addition, seasonal variations of the predictions are examined and the underlying physical processes responsible for the changes are investigated and compared with those identified in Zhang et al. (1996a, 1996b) for the CCM1-OZ model.

12.4.1 Geographical distribution of surface climate changes in BMRC GCM simulation

Amazon Basin

Figure 12.3 shows the distribution of changes in surface temperature (Ts) and the monthly mean rainfall over South America and the signal-to-noise ratio of these changes in January and July in the BMRC simulation. The shaded areas having a value of the signal-to-noise ratio greater than 2.0 indicate that such changes can be taken to be the result of the imposed deforestation.

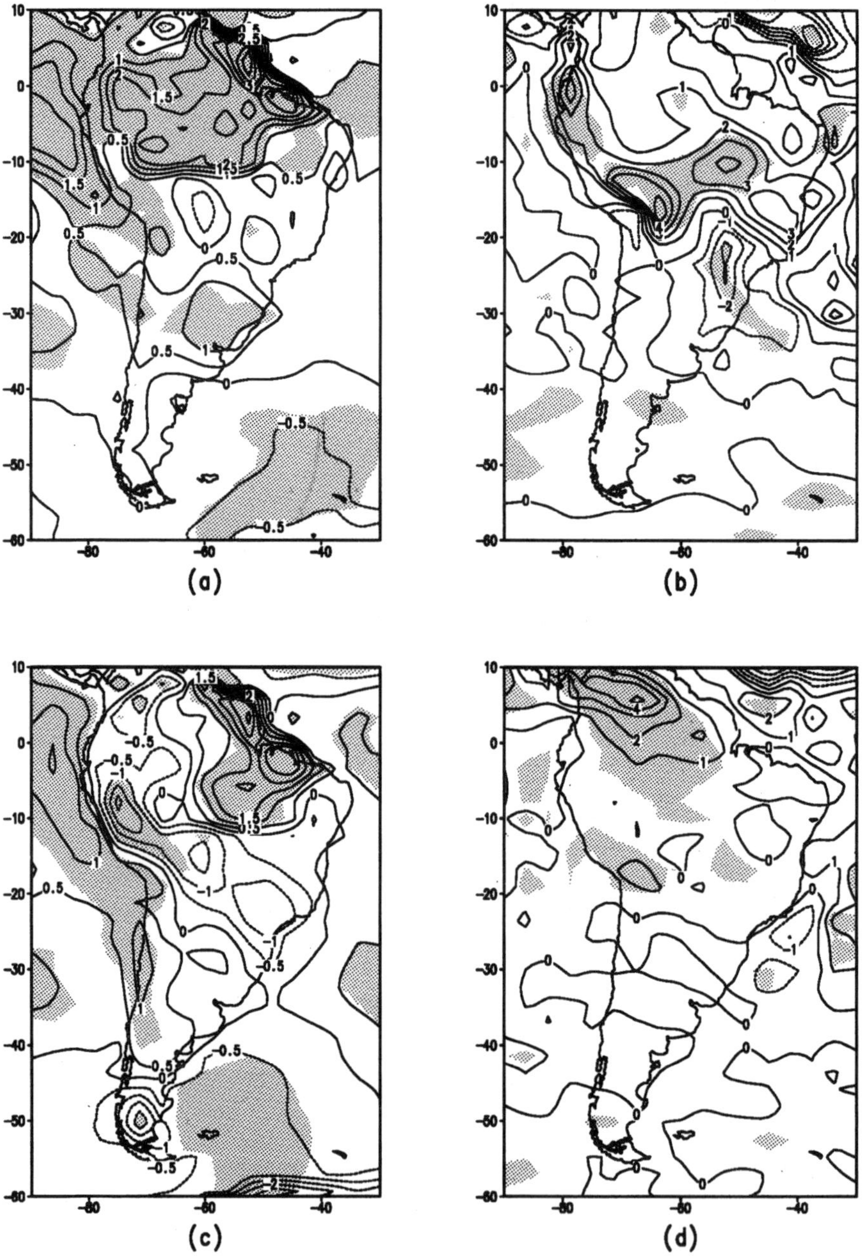

Figure 12.3 Changes of surface temperature and precipitation in January and July over South America in BMRC GCM: (a) changes in surface temperature in January with a contour interval of 0.5 K (shaded areas indicate the signal-to-noise ratio of these changes is greater than 2.0), (b) as (a) but for precipitation with a contour interval of 1 mm d^{-1}, (c) as (a) but in July, and (d) as (b) but in July.

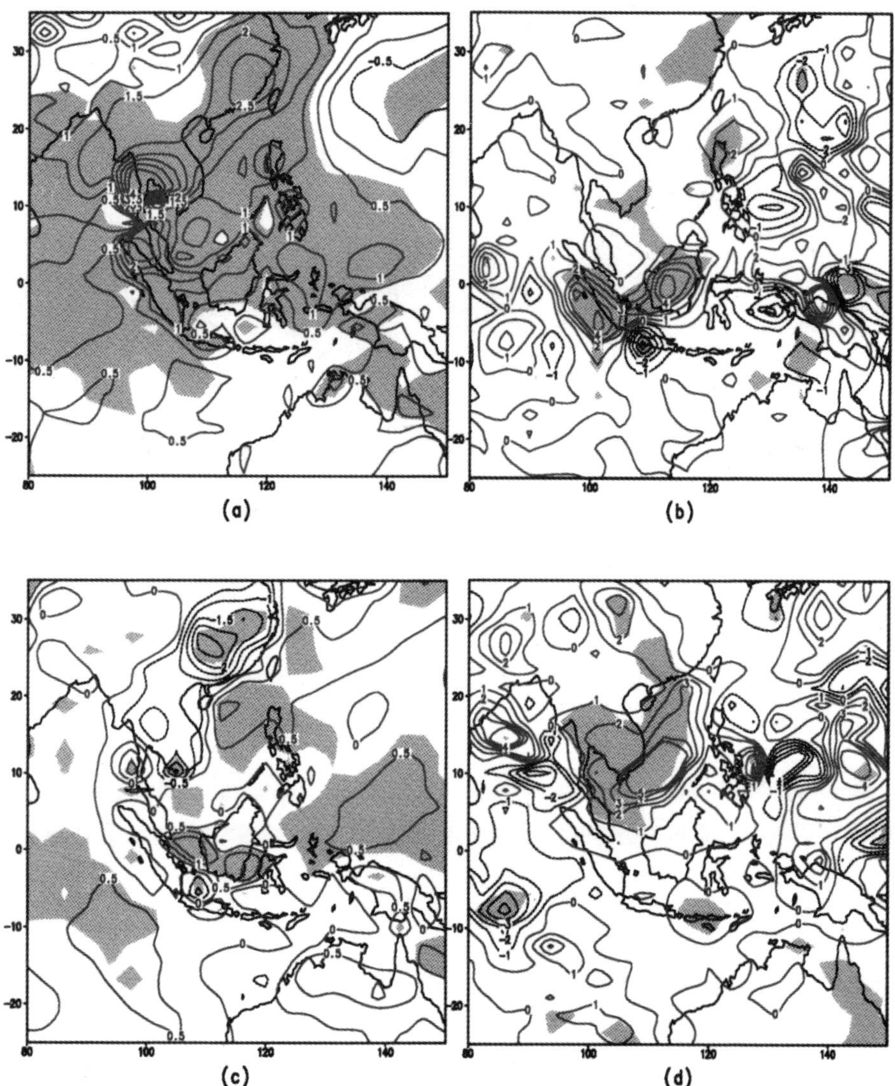

Figure 12.4 Changes of surface temperature and precipitation in January and July over Southeast Asia in BMRC GCM: (a) changes in surface temperature in January with a contour interval of 0.5 K (shaded areas indicate the signal-to-noise ratio of these changes is greater than 2.0), (b) as (a) but for precipitation with a contour interval of 1 mm d^{-1}, (c) as (a) but in July, and (d) as (b) but in July.

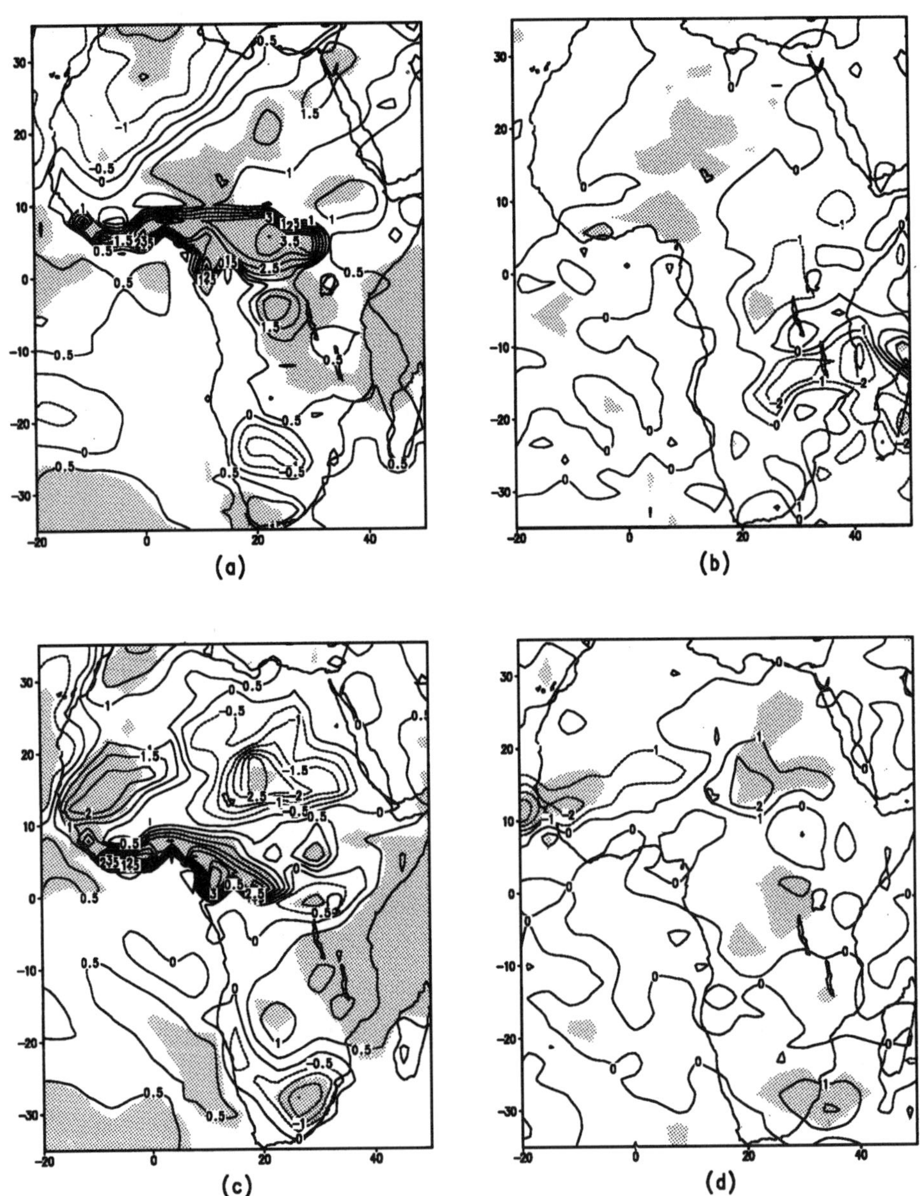

Figure 12.5 Changes of surface temperature and precipitation in January and July over Africa in BMRC GCM: (a) changes in surface temperature in January with a contour interval of 0.5 K (shaded areas indicate the signal-to-noise ratio of these changes is greater than 2.0), (b) as (a) but for precipitation with a contour interval of 1 mm d^{-1}, (c) as (a) but in July, and (d) as (b) but in July.

Over the deforested Amazon Basin, surface temperature in January (Figure 12.3(a)) is significantly increased, particularly over the northeast coast of South America. The increment is around 2.5 K over the central Amazon Basin and the largest increase occurs over the Guiana Highlands. In general, more rainfall is received over the Amazon Basin following deforestation (Figure 12.3(b)) and the maximum increase in rainfall is located around 15°S where heavy rainfall is simulated in the control experiment. The signal-to-noise ratio suggests that the increase in precipitation over the Amazon Basin is mainly due to the process of the deforestation and thus that the underlying physical processes would benefit from investigation. In July, changes in surface temperature are heterogeneous (Figure 12.3(c)): decreases over the west of the Amazon Basin but large increases in the east with the maximum increase in surface temperature (about 3 K) located over the northeast coast. Due to the northward shift of the ITCZ, precipitation shows only small increases over the Amazon Basin (Figure 12.3(d)) compared with that in January and the major increases occur over the north Amazon Basin. The heterogeneous feature of the changes in surface temperature is difficult to understand although there is a suggestion that the increase of evapotranspiration (not shown here) is larger over the west Amazon Basin which could mean that more energy is lost to the atmosphere from the surface which may result in the decrease of surface temperature. The mechanism for the increase in precipitation will be discussed in Section 12.5.

Southeast Asia and tropical Africa

Following deforestation, surface temperatures over Southeast Asia are increased by around +1 K over the Sumatra and Borneo island and Indochina peninsula in January (Figure 12.4(a)). Changes in monthly total precipitation in January (Figure 12.4(b)) do not have systematic spatial patterns, and only the increase over Sumatra and Borneo islands and the decrease over the west Pacific Ocean (around 160°E) can be distinguished above the model noise. This suggests that the monsoon precipitation over Southeast Asia as a whole is little affected by the imposed disturbance. In the summer monsoon season (July), surface temperature is increased over tropical Southeast Asia but it is decreased over the East Asian continent (Figure 12.4(c)). In addition, surface temperatures over west Pacific Ocean show some slight warming. Changes in the July precipitation (Figure 12.4(d)) show that the summer monsoon precipitation is significantly strengthened over the Indochina peninsula and in some regions of the East Asian continent following deforestation.

Over tropical Africa, deforestation results in a large increase in January surface temperature by about +3 K (Figure 12.5(a)). Changes in precipitation (Figure 12.5(b)) are very small over tropical Africa although a large reduction in precipitation occurs over southern Africa. The surface warming over tropical Africa is also seen in July (Figure 12.5(c)) but at the same time the surface temperature is decreased over central Africa by about 2 K. The reduction of precipitation in July (Figure 12.5(d)) is located over the subtropical regions around 10°N where the highest precipitation is simulated in the control experiment.

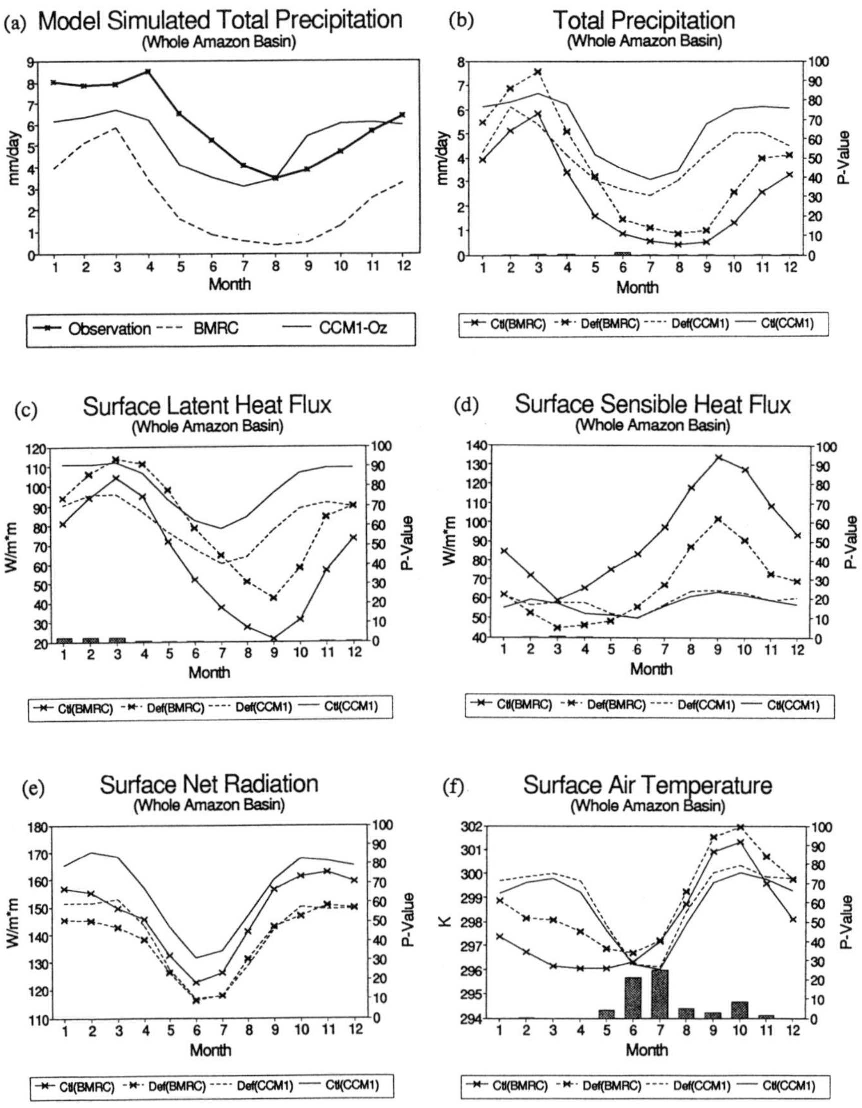

Figure 12.6 Regional averaged surface climate changes over the Amazon Basin with Student's t-test (the vertical bar represents the P-value of the Student's t-test): (a) comparison of regional averaged land surface precipitation over the Amazon Basin (mm d^{-1}) with the observation (Legates and Willmott 1990), (b) total precipitation (mm d^{-1}), (c) surface latent heat flux (W m^{-2}), (d) surface sensible heat flux (W m^{-2}), (e) upward surface net radiation (W m^{-2}), and (f) surface air temperature (K).

12.4.2 Seasonal variation of surface climate changes

As pointed out in Nobre et al. (1991) and Zhang et al. (1996a, 1996b), changes in surface climate's seasonality may impact the local and regional ecosystems. These changes are evaluated here from the BMRC model and compared with those from CCM1-OZ as described by Zhang et al. (1996a). In addition to the seasonal cycles the significance of the changes produced by the BMRC model are shown as vertical bars. If these 'P-values' lie between 0% and 5% the differences are likely to be significant.

Amazon Basin

Figure 12.6(a) display the regionally averaged precipitation over the Amazon Basin (15°S to 5°N and 80°W to 50°W) in the observations (Legates and Willmott 1990) and in the two control experiments. In the BMRC GCM, the local precipitation has a large seasonal variation with maximum precipitation (about 4 mm d^{-1}) in January–February–March, while in the rest of the months the rainfall is lower than 2 mm d^{-1}. Comparison with observations indicates that wet season precipitation is about 3 mm d^{-1} too low and that this season is too short in BMRC GCM. In addition, the dry season over the Amazon Basin is about 3.5 mm d^{-1} drier and longer in the control simulation than observed.

The seasonal variation of surface evapotranspiration (Figure 12.6(c)) exhibits features similar to those of the local precipitation in the BMRC GCM. The simulated evapotranspiration is lower in the months from September to December than that from January to April, presumably because the underestimation of precipitation in this season limits surface moisture. Over the Amazon Basin, surface net radiation has its minimum in the months of May–June–July and high values in the rest of the months (Figure 12.6(e)).

Following deforestation, monthly total precipitation increases in all the months in BMRC model (Figure 12.6(b)): about 2 mm d^{-1} in the wet season and 0.3 mm d^{-1} in the dry season. The increased rainfall acts to alleviate the insufficiency in surface moisture. Therefore, more energy is lost to the atmosphere by evapotranspiration which is increased in all the months by about +20 W m^{-2}. This also leads to a large reduction of surface sensible heat flux by about –27 W m^{-2}. The reduction of the surface net radiation in Figure 12.6(e) (about –10 W m^{-2}) is the direct result of the increase in surface albedo in the 'deforested' land surface and could be part of the reason for the large reduction of the surface sensible heat flux. As a combined result of the changes in surface energy balance, surface temperature is increased throughout the year (Figure 12.6(f)): the largest increase temperature (about 2 K) is seen during the wet season, while changes of temperature in June and July are not statistically significant. All the BMRC predicted changes in surface net radiation, surface latent heat flux and sensible heat flux are found to be statistically significant when evaluated by the Student's t-test.

The seasonal variation of surface climate changes over the Amazon Basin predicted by the BMRC are quite different from CCM1-OZ. The former shows increased rainfall and warmer temperature while the latter predicts less precipitation and evapotranspiration and only small changes in surface temperature. The seasonal variation of the surface climate in

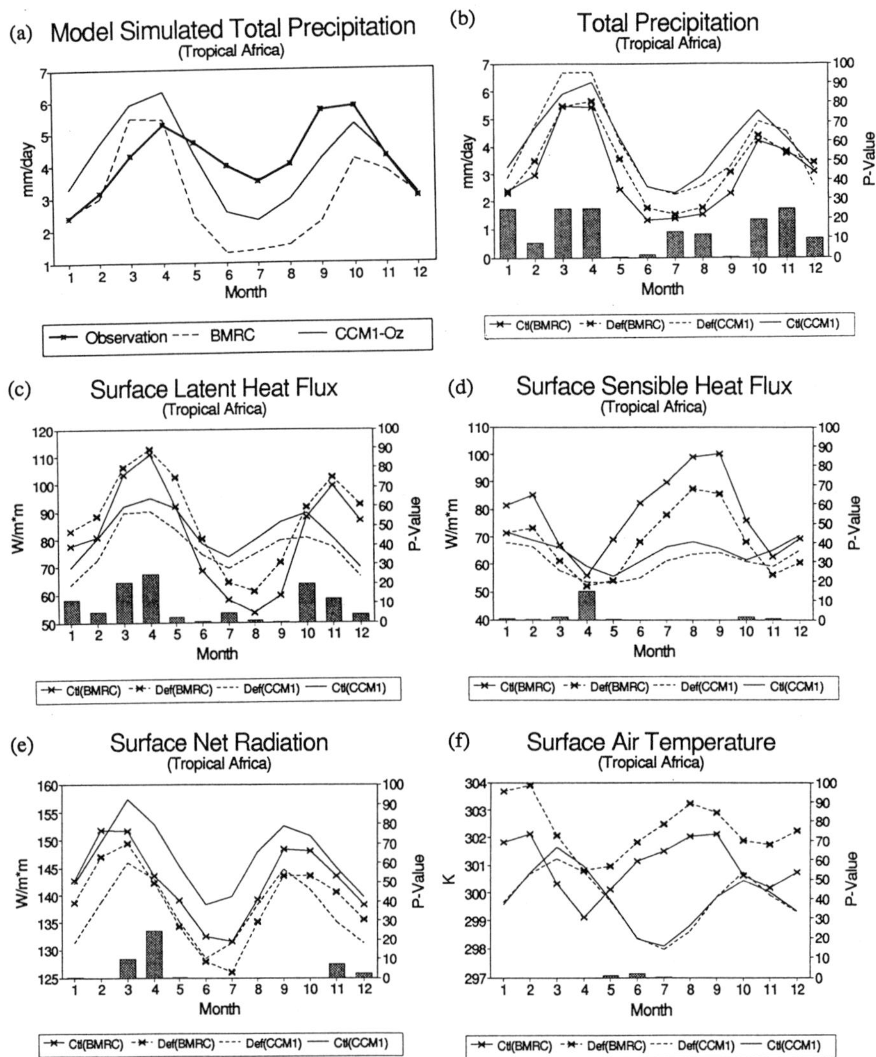

Figure 12.7 Regional averaged surface climate changes over the Southeast Asia with Student's t-test (the vertical bar represents the P-value of the Student's t-test): (a) comparison of regional averaged land surface precipitation over Southeast Asia (mm d^{-1}) with the observation (Legates and Willmott 1990), (b) total precipitation (mm d^{-1}), (c) surface latent heat flux (W m^{-2}), (d) surface sensible heat flux (W m^{-2}), (e) upward surface net radiation (W m^{-2}), and (f) surface air temperature (K).

the Amazon Basin suggests that failure of the BMRC model to simulate an adequate climate prior to the imposed disturbance could be the predominant reason for the differences between the two models' predictions following deforestation.

Southeast Asia and tropical Africa

Simulated precipitation over the land-points of Southeast Asia (15°S to 20°N and 90°E to 150°E) is displayed in Figure 12.7(a) together with the observed values. BMRC rainfall over the marine continent of Southeast Asia is far less than observed and no monsoon signal is distinguishable. In contrast, CCM1-OZ has a reasonable simulation of the local rainfall seasonal variation, with precipitation being overestimated by about 1.5 mm d^{-1}.

As discussed in the analysis of the Amazon Basin results, the low precipitation over the land surface of Southeast Asia would certainly affect the model simulation of climate changes following deforestation. Deforestation in the BMRC model leads to an increase of precipitation (about +0.5 mm d^{-1}) throughout the year. This increase in precipitation results in the increase of the surface evapotranspiration: the largest increment about +12 W m^{-2} occurs in June–July–August. The surface sensible heat flux is decreased in the BMRC model: by –15 W m^{-2} in all the months. The effect of the imposed increase of surface albedo leads to the reduction of the surface net radiation of about –8 W m^{-2} in both models. The differences in the responses of the two models to deforestation seem, again, to be due to the different control simulations of precipitation.

Over tropical Africa, the double-phase of local precipitation (March–April and September –October–November) is represented in the BMRC GCM control simulation (Figure 12.8(a)) but the local dry season (May to September) has lower precipitation than observed. CCM1-OZ overestimates the precipitation from December to April but underestimates it in other months. In contrast to the results from the Amazon Basin and Southeast Asia, local precipitation over tropical Africa is little affected by the deforestation (Figure 12.8(b)). Although there are some small changes in most months, only the changes in May, June and September in the BMRC model are statistically significant. Changes in the surface energy allocation have the same directions as that over the other regions: surface evaporation increases in all the months (+6 W m^{-2}) and the sensible heat flux decreases (–10 W m^{-2}) in the BMRC model but evaporation is reduced in CCM1-OZ although the surface sensible heat flux also decreases following deforestation in the CCM1-OZ simulation.

12.5 A search for the causes of uncertainty in model predictions

Changes in the surface and atmospheric energy budgets and in the local hydrological processes are analysed in detail to try to understand the differences in the simulated regional responses. As changes in surface climate simulated in the BMRC GCM and CCM1-OZ models (Zhang et al. 1996a, 1996b) in Section 12.4 show similar features (though the magnitudes are different) over all three deforested regions, the Amazon Basin responses are analysed as representative.

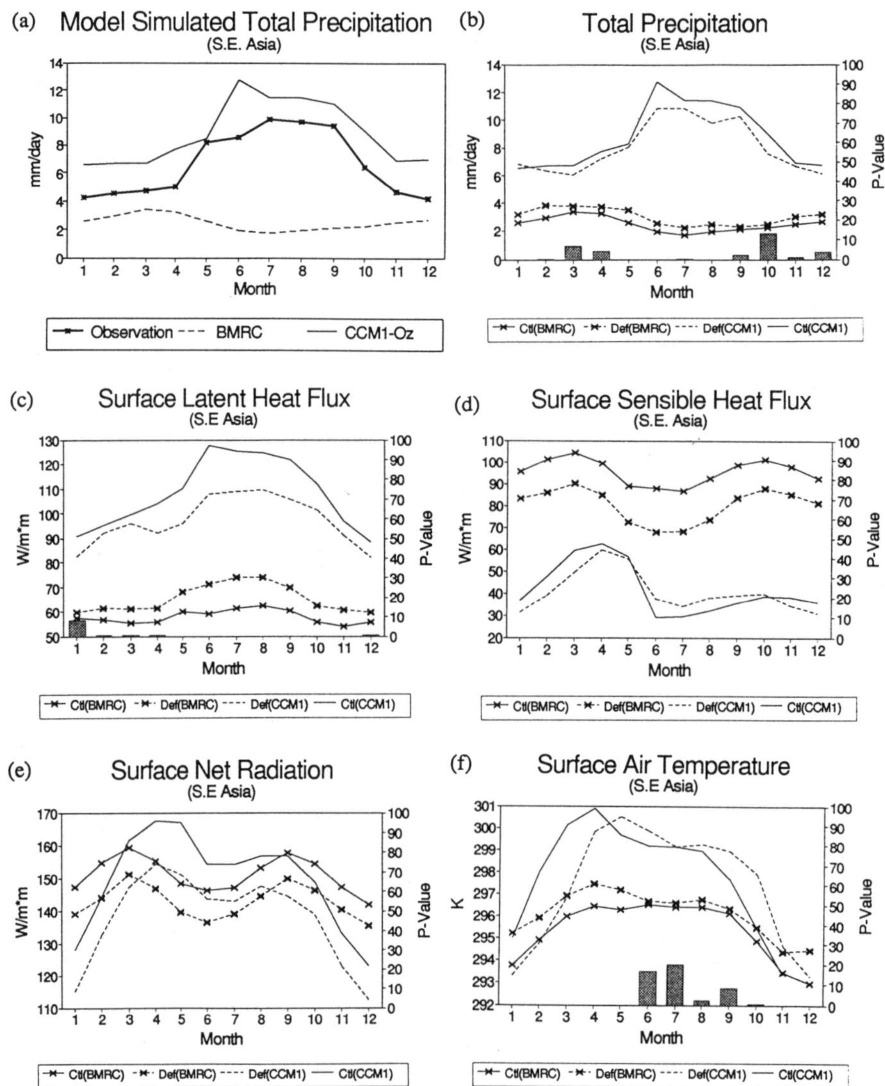

Figure 12.8 Regional averaged surface climate changes over tropical Africa with Student's t-test (the vertical bar represents the P-value of the Student's t-test): (a) comparison of regional averaged land surface precipitation over tropical Africa (mm d^{-1}) with the observation (Legates and Willmott 1990), (b) total precipitation (mm d^{-1}), (c) surface latent heat flux (W m^{-2}), (d) surface sensible heat flux (W m^{-2}), (e) upward surface net radiation (W m^{-2}), and (f) surface air temperature (K).

12.5.1   Changes in surface and atmospheric energy budgets

Figure 12.9 shows the changes in the components of the land-surface and atmospheric energy budgets over the Amazon Basin in both model simulations. In the BMRC control experiment (Figure 12.9(a)), 39.2 W m^{-2} of the solar radiation incident at the land-surface is reflected back to the space and 206.4 W m^{-2} net shortwave radiation heats the land surface. The net outgoing longwave radiation at the land surface is 59.2 W m^{-2} and consequently there is 147.1 W m^{-2} net radiative energy gain which is balanced by latent and sensible heat fluxes, heating the overlying atmosphere. In the BMRC model the surface sensible heat flux (93.2 W m^{-2}) is larger than the latent heat flux (62.2 W m^{-2}).

Following deforestation, the net absorbed solar radiation at the top of the atmosphere (TOA) is reduced in BMRC model by –13.1 W m^{-2} partly due to the increase in the cloud cover which is seen in this model and the increase of surface albedo. Increased cloud amount also contributes to the reduction (–5.5 W m^{-2}) in outgoing longwave radiation at the TOA. Because of the increase in surface albedo, an extra 7.8 W m^{-2} shortwave radiation is reflected from the surface over the deforested landscape. The net solar radiation lost to the land surface after deforestation is –17.1 W m^{-2}. From the energy 'supply and demand' point of view, both the latent and sensible heat flux could be weakened. However, in the BMRC model the soil moisture 'demand' dominates and latent heat flux is increased. The combined effect of more energy being allocated to latent heat flux (+21.3 W m^{-2}) and the net radiative energy being reduced results in a large reduction in the sensible heat flux (–27.3 W m^{-2}).

For comparison, Figure 12.9(b) shows the Amazon surface and atmospheric energy budgets from the CCM1-OZ model. The major changes in the CCM1-OZ simulation lie in the shortwave radiation terms: incident solar radiation at the land-surface and the reflected solar radiation by the land-surface. Following deforestation, the increased surface albedo results in more solar radiation being reflected back to the atmosphere and space (+17.1 W m^{-2}), similar to, but larger than the changes in BMRC GCM. However, in CCM1-OZ, more solar radiation reaches the land-surface and the increased incident solar radiation (+10.4 W m^{-2}) is primarily due to the decrease of cloud cover (–4%). The effect of increased surface albedo is mitigated by the increase in incident solar radiation and there is only a decrease of –6.7 W m^{-2} in net solar radiation at the land-surface. It is clear that changes in cloud cover are of importance in determining the surface radiative energy following deforestation in the CCM1-OZ simulation.

12.5.2   Changes in the low-level dynamics and moisture transport

Removal of tropical rainforest has the potential to modify the low-level atmospheric dynamics since reducing surface roughness can result in an enhanced surface wind speed (Sud et al. 1988). In the BMRC simulation, reducing the surface friction leads to a large increase in the low-level wind speed (Figure 12.10(a)). Reduction of surface roughness length from the 1.1 m to 0.1 m roughly corresponds with the reduction of the surface aerodynamic drag coefficient (C_D) from 0.0065 to 0.0040. However, the wind speed at the

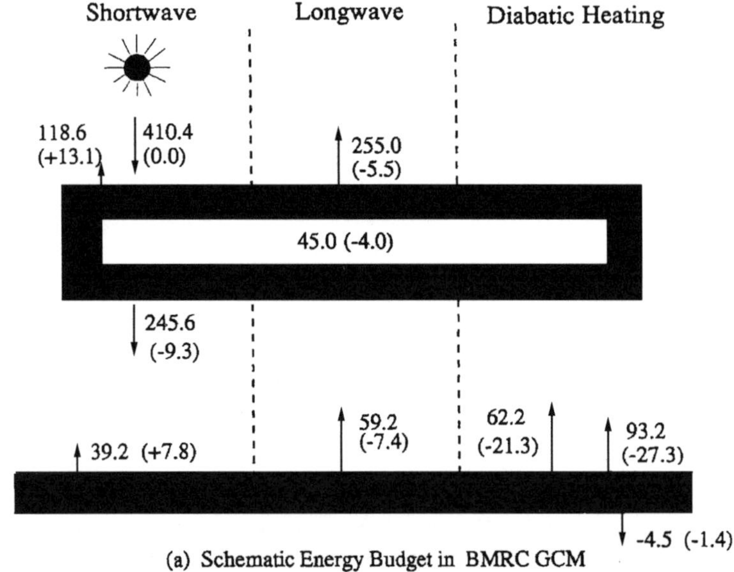

(a) Schematic Energy Budget in BMRC GCM

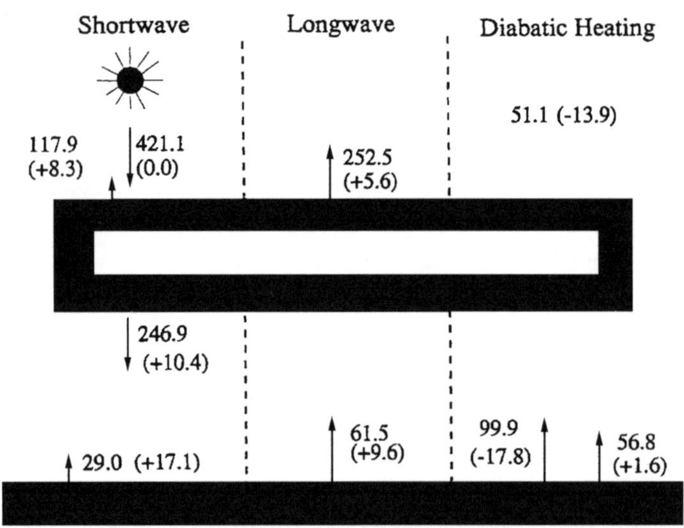

(b) Schematic Energy Budget in CCM1-Oz

Figure 12.9 Schematic energy budgets in (a) BMRC GCM and (b) CCM1-OZ in the model control experiment and changes following deforestation (in parenthesis). All units are W m^{-2}.

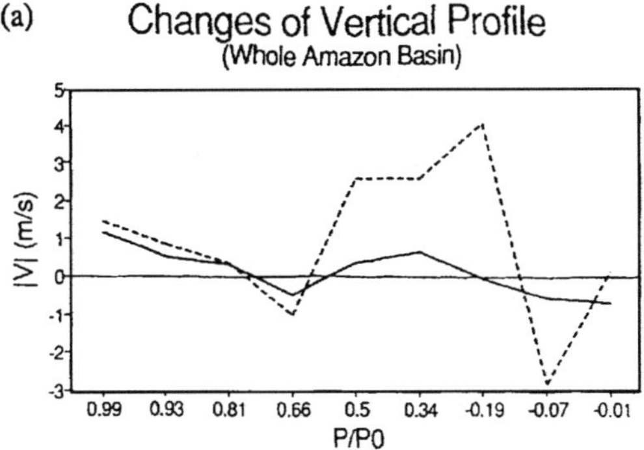

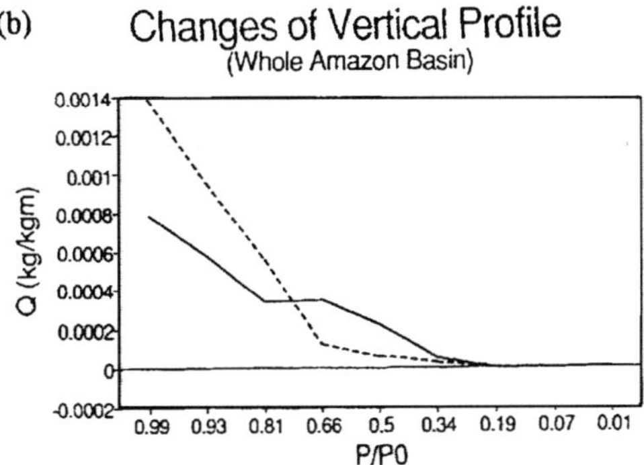

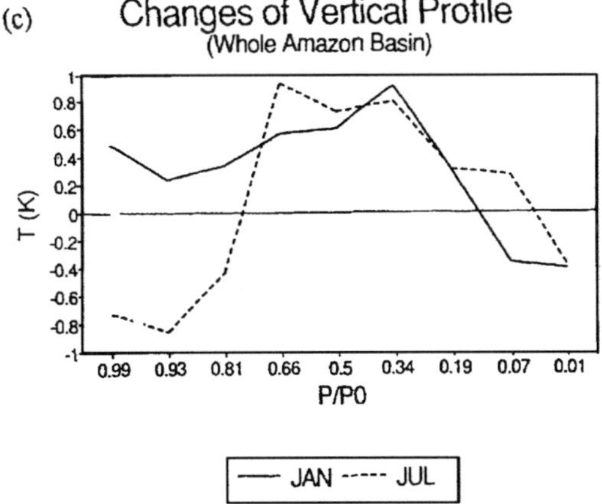

Figure 12.10 Changes in the vertical profiles over the Amazon Basin in January and July in BMRC deforestation experiment: (a) wind speed (m s^{-1}), (b) specific humidity (kg^{-1} kg), and (c) temperature (K).

lowest atmospheric level increases from 0.82 m s^{-1} to 2.0 m s^{-1}. The effect of increased wind speed dominates the reduction of C_D and the total effect is that $|V| * C_D$, which controls the dynamic conditions for evaporation and sensible heat flux, is actually increased. In addition, Figure 12.10(b) shows that the specific humidity in the low levels of atmosphere is increased. Thus the latent heat flux is increased in BMRC model simulation as a result of the more favourable dynamical and low-level moisture conditions.

Changes in the low-level dynamics in the BMRC GCM are quite different from those of CCM1-OZ which are described in Zhang et al. (1996b). In the CCM1-OZ deforestation experiment, even though the increase in the surface wind speed is represented (from 1.5 m s^{-1} to 2.1 m s^{-1}), the effect of the reduction of surface roughness length dominates the effect of the increased surface wind speed. Thus the dynamical term of $|V| * C_D$ is weakened from 0.0985 to 0.0100 m s^{-1} following deforestation in CCM1-OZ and more reduction of evapotranspiration is induced.

Figure 12.11 shows changes in total precipitation and the horizontal moisture advection, $q * V$, at the lowest model level over the Amazon Basin and the nearby regions in January and July in both models. A significant feature illustrated in Colour Plate 12.1 is that two models have different low-level responses to the imposed deforestation disturbances. In BMRC model (Colour Plates 12.1(b) and (d)) as the reduction of surface roughness after deforestation, much more water is transported from the tropical Atlantic Ocean into the Amazon Basin, especially in January. Following the methodology used (Zhang et al. 1996b) to analyse the CCM1-OZ deforestation experiment, changes of the incoming moisture transport at the boundaries of the Amazon Basin at different levels are calculated in January (Figure 12.11(b)). Over the northern and eastern boundaries, more moisture is delivered into the Amazon Basin and the incoming moisture over the whole boundaries is greatly increased in the lower atmosphere. Therefore, there is more water vapour over the Amazon Basin and local precipitation may be increased due to this process.

Changes in the low-level moisture transport over the Amazon Basin in CCM1-OZ (Colour Plates 12.1(a) and (c)) are quite different from those seen in the BMRC simulation. Instead of the enhanced moisture convergence from the nearby oceans seen in BMRC model, the responses from CCM1-OZ to the imposed disturbances result in less water vapour being delivered from the nearby regions into the Amazon Basin and the smoothed land results in more water being transported out of the basin. Calculations of the incoming water vapour at the boundaries of the Amazon Basin (Figure 12.11(a)) show that the largest changes in water vapour transport in CCM1-OZ occur at the western boundary of the Amazon Basin where more water vapour is blown out of the basin by the enhanced low-level flow. In addition, there is a decrease of incoming water vapour at the northern boundary and the combined effect is that the total incoming water vapour over the Amazon Basin is decreased following deforestation in CCM1-OZ.

Overall, changes of surface climate over the Amazon Basin simulated by this pair of GCMs can be explained in terms of the changes in the land-surface and atmospheric energy budgets and changes in dynamical process controlling the regional moisture budget.

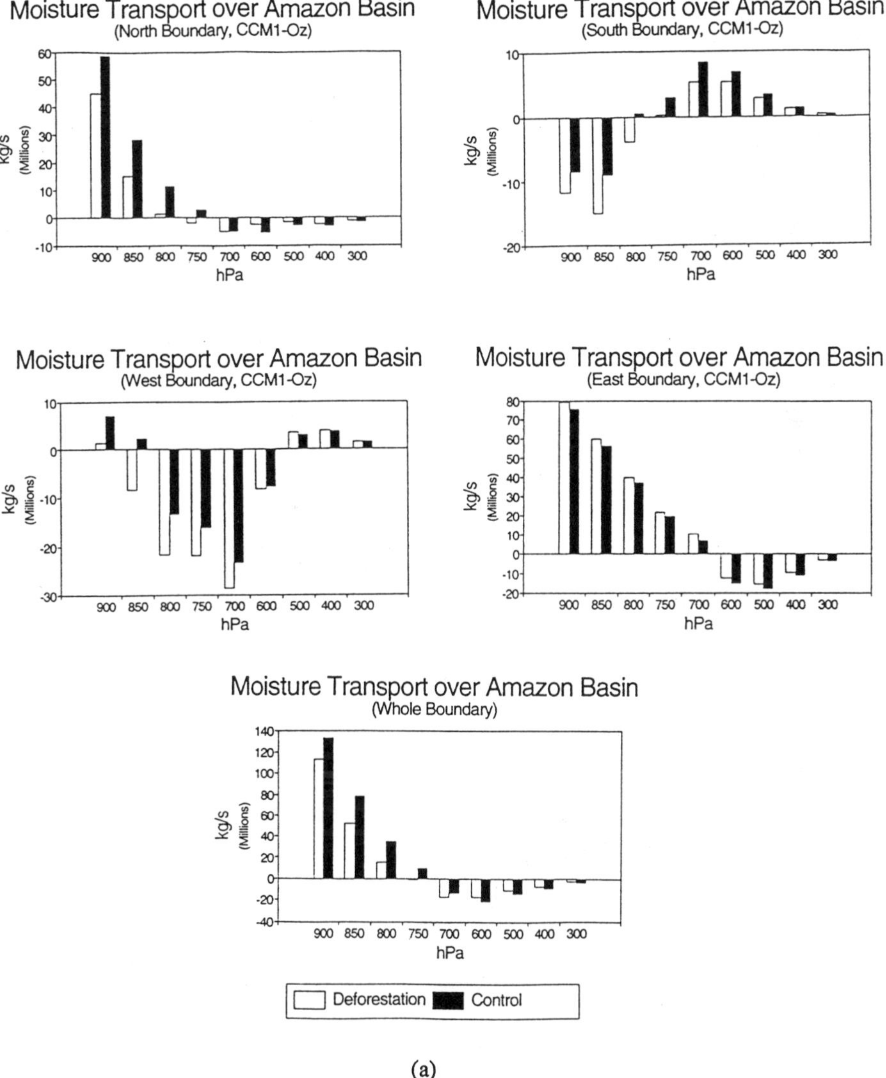

Figure 12.11 Incoming moisture over the boundaries of the Amazon Basin (kg s^{-1}) in January in the model control and deforestation experiments: (a) CCM1-OZ and (b) BMRC GCM.

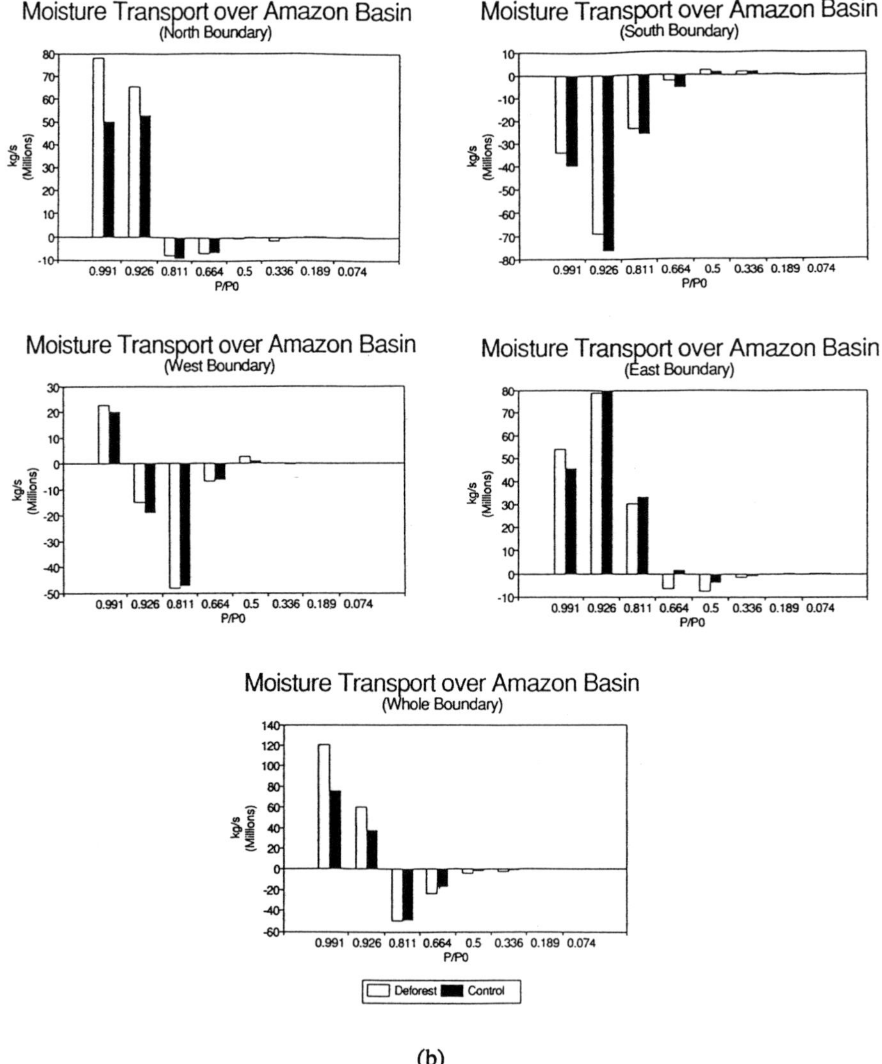

(b)

Figure 12.11 *Continued*

Table 12.3 Cloud radiative forcing and its changes following deforestation in the BMRC and CCM1-OZ simulations of tropical deforestation over the Amazon Basin

	Experiment	BMRC	CCM1-OZ
Clear-sky OLR	Deforestation	295.8	295.1
	Control	297.0	294.2
Clear-sky net TOA solar radiation	Deforestation	330.9	345.2
	Control	341.5	362.9
Cloudy-sky OLR	Deforestation	295.8	258.1
	Control	297.0	252.5
Cloudy-sky net TOA solar radiation	Deforestation	295.8	294.9
	Control	297.0	303.2
Longwave cloud radiative forcing	Deforestation	295.8	+37.0
	Control	297.0	+41.7
Shortwave cloud radiative forcing	Deforestation	−52.3	−50.4
	Control	−49.7	−59.7
Net cloud radiative forcing	Deforestation	−6.0	−13.3
	Control	−7.7	−17.9

Note: OLR is the outgoing longwave radiation and TOA is the top-of-atmosphere. All units are $W\,m^{-2}$ and the incoming radiation to the atmosphere (warming effect) is defined as positive.

However, the sequence and direction of changes differs between the two models. In the BMRC simulation, the modified regional atmospheric circulation produces more favourable condition for precipitation. The resulting increase in local rainfall affects the land-surface energy partitioning and hence results in a large increase in latent heat flux. However, as shown in Zhang et al. (1996b), changes in the regional atmospheric circulation over the Amazon Basin in CCM1-OZ result in less water vapour to be delivered into the Amazon Basin and consequently a larger reduction in precipitation than in local evapotranspiration is predicted.

12.5.3 Changes in the cloud radiative forcing

Cloud radiative forcing and its changes following a climate perturbation are very important components of overall climate model uncertainty (e.g. Cess et al. 1990; Cess et al. 1995). This is analysed in this section and the BMRC simulation is compared with that of CCM1-OZ.

The cloud radiative forcing for the whole atmosphere and land-surface system is calculated as the difference between the radiative flux under a cloudy sky and in clear sky conditions (cf. Zhang et al. 1996a). Table 12.3 shows that in both the BMRC GCM control and deforestation experiments, cloud radiative forcing plays an important role in the whole system longwave and shortwave radiative energy budgets over the tropical rainforest regions. The longwave cloud radiative forcing has a warming effect on the system of about 40 to 50 W m^{-2}, due to the absorption of the longwave radiation from the surface to the atmosphere. At the same time, cloud reflects more solar radiation back to space and has a cooling effect of about 45 to 55 W m^{-2} over tropical regions. The net radiative forcing is small due to the opposing effects of the long and shortwave forcing.

The longwave cloud radiative forcing is larger in the BMRC control simulation (46.3 W m^{-2}) than in that of CCM1-OZ (41.7 W m^{-2}) but the shortwave cloud radiative forcing is larger in the CCM1-OZ (–59.7 W m^{-2}) than in BMRC GCM (–49.7 W m^{-2}). Consequently, the cloud forcing has a larger influence in the CCM1-OZ with net effect of –17.9 W m^{-2} over the Amazon Basin compared with that of BMRC (–6.0 W m^{-2}).

Following deforestation, cloud radiative forcing does not display large changes in the BMRC simulation. The increase of cloud cover over the Amazon Basin leads to the increase of the cooling effect of shortwave cloud forcing by –2.7 W m^{-2}. This is offset by the increased warming effect of the longwave cloud forcing (+4.3 W m^{-2}) and the net effect is that the whole system gains +1.7 W m^{-2} radiative energy due to the reduction of the cloud cover over the Amazon Basin. This effect is also weaker than that in CCM1-OZ (+4.6 W m^{-2}).

The results presented here underline that cloud radiative forcing is an important component of the overall atmospheric radiative processes and that changes in the cloud radiative forcing can affect the changes in the energy budgets of the whole atmospheric and the land-surface systems over the deforested regions. From this analysis, it seems that changes in the cloud radiative forcing in the BMRC simulation of tropical deforestation are not as large as those in the CCM1-OZ experiment (Zhang et al. 1996a). However, the differences in the cloud radiative forcing in the control and disturbed climates between these two models adds considerably to the differences between their predictions of the impacts of tropical deforestation.

12.6 Large-scale or global impacts of tropical deforestation

The importance of tropical rainforest in the maintenance of the tropical and/or global atmospheric circulation is not certain either for the individual numerical models or for the real world. Some researches (e.g. Numaguti 1993; Milly & Dunne 1994) stress the role of tropical evaporation and the changes of land surface conditions in supporting the Hadley circulation over tropical regions. Modifying the surface and boundary layer conditions by imposing a tropical deforestation may affect these dynamics.

Results from CCM1-OZ and some other GCM simulations (e.g. McGuffie et al. 1995; Sud et al. 1996; Zhang et al. 1996b) state that tropical deforestation may affect the climate beyond the tropical deforested regions. Changes in the Walker circulation, the meridional Hadley circulation and the propagation of stationary Rossby waves forced by the anomalous tropical heating were identified by Zhang et al. (1996b) as possible dynamic mechanisms responsible for the climatic responses in middle and high latitudes following tropical deforestation. This hypothesis is re-examined here in the context of the BMRC simulation.

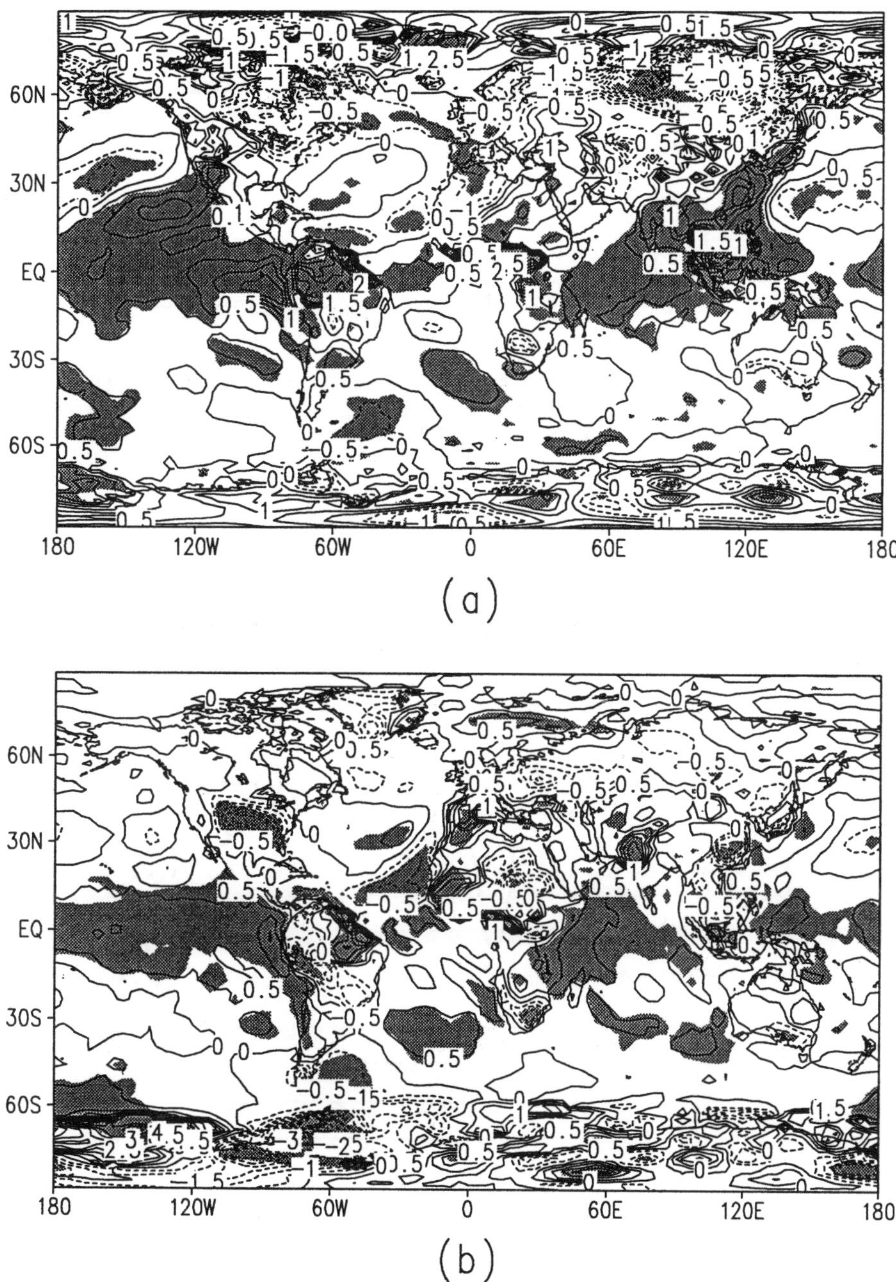

Figure 12.12 Global changes in surface temperature following tropical deforestation in BMRC GCM in (a) January and (b) July with contour intervals of 0.5 K. Shaded areas indicate the signal-to-noise ratio of these changes is greater than 2.0.

Meridional Energy Transport (BMRC+BEST)
(Control, Jan)
J*m/(kg*s)
(Thousands)
40
30
20
10
0
-10
-20
88.3 70.4 52.5 34.7 16.8 -1.1 -16.8 -34.7 -52.5 -70.4 -88.3
Latitude
Lq   CpT   KE   gz

Meridional Energy Transport (BMRC+BEST)
(Changes, Jan)
J*m/(kg*s)
4000
2000
0
-2000
-4000
-6000
-8000
88.3 70.4 52.5 34.7 16.8 -1.1 -16.8 -34.7 -52.5 -70.4 -88.3
Latitude
Lq   CpT   KE   gz

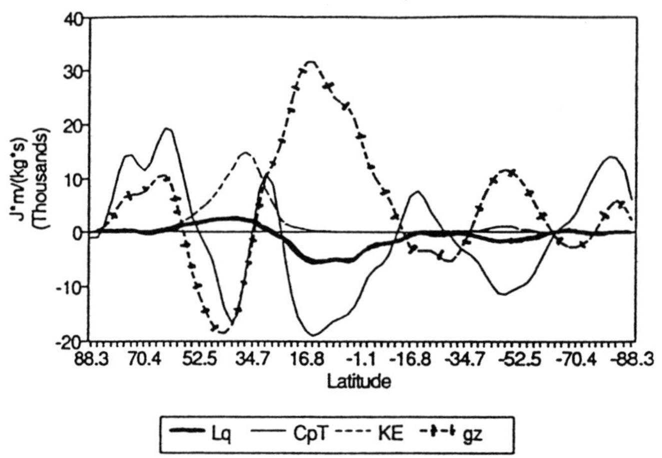

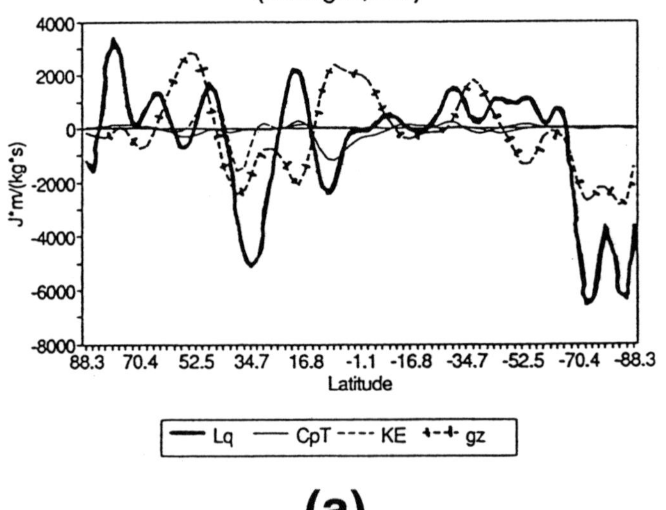

(a)

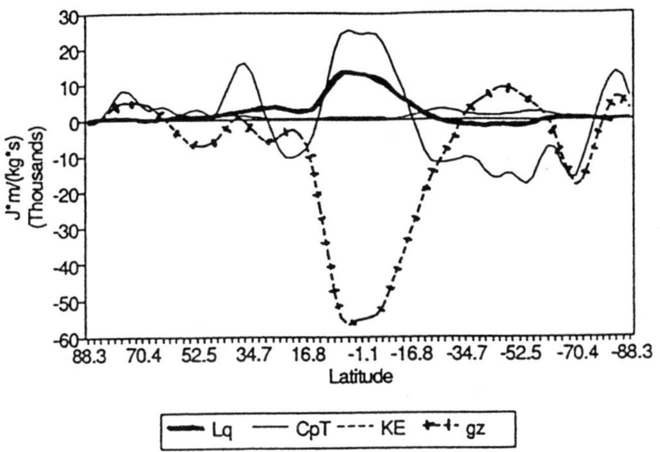

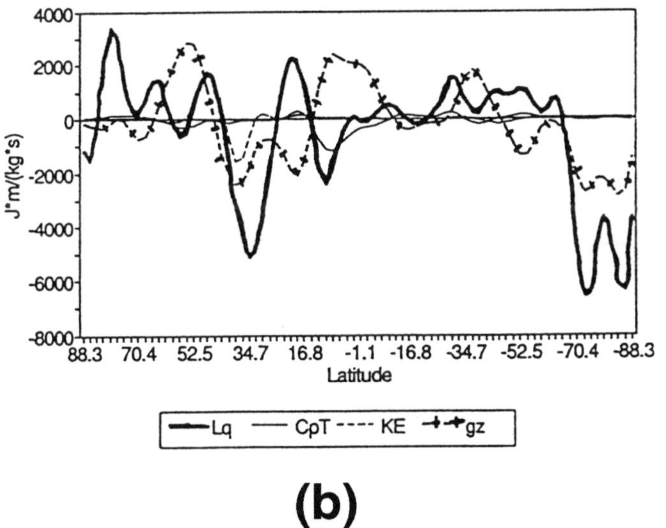

(b)

Figure 12.13 Vertical integrated meridional energy transport (northward) following tropical deforestation in BMRC GCM in (a) January and (b) July.

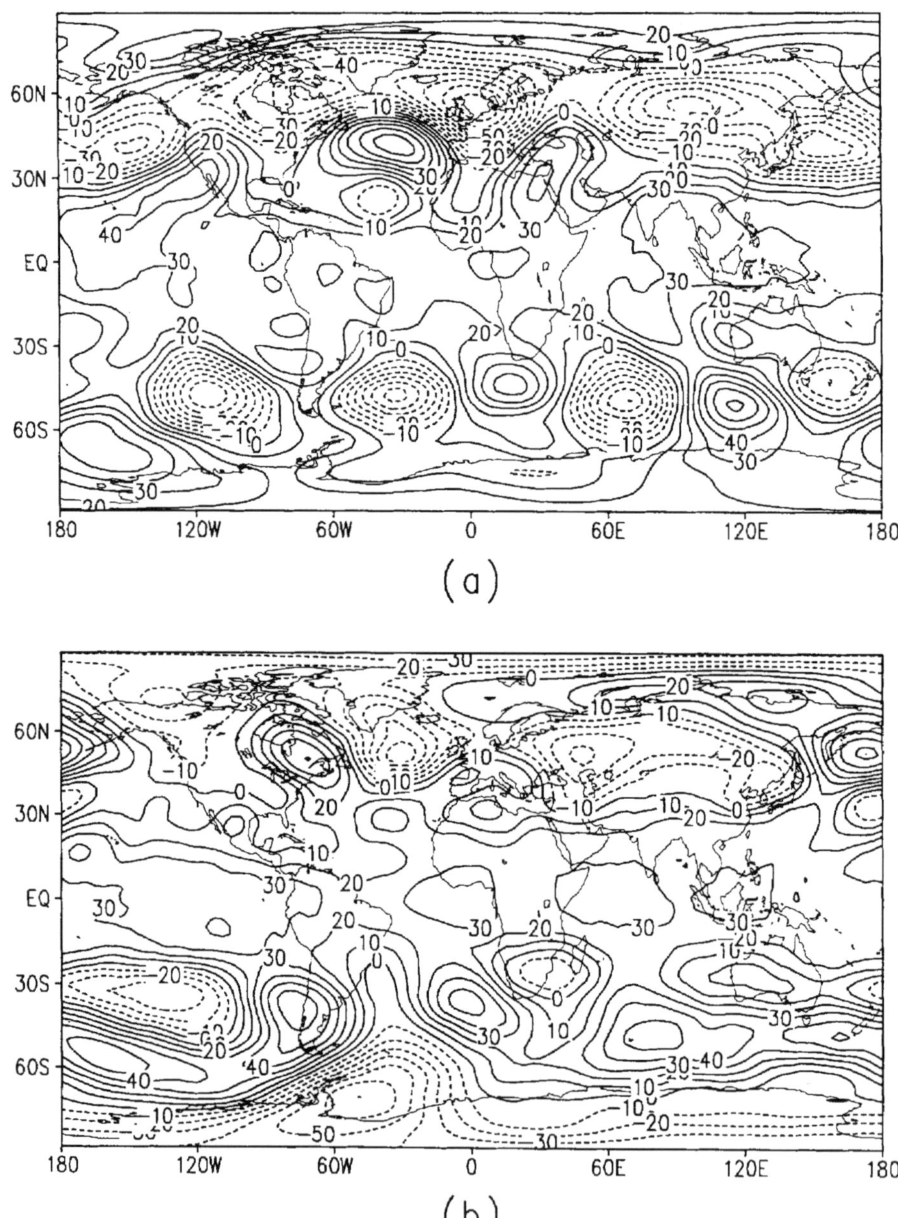

Figure 12.14 Changes in the geopotential height at 200 hPa in BMRC GCM tropical deforestation experiment in January and July: (a) 200 hPa in January with contour interval of 10 m and (b) as (a) but for July.

Changes in the surface temperature in January and July are displayed in Figure 12.12. As seen in the results from CCM1-OZ, large changes in surface temperature do occur in middle and high latitudes. Even though the natural variability of the model climate in the middle and high latitudes is much greater than in the tropical regions, the signal-to-noise ratios demonstrate that some of these changes are larger than the inherent model variation and might, therefore, be expected to be the result of tropical deforestation. Changes of precipitation are also simulated in this model (not shown here) but the geographical distribution pattern is not as large as for the surface temperature, nor are there as many regions that could be significant.

As the tropics are the source of moisture and energy for the global climate system, an important question is whether the changes in energy transport into the middle and high latitudes are affected by tropical deforestation. Transports of potential energy (gz), latent heat (Lq), sensible heat (CpT) and kinetic energy in the control experiment are discussed by Colman et al. (1994) who showed the basic features captured in BMRC GCM. As shown in Figure 12.13, changes in the meridional energy transport following tropical deforestation are simulated over all latitudes, with large changes being identified over the middle and high latitudes in both January and July.

Rossby wave propagation is another means by which the effects of tropical disturbances can be transported into the extra-tropical regions (Zhang et al. 1996b). Wave patterns in the changes of geopotential height found in the CCM1-OZ simulation of tropical deforestation also occur in the geopotential height at 200 hPa from the BMRC GCM simulation (Figure 12.14). In January large scale wave patterns are located in the Northern Hemisphere, while they are simulated in the Southern Hemisphere in July: a characteristic explained by the dependence of meridional Rossby wave propagation on the background zonal wind field. Stronger westerly zonal wind in the winter hemisphere is more favourable for the propagation of Rossby waves. Results from the BMRC simulation agree with those from CCM1-OZ, increasing the possibility that Rossby wave propagation is an effective dynamical process by means of which the disturbance of tropical deforestation have global impact.

12.7 Summary

Results of two MECCA-sponsored simulations of the impacts of tropical deforestation on the local, regional and large scale climate have been analysed and compared. The changes in the surface climate over the tropical deforested regions identified in the BMRC predictions are quite different from the earlier simulations and from those of CCM1-OZ. The BMRC results include an increase in the local precipitation, an increase in the surface latent heat flux and a large decrease in the surface sensible heat flux in contrast to large reductions in precipitation and evapotranspiration in the CCM1-OZ and other earlier results (Table 12.1).

Both the BMRC and CCM1-OZ models underestimate the local precipitation over the Amazon Basin. However, further analysis shows that even though the rainfall over the

Table 12.4 Comparison of the surface climate regimes over the Amazon Basin in the BMRC GCM and CCM1-OZ in January–February–March and June–July–August

	BMRC GCM		CCM1-OZ	
	J–F–M	J–J–A	J–F–M	J–J–A
Surface net radiation (W m^{-2})	153.9	130.1	167.9	137.7
Maximum evaporation (mm d^{-1})	5.3	4.5	5.8	4.8
Precipitation (mm d^{-1})	5.0	0.6	6.4	3.4

Amazon Basin is less than observed during the local wet season in the CCM1-OZ model control simulation, the model simulates enough precipitation so that the surface energy allocation occurs in a 'wet' climate fashion. In contrast, the very low rainfall in the BMRC simulation results in the surface energy allocation over the Amazon Basin in the wet season resembling a 'dry' climate regime. This is illustrated in Table 12.4 which gives the three-month (January–February–March and June–July–August) averaged surface net radiation, 'maximum evaporation' and precipitation over the Amazon Basin in the BMRC and CCM1-OZ control simulations. 'Maximum evaporation' is defined here as the amount of water which could be evaporated using all the simulated surface net radiation.

Following this definition, if the simulated local precipitation is larger than the maximum evaporation, the region has a rainfall supply such that the evapotranspiration is controlled by the surface net radiative energy: the surface energy allocation belongs to the 'wet' climate regime. Otherwise, if the simulated local precipitation is less than the maximum evaporation, there is not enough surface moisture to satisfy the evaporative demand and the evaporation is dominated by the local rainfall and it belongs to the 'dry' climate regime.

The surface net radiation over the Amazon Basin in the BMRC control experiment is 153.9 W m^{-2} during the wet season, equivalent to a 5.3 mm d^{-1} maximum evaporation (Table 12.4). However, as the model simulated precipitation is only 5.0 mm d^{-1}, there is not sufficient surface moisture for evaporation and the surface energy allocation happens in such a way that a large part of the radiative energy is lost as sensible heat flux and the amount of surface evaporation depends upon the local precipitation. In contrast, in the CCM1-OZ control simulation, even though the local precipitation (6.4 mm d^{-1}) is less than the observation (7.9 mm d^{-1}) the model simulated precipitation is larger than the maximum evaporation of the surface net radiation. Thus the surface climate in JFM in the CCM1-OZ control simulation has 'wet' climate characteristics: a large part of the surface radiation is allocated to evaporation and the surface evaporation is dominated by the surface radiation. During the local dry season (June–July–August), the surface climates of both GCMs belongs to the 'dry' regime with less precipitation than the maximum evaporation. However, Table 12.4 shows that the surface precipitation (0.6 mm d^{-1}) in the BMRC simulation is far below the observation (4.3 mm d^{-1}) but this 'feature' is much less extreme in the CCM1-OZ simulation (3.4 mm d^{-1}).

Analyses of the land surface and atmosphere energy and moisture budgets of these two model simulations suggest that useful predictions of the climatic impacts depends first, on the model's capacity to simulate the current climate. The excessively dry climate in the BMRC control simulation seems to be primarily responsible for the different response to the imposed land surface disturbance of deforestation. Also the boundary layer processes included in the model influence its response to the prescribed disturbance of tropical deforestation. In the BMRC GCM, the decrease of surface roughness results in a large increase in the moisture advection from the nearby oceans into the Amazon Basin, Southeast Asia and tropical Africa. This seems to be the major reason for the increased precipitation following deforestation; the result which contrasts most strongly with those of previous simulations.

It should be noted that efforts have been made to improve the BMRC control simulation by studying the impacts of changes in convective parameterization on surface-atmosphere interactions and the sensitivity of model climate to changes in convective parameterizations and horizontal resolution (cf. McAvaney et al. 1995; Colman & McAvaney 1995; Colman et al. 1995). Improvements have been seen in the recent model simulations (McAvaney 1995, pers. comm.) and deforestation experiment with improved model parameterizations will be undertaken in the future to further study the uncertainty sources in GCM simulations of tropical deforestation.

Acknowledgments

This research is sponsored by the Model Evaluation Consortium for Climate Assessment. Huqiang Zhang acknowledges funding as a National Greenhouse Advisory Committee Scholar. Ann Henderson-Sellers and Andrew J. Pitman acknowledge support from the Australian Research Council. This paper is CIC contribution number 95/27.

References

Cess, R.D., Potter, G.L., Blanchet, J.P., Boer, G.J., Del Genio, A.D., Déqué, M., Dymnikov, V., Galin, V., Gates, W.L., Ghan, S.J., Kiehl, J.T., Lacis, A.A., Le Treut, H., Li, Z.-X., Liang, X.-Z., McAvaney, B.J., Meleshko, V.P., Mitchell, J.F.B., Morcrette, J.-J., Randall, D.A., Rikus, L., Roeckner, Royer, J.F., E., Schlese, U., Sheini, D.A., Slingo, A., Sokolov, A.P., Taylor, K.E., Washington, W.M., Wetherald, R.T., Yaga, I. and Zhang, M.-H. (1990) Intercomparison and interpretation of climate feedback processes in 17 atmospheric general circulation models. *Journal of Geophysical Research*, **95**, 16601–16615.

Cess, R.D., Zhang, M.H., Minnis, P., Corsetti, L., Dutton, E.G., Forgan, B.W., Garber, D.P., Gates, W.L., Hack, J.J., Harrison, E.F., Jing, X., Kiehl, J.T., Long, C.N., Morcrette, J.-J., Potter, G.L., Ramanathan, V., Subasilar, B., Whitlock, C.H., Young, D.F. and Zhou, Y. (1995) Absorption of solar radiation by clouds: Observations versus models. *Science*, **267**, 496–499.

Colman, R.A., McAvaney, B.J., Fraser, J.R. and Power, S.B. (1994) Annual mean meridional energy transport modelled by a general circulation model for present and $2 \times CO_2$ equilibrium climates. *Climate Dynamics*, **10**, 221–229.

Colman, R.A. and McAvaney, B.J. (1995) The sensitivity of the climate response of an atmospheric circulation model to changes in convective parameterization and horizontal resolution. *Journal of Geophysical Research*, **100(D2)**, 3155–3172.

Dickinson, R.E. and Kennedy, P. (1992) Impacts on regional climate of Amazon deforestation. *Geophysical Research Letters*, **19**, 1947–1950.

Dickinson, R.E. and Henderson-Sellers, A. (1988) Modelling tropical deforestation: A study of GCM land-surface parameterizations. *Quarterly Journal of the Royal Meteorological Society*, **114**, 439–462.

Dickinson, R.E., Henderson-Sellers, A., Kennedy, P.J. and Giorgi, F. (1993) Biosphere-atmosphere transfer scheme (BATS) version 1e as coupled to the NCAR community climate model. NCAR Technical Note, NCAR/TN-387+STR, Boulder, Colorado.

Dirmeyer, P.A. (1992) GCM studies of the influence of vegetation on the general circulation: The role of albedo in modulating climate change. *Ph.D thesis*, Department of Meteorology, University of Maryland, USA.

Fels, S.B. and Schwarzkopf, M.D. (1975) The simplified exchange approximation: A new method for radiative transfer calculations. *Journal of Atmospheric Science*, **32**, 1475–1488.

Gates, W.L. (1992) AMIP: The atmospheric model intercomparison project. *Bulletin of the American Meteorological Society*, **73**, 1962–1970.

Hart, T.L., Bourke, W., McAvaney, B.J., Forgan, B.W. and McGregor, J.L. (1990) Atmospheric general circulation simulations with the BMRC global spectral model: The impacts of revised physical parameterization. *Journal of Climate*, **3**, 436–459.

Henderson-Sellers, A., Yang, Z.-L. and Dickinson, R.E. (1993a) The project for intercomparison of land-surface parameterization schemes. *Bulletin of the American Meteorological Society*, **74**, 1335–1349.

Henderson-Sellers, A., Durbidge, T.B., Pitman, A.J., Dickinson, R.E., Kennedy, P.J. and McGuffie, K. (1993b) Tropical deforestation: Modelling local to regional-scale climate change. *Journal of Geophysical Research*, **98**, 7289–7315.

Henderson-Sellers, A., Howe, W. and McGuffie, K. (1995a) The MECCA analysis project. *Global and Planetary Change*, **10**, 3–21.

Henderson-Sellers, A., Pitman, A.J., Love, P.K., Irannejad, P. and Chen, T.H. (1995b) The project for intercomparison of landsurface parameterization schemes (PILPS): Phase 2 & 3. *Bulletin of the American Meteorological Society*, **76**, 489–503.

Kiehl, J.T., Wolski, R.J., Briegleb, B.P. and Ramanathan, V. (1987) Documentation of radiation and cloud routines in the NCAR community climate model (CCM1). NCAR Technical Note, NCAR/TN-288+IA, Boulder, Colorado.

Lean, J. and Warrilow, D.A. (1989) Simulation of the regional climatic impacts of Amazon deforestation. *Nature*, **342**, 411–413.

Lean, J. and Rowntree, P.R. (1992) *A GCM Simulation of the Impacts of Amazonian Deforestation on Climate Using an Improved Canopy Representation*, Climate Research Technical Note, No. 26, United Kingdom: Hadley Centre for Climate Prediction and Research, Meteorological Office.

Legates, D.R. and Willmott, C.J. (1990) Mean seasonal and spatial variability in gauge-corrected global precipitation. *International Journal of Climatology*, **10**, 111–127.

Louis, J.-F. (1979) A parametric study of vertical eddy fluxes in the atmosphere. *Boundary Layer Meteorology*, **17**, 187–202.

Manabe, S., Smagorinsky, J. and Strickler, R.F. (1965) Simulated climatology of a general circulation model with a hydrologic cycle. *Monthly Weather Review*, **93**, 769–798.

McAvaney, B.J., Dahni, R.R., Colman, R.A. and Fraser, J.R. (1995) The dependence of the climate sensitivity on convective parameterization: Statistical evaluation. *Global and Planetary Change*, **10**, 181–200.

McGuffie, K., Henderson-Sellers, A., Zhang, H., Durbidge, T.B. and Pitman, A.J. (1995) Global climate sensitivity to tropical deforestation. *Global and Planetary Change*, **10**, 97–128.

Milly, P.C.D. and Dunne, K.A. (1994) Sensitivity of the global water cycle to the water-holding capacity of land. *Journal of Climate*, **7**, 506–526.

Moran, E.F., Brondizio, E., Mausel, P. and Wu, Y. (1994) Integrating Amazonian vegetation, land-use, and satellite data. *Bioscience*, **44**, 329–338.

Nobre, C.A., Sellers, P.J. and Shukla, J. (1991) Amazonian deforestation and regional climate change. *Journal of Climate*, **4**, 957–988.

Numaguti, A. (1993) Dynamics and energy balance of the Hadley circulation and the tropical precipitation zones: Significance of the distribution of evaporation. *Journal of Atmospheric Science*, **50**, 1874–1887.

Pitman, A.J., Yang, Z.-L., Cogley, J.G. and Henderson-Sellers, A. (1991) Description of bare essentials of surface transfer for the Bureau of Meteorology Research Centre AGCM, BMRC Research Report No. 32.

Polcher, J. and Laval, K. (1994) A statistical study of the regional impact of deforestation on climate in the LMD GCM. *Climate Dynamics*, **10**, 205–219.

Salati, E. and Vose, P.B. (1984) Amazon Basin: A system in equilibrium. *Science*, **225**, 129–137.

Salati, E. (1987) The forest and the hydrological cycle. *The Geophysiology of Amazonia*, edited by R.E. Dickinson, pp. 273–296. John Wiley & Sons.

Slingo, A. (1989) A GCM parameterization for the shortwave radiative properties of water clouds. *Journal of Atmospheric Science*, **46**, 1419–1427.

Sud, Y.C., Walker, G.K., Kim, J.-H., Liston, G.E., Sellers, P.J. and Lau, K.-M. (1996) Biogeophysical effects of a tropical deforestation scenario: A GCM simulation study. *Journal of Climate*, **12(9)**.

Tiedtke, M. (1984) The effect of penetrative cumulus convection on the large-scale flow in a general circulation model. *Beitr. Phys. Atmos.*, **57**, 216–239.

Tiedtke, M. (1989) A comprehensive mass flux scheme for cumulus parameterization in large-scale models. *Monthly Weather Review*, **117**, 1779–1800.

Williamson, D.L., Kiehl, J.T., Ramanathan, V., Dickinson, R.E. and Hack, J.J. (1987) Description of NCAR community climate model (CCM1). NCAR Technical Note, NCAR/TN-285+STR, Boulder, Colorado.

Zhang, H., Henderson-Sellers, A. and McGuffie, K. (1996a) Impacts of tropical deforestation I: Process analysis of local climatic change. *Journal of Climate*, **9**, 1497–1517.

Zhang, H., McGuffie, K. and Henderson-Sellers, A. (1996b) Impacts of tropical deforestation II: The role of large scale dynamics. *Journal of Climate*, **9**, 2498–2521.

CHAPTER 13

THE CLIMATIC EFFECT OF DOUBLED STOMATAL RESISTANCE

David Pollard and Starley L. Thompson

13.1 Introduction

As atmospheric CO_2 concentrations increase in the next century, it is possible that many plant species will respond by increasing their stomatal resistance (Strain & Cure 1985; Eamus & Jarvis 1989; Bazzaz 1990; Bazzaz & Fajer 1992). In a CO_2-rich atmosphere, a plant can maintain the same intake of CO_2 for photosynthesis by reducing its stomatal openings, which also has the effect of reducing water loss by transpiration, i.e. its water-use efficiency increases. This response would therefore be beneficial in natural environments where water availability is the primary constraint on plant growth. In addition, smaller stomatal openings could improve the health of certain plants by limiting the entrance of air pollutants such as sulphur dioxide (Bazzaz & Fajer 1992). Even where water and nutrients are not limiting factors, plants in many natural environments may still not be able to incorporate CO_2 at faster rates (CO_2 'fertilization') due to internal biological constraints or inter-species competition (Bazzaz & Fajer 1992), so stomatal resistances might still increase in order to prevent the intracellular CO_2 concentration from rising. Direct experiments with isolated acclimatised temperate trees have shown that stomatal resistance does increase with elevated CO_2 levels in many species by factors of up to two or three, but the response varies widely and may depend on temperature, light levels, water availability and other factors (Eamus & Jarvis 1989, Table 13.2). Increases in both stomatal resistance and growth rate with elevated CO_2 have also been observed in controlled crop experiments (e.g. Cure 1985), with more stomatal closing in C_4 plants and greater increases in photosynthetic rates in C_3 plants (Rosenberg 1981).

Widespread reductions in transpiration caused by increases in stomatal resistance have the potential to significantly alter climate on regional and even global scales. General circulation models with explicit vegetation components have shown that in the tropics and in the northern boreal forests in summer, transpiration can be a large fraction of the total evaporation to the atmosphere (Nobre et al. 1991). The direct local effects would be to dry the lower atmosphere, warm the canopy due to reduced latent heat cooling, increase the upward sensible heat flux and the Bowen ratio, and increase soil moisture due to less water uptake for transpiration. However, a general circulation model with explicit physical models of vegetation and soil is needed to predict interactions with the atmosphere and the global climate system. Recently, Henderson-Sellers et al. (1995) have compared the effects of doubling stomatal resistance and doubling CO_2 in a global climate model, and discussed similarities and differences with our results.

The following sections describe a preliminary sensitivity test using a global climate model in which the stomatal resistance parameterization of all plants is uniformly doubled at all locations and at each time step. The global model used here, GENESIS version 1.02, is described in Thompson and Pollard (1995a). The GENESIS model has been under development for several years at NCAR for the purpose of performing greenhouse and paleoclimatic experiments. The model consists of an atmospheric general circulation model, multi-layer models of snow, soil and sea ice, sea-ice dynamics and a slab ocean mixed layer. As a prerequisite for the current experiment, it includes a land-surface-transfer scheme (LSX) that explicitly models the physical effects of vegetation including stomatal resistance and transpiration (Pollard & Thompson 1995).

Our choice of a factor of two increase in stomatal resistance is guided mainly by the observed values in Eamus and Jarvis (1989, Table 13.2), since they represent the net biological response of real plants. Although oversimplified, it is also instructive to consider the choice using an idealised resistance model (e.g. Gates 1980). In such a model a doubling of stomatal resistance would correspond to a doubling of atmospheric CO_2 only if: (1) plants act to maintain today's CO_2 intake rates; (2) stomatal resistance is much larger than intercellular and leaf boundary-layer resistances; and (3) CO_2 concentration in the chloroplasts is much less than in the ambient air. The latter is a good assumption, but the second is not: intercellular resistances are in fact somewhat larger than stomatal resistances (Rosenberg 1981). Taken at face value that would require even greater multiples of stomatal closure in order to keep the CO_2 intake rate constant.

Of course, increased stomatal resistance will not be the only response of global vegetation to elevated CO_2 levels and other climatic changes in the next century. For instance, CO_2 fertilization in areas with plentiful water and nutrients may cause an increase in net primary productivity and leaf area index and a net *increase* in transpiration (e.g. Rosenberg 1982). Furthermore, the geographic boundaries and/or species compositions of forests and other unmanaged ecosystems may change in perturbed climates (Botkin et al. 1973; Emanuel et al. 1985; Pastor & Post 1988; Prentice & Fung 1990; Smith et al. 1992). On somewhat longer time scales, changes in the vegetation can alter climatically important soil properties such as texture and albedo. However, in the experiment described below we isolate just one possible response in order to test its significance for the large-scale climate in our global model. For clarity in interpreting the results we will leave the model's prescription of plant distributions, leaf area indices, soil properties and the atmospheric CO_2 concentration unchanged. This will allow a clearer comparison of the magnitude of the changes caused by increased stomatal resistance with those of separate CO_2 doubling experiments.

In LSX, transpiration from green leaves to the canopy air enters into the main LSX equations at each time step, with stomatal resistance parameterized as a function of plant species, leaf temperature, visible sunlight incident on the leaves, and vapour pressure deficit. It is also constrained by a maximum soil-water uptake rate that depends on soil moisture in each soil layer (see Pollard & Thompson 1995, Appendix A7). Thus the instantaneous stomatal resistance is a function of several local environmental variables. For the perturbed experiment described below, we simply inserted a factor of 2 in the resistance parameterization, which does not necessarily result in doubled transpiration or even doubled stomatal resistances compared to the control case since the local environmental variables also change due to interactions within the climate system. However, we will use the phrase 'doubled stomatal resistance' loosely to refer to this factor of 2.

In Pollard and Thompson (1995), the response of LSX to doubling stomatal resistance is described for a few single points driven by prescribed meteorological forcing. Those results are not shown here for brevity, but they confirm the anticipated direct effects of doubling stomatal resistance in a dense canopy: decreases in transpiration and total latent heat flux of $\sim$50 to 100 W m^{-2} at midday, slightly smaller increases in upward sensible heat flux (i.e. increased Bowen ratios), and increases in surface-air temperatures of $\sim$0.5°C. Another effect was an increase in rooting-zone soil moisture (fractional amount relative to saturation) of $\sim$0.05. The results also suggested that the changes are more significant for dense (high LAI) forests than for grasslands. Flux perturbations of this magnitude would of course drive the regional atmosphere away from the prescribed conditions used for those single point results, and motivate the use of a fully coupled global GCM as described in the next section.

13.2 Global results

Results of two simulations with the global climate model are presented below, divided into subsections for zonal mean fields, global surface fields and atmospheric cross-sections. The control run is a present-day simulation using GENESIS version 1.02 with present-day stomatal resistances (Thompson & Pollard 1995a), previously spun up for several decades to achieve climatic equilibrium. The only change for the perturbed run is the additional factor of 2 in the stomatal resistance parameterization, initialised to the control climate and spun up for a further decade. All results shown here are based on monthly means from ten additional years of integration for both runs.

13.2.1 Zonal-mean surface fields

Figures 13.1 to 13.4 show July monthly mean changes due to doubled stomatal resistance, averaged over land only. Although there are significant changes over ocean (section 13.2.2), the changes over land are larger and more relevant to human society. Throughout this section we concentrate on the month of July since the largest changes in mid and high latitudes occur in the northern hemisphere growing season, and changes in the tropics are relatively constant year round.

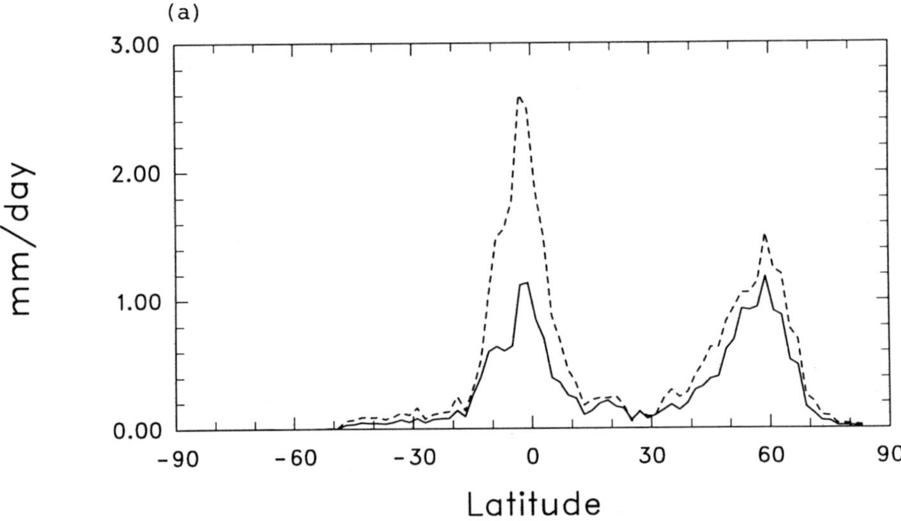

Figure 13.1(a) Zonal-mean transpiration rate in global model simulations, for July and over land surfaces only, in mm day^{-1}. Dashed line: control (present-day stomatal resistances); solid line: doubled stomatal resistance parameterization.

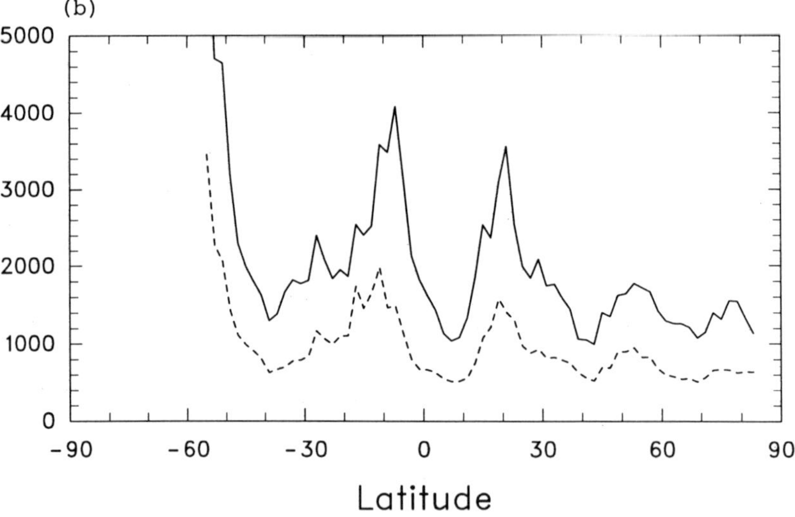

Figure 13.1(b) Zonal-mean stomatal resistance, for July and over land surfaces only, in sm^{-1}. Dashed line: control; solid line: doubled stomatal resistance parameterization. This is basically the quantity r given by Equation A42 of Pollard and Thompson (1995). First the stomatal conductances r_u^{-1} and r_l^{-1} were first weighted by the respective $f \times$ LAI, and the zonal mean of the sum over vegetated points only was taken. The inverse of the result at each latitude is plotted.

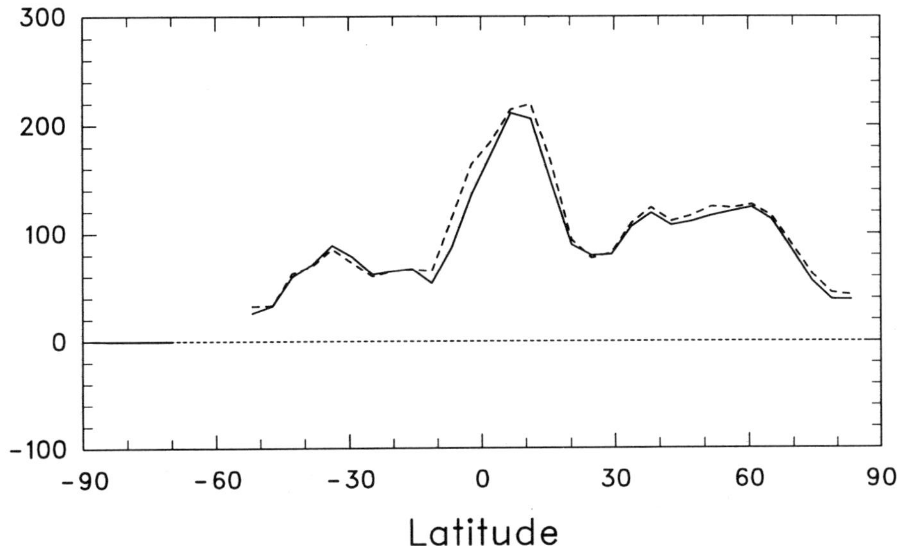

Figure 13.2 Zonal-mean net upward latent heat flux (evapotranspiration), for July and over land surfaces only, in W m^{-2}. Dashed line: control; solid line: doubled stomatal resistance parameterization.

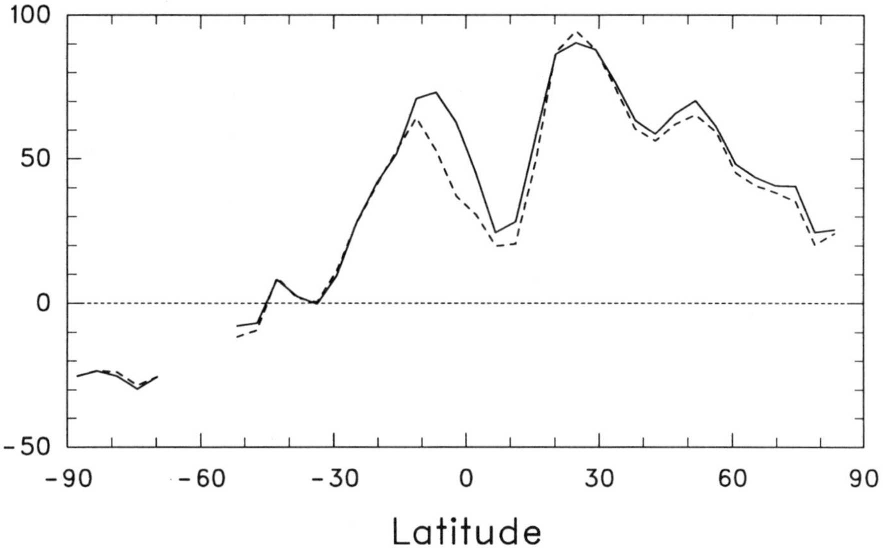

Figure 13.3 Zonal-mean net upward sensible heat flux, for July and over land surfaces only, in W m^{-2}. Dashed line: control; solid lines: doubled stomatal resistance parametrization.

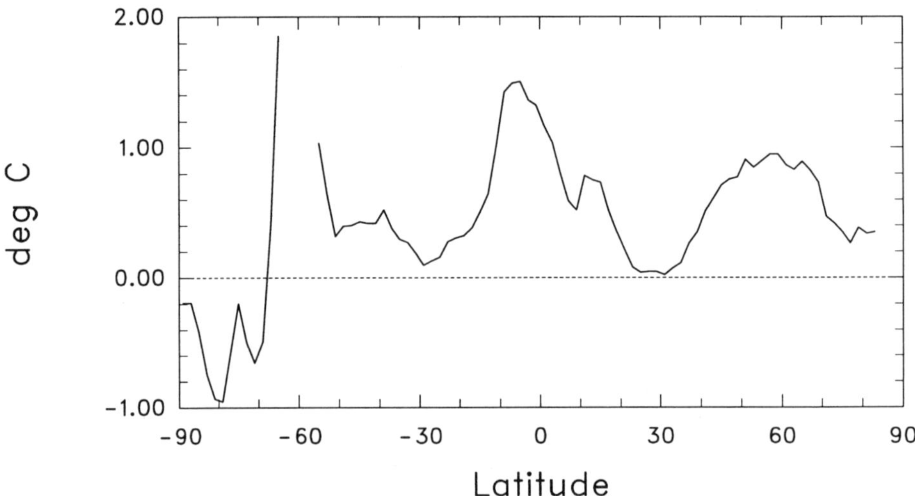

Figure 13.4 Zonal-mean *change* in 2 m surface-air temperature due to doubling the stomatal resistance parameterization, for July and over land surfaces only, in °C. Dashed line: control; solid line: doubled stomatal resistance parameterization.

Figure 13.1(a) confirms that transpiration is reduced everywhere. In the tropics (dominated by tropical forests in Amazonia, the Congo and the Far East) transpiration has been reduced by about a half, consistent with the single point results in Pollard and Thompson (1995). However, in the northern mid and high latitudes including the boreal forests, transpiration has only been reduced by approximately one quarter, more like the single point results for grassland. As mentioned above, the factor of 2 inserted in the stomatal resistance parameterization does not necessarily cause doubled transpiration or even doubled stomatal resistance, due to changes in the local environmental variables, especially soil moisture stress and leaf-air vapour gradients. Figure 13.1(b) shows that the zonal mean stomatal resistance does approximately double at most latitudes, suggesting that the lesser reduction in boreal forest transpiration rates in Figure 13.1(a) is due more to changing leaf-air vapour gradients.

The differences in the latent and sensible heat flux curves in Figures 13.2 and 13.3 are largest in the tropics ($\sim$20 W m^{-2}) and in the boreal forests around $\sim$60°N ($\sim$7 W m^{-2}). The changes are the same sign but smaller than those in single point results (Pollard & Thompson 1995). However, the zonal mean increases in surface-air temperature in Figure 13.4 are $\sim$1°C in both the tropics and high northern latitudes, about double the increase found in the single point results. This indicates that significant feedbacks are operating in the surface-atmosphere system to amplify the surface warming. Although we have not done enough analysis to conclusively identify the most important mechanisms, a likely candidate is the decrease in cloud amounts due to drying of the atmosphere (Figure 13.9).

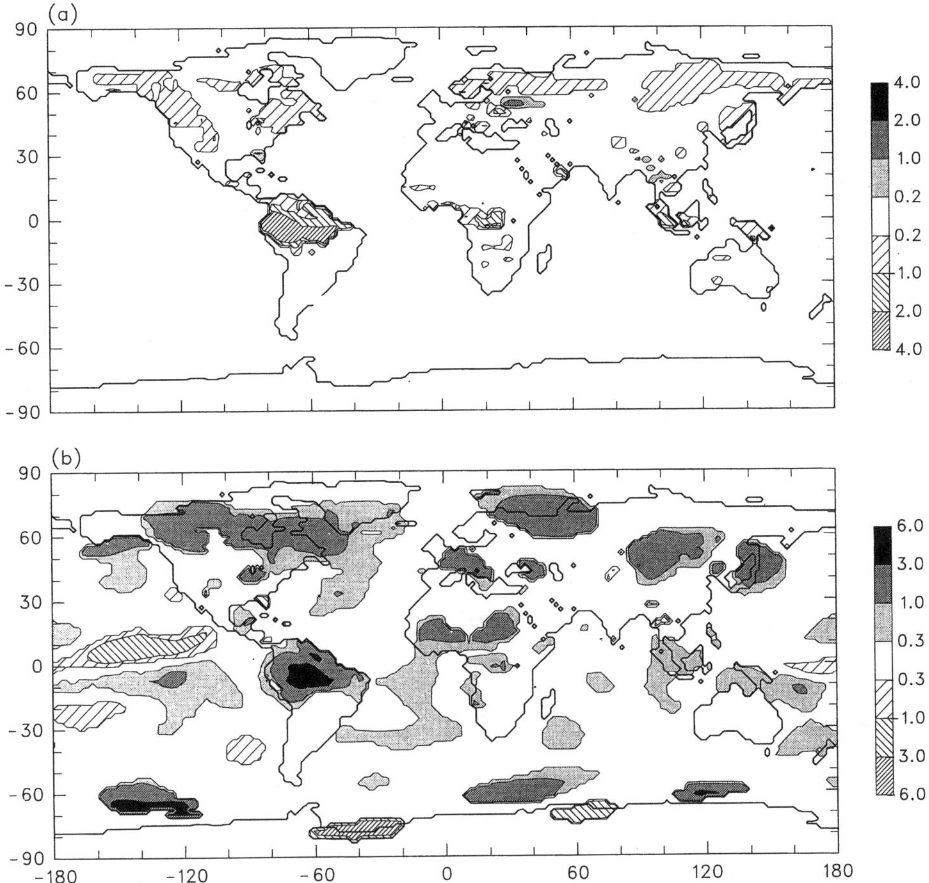

Figure 13.5 Changes in July monthly means due to a uniform doubling of the stomatal resistance parameterization, in global model simulations. Only points with Student's t-test significance level of $\leq 10\%$ are shown. (a) Change in transpiration rate in mm day^{-1} and (b) change in 2 m surface-air temperature, in °C.

As discussed further below, the interannual variability of GENESIS surface temperatures is very high in the south polar region, and the changes in Figure 13.4 poleward of $\sim 70°$S reflect this spurious signal; they are very different in the equivalent plot for five earlier independent years of model integration. However, the zonal mean changes at all other latitudes and for the other variables in Figures 13.1 to 13.4 are highly repeatable and essentially unaffected by model interannual variability.

13.2.2 Global surface fields

Global maps of the changes in various near surface fields are shown for July in Figures 13.5 to 13.7, averaging over the ten Julys of the model runs. Since the changes in these ten

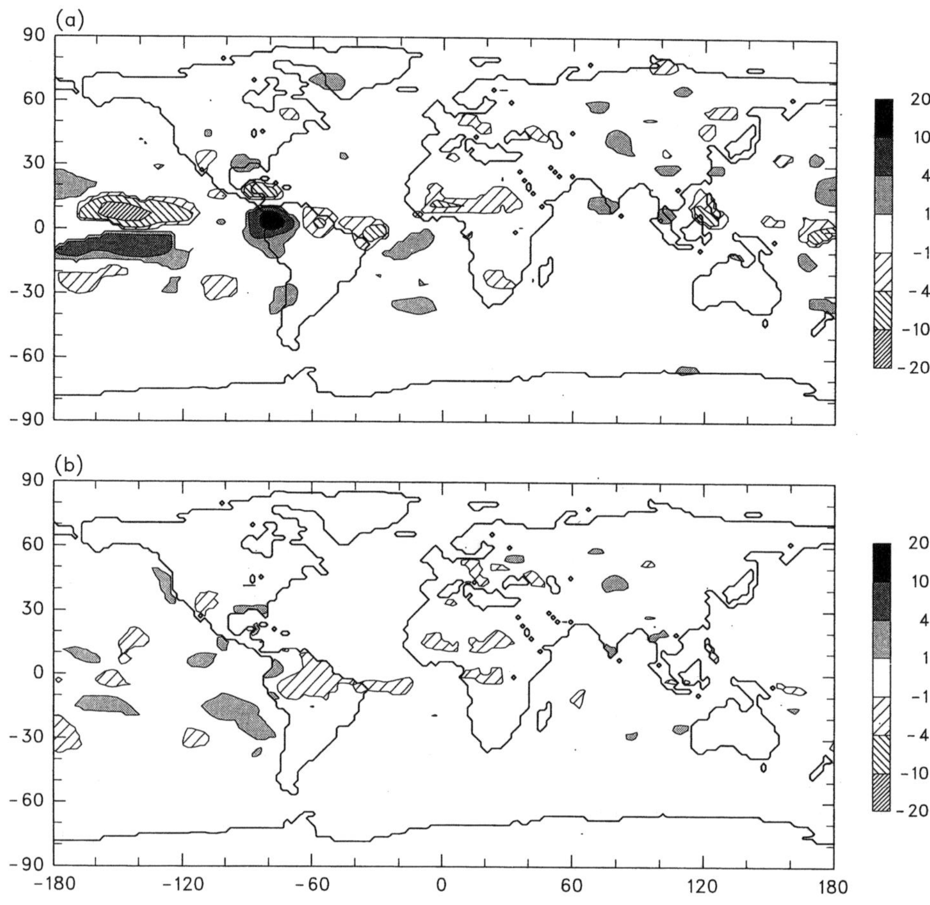

Figure 13.6 Changes in July monthly means as in Figure 13.5 except for: (a) change in precipitation in mm day^{-1} and (b) change in evapotranspiration in mm day^{-1}.

year means are quite small in some regions compared to the natural interannual variability, the statistical significance of the results is an important concern. Following Chervin and Schneider (1976) we have calculated a Student's t-test variate at each point by dividing the change in the ten year means (the signal) by the combined standard deviation of the individual July means (the noise), and rejected points from Figures 13.5 to 13.7 at the 10% significance level (i.e. $\geq 10\%$ probability that the change could be due to random interannual variability alone). We set the significance level at 10% and chose the minimum contour value for each plot in order to evoke only the large scale features; that is, if larger significance levels and/or smaller contour values were used, more small scale 'noise' would appear but no other major features would emerge.

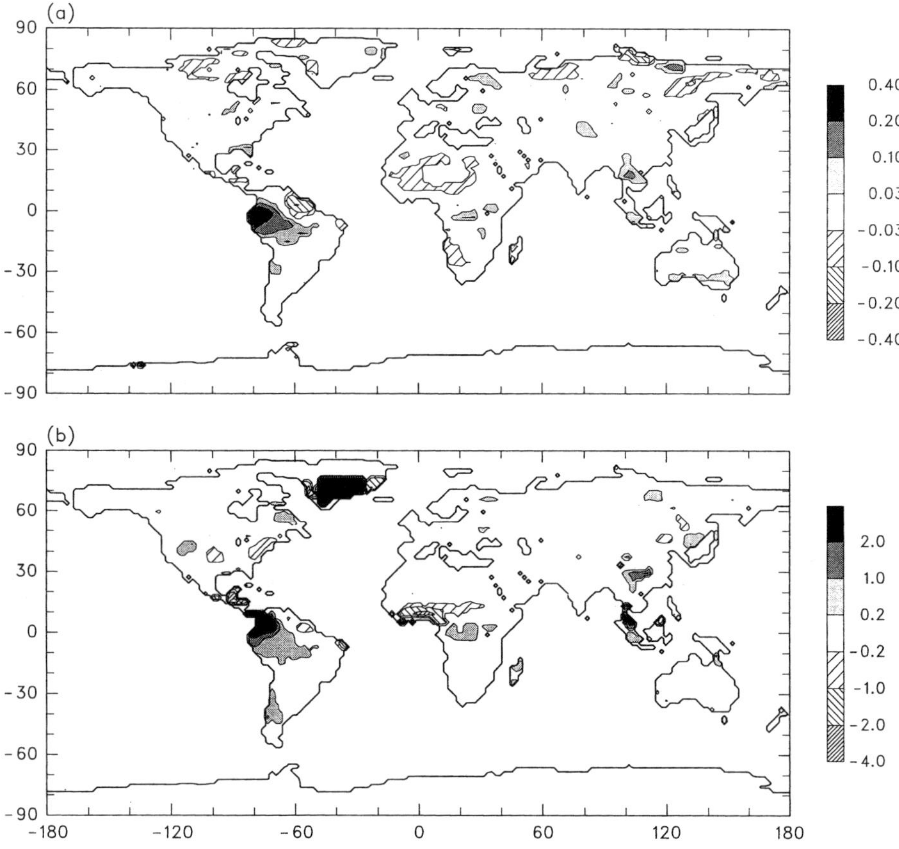

Figure 13.7 Changes in July monthly means as in Figure 13.5 except for: (a) change in fractional soil moisture (liquid plus ice content relative to saturation) averaged over the top four soil layers, i.e. 0 to 75 cm depth, and (b) change in surface runoff plus bottom drainage, in mm day^{-1}.

The largest reduction in transpiration in Figure 13.5(a) occurs in tropical South America, with widespread lesser reductions in the northern hemispheric boreal forests. As expected from the zonal mean results above, surface-air temperatures over land increase in roughly the same areas (Figure 13.5(b)), with warming up to ~4°C in Amazonia and ~2°C in boreal forest regions. (The boreal forest patches in Figures 13.5(a) and 13.5(b) do not coincide exactly, but the apparent displacements are exaggerated by masking of adjacent regions due to the 10% significance criterion). A similar but smaller scale pattern of warming and reduced transpiration also occurs in the West African equatorial rainforest. Since the changes in precipitation in these regions are quite small (Figure 13.6(a)), we infer that the reduction in transpiration directly causes the surface warming as in the single point runs in Pollard and Thompson (1995).

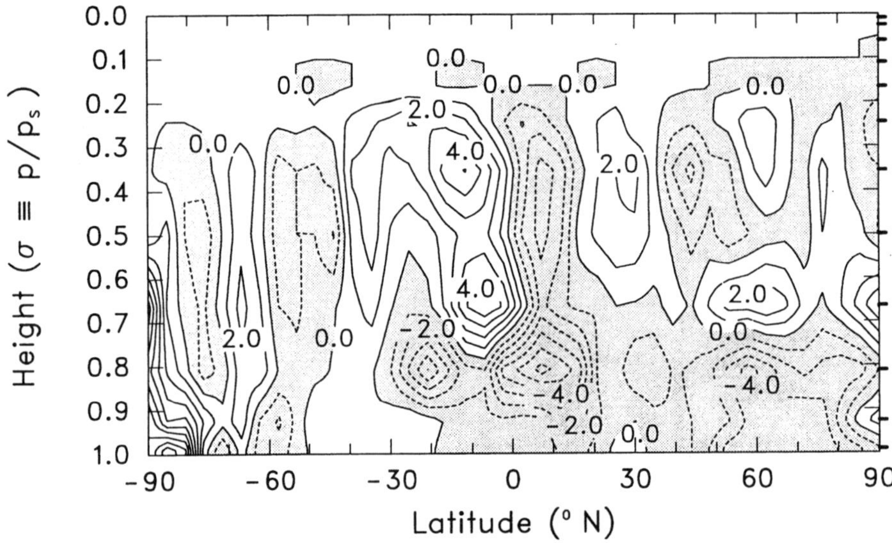

Figure 13.8 Changes in the zonal mean relative humidity for July versus latitude and height due to a uniform doubling of the stomatal resistance parameterization, in %. Regions with negative changes are shaded. σ = pressure/surface pressure is the AGCM vertical coordinate, and the thick tick marks show the centres of the 12 model layers.

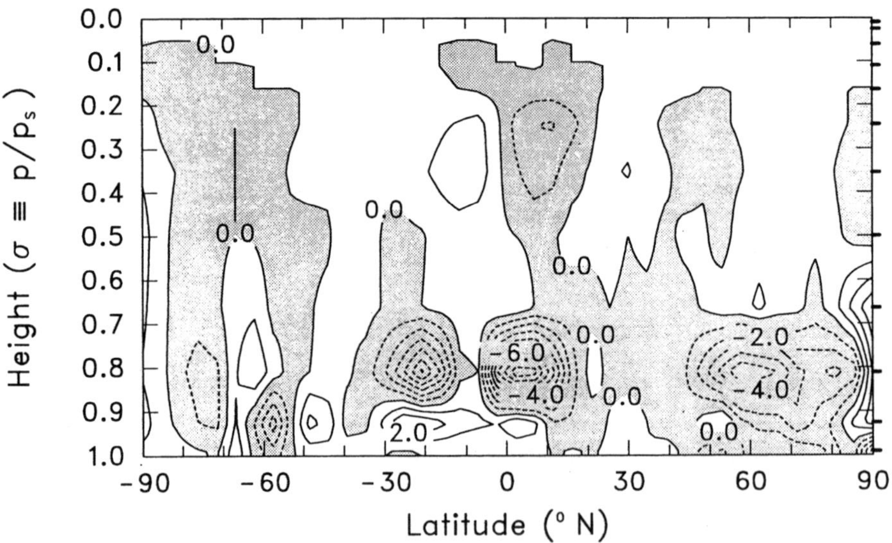

Figure 13.9 Changes in the zonal mean cloud fraction for July, in %.

The stomatal doubling perturbation causes a large change in the July precipitation pattern in the eastern tropical Pacific ocean (Figure 13.6(a)), where the pronounced north–south dipole pattern indicates the ITCZ has shifted southwards. We have noticed this same pattern in other sensitivity experiments with GENESIS, for instance doubled atmospheric CO_2, and it may reflect a canonical jump of the model tropical circulation between two quasi-stable modes. (Since our slab ocean model has no dynamics or vertical structure this effect cannot be directly related to the traditional El Niño.) There is a large increase in precipitation in Colombia, probably induced by the ITCZ displacement in the Pacific, and large scale decreases in parts of northern subtropical Africa around $\sim$10–15°N. Apart from those two regions there are no widespread changes in precipitation over land. One would expect the reductions in transpiration in Figure 13.5(a) might be reflected in net evaporation changes in Figure 13.6(b), but this is true only in Amazonia and the equatorial African rainforest.

One would also expect soil moisture in the rooting zone to increase in regions with reduced transpiration, but this is not born out to any significant degree in Figure 13.7(a). The only large-scale increase occurs in western equatorial South America, coincident with a significant increase in total runoff (Figure 13.7(b)). This feature is likely caused directly by the increase in precipitation centred over Colombia in Figure 13.6(a), although its extent to the east may be helped by the lesser transpiration rates in the Amazon. A smaller patch of increased moisture and runoff in the West African rainforest is also probably due directly to reduced transpiration. In other land regions there are no widespread large changes in the soil hydrology, with the exception of soil drying in parts of the Sahel due to reduced precipitation noted above.

In general, these global maps show that large and statistically significant changes in surface climate in July occur in tropical South America, with smaller but still significant changes in surface-air temperature in some regions of Canada, northern Russia and northern Asia. These locations coincide with widespread dense forests and so it is not surprising the largest responses occur there. However, the only appreciable changes to surface hydrology (soil wetness, runoff) occur in western equatorial South America and subtropical Africa, primarily due to shifts in ITCZ precipitation.

13.2.3 Atmospheric zonal cross-sections

The zonal cross-sections in Figures 13.8 to 13.11 show some of the changes induced in the troposphere in July by doubling stomatal resistance. Note that these zonal means include land and ocean, so the changes are probably smaller than those over land alone. We are confident that nearly all of the features in these figures are true long-term climatic changes and are not contaminated by interannual variability, since they are very similar to equivalent figures for five earlier independent years of model integration. The exceptions are the polar regions, poleward of $\sim$70°S and $\sim$80°N, where the model interannual variability tends to be very high and the two sets of figures differ wildly.

As expected, the greatest reductions in relative humidity (Figure 13.8) occur over the tropics and mid and high northern latitudes. Although the drying is largest in the

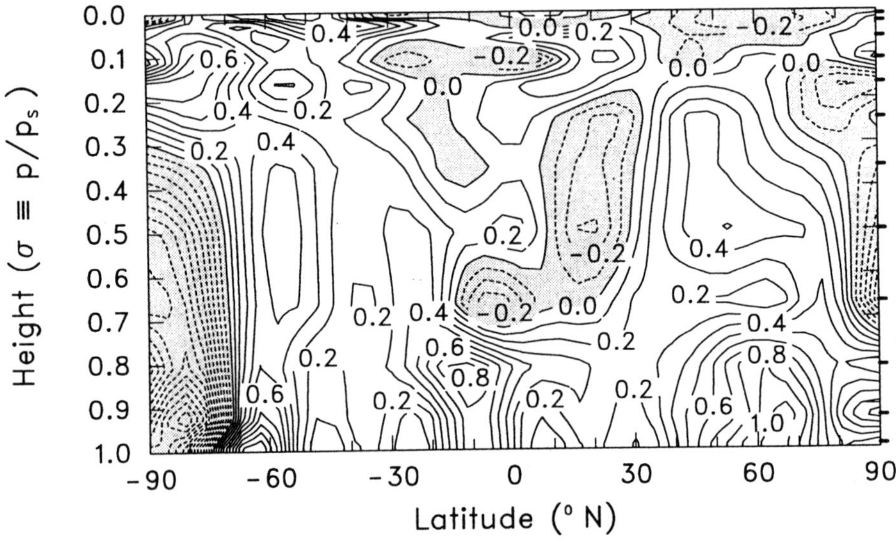

Figure **13.10** Changes in the zonal mean air temperature for July, in °C.

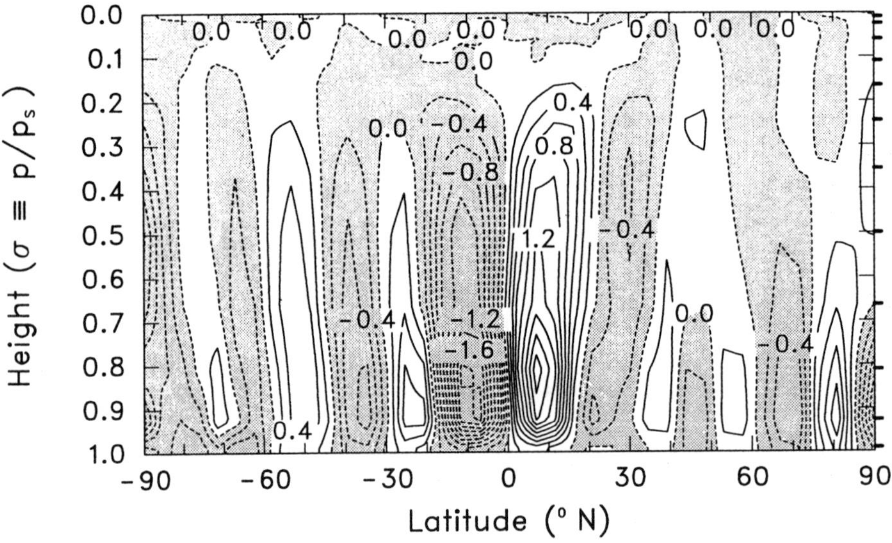

Figure **13.11** Changes in the zonal mean vertical velocity dp/dt for July, in 10^{-2} Pa s^{-1}.

lowest $\sim$3 km (below $\sim$700 mb), changes as large as 2% occur throughout the troposphere. Figure 13.9 shows similar reductions in cloudiness that are well correlated with the relative humidity changes below $\sim$700 mb. These cloudiness reductions are possibly the main cause of the $\sim$0.5°C warming in zonal mean surface-air temperatures in Figure 13.4; the zonal mean of total cloudiness over land decreases by about .05 fractional area at most latitudes (not shown). As shown in Figure 13.10, this warming does extend throughout much of the troposphere, except for the cooling at higher levels at $\sim$25°N in the descending northern branch of the Hadley Cell. This coincides with a region of increased cloudiness which in turn may be associated with a general weakening of the Hadley circulation itself, as shown by the changes in the vertical velocity field in Figure 13.11.

Except for specific humidity (not shown) and temperature, the atmospheric changes for July are about the same size and generally opposite in sign to the annual mean changes induced in the same model by doubled atmospheric CO_2 (Thompson & Pollard 1995b). In doubled CO_2 scenarios both temperature and specific humidity increase, with near cancelling effects on relative humidity and stratus cloud amounts. In contrast, increased stomatal resistance causes relatively small increases in atmospheric temperature and *decreases* in specific humidity ($\sim$ $1/4$ to $1/10$ of the doubled CO_2 changes), which combine to give relatively large decreases in relative humidity and cloudiness. If both perturbations were applied simultaneously, we expect that the larger CO_2-induced changes in temperature and specific humidity would dominate, but with important modulating influences due to stomatal resistance especially over the tropics and northern high latitude boreal forests in summer.

13.3 Summary

The climatic effects of a uniform doubling of stomatal resistance in our global climate model are largest in Amazonia, the West African equatorial rainforest and in parts of the northern hemispheric boreal forests in summer in Canada, northern Russia and Siberia. These regions experience decreases in transpiration of 20% to 50% which directly causes surface warming of $\sim$2 to 5°C. There are significant feedbacks on the climate, notably north–south shifts in the ITCZ which cause large increases in precipitation, soil moisture and runoff in western tropical South America and decreases in these variables in northern subtropical Africa. These changes are statistically significant compared to natural interannual variability at the 90% confidence level (and almost all of them are still significant at the 95% level, not shown).

Changes in the atmosphere are largest in the lowest $\sim$3 km but do not extend throughout the troposphere. The main drying effects lower relative humidity and clouds, and a weakening of the Hadley Cell, are generally of comparable magnitude and opposite to those associated with doubled atmospheric CO_2, at least in July. However, the primary changes in atmospheric temperature and specific humidity are considerably smaller than those due to doubled CO_2, so that CO_2 greenhouse climate change should still dominate on global scales when both perturbations are applied simultaneously (cf. Henderson-Sellers et al. 1995). (The global mean surface warming due to increased stomatal resistance alone in July is a few tenths of a $°C$, partly due to the increased Bowen ratio and partly due to the feedback of reduced cloudiness.)

As discussed in the introduction, widespread increases in plant stomatal resistance are firstly not certain to occur in all species and ecosystems, and secondly by no means the only important changes in vegetation likely to be induced by future climatic variations. Nevertheless, our results suggest that in at least one instance the plant response to a large scale environmental perturbation can have significant regional and even global feedbacks on the physical climate. Our experiment has dealt with one possible response of stomatal resistances to increasing CO_2 levels. Other GCM studies involving climate-vegetation interactions include the geographical redistribution of natural life forms in response to global warming (Friedlingstein et al. 1992), regional changes due to Amazonian deforestation (e.g. Nobre et al. 1991) and northern hemispheric cooling due to boreal forest deforestation (Thomas & Rowntree 1992; Bonan et al. 1992), all of which have found significant climatic effects on regional or global scales. As GCM predictions of future climatic change continue to improve in the coming years, it will become increasingly important to include the biological response of vegetation and subsequent physical interactions with the climate.

Acknowledgments

The development of the GENESIS earth systems model at NCAR is supported in part by the US Environmental Protection Agency Interagency Agreement No. DW49935658-01-0. The National Center for Atmospheric Research is sponsored by the National Science Foundation.

References

Bazzaz, F.A. (1990) The response of natural ecosystems to the rising global CO_2 levels. *Annual Review of Ecological Systems*, **21**, 167–196.

Bazzaz, F.A. and Fajer, E.D., 1992. Plant life in a CO_2-rich world. *Scientific American*, **266**, 1, 68–74.

Bonan, G.B., Pollard, D. and Thompson, S.L. (1992) Effects of boreal forest vegetation on global climate. *Nature*, **359**, 716–718.

Botkin, D.B., Janek, J.F. and Wallis, J.R. (1973) Estimating the effects of carbon fertilization on forest composition by ecosystems simulation. In *Carbon and the Biosphere*, edited by G.M. Woodwell and Pecan E.V., pp. 328–344. US Atomic Energy Commission, Symposium Series 30.

Chervin, R.M. and Schneider, S.H. (1976) On determining the statistical significance of climate experiments with general circulation models. *Journal of Atmospheric Science*, **33**, 405–412.

Cure, J.D. (1985) Carbon dioxide doubling responses: a crop survey. In *Direct Effects of Increasing Carbon Dioxide on Vegetation*, edited by B.R. Strain and Cure J.D., pp. 99–116. US Department of Energy Report DOE/ER-0238, Washington, DC.

Eamus, D. and Jarvis, P.G. (1989) The direct effects of increase in the global atmospheric CO_2 concentration on natural and commercial temperate trees and forests. *Advances in Ecological Research*, **19**, 1–55.

Emanuel, W.R., Shugart, H.H. and Stevenson, M.P. (1985) Climatic change and the broad-scale distribution of terrestrial ecosystem complexes. *Climatic Change*, **7**, 29–43.

Friedlingstein, P., Delire, C. , Muller, J.F. and Gerard, J.C. (1992) The climate induced variation of the continental biosphere: a model simulation of the last glacial maximum. *Geophysical Research Letters*, **19**, 897–900.

Gates, D.M. (1980) *Biophysical Ecology*. Springer-Verlag Publishers.

Henderson-Sellers, A., McGuffie, K. and Gross, C. (1995) Sensitivity of global climate model simulations to increased stomatal resistance and CO_2 increases. *Journal of Climate*, **8**, 1738–1756.

Nobre, C.A., Sellers, P.J. and Shukla, J. (1991) Amazonian deforestation and regional climate change. *Journal of Climate*, **4**, 957–988.

Pastor, J. and Post, W.M. (1988) Response of northern forests to CO_2-induced climate change. *Nature*, **344**, 55–58.

Pollard, D. and Thompson, S.L. (1995) Use of a land-surface-transfer scheme (LSX) in a global climate model: The response to doubling stomatal resistance. *Global Planetary Change*, **10**, 129–161.

Prentice, K.C. and Fung, I.Y. (1990) The sensitivity of terrestrial carbon storage to climate change. *Nature*, **346**, 48–51.

Rosenberg, N.J. (1981) The increasing CO_2 concentration on the atmosphere and its implications on agricultural productivity. *Climatic Change*, **3**, 265–279.

Rosenberg, N.J. (1982) The increasing CO_2 concentration on the atmosphere and its implications on agricultural productivity. II. Effects through CO_2-induced climatic change. *Climatic Change*, **4**, 239–254.

Smith, T.M., Leemans, R. and Shugart, H.H. (1992) Sensitivity of terrestrial carbon storage to CO_2-induced climate change: Comparison of four scenarios based on general circulation models. *Climatic Change*, **21**, 367–384.

Stössell, A., Lemke, P. and Owens, W.B. (1990) Coupled sea ice—mixed layer simulations for the southern ocean. *Journal of Geophysical Research*, **95**, 9539–9555.

Strain, B.R. and Cure, J.D. (1985) *Direct Effects of Increasing Carbon Dioxide on Vegetation*. US Department of Energy Report DOE/ER-0238, Washington, DC.

Thomas, G. and Rowntree, P.R. (1992) The boreal forests and climate. *Quarterly Journal of the Royal Meteorological Society*, **118**, 469–497.

Thompson, S.L. and Pollard, D. (1995a) A global climate model (GENESIS) with a land-surface-transfer scheme (LSX). Part 1: Present-day climate. *Journal of Climate*, **8**, 732–761.

Thompson, S.L. and Pollard, D. (1995b) A global climate model (GENESIS) with a land-surface-transfer scheme (LSX). Part 2: CO_2 sensitivity. *Journal of Climate*, **8**, 1104–1121.

PART 5

IMPACTS OF CLIMATE CHANGE

CHAPTER 14

COMMENTARY

Sascha Schubert

14.1 Introduction

Recently rising public discussion about future anthropogenic induced climate change has raised the concern about the consequences those changes might have for the natural environment and human society. The identification of many human and natural systems which are directly or indirectly influenced by climate has been an important element in the fast development of climate impact research. Assessments of climate impacts, however, require information about climate change with spatial and temporal resolutions corresponding to the scale of the investigated system. While hydrological models for estimating area, precipitation and runoff need climate data on a regional scale (several 100 km^2) (Dooge 1992), the simulation of soil erosion processes requires local (point to 1 km^2) climate information (Favies-Mortlock 1994).

The most severe impacts of climate are often caused by extreme events (storms, tropical cyclones, droughts, floods, frost, heat waves) (Tonkin et al., Chapter 17), which occur on a time scale of days to weeks (Katz & Brown 1992). Climate impact assessments, therefore, require information not only on a fine spatial scale but also with a high temporal resolution. Furthermore, climate vulnerable systems are affected by several climate parameters which are related to each other. For instance, storms are often accompanied by flooding rain while droughts are often accompanied by heat waves. Information provided for climate impact research has to reproduce these relationships accurately in order to give a meteorologically consistent picture of future climate change.

The most fully developed tools for the estimation of future climate change are global climate models (GCMs). In other chapters of this book much has been written about the strengths and the weaknesses of these models. Regarding climate impact research, the most important limitation of the GCMs is their coarse horizontal resolution, which is currently a few hundred kilometres. At present limited computer resources prevent the GCMs from being run at the required high horizontal resolutions. Even if the models could be run on much higher resolutions there remain problems to be solved, for example, the problem of parameterization. This term describes the technique developed for the presentation of processes on a scale smaller than the model resolution. The current parameterization schemes developed for the coarse-resolution models may not be suitable for high-resolution models (cf. Tanaka & Hasegawa, Chapter 4).

The coarse horizontal resolution of the GCMs is thought to have two effects. Firstly, the models do not resolve high resolution features of the general atmospheric circulation and, thus, they exhibit rather poor performance in simulating regional present-day climate (cf. Henderson-Sellers & Howe, Chapter 18). This is mainly caused by the strict limitations the coarse horizontal resolution sets on the size of eddies which can be resolved by the model. All energy exchange by eddies smaller than the resolution of the GCM is not explicitly simulated in the model but described in a very general way by large-scale parameters. Secondly, the failure to capture smaller scale forcing can contribute to errors in the simulations, even at the global scale, because the feedback of sub grid scale circulation systems to the global flow is not modelled realistically.

However, it should be noted here that these limitations have only become important with the changing role of the GCMs over the last few years. Initially designed as a tool for the study and better understanding of the climate system the GCMs have been used more and more for producing estimations or predictions of future climate change. While many GCMs reliably simulate the global scale climate, their original design goal, their performance on a regional scale, remains poor. The same statement is true for the reliability of the GCMs on a temporal resolution higher than monthly averages (Cao et al. 1992; Whetton et al. 1994). Unfortunately, these are the temporal and spatial dimensions of the climate information requested by the climate impact community.

14.2 Methods to downscaling global climate model results

This scale mismatch or inconsistency (Storch 1994) between GCM-produced climate change predictions and the climate impact models has led to the development of several approaches to derive more detailed regional climate change scenarios on the basis of GCM simulations (Kim et al. 1984; Mearns et al. 1984; Wilks 1989; Wigley et al. 1990; Karl et al. 1990; Giorgi et al. 1990; Hay et al. 1992; Storch et al. 1993; Hewitson & Crane 1992; Giorgi et al. 1994; Frey-Buness et al. 1995; Heyen & Storch 1995; Déqué & Piedelievre 1995). These techniques are generically termed 'downscaling'.

The basic idea of all downscaling approaches is that the regional climate variability is driven by forcing on different spatial scales; the macroscale (1000 km to global) and the mesoscale (1–100 km) (Giorgi & Mearns 1991). The large-scale forcing (jet stream, semipermanent pressure systems) regulate the general circulation which in turn determines the succession of weather events in a certain region. Mesoscale forcing, such as topographic features, surface characteristics, and sea-surface temperature anomalies, modify the structure of the weather events and thus contribute to the regional distribution of the climate variables. Building on this concept of scale interaction the GCMs are used to estimate the consequences of changing external forcing (e.g. the increasing CO_2 content of the atmosphere) to the large-scale general circulation. The regional modification of the changed circulation regime can then be calculated using either modelling or statistical techniques.

14.2.1 Modelling approaches

Modelling techniques attempt to describe explicitly the effect of mesoscale forcing by embedding a model with higher horizontal resolution in the coarse GCM over an area of interest. These Limited Area Models (LAMs) compute the impact of the mesoscale forcing to the large-scale flow simulated by the GCM (Giorgi 1990; Giorgi et al. 1994; Déqué & Piedelievre 1995).

The way of interaction between the two models can be realised as one-way or two-way interaction. In the case of one-way interaction information from the coarse GCM is given to the smaller scale LAM in the form of boundary conditions. The smaller scale model, however, does not affect the larger scale atmospheric processes. The two-way interaction allows for the feedback of the smaller scale circulation systems on the large-scale flow. However, the problems connected with the two-way interaction have not been satisfactorily solved for regional climate simulations, though only one-way interactive model simulations have been investigated so far (Giorgi 1995). The LAMs require huge computer resources. For this reason they are mostly run over time periods of a few months or years, far too short for investigating changes in climate variability.

The more detailed representation of the topography in the LAM yields a more realistic simulation of some of the regional structures of the climate variables, particularly orographically caused precipitation. However, the one-way LAMs are not able to improve the performance of the GCMs. Unrealistic large-scale boundary conditions will therefore produce an unrealistic mesoscale climate (Machenhauer 1995).

Another way of downscaling using dynamical models is the approach introduced by Tonkin, Holland, Landsea and Li in Chapter 17, perhaps better described as 'upscaling'. Here climate impact models are established, driven by parameters in a spatial scale comparable to the horizontal resolution of the GCMs. For example, the Holland model (Holland 1995) calculates the maximum potential intensity (MPI) of tropical cyclones based on the sea surface temperature and the vertical temperature profile. Since these parameters are spatially rather invariant in the tropics, the GCM simulated temperatures can be directly used as input for the impact model. The application to $1 \times CO_2$ control runs and $2 \times CO_2$ scenarios of six MECCA-sponsored GCMs yields a general agreement of potentially more intense tropical cyclones in the north-west Pacific in a greenhouse warmed climate.

14.2.2 Statistical approaches

Instead of describing the mesoscale forcing with dynamic models the interactions between the scales can alternatively be simulated by statistical means. This approach has enjoyed a rapid development over the last few years, during which the applicability of many different statistical methods; multiple regression (e.g. Karl et al. 1990; Wigley et al. 1990; Hewitson

& Crane 1992), canonical correlation (e.g. Storch et al. 1993; Heyen & Storch 1995) and stochastic simulation (e.g. Bardossy & Plathe 1992; Matyasovszky et al. 1994; Schubert 1994) have been explored. All methods, however, translate the large-scale GCM information into a high resolution climate distributions based on empirically derived relationships (Figure 14.1).

A convincing example of the simplest downscaling method and its effect on the assessment of climate impact is given by Ciret in Chapter 16. It is shown that the results of climate impact models driven by GCM-simulated input parameters are strongly sensitive to the spatial pre-processing of the input. In the case presented, a vegetation model was forced either directly by GCM output or by pre-processed GCM results. In the latter, the change in temperature, precipitation and cloudiness between $1 \times CO_2$ and $2 \times CO_2$ GCM experiments was appended to a baseline climate, itself derived from observed climate records. Even though the relative changes of the input parameters due to an enhanced greenhouse effect are the same in these two climate change scenarios, the estimated variations in the vegetation distribution differ considerably because of the non-linear response of the vegetation models to climate forcing. Since the results derived with the pre-processed input parameters seem more realistic a direct use of GCM output as input for impact models is discouraged.

Thus, besides the uncertainties related to the GCM predictions, the use of different downscaling methods for constructing regional climate change scenarios is expected to add another component to the uncertainty 'explosion' (Henderson-Sellers 1993) on the way to climate impact assessments.

Stochastic simulation is applied in order to overcome the underestimation of extreme events when regular statistical methods (e.g. regression) are used. By explicitly including the variability of the observed regional or local climate long time series of climatic variables with more realistic statistical properties (mean, variances, occurrence of extreme events) can be generated. There are two different approaches to producing stochastically-derived climate change scenarios. Firstly, the statistical properties of the distributions of the climate variables are calculated from observations. According to the changes in the GCM experiments these parameters are altered and used for the stochastic generation of the 'new' time series (Mearns et al. 1984; Wilks 1992).

Alternatively, the stochastic model is linked to large-scale circulation parameters. Changes in the frequencies of these circulation patterns, as calculated by a GCM experiment, are then used to simulate time series of climate variables which reflect the new conditions of the large-scale circulation (Bardossy & Plathe 1992; Hay et al. 1992; Schubert 1994).

The most widely used statistical downscaling tool however is perfect prognosis (Perfect Prog), a method known from numerical weather predictions. Here, the relationships between observed large-scale and observed regional climate are used to establish the statistical model which is later applied to the GCM-simulated large-scale parameters (e.g. Karl et al. 1990; Wigley et al. 1990; Storch et al. 1993; Hewitson & Crane 1992; Heyen & Storch 1995).

In order to successfully apply statistical downscaling techniques several conditions have to be fulfilled according to Storch et al. (1993):

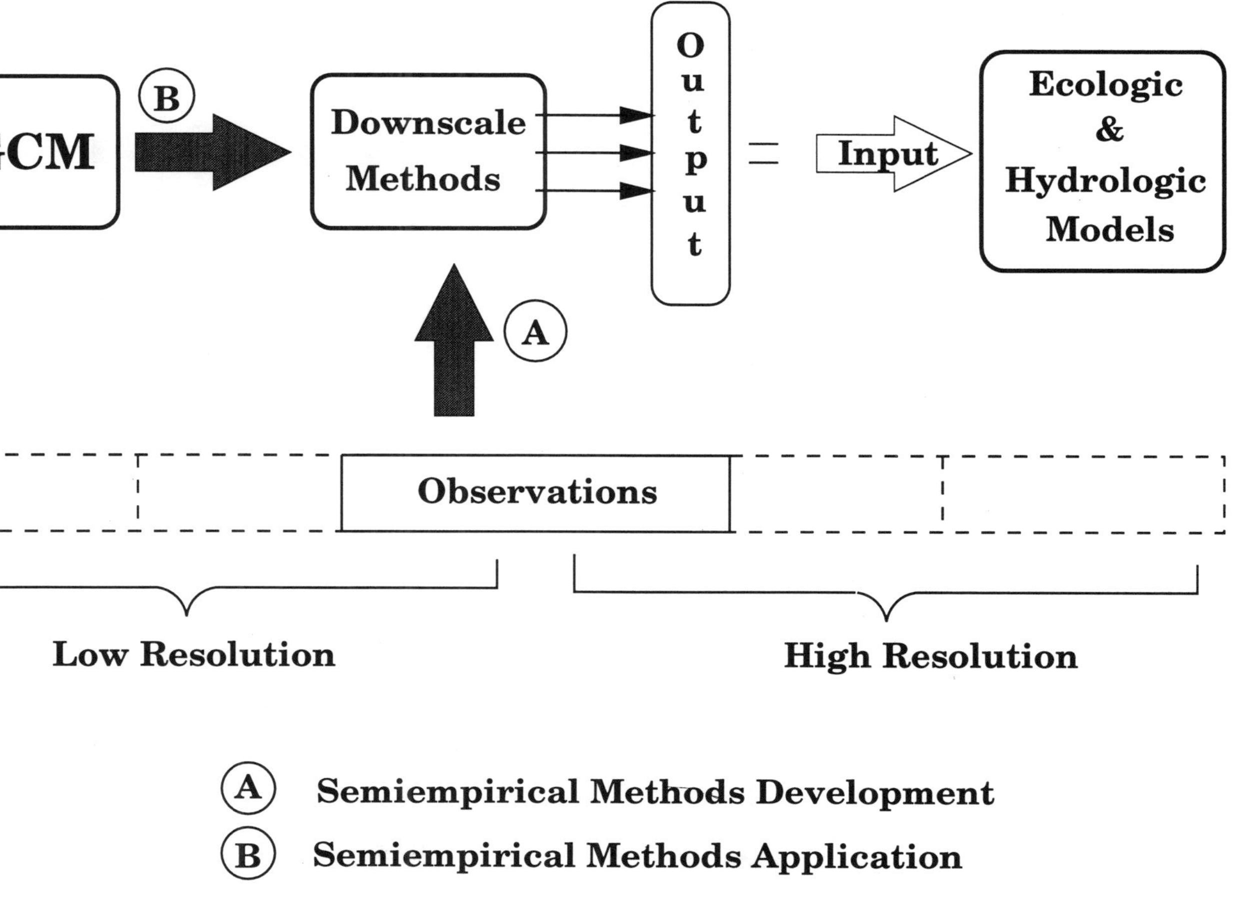

Figure 14.1 Conceptual representation of statistical downscaling approaches (BAHC 1993).

- macroscale variables have to be found that are strongly linked to the regional climate;
- the macroscale variables, acting as predictors in the statistical model, have to be well simulated by the GCMs; and
- future changes of the macroscale variables have to be small to ensure that the statistical model is valid under changed climate conditions.

Statistical models are not very computer intensive and, therefore, it is possible to produce long time series of climate variables for climate impact studies. Once the methodology has been established and tested it can be applied quite easily to other GCMs, regions and parameters. Assuming observed data are available, a statistical model could also be used to derive parameters which are not explicitly simulated by the GCM. For instance, many GCMs still do not simulate a diurnal cycle. Parameters such as daily maximum and minimum temperatures, which are very important for climate impact assessments, are not produced by those models. However, if a reliable relationship between parameters well simulated by the models, like daily pressure fields, and local extreme temperatures can be found, the statistics can be used to derive these temperatures.

A project conducted by MAT demonstrated that local extremes of daily surface air temperatures in southeastern Australia are substantially driven by the synoptic scale circulation. Figure 14.2 shows the daily maximum temperature anomalies from the long-term monthly mean (1979–1988) observed and predicted by a multiple regression model for January 1983 and 1984. The multiple regression model was derived from a principal component analysis (PCA). Nine important synoptic scale circulation modes were extracted from 10 years (1979–1988) of daily analysed mean sea level pressure fields over the Australian region. The temporal behaviour of these primary circulation modes, which explain more than 88% of the observed spatial pressure variability, is reflected by the time series of the scores of the principal components.

Using these scores as multivariate parameters of the daily synoptic circulation its influence on the local temperature variability was quantified by stepwise multiple regression. Since daily maximum temperature exhibits a strong temporal autocorrelation the inclusion of the PC scores of successive three-day sequences considerably improved the predictive skill of the model. The model was established over a reference period, which comprised the years 1979–1982 and 1985–1988. In the remaining independent years 1983 and 1984 the model was validated by estimating the daily local maximum temperature anomalies from the PC scores and comparing them to the observed values (see Figure 14.2). In general, the oscillations and the persistence of the observed and the predicted time series are similar. The ability of the model to follow day-to-day temperature changes of several degrees is very satisfactory.

Following the normal 'recipe' for statistical downscaling, the next step would be the application of the transfer functions between synoptic scale circulation and local temperature to the GCM simulated circulation fields. However, the comparison of the frequency and intensity of synoptic scale circulation features simulated by five MECCA-sponsored GCMs (BMRC, CCM1, CCM-1OZ, CCM0 and CCM1W) with observations revealed significant

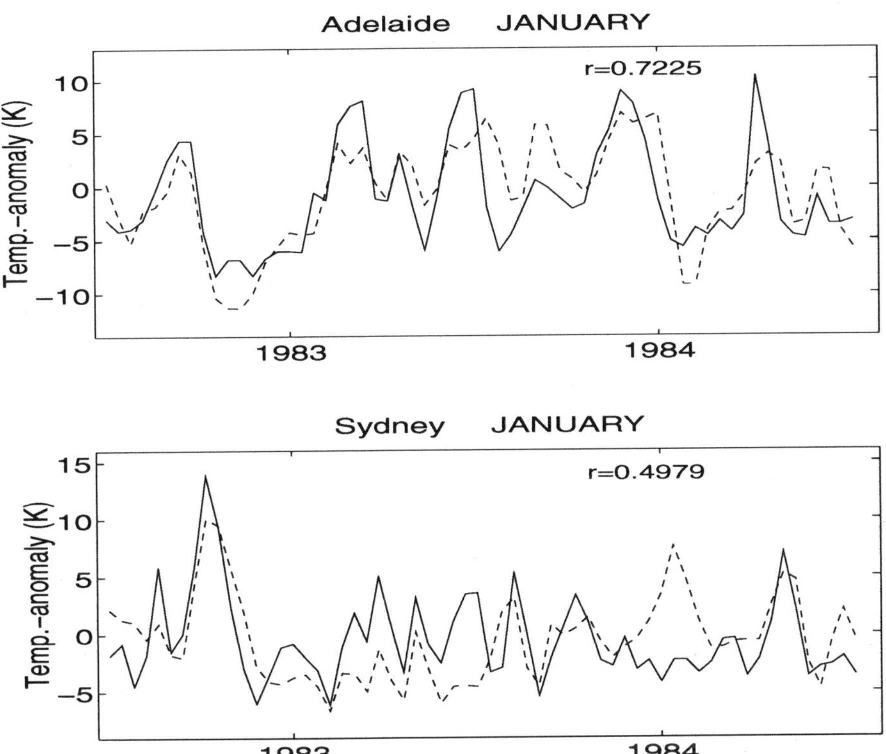

Figure 14.2 Daily maximum temperature anomalies (from long term mean 1979–1988) at two stations in southeastern Australia in January 1983 and 1984, observed (solid) and predicted (dashed) using a multiple regression model.

differences from the current climate (Schubert 1995). Hence, the second of the conditions mentioned above is not fulfilled and downscaling does not seem to offer much potential.

This result underlines the fact that statistical downscaling methods are strongly dependent on the performance of the GCM in simulating the driving large-scale parameters. Furthermore, statistical models always need long and homogeneous time series of observations for the fitting and testing. Therefore, in data sparse regions, this technique will not be suitable. Another disadvantage arises from the assumption that the underlying scale interactions will remain constant under changed climate conditions. However, changing local surface characteristics, such as different land use, could change the way large-scale weather events are modified by mesoscale forcing. The uncertainty associated with this problem is hard to estimate and generally depends on the physical processes involved. Only very recently has this issue started to be addressed in relation to downscaling (Matyasovszky et al. 1994).

Since both methods — dynamical and statistical downscaling — exhibit advantages and disadvantages, a combination of the two might produce more reliable results as suggested by Mearns and Rosenzweig in Chapter 15. They present a careful approach to the development of high resolution climate change scenarios that include changes in both mean and second order moments of daily temperature and precipitation. Firstly, a regional GCM (RegGCM) of Giorgi et al. (1994) is used to produce high resolution climate simulations over the continental USA for $1 \times CO_2$ and $2 \times CO_2$ conditions. Based on these simulated time series statistical parameters are calculated for the generation of long synthetic time series using a stochastic weather generator.

By changing the appropriate parameters in the weather generator according to the changes in the RegGCM, two different regional climate scenarios are produced: one with changes in the means only and one with changes in the mean and the variability. The differences in the impact of climate change arising from the differences in the scenario construction is assessed by applying both scenarios to a wheat crop model. Mearns and Rosenzweig draw the conclusion that 'substantial differences in the evaluation of the impact of a climate change on crop yields can evolve depending upon how the scenario is formed and to what degree changes in the variability are taken into account'.

The linkage of dynamical and statistical downscaling methods unavoidably increases the uncertainty associated with the scenario construction because both methods have their limitations. For example, the short length of the regional climate simulations (only $3 \, ^1/_2$ years) affects the stability of the statistical parameters needed for the weather generator. However, it provides a promising way to a more systematic analysis of how sensitive several resource systems are to changes in different climatic (e.g. temperature, precipitation, wind) and/or statistical (e.g. means, variability, extremes) parameters.

14.3 Summary

Studying the impacts of climate change on natural and human systems has identified the need for reliable regional climate change predictions as the primary pre-condition for climate impact assessments. Although there has been considerable research devoted to developing downscaling techniques, over the last few years an optimal and universal interface between GCMs and climate impact models has not yet been found. The often poor performance of the GCMs in reproducing the larger scale boundary conditions for both dynamical and statistical downscaling causes very significant problems (Landsea 1995; Machenhauer 1995). The identification of this limitation on downscaling methods was one important achievement of the MECCA project.

There have also been many achievements by the downscaling community. The development of downscaling techniques, which will without doubt also be necessary in future, has usually been accompanied by an extensive evaluation of the GCMs. In this way systematic errors in those GCM-simulated large-scale climate parameters which are essential for downscaling models (i.e. the synoptic circulation) can be identified (Hewitson & Crane 1992; Hay et al. 1992; Storch et al. 1993). Future research then can be focused on the improvement of these parameters in order to provide more reliable large-scale boundary conditions for regional climate change constructions and consequently for climate impact applications.

The problem of spatial scale mismatch does not appear to be amenable to one general or generic solution. As stated in Chapters 15 and 16 different downscaling techniques give different regional climate change scenarios and these in turn different climate impact responses. The systematic comparison of downscaling approaches, thus, is helpful to assess the uncertainties related to the manner in which regional climate change information is provided.

The answer to the question, how is climate change going to affect natural and human systems?, will always be uncertain. The MECCA project has very significantly contributed to identifying and underlining the fact that political response strategies to climate change have to incorporate these unavoidable uncertainties.

Acknowledgments

The author wishes to thank the Australian Coal Association Research Program/Electricity Supply Association of Australia (ACARP/ESAA) for funding this work and MECCA for providing the necessary data. This project is CIC contribution number 95/32.

References

BAHC (1993) Biospheric Aspects of the Hydrological Cycle, Focus 4: The weather generator project, Project report 1993.

Bardossy, A. and Plathe, E. (1992) Space-time model for daily rainfall using atmospheric circulation patterns. *Water Resources Research*, **28**, 1247–1255.

Cao, H.X, Mitchell, J.F.B. and Lavery, J.R. (1992) Simulated diurnal range and variability of surface temperature in a global climate model for present and doubled CO_2 climates. *Journal of Climate*, **5**, 920–943.

Déqué, M. and Piedelievre, J.Ph. (1995) High resolution climate simulation over Europe. *Climate Dynamics*, **11**, 321–339.

Dooge, J.C.I. (1992) Hydrologic models and climate change. *Journal of Geophysical Research*, **97**(D3), 2677–2686.

Favies-Mortlock, D. (1994) The use of synthetic weather for soil erosion modelling. *Agricultural and Forest Meteorology*, **52**, 321–337.

Frey-Buness, A., Heimann, D. and Sausen, R. (1995) A statistical-dynamical downscaling procedure for global climate simulations. *Theoretical Applied Climatology*, **50**, 117–131.

Giorgi, F. (1990) Simulation of regional climate using a limited-area model nested in a general circulation model. *Journal of Climate*, **3**, 941–963.

Giorgi, F. and Mearns, L.O. (1991) Approaches to the simulation of regional climate change: A review. *Review in Geophysics*, **29**(2), 191–216.

Giorgi, F., Shields-Brodeur, C. and Bates, G.T. (1994) Regional climate change scenarios over the United States produced with a nested regional model. *Journal of Climate*, **7**, 375–399.

Giorgi, F. (1995) Perspectives for regional earth system modelling. *Global and Planetary Change*, **10**, 23–42.

Hay, L.E., McCabe, G.J. , Wolock, D.M. and Ayers, M.A. (1992) Use of weather-types to disaggregate general circulation model predictions. *Journal of Geophysical Research*, **97**(D3), 2781–2790.

Heyen, H., Zorita, E. and Storch, H.V. (1994) Statistical downscaling of winter monthly mean North Atlantic sea-level pressure to sea level variations in the Baltic Sea. Max-Planck-Institut fur Meteorologie-Report, No. 150.

Henderson-Sellers, A. (1993) An antipodean climate of uncertainty? *Climatic Change*, **25**, 203–224.

Hewitson, B. and Crane, R.G. (1992) Regional climate prediction from the GISS GCM. *Palaeogeography, Palaeoclimatology, Palaeoecology* (Global and Planetary Change Section), **97**, 249–267.

Holland, G.J. (1996) The maximum potential intensity of tropical cyclones. Submitted to *J. Atmos, Sci.*

Karl, T.R., Wang, W.C., Schlesinger, M.E., Knight, R.W. and Portman, D. (1990) A method of relating general circulation model simulated climate to the observed local climate. Part I. Seasonal statistics. *Journal of Climate*, **3**, 1053–1079.

Katz, R.W. and Brown, B.G. (1992) Extreme events in a changing climate: Variability is more important than averages. *Climatic Change*, **21**, 289–302.

Kim, J.W., Chang, N.L., Baker, N.L., Wilks, D.S. and Gates, W.L. (1984) The standard problem of climate inversion: Determination of the relationship between local and large scale climate. *Monthly Weather Review*, **112**, 2069–2077.

Landsea, C.W. (submitted) Comments on multi-year present day and $2 \times CO_2$ simulations of monsoon climate over eastern Asia and Japan with a regional climate model nested in a GCM. *Journal of Geophysical Research*.

Machenhauer, B. (1995) Dynamical downscaling of climate simulations. In *Abstracts of the XXI General Assembly of the International Union of Geodesy and Geophysics*. Boulder, Colorado, 2–14 July, B279.

Matyasovszky, I., Bogardi, I., Bardossy, A. and Duckstein, L. (1993) Space-time precipitation reflecting climate change. *Hydrological Science Journal*, **38**, 539–558.

Mearns, L.O., Katz, R.W. and Schneider, S.H. (1984) Extreme high-temperature events. Change in their probabilities with changes in mean temperature. *Journal of Climate Applications in Meteorology*, **23**, 1601–1613.

Schubert, S. (1994) A weather generator based on the European 'Grosswetterlagen'. *Climate Research*, **4**, 191–202.

Schubert, S. (1995) Synoptic scale circulation in the MECCA general circulation models: A comparison with observations. In *Abstracts of the XXI General Assembly of the International Union of Geodesy and Geophysics*, Boulder, Colorado, 2–14 July, B259.

Storch, H.V., Zorita, E. and Cubasch, U. (1993) Downscaling of global climate change estimates to regional scales: an application to Iberian rainfall in wintertime. *Journal of Climate*, **6**, 1161–1171.

Storch, H.V. (1994) Inconsistencies at the interface of climate impact studies and global climate research. Max-Planck-Institut fur Meteorologie report, No. 122.

Whetton, P.H., Rayner, P.J., Pittock, A.B. and Haylock, M.R. (1994) An assessment of possible climate change in the Australian region based on the intercomparison of general circulation model results. *Journal of Climate*, **7**, 441–463.

Wigley, T.M.L., Jones, P.D., Briffa, K.R. and Smith, G. (1990) Obtaining sub-grid-scale information from coarse resolution general circulation model output. *Journal of Geophysical Research*, **95**(D2), 1943–1953.

Wilks, D.S. (1989) Statistical specification of local surface weather elements from large scale information. *Theoretical Applied Climatology*, **40**, 119–133.

CHAPTER 15

CLIMATE CHANGE SCENARIOS INCORPORATING CHANGES IN DAILY CLIMATE VARIABILITY

Linda O. Mearns and Cynthia Rosenzweig

15.1 Introduction

While most climate change agricultural impact studies have analysed the effects of mean changes of climate variables on crop production (Smith & Tirpak 1989; Rosenzweig 1990; Rosenzweig & Parry 1994), impacts of changes in climate variability (on daily to annual time scales) have been studied much less (Mearns 1989a). The focus on the effects of mean climate change has provided only limited information on how future climate could affect agriculture. Additional information on changed variability effects is needed to further elucidate uncertainties in our knowledge of possible impacts of climate change.

Very few climate change scenarios used for climate change impact analysis have included detailed, explicit changes in variability (Panturat & Eddy 1989; Wilks 1992; Mearns et al. 1992). This is partially because there is considerable uncertainty regarding how climate variability may change in the future, due to greenhouse warming or any other cause (Houghton et al. 1990, 1992). There is also quantitative uncertainty concerning how agricultural crops respond to changes in climate variability, but it is known qualitatively that changes in variability can have serious effects. One of the main means by which crops would be affected would be through changes in the frequency of extreme climate events (e.g. heat waves). It has been demonstrated that changes in variance have a greater effect on the frequency of extremes than changes in the mean (Katz & Brown 1992).

The main focus of our MECCA project was the development of climate change scenarios that include changes in both mean and second order moments of the variables of interest and application of these scenarios to crop climate models. The study also included a series of preliminary research steps that are also reviewed in this chapter: analysis of the sensitivity of crop models to variability change, and evaluation of change in climatic variability in the perturbed climate runs of interest. In this study we use the control and doubled CO_2 runs from the regional climate model (RegCM) nested within a general circulation model (GCM) over the continental USA (Giorgi et al. 1994). We also include an inchoate method for quantifying uncertainties in the climate change scenario based on the relative errors in the control run climate, and how these errors affect the crop models.

15.2 Sensitivity of impact models to changes in daily variability

Recently a series of studies have analysed the effect of daily climate variability change on simulated crop yields, by adjusting the parameters of weather generators to create time series of weather variables that exhibit changes in daily variability and then using these series to drive crop models (Mearns 1995; Semenov & Porter 1995; Mearns et al. 1996; Riha et al. 1995). Here we briefly review results of Mearns et al. 1995 (hereafter referred to as MRG), which research formed part of our MECCA project. We next describe the weather generator used in MRG and for the scenario formation study described in section 15.4.

15.2.1 Description of the Richardson model

Richardson's (1981) stochastic weather generator simulates daily times series of maximum and minimum temperature, incident solar radiation, and precipitation. Daily precipitation occurrence is represented by a two-state first-order Markov chain model. It accounts for the stochastic dependence of the series of wet and dry days. Parameters estimated are two transition probabilities: P_{11} and P_{01}, the probability of a wet day following a wet day, and the probability of a wet day following a dry day. Rainfall amounts (x) are simulated for rain days using the gamma distribution:

$$f(x) = x^{\alpha-1}e^{(-x\beta-1)}/(\beta^{\alpha}\Gamma(\alpha)), \quad x \geq 0, \tag{1}$$

where:

α = the shape parameter;
β = the scale parameter; and
$\Gamma(\alpha)$ = the gamma function of α.

The mean μ of the distribution is $\alpha\beta$ and the variance σ^2 is $\alpha\beta^2$.

Maximum and minimum temperature, and solar radiation are modelled as a multivariate first-order autoregressive process:

$$x_t(j) = Ax_{t-1}(j) + B\varepsilon_t(j), \tag{2}$$

where:

$x_t(j)$ = 3×1 matrix for day t for $j = 3$ elements, which are standardised values of maximum temperature ($j = 1$), minimum temperature ($j = 2$) and solar radiation ($j = 3$);
$\varepsilon_t(j)$ = 3×1 matrix for day t for j elements, of independent random normal components; and

$A, B = 3 \times 3$ matrices constructed from matrices of lag 0 and lag 1 correlations among the three j elements.

$A \cong$ time dependence,

$B \cong$ simultaneous correlations among the j elements.

Then the actual daily values of the j elements X_t are determined for $j = 1$, or $j = 3$ by:

$$X_{ti}(j) = x_{ti}(j) \times s_{ti}(j) + m_{ti}(j), \tag{3}$$

where:

$X_{ti}(j)$ = daily value of variable j on day t for precipitation occurrence state i, $i = 1$ for a wet day, $i = 0$ for a dry day;

$s_{ti}(j)$ = standard deviation of variable j on day t for state i;

$m_{ti}(j)$ = mean of variable j on day t for state i.

The seasonal cycle for the means and standard deviations of the j elements is determined by two-harmonic Fourier series. Since maximum temperature and solar radiation are conditioned on the occurrence of precipitation, separate models (including different harmonics, means, and variances) are used for values occurring on rain days and dry days, indicated by the i index. For $j = 2$, minimum temperature, there is no conditioning, and the index i in Equation (3) is not used.

The Richardson model, or models very similar to it, have been evaluated as to how well they reproduce observed climate characteristics of particular locations (MRG; Johnson et al. 1995; Katz 1995). The model tends to underestimate interannual variability of precipitation, as well as the autocorrelation structure of daily temperature. However, we found that the model performed reasonably well for our purposes.

15.2.2 Relationship between annual and daily variability

Based on the stochastic model for precipitation occurrence and intensity, the variance of the monthly total precipitation is related to the characteristics of daily precipitation according to the following:

$$\sigma_I^2 \cong N\pi\alpha\beta^2 \left[1 + \alpha(1 - \pi)\frac{1 + d}{1 - d} \right], \tag{4}$$

where:

σ_I^2 = year-to-year variance of monthly precipitation;

N = number of days in the time series;

π = unconditional probability of a wet day ($P_{01}/(P_{10} + P_{01})$);

α, β = shape, scale parameters of the gamma distribution;

d = the persistence parameter for a first order Markov chain of precipitation occurrence ($d = P_{11} - P_{01}$).

The expression $N\pi\alpha\beta$ is the mean monthly total precipitation.

The relationship between interannual and daily variance of temperature is the following:

$$Var(T) \cong \frac{\sigma_d^2}{N}\left(1 + 2\sum_{k=1}^{\infty}\rho_k\right), \tag{5}$$

where:

$Var(T)$ = interannual variance of monthly temperature ($^\circ$C^2);
σ_d^2 = variance of daily temperature ($^\circ$C^2);
ρ_k = autocorrelation coefficient of order k.

Since temperature is modelled as a first order autoregressive process Equation (5) is simplified to consider only ρ_1. It is important to note that these relationships are approximations.

15.2.3 Crop model sensitivity

MRG analysed the effect of daily variability change on the CERES-wheat model (Otter-Nacke et al. 1985; Richie & Otter 1985) at two locations in the central Great Plains, one where moisture is limiting and summer fallow production is common (Goodland) and one farther east where soil moisture is not limiting and continuous rainfed cropping is used (Topeka). Daily temperature variability (on a monthly basis) was changed by factors of 0.33, 0.5, 2, and 3 $\times$ the baseline (observed) variance. At both locations (Topeka & Goodland), large changes in the mean and variability of simulated yield resulted (refer to Figure 15.1 for Topeka). With increasing temperature variance, mean yields decreased and the relative variability of yield increased. The major cause of these yield changes was crop failure and damage due to extreme winter temperatures, which resulted in high incidence of winter kill. For example, at Goodland, the probability of crop failure (assuming summer fallow production methods) increased from the base value of 0.05 to 0.3 with a doubling of daily variance.

For precipitation variance change MRG constructed two different types of variance change. In both cases the mean total monthly precipitation ($N\pi\alpha\beta$) remained constant. In the first case, the frequency of precipitation and scale parameter of the gamma distribution (by which precipitation amounts are modelled, Equation (1)) were changed; and in the second only the persistence of precipitation occurrence (d) was changed. These parameters were changed such that interannual variability was changed by the factors 0.3, 0.5, 2, and 3. For the first change β is changed approximately by the factor and π by the inverse of the factor to bring about the desired change. Interesting interactions of the precipitation variability changes with the contrasting base climates were found at the two locations. At Topeka, mean yield decreased and variability of yield increased with increasing precipitation variability (Figure 15.2), whereas mean yields increased at Goodland, where soil moisture is limiting and fallow production methods are used (Figure 15.3). Yield changes were similar for the two different types of precipitation variability change investigated. Changes in frequency

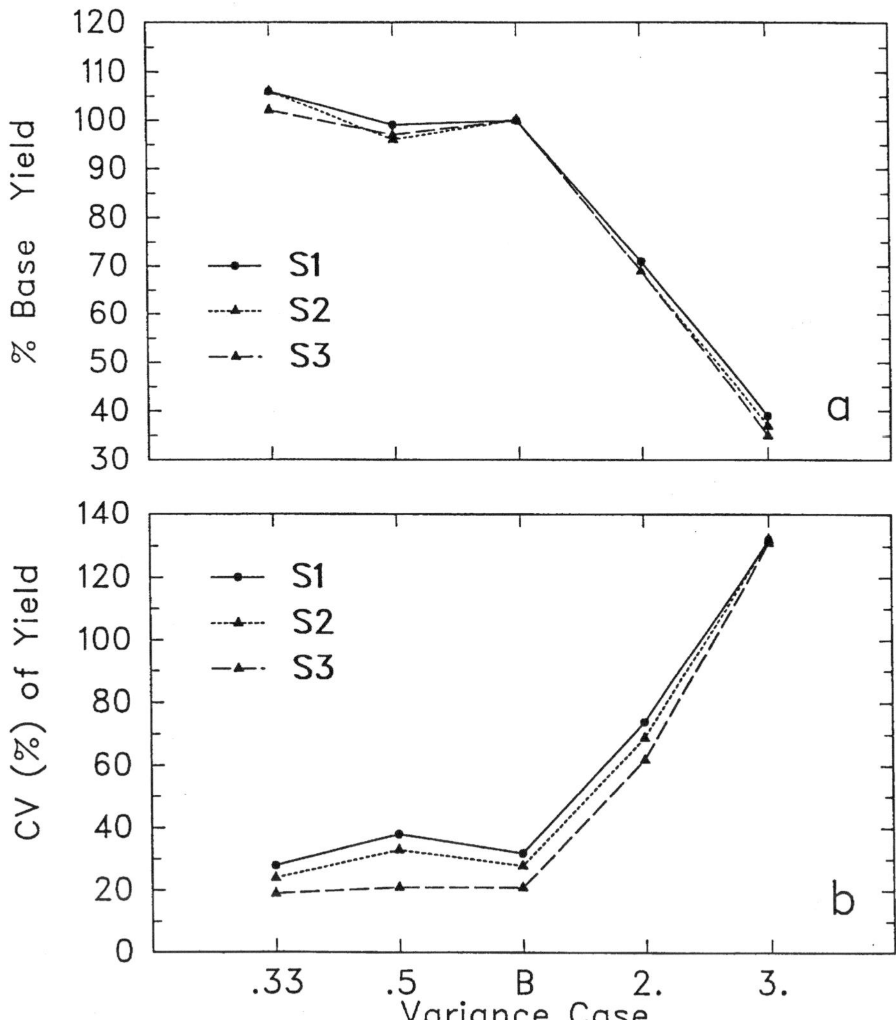

Figure 15.1 Topeka rainfed (a) % base yields and (b) coefficient of variation (CV) for temperature variance change cases. S1, S2, and S3 refer to the three soil types. B on x-axis refers to the base case (no variance change).

proved particularly important in the first type of change. With variance decrease (when frequency increases) the amount of evaporation from the soil is high at both locations. At Goodland, the plants are stressed throughout the crop season. At Topeka, there is a sudden onset of stress late in the season, which strongly affects final yield, even though a high level of dry matter production is maintained during the growing season. These results are similar to those found by Mearns et al. (1992), when monthly observed time series were mechanically altered to bring about changes in (mainly) interannual variability of precipitation, but results in MRG were more extreme.

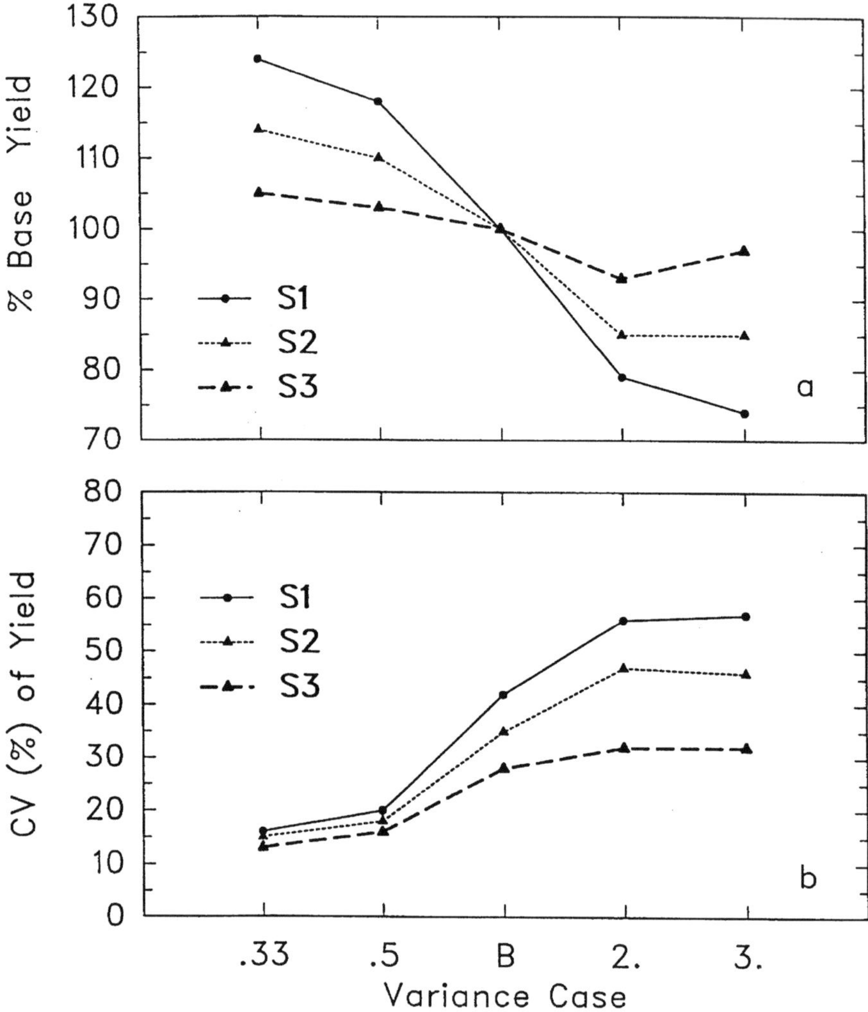

Figure 15.2 Topeka (a) % base yields and (b) CV for precipitation variance (PV) change cases.

15.3 Changes in variability in the RegCM

Giorgi et al. (1994) used a RegCM to generate two continuous $3\frac{1}{2}$ year-long high resolution climate simulations over the continental United States, one for present-day conditions and one for conditions under doubled carbon dioxide concentration. The regional model was nested within the GENESIS version of the NCAR Community Climate Model (CCM) (Thompson & Pollard 1995a, 1995b). The regional model was run at 60 km grid point spacing.

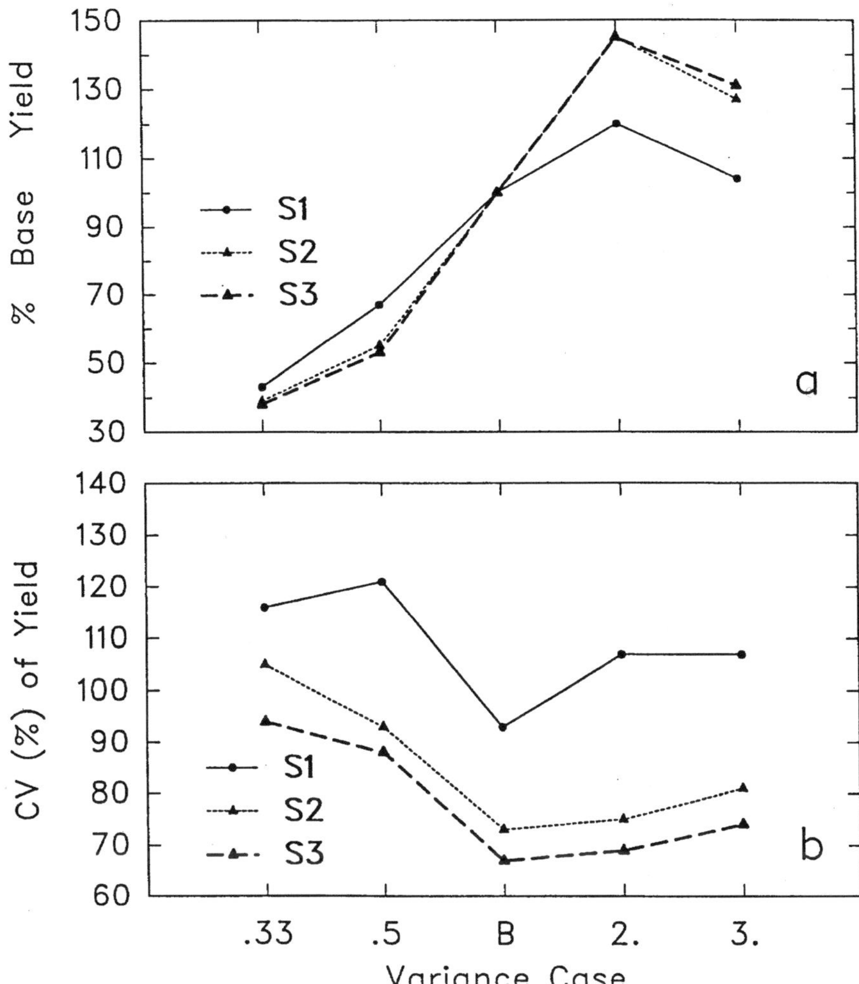

Figure 15.3 Goodland (a) % base yields and (b) CV for precipitation variance (PV) change cases.

As an earlier part of our MECCA project, detailed analysis of the regional climate model runs was performed in order to evaluate the quality of the control run and to determine the magnitude and direction of variability change. What we summarise briefly here can be found in detail in Giorgi et al. (1994) and Mearns et al. (1995a). The regional model control run had some serious errors. There was a serious bias in mean precipitation such that the estimated precipitation was too low in the eastern half of the domain and too high in the western half. In addition there was a consistent overestimation of the frequency of precipitation, throughout the domain, even when the mean precipitation was underestimated. Smallest errors were found in the northwest portion of the domain, where even variability of precipitation was well reproduced. Mean daily temperature was seriously overestimated in the Great Lakes region. In the central plains it was relatively well reproduced. The diurnal range, however, was consistently underestimated as was daily variance (of both maximum and minimum temperature) throughout the domain.

Changes in the variance of daily temperature, comparing the three-year control run to the three year doubled CO_2 run were quite substantial. Large decreases in variance occurred throughout the winter and early spring. In late spring and early summer some regions (such as the central Plains) exhibited sharp increases in variance. This latter result is particularly interesting, since it indicates that increases in the frequency of extreme high temperatures in summer might be larger than assumed when only considering mean increases in temperature.

Changes in precipitation varied a great deal seasonally and regionally. In the central Plains largest changes occurred in the summer, particularly in July, when mean precipitation increased, intensity increased, but frequency often decreased.

15.4 Variability uncertainty in impacts assessment

How different is simulated wheat yield generated from a climate change scenario that incorporates only mean changes in climate compared to yields from a climate change scenario that includes changes in mean and daily variability of climate?

The significant results from the crop model sensitivity study discussed in section 15.2 motivated us to go on to construct scenarios that included changes in daily variability of climate, calculated from the regional climate model results discussed in section 15.3. We formed two scenarios: one incorporating only mean changes in climate variables and the other incorporating both mean and variability changes. It should be noted that the results of these climate model experiments, although quite interesting, do not necessarily contain information regarding climate change that is of any higher quality than other climate model experiments. However, the scenario is spatially detailed, and the time step used in the model for dynamic calculations (e.g. six minutes) was of an order such that some greater confidence may be placed in the output on a daily time scale.

15.4.1 Methods for forming the scenarios

To form the scenario incorporating only mean changes in relevant climate variables we applied what can be called the classical scenario formation technique (e.g. Smith & Tirpak 1989). Differences ($2 \times CO_2$ – control) in monthly mean maximum and minimum temperature, were calculated for the grid locations roughly corresponding to the locations Goodland and Topeka (Table 15.1). These monthly mean differences were then added to the daily observational time series of maximum and minimum temperature for the 30 years of observed data. These procedures result in a change only in the mean of the temperature—daily (and interannual) variability remains the same as for the observations.

We calculated the ratios (r) of $2 \times CO_2$ to control mean monthly precipitation from the regional model output for the two grids, and then the observed daily precipitation values

Table 15.1 Changes from $2 \times CO_2$ climate for Topeka

	Month											
	1	2	3	4	5	6	7	8	9	10	11	12
Max T	4.33	4.23	6.77	3.23	2.13	4.27	2.97	3.43	4.43	4.90	4.87	5.87
Min T	4.67	4.20	7.00	2.93	3.20	4.17	3.07	4.00	4.73	3.70	4.20	4.80
Precip	0.75	0.42	3.43	0.65	0.82	1.60	0.78	1.41	0.86	0.74	0.89	0.86
SR	1.01	1.02	0.93	1.01	0.99	0.99	1.05	0.99	1.02	0.99	0.99	1.00

Max T = Maximum temperature (difference between $2 \times CO_2$ and control from RegCM)
Min T = Minimum temperature (difference between $2 \times CO_2$ and control from RegCM)
Precip = Precipitation (ratio of $2 \times CO_2$ to control from RegCM)
SR = incident solar radiation (ratio of $2 \times CO_2$ to control from RegCM)

were multiplied by these ratios. The mean monthly total precipitation is thus changed, but so is the daily intensity and its variability. The frequency of precipitation (unconditional probability of precipitation, π), generally remains as in the observations, except when the application of the ratio results in a daily precipitation amount less than 0.1 mm. Under such circumstances small changes in frequency result. The mean intensity changes by the ratio (r), and the variance of the intensity by r^2. This is basically equivalent, in the Richardson stochastic weather generator, to changing the scale parameter β by the ratio and leaving α constant. For solar radiation, we also used the ratio approach. In general changes in solar radiation were small (Table 15.1), and the CERES-wheat model is relatively insensitive to small changes in solar radiation compared to the changes in other climate variables. Hence, we do not discuss further changes in solar radiation. We then calculated the parameters for the weather generator for this mean altered dataset and generated several realisations of different lengths (from 30 to 90 years). This became the 'mean change only' climate, henceforth referred to as MC, for mean change. Due to sampling errors, the characteristics of the simulated dataset vary slightly from those of the altered observational dataset. Table 15.1 presents the ratios and differences used to construct the mean change scenario for Topeka. Maximum and minimum temperatures increase each month, whereas precipitation exhibits contrasting increases and decreases. Changes in solar radiation are quite small.

To generate the scenario with mean and variability changes, we calculated the parameters for the weather generator for the control and $2 \times CO_2$ output from the regional model, and then took the ratio or difference of these variables, what ever was appropriate for the parameter considered. Mainly this involved ratios. Wilks (1992) recommended using changes in interannual variability and then selecting among possible alternatives for how to change daily parameters. Here we follow an approach slightly modified from that suggested by Mearns (1989b), and briefly presented in Mearns and Rosenzweig (1994), wherein changes in daily parameters are determined and these then impose changes in interannual variability according to Equations (4) and (5). For precipitation, the ratios of the αs and βs were calculated, and then αs and βs calculated from the observed climate were multiplied by the ratios to form new parameters. Ratios of π and d were similarly calculated and new parameters determined again from multiplying the parameters generated from the observed data by these ratios. From the new π and d, new transition probabilities P_{11} and P_{01} were determined according to the relations:

$$\pi = P_{01}/(P_{01} - P_{11} + 1);$$

and

$$d = P_{11} - P_{01}.$$

This method differs from that described in Mearns and Rosenzweig (1994) in how the new transition probabilities are determined. In the earlier work, ratios of the transition probabilities themselves were calculated from the control and $2 \times CO_2$ values, and these ratios were then used directly to modify the observed transition probabilities. The difference here is that how transition probabilities are changed in the current study results in the mean monthly precipitation change being the same for the two scenarios, that is, for each month

$$\pi'\alpha'\beta' = \pi\alpha\beta,$$

where the prime indicates parameters for the variance change precipitation and the non prime are those for the mean change precipitation. When the transition parameters are changed directly, this relation is not maintained and mean precipitation in the two scenarios can differ.

Given the shortness of the RegCM model runs (three years control and three years $2 \times CO_2$), values of the ratios of these parameters are not very stable. This exercise should be viewed as only a test application of our method and in no way as a scenario that has significant predictive meaning. It might be more prudent to change only the unconditional probability of precipitation and leave persistence constant. However, for our first experiments we wanted to make the most direct changes we could in relation to the parameters for the weather generator.

For temperature, the means were changed according to the differences in $2 \times CO_2$ and control runs. The Fourier series (first harmonic) was altered by transforming the harmonics to their trigonometric forms (of amplitudes and phase angles) and then ratios ($2 \times CO_2$/control) of amplitudes and phase angles were taken. The second harmonics were held constant. The amplitudes and phase angles from the observations were then multiplied by the ratios. Then the resulting new amplitudes and phase angles were transformed back to the Fourier series form used in the Richardson weather generator. This second scenario is henceforth referred to as MVC, for mean and variance change. We then compared the MC and the MVC climates for the two locations.

Table 15.2 Comparison of simulated observed, mean change, and mean plus variance change climates for Topeka

	Month			
	1	4	7	10
(a) Mean temperature ($^\circ$C)				
OBS				
Mean	−3.6	12.2	25.8	14.0
Variance	40.0	28.6	10.5	28.0
MC				
Mean	0.6	15.3	28.7	18.3
Variance	34.3	29.0	10.6	29.1
MVC				
Mean	1.0	15.4	28.6	18.3
Variance	19.5	25.6	12.5	19.0
(b) Precipitation (mm)				
OBS				
Mean Daily	0.72	2.57	3.41	2.45
Mean Intensity	3.9	8.7	12.5	12.2
S.D.	4.2	10.9	15.4	14.8
Frequency	0.18	0.30	0.27	0.20
MC				
Mean Daily	0.49	1.71	2.47	1.64
Mean Intensity	2.8	5.7	9.6	8.2
S.D.	3.3	6.4	11.1	9.0
Frequency	0.18	0.30	0.26	0.20
MVC				
Mean Daily	0.55	1.60	2.34	1.40
Mean Intensity	4.1	7.8	10.0	6.6
S.D.	4.2	9.3	12.0	6.2
Frequency	0.14	0.21	0.23	0.21

OBS = 90 year simulated current climate time series
MC = 90 year simulated climate with mean changes only
MVC = 90 year simulated climate with mean and variance changes
Mean Intensity = Mean daily intensity of precipitation
S.D. = Standard deviation of intensity
Frequency = Frequency of precipitation occurrence (probability)

Table 15.2 shows contrasts between the key variables for four months for simulated time series for the current (OBS) climate, the MC climate and the MVC climate for Topeka.

In the MVC scenario, for Topeka, in general there were temperature variance decreases in the winter and increases in the summer. Regarding precipitation, changes were mixed seasonally. Of the four months presented in Table 15.2, largest contrasts in the statistics for precipitation are seen in April, when the ratio r from Table 15.1 is 0.65; and the ratio of MC and OBS for daily precipitation is indeed close to that, 0.68. Although the mean daily precipitation amounts for the two scenarios are close as they are constrained to be, the frequency and intensity statistics differ.

These differences can result because there is no direct relationship between the ratio r (0.65 for April) and the ratios of the individual parameters α, β, π and d, that determine the new precipitation parameters for the MVC case. In the case of MC, there is no change in π, and both the mean daily precipitation and intensity are changed by the single ratio. Essentially the mean ratio change method is equivalent to changing β by the ratio r (0.65 for April). In the MVC method, for this particular month, β does not change since the ratio of $2 \times CO_2$ to control the βs from the RegCM output is 1.0; and α is changed very little, by 0.91. The probability of precipitation, π, is changed by the ratio 0.7. Hence, in MVC mean intensity changes by 0.91, the change in α, as does the variance, since there is no change in β. In the MC case mean intensity changes by r, 0.65, and the variance by r^2 (0.42). As noted earlier there is no change in π in the MC case. Mean daily precipitation in MVC changes by $0.91 \times 1.0 \times 0.70 = 0.64$, approximately the same change as in the MC case.

15.4.2 Contrasting effects on simulated yields

The contrasting effect of the scenarios on simulated wheat yield for both Topeka and Goodland are presented in Table 15.3.

At Goodland there are large differences in the yields generated from the two different climate change scenarios, but little contrast in yields is obtained at Topeka. At Goodland, where soil moisture in particular is limiting, the crop is more sensitive to changes in precipitation variability, in particular. At Topeka, changes in precipitation variability are less extreme and the crop is less sensitive to such changes since soil moisture is not an important limiting factor. Thus, substantial differences in the evaluation of the impact of climate change on crop yields can evolve depending upon how the scenario is formed, to what degree changes in variability are taken into account, and the nature of the base climate for a particular location.

Further development in the scenario formation techniques described above are necessary. For example, it would be useful to determine to what degree changes in temperature variability as opposed to precipitation variability are responsible for the differences in yield. We plan to run certain trials, separating the temperature and precipitation change scenarios. Results from fully irrigated runs suggest that, at the two sites discussed here, changes in precipitation variability are more important. We will also determine whether the temperature and precipitation variability changes augment or cancel out each other's effect on yield.

At this stage we are not suggesting that one scenario method is in some fundamental way better than the other. We are asserting that how a scenario is formed makes up another of the myriad uncertainties associated with determination of climate change impacts on resource systems. We do suggest, however, that the 'classic' scenario formation method has not been sufficiently examined. For example, in no impacts papers we surveyed was the fact that the variance of mean intensity of precipitation changes with the MC method recognised and discussed.

Table 15.3 Simulated yields (Kg/ha) from different climate change scenarios (Soil 2)

	CASE		
	Base	MC	MCV
(a) Goodland Fallow			
Mean	1266	2121	1744
S.D.	947	1098	1263
CV	75	52	72
(b) Topeka Rainfed			
Mean	4208	4038	4014
S.D.	1275	1322	1366
CV (%)	30	33	34

Base = Base yields simulated with simulated current climate (kg/ha)
MC = Yields simulated with MC climate
MVC = Yields simulated with MVC climate
S.D. = Standard deviation (kg/ha)
CV = Coefficient of variation (%)

Table 15.4 Uncertainty measures for impacts

Yield ratio	Location				
(a) Integrated climate model validation using crop models					
	Eugene	Spokane	Goodland	Topeka	Indianapolis
obs sim/control sim	0.81	0.48	0.78	14.1	7.1
(b) Yield Uncertainty Parameter					
	Goodland		Topeka		
$2 \times$ param – obs sim	1.5		2.0		
$2 \times CO_2$ – control sim					

15.5 Determination of uncertainty in impacts

In numerous climate change impacts analyses, authors comment on the problem of errors
in the control run. Indeed, the errors in control runs is the very reason that the direct use of
output from climate models is not feasible. Yet, very little progress has been made at trying
to quantify this uncertainty. Here we make a humble beginning in this direction by forming
certain indexes which can give crude indications of the quantitative impacts of quantitative
error in the control run. The first approach is to provide what we call an integrated validation
of the control run by making use of the direct output from the climate model. The simple idea
is to determine the degree of difference in the simulation of yield using observed climate
data compared to yield simulated from using the control run of the climate model. A ratio

of these two yields is formed (Table 15.4). This we consider an integrated validation in that the total effect of the errors in the control run on the impact model can be determined. We include locations in Table 15.4(a) in addition to Goodland and Topeka to give the reader some indication of the effect of control run errors across the entire domain. The farther the ratio is from 1.0, the greater is the effect of the control run errors on the impacts model. Note the very large difference between the ratios at Goodland and Topeka. Between these locations the error in underestimating precipitation increases very quickly, and this mainly accounts for the differences in the ratios. This method is certainly very simple and relatively crude. It also of course cannot account for compensating errors in the control run that might lead to the impact model results looking 'good' even though the climate model climatology is quite bad. Another important consideration is whether the observed data being used truly represents the control run model, which usually does not correspond to any specific series of years of observations.

The second part of Table 15.4, presents what we call 'yield uncertainty parameters' (YUP), which attempts to measure the effect of the climate model errors on the uncertainty of the climate change scenario and hence, the uncertainty of assessment of the climate change impact. Again, a simple ratio is formed incorporating the integrated validation ratio with estimates of the climate change impact. The numerator is formed by the difference in yield determined from the classic climate change scenario formation method (see section 15.4), and yield generated from the actual observed climate (i.e. the numerator of the integrated validation ratio). The denominator consists of the yield simulated by using the actual output of the regional climate model from the $2 \times CO_2$ run and the control run. Thus, the numerator and denominator represent two different ways of calculating differences in yield between a perturbed climate and a current climate. Again the farther the ratio is from 1.0, the larger is the uncertainty in the impact assessment. For both locations YUP is greater than 1.0, double that in the case of Topeka, which indicates substantial uncertainty in the impact.

Again it must be noted that this parameter is quite crude. Arguments can be made that errors in the absolute values of climate variables compared to point observations is not the best way to determine the quality of a climate model. The overall spatial pattern correlation for a large area or the entire earth may be more important than absolute value errors at one location. We are only providing this parameter as a simple initial step in attempting to quantify uncertainty in climate change scenarios.

15.6 Summary

In our MECCA project, we attempted to advance the state of art of climate change scenario formation by evaluating the difference incorporation of variability change has on resulting assessment of climate change impacts in an agricultural context. Obviously, these type of scenarios would be useful in evaluating the effect of climate change in any number of contexts, such as human mortality and morbidity (where daily extreme events are quite critical), water resources, and unmanaged ecosystems. Our method of including variability change is only one of several possiblities. One aspect of variability change that our method does not consider is changes in the persistence of important interannual events such as El Niño and La Niña events; and the importance of these events for agricultural production has been demonstrated (e.g. Adams et al. 1995). Since the weather generator does not reproduce the autocorrelation structure of the variables on interannual time scales, changes in the persistence of these important events are not captured by our method. One way of capturing these autocorrelations and then changes in persistence is to condition the weather generator on different states (e.g. El Niño versus normal years). Then, by superimposing a first (or higher order) Markov chain on the weather generator, different sequences of states (with different daily characteristics) could be generated. The success of conditioning to improve the simulation of current climate has been demonstrated (Wilks 1989; Wang & Conner 1995).

It is striking in the evolution of climate change impacts research and its recent incorporation in the framework of integrated assessment how little attention has been given to improvement of climate change scenario development, especially from climate models, and efforts to analyse the effects of climate model errors on scenario formation. For example, much attention (rightfully) has recently been focussed on issues such as possible adaptation (of agricultural systems) to climate change (Mendelsohn et al. 1994; Rosenzweig & Parry 1994). Yet what type of climate agricultural systems will be confronted with in the future remains very vague and uncertain, and more research efforts should be aimed at scenario development. We trust our MECCA project has served as a useful start in rectifying this imbalance in integrated assessment and climate change impacts research in general.

Acknowledgments

We wish to thank Richard Goldberg for computing assistance in this project. We also thank MECCA for providing the necessary computer time. Part of this research was also funded by a grant from the EPA Global Change Program.

References

Adams, R.M., Bryant, K.J., McCarl, B.A., Legler, D.M., O'Brian, J., Solow, A. and Weiher, R. (1995) Value of improved long-range weather information. *Contemporary Economic Policy*, **13**, 10–19.

Giorgi, F., Shields Brodeur, C. and Bates, G.T. (1994) Regional climate change scenarios over the United States produced with a nested regional climate model: Spatial and seasonal characteristics. *Journal of Climate*, **7**, 375–399.

Johnson, G.L., Hanson, C.L. and Ballard, E.B. (1996) Stochastic weather simulation: Overview and analysis of two commonly used models. *Journal of Applied Meteorology*, **35**, 1878–1896.

Katz, R.W. (1996) The use of stochastic models to generate climate change scenarios. *Climatic Change*, **32**, 237–255.

Katz, R.W. and Brown, B.G. (1992) Extreme events in a changing climate: Variability is more important than averages. *Climatic Change,* **21**, 289–302.

Mearns, L.O. (1989a) Climate variability. In *The Potential Effects of Global Climate Change on the United States*, edited by J. Smith and D. Tirpak, pp. 29–54. US EPA, Report to Congress No. 230-05-61-050, Washington, DC.

Mearns, L.O. (1989b) The simulation of meteorological time series for climate change scenario development, paper presented at the US EPA Meeting on Climate Change Scenario Development, Boulder, Colorado, 31 August. (Summarised in Scenarios Advisory Meeting Summary Report, December 1989, US EPA, Washington, DC.)

Mearns, L.O. (1995) Research issues in determining the effects of changing climatic variability on crop yields. In *Climate Change and Agriculture: Analysis of Potential International Impacts*, edited by C. Rosenzweig, pp. 123–146. American Society of Agronomy Special Publication No. 59, Madison: ASA.

Mearns, L.O. and Rosenzweig, C. (1994) Use of a nested regional climate model with changed daily variability of precipitation and temperature to test related sensitivity of dynamic crop models. In *Preprints of the AMS Fifth Symposium on Global Change Studies*, pp. 142–155. 23–28 January. American Meteorological Society: Boston.

Mearns, L.O., Rosenzweig, C. and Goldberg, R. (1992) The effect of changes in interannual climatic variability on CERES-wheat yields: Sensitivity and $2 \times CO_2$ studies. *Journal of Agricultural and Forest Meteorology*, **62**, 159–189.

Mearns, L.O., Giorgi, F., Shields Brodeur, C. and McDaniel, L. (1995a) Analysis of the variability of daily precipitation in a nested modeling experiment: Comparison with observations and $2 \times CO_2$ results. *Global and Planetary Change*, **10**, 55–78.

Mearns, L.O., Giorgi, F., Shields, C. and McDaniel, L. (1995b) Analysis of the diurnal range and variability of daily temperature in a nested modeling experiment: Comparison with observations and $2 \times CO_2$ results. *Climate Dynamics*, **11**, 193–209.

Mearns, L.O., Rosenzweig, C. and Goldberg, R. (1996) The effect of changes in daily and interannual climatic variability on CERES-wheat: A sensitivity study. *Climatic Change*, **32**, 257–292.

Mendelsohn, R., Nordhaus, W.D. and Shaw, D. (1994) The impact of global warming on agriculture: A Ricardian analysis. *The American Economic Review*, **84**, 753–771.

Otter-Nacke, S., Goodwin, D.C. and Ritchie, J.T. (1986) *Testing and Validating the CERES-Wheat Model in Diverse Environments*, AGRISTARS YM-15-00407 JSC 20244. Houston: Johnson Space Center.

Riha, S.J., Wilks, D.W. and Simoens, P. (1996) Impact of temperature and precipitation variability on crop model predictions. *Climatic Change*, **32**, 293–311.

Ritchie, J.T. and Otter, S. (1985) Description and performance of CERES-wheat: A user-oriented wheat yield model. In *ARS Wheat Yield Project*, edited by W.O. Willis, pp. 159–175. ARS-38, US Department of Agriculture-Agricultural Research Service.

Richardson, C.W. (1981) Stochastic simulation of daily precipitation, temperature, and solar radiation. *Water Resources Research*, **17**, 182–190.

Rosenzweig, C. (1990) Crop response to climate change in the Southern Great Plains: A simulation study. *The Professional Geographer*, **42**, 20–39.

Rosenzweig, C. and Parry, M.L. (1994) Potential impact of climate change on world food supply. *Nature*, **367**, 133–137.

Semenov, M.A. and Porter, J.R. (1995) Climatic variability and modelling of crop yields. *Agricultural and Forest Meteorology*, **73**, 265–283.

Smith, J.B. and Tirpak, D.A. (eds.) (1989) *The Potential Effects of Global Climate Change on the United States*, US EPA, Report to Congress No. 230-05-61-050, Washington, DC: US Environmental Protection Agency.

Thompson, S.L. and Pollard, D. (1995a) A global climate model (GENESIS) with a land-surface-transfer scheme (LSX). Part I: Present climate simulation. *Journal of Climate*, **8**, 732–761.

Thompson, S.L. and Pollard, D. (1995b) A global climate model (GENESIS) with a land-surface-transfer scheme (LSX). Part II: CO_2 sensitivity. *Journal of Climate*, **8**, 1104–1121.

Wang, Y.P. and Conner, D.J. (1996) Simulation of optimal development for spring wheat at two locations in southern Australia under present and changed climate conditions. *Agricultural and Forest Meteorology*, **79**, 9–28.

Wilks, D.S. (1989) Conditioning stochastic daily precipitation models on total monthly precipitation. *Water Resources Research*, **25**, 1429–1439.

Wilks, D.S. (1992) Adapting stochastic weather generation algorithms for climate change studies. *Climatic Change*, **22**, 67–84.

CHAPTER 16

GLOBAL CLIMATE MODELS AND GLOBAL VEGETATION MODELS

Catherine Ciret

16.1 Introduction

The prospect of a climate change due to an increase in greenhouse gases demands that the assessment of the potential climate change and its impact on various systems be undertaken. One of the main tools available for such assessments are numerical models, since no experiment of this sort can be done in a laboratory. Global Climate Models (GCMs) simulate the climate but have, to date, incorporated a simplified representation of the vegetation–climate subsystems on the land surface. Numerous vegetation models exist whose aims are either to simulate the vegetation processes (photosynthesis, growth, reproduction) or to predict the vegetation distribution as a function of climate. Both GCMs and vegetation models have been widely used in the assessment of future climates and climatic impacts. There are a number of reasons why climate models and vegetation models might be linked: (i) the vegetation models (e.g. global 'static' models) are used as impact models when linked into a simulated changed climate model, (ii) the vegetation models are used with current climate simulation as a diagnostic tool (e.g. Claussen & Esch 1994; McAvaney 1995, pers. com.) to evaluate the current climate simulated by the GCMs, and finally (iii) the vegetation models are coupled in a two-way mode to climate models to see if the quasi-interactive vegetation influences the simulated climate.

The aim of this chapter is to evaluate the uncertainties related to the use of climate models and global vegetation models when they are linked together, particularly as in (i). There are numerous problems in using GCMs and global static vegetation models, such as problems of mismatch of spatial and temporal scales, inaccuracy of the climate model's prediction at the regional scale and perhaps more importantly, inadequacy of the use of static vegetation models and equilibrium climate experiments to represent two intrinsically dynamic systems (i.e. climate and ecosystems). Ideally, a dynamical global vegetation model fully coupled to a GCM that simulates a transient increased greenhouse gases experiment should be used to assess the impacts of climate change on the vegetation. However, these global change models are not yet available. The equilibrium vegetation models and equilibrium climate experiments provide, nevertheless, some valuable information such as the evaluation of the sensitivity of the different vegetation types to the input climate and to coupling techniques, for example, temporal resolution of linkage, spatial resolution, methodology chosen in using scenarios, sensitivity to different datasets of observed climate and to changes in parameters. Furthermore, the static vegetation models can be used as diagnostic tools to evaluate the ability of the GCMs to simulate the bioclimatic variables required by the vegetation models.

The models employed (vegetation models and climate models) are presented in section 16.2. The impacts of the GCM climates on the simulation of the different bioclimatic variables are analysed in section 16.3 and some preliminary analysis about the sensitivity of the vegetation models to the method chosen in constructing climate scenarios are presented in section 16.4.

16.2 Vegetation models and climate models

16.2.1 Vegetation models

It has been recognised for centuries that there is an apparent correspondence between the climate and vegetation distributions. Many early plant geographers tried to correlate the distribution of flora with climatic factors, and the most widely known classifications of climate and vegetation are Köppen (1936), Holdridge (1947) and Thornthwaite (1948). These schemes all have in common the fact that they are correlative: it is the observed spatial correlation between plants and climate which determines the limits given to the different categories, whether those categories are called vegetation types or climate types.

These correlative schemes have been compared in the literature (e.g. Prentice & Solomon 1991; Claussen & Esch 1994) to recent models which are based on more mechanistic considerations such as the Woodward model (Woodward 1987) or the BIOME model (Prentice et al. 1992). In these models, most of the environmental constraints delimiting the different vegetation types are physiologically based, in contrast to the schemes described earlier in which the upper and lower climatic limits are based on the empirical correlation between present-day vegetation and present-day climate. The difference in these approaches is relevant if the vegetation models are meant to predict the distribution of natural vegetation under different climate conditions (e.g. Quaternary climate, future climate).

The vegetation schemes employed in this study are the BIOME model of Prentice (1992) and a version of the Holdridge Life Zone classification (Holdridge 1947). The Holdridge scheme is merely a climatic classification, defining life zones with the use of simple climatic indices: the mean annual biotemperature (defined as the mean of the daily or monthly temperatures above $0°C$) and the total annual precipitation. The BIOME model is a more sophisticated prediction of plant functional types based on environmental constraints. These functional types are combined into 17 biomes. The environmental limits are expressed in terms of cold tolerance, heat requirement and soil moisture availability (see Figure 16.1).

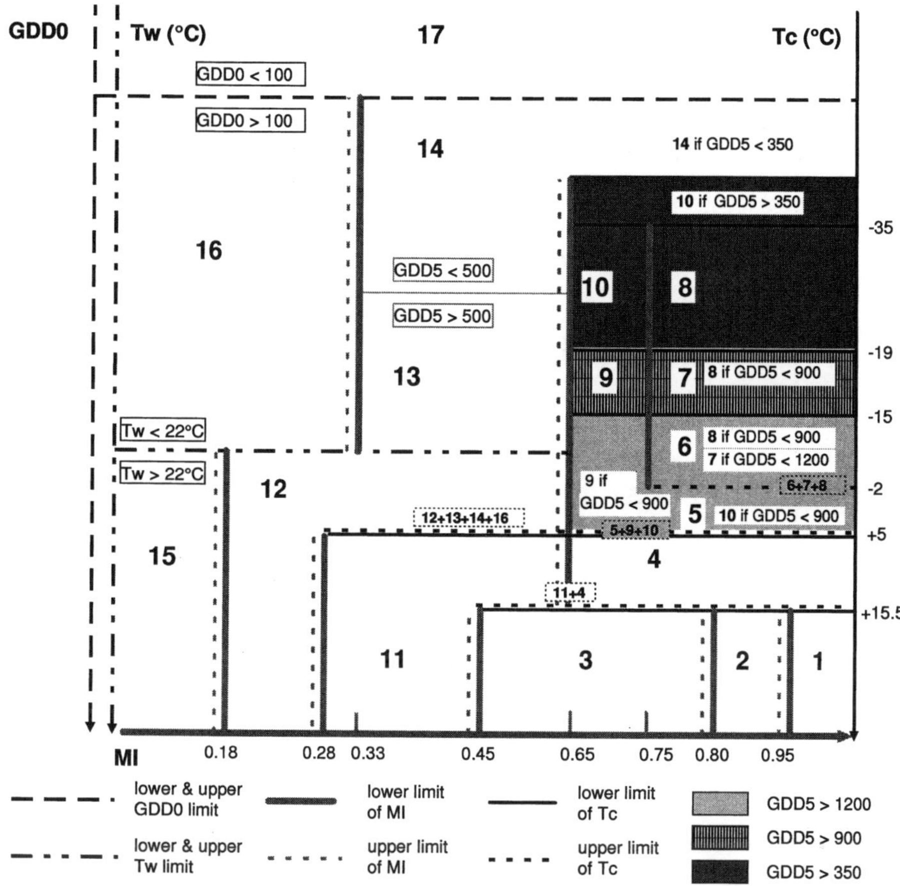

Figure 16.1 Environmental constraints imposed to 17 biomes (bold numbers) in the BIOME model (see Table 16.2 for the list of biomes). Each biome in the diagram has one or more environmental limits: moisture index MI, temperature of the coldest month Tc, growing degree day base 5°C and 0°C, GDD5 and GDD0, respectively, and temperature of the warmest month Tw. The biomes are limited by lower and/or upper environmental limits, but there is not necessarily one lower and one upper environmental limit allocated to each biome. The dotted boxes with bold numbers represent the biomes for which there is an upper Tc limit (e.g. the upper Tc limit of biomes 6, 7 and 8 is –2°C).

The primary input variables are monthly precipitation, surface temperature, the total amount of cloud and soil data. In this study the soil water capacity is set to 15 cm and so no additional soil data are employed. The Holdridge model was tuned (Figure 16.2) in order to match the distribution of vegetation given by the BIOME model when observed climate data are used. This tuning was done using a high resolution dataset of climate (i.e. 0.5° by 0.5°).

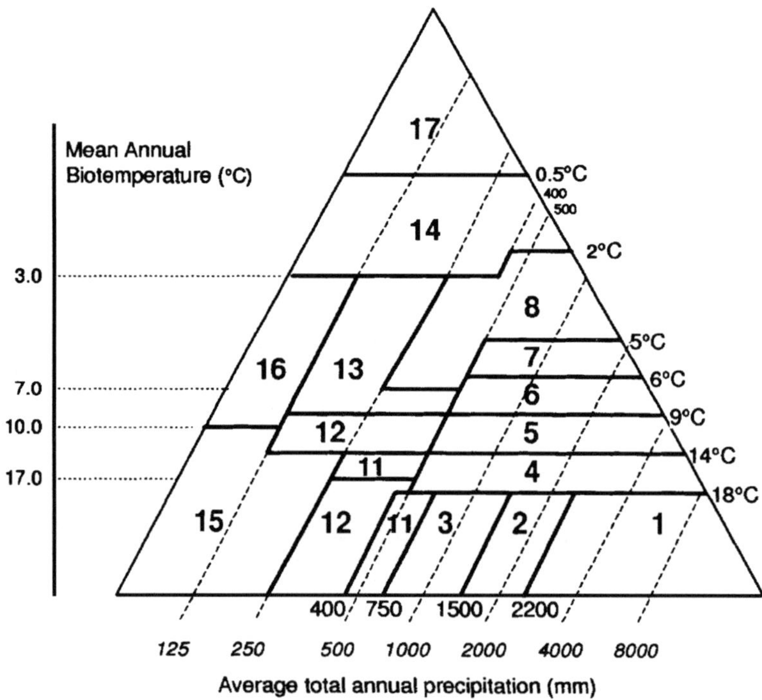

Figure 16.2 Holdridge model tuned to match the distribution of vegetation predicted by the BIOME model using observed climate data, (bold numbers refer to the biome names listed in Table 16.2).

The aim was to obtain two distributions of vegetation with roughly the same number of vegetation types, and as far as possible, similar geographical distributions. These models carry with them the assumption of the existence of an equilibrium between mature vegetation and climate, i.e. they predict the potential natural vegetation types (plant functional types or biomes) which are in equilibrium with the climate. They are *static*, or time-independent models. There are several problems in using *static* global vegetation model as impacts model (Malanson 1993). The shift in the geographical ranges of the vegetation types is simulated without taking into account the spatial mechanisms of vegetation changes. The processes of competition and dispersal are not included, nor are the potential inertia of vegetation, or the potential extinction of species. Moreover, the rate of climatic change cannot be accounted for in these projections. The sort of information these models provide is about the potential future or past favourable conditions for the natural ecosystems.

16.2.2 Climate models

The control climate integrations from six Global Climate Models were used: BMRC (Colman & McAvaney 1995), CCM1 (Oglesby & Saltzman 1992), CCM0 (Washington

& Meehl 1993), CCM1W (Wang et al. 1992), CCM1-OZ (Henderson-Sellers et al. 1993) and CCM2 (Williamson et al. 1995). The disturbed climate integrations (i.e. 'greenhouse' effect experiments) were available for five models only, that is, BMRC, CCM1, CCM0, CCM1W and CCM1-OZ.

The experiments are fully described in the papers mentioned above, and a summary of the main characteristics of the models and experiments are summarised in Table 16.1. All disturbed climate simulations are instantaneous increase greenhouse gases experiments.

It must be noted that most GCMs employed in this study have a coarse horizontal resolution (i.e. R15). The exceptions are BMRC at R21 and CCM2 for which a range of resolutions are available. To allow comparison with the other models, the results used in this section are from the R15 version of CCM2. In addition, it was decided to aggregate spatially the BMRC results to the R15 grid as it would be too difficult to compare the results obtained by the R21 version of the BMRC model with the other model results.

16.3 Evaluation of the ability of the GCMs to simulate bioclimatic variables

The comparison of simulated climates generated by several GCMs shows that the simulated climates are often different among each other and are also different to the observed climate. The intercomparison of climate models undertaken in projects like AMIP (Gates 1992), for instance, provides invaluable information concerning the discrepancies between the observed climate and the simulated current climate. However, this information might not be particularly relevant for the impact modelling community. The impact modelling approach is based on the assumption of a cause and effect relationship between the climate and an 'exposure unit' (e.g. natural vegetation or water resource) (Kates 1985). Only certain characteristics of the climate are important for a given exposure unit (for instance the regime of precipitation, the amount of solar radiation or a certain temperature threshold). Whether a variation of climate is significant or not depends upon the exposure unit considered.

From a modelling perspective, the importance of the variation of climate depends upon the impact model employed. If the output of the model are discrete categories, (e.g. the vegetation classes from the models presented in this chapter), the variations, or differences, of climate are significant only when the climatic variables cross the thresholds imposed by the impact model. The climatic variations are seen through a 'filter' imposed by the impact model. If simulated climates are evaluated from the perspective of investigating effects on potential vegetation, it is essential to look at the climate variables once they have been 'filtered' through the vegetation model.

This methodology assumes that the limits imposed by the vegetation models are themselves significant. If the limits are physiologically based as in BIOME, such an assumption could be justified. A variation, however small, around a temperature threshold of say $-15°C$, would definitely have an impact on some woody plants. If the limits are empirically based, as in the Holdridge scheme, this assumption can be less confidently asserted.

Table 16.1   Global climate model characteristics

Model	Horizontal resolution	Vertical resolution (layers)*	Diurnal cycle	Soil moisture	Convective scheme	SSTs	CO_2 (in ppmv) Control (C)	CO_2 (in ppmv) Increased (I)	ΔT (K)	Number of years in the integrations
BMRC	R21 ~5.6° Ion by lat 3.2°	9	Yes	Bucket	Mass Flux + cirrus albedo	Mlo 50m Q-flux	330	660	2.1	C: 10 I: 10
CCM0	R15 ~7.5° Ion by lat 4.5°	9	No	Bucket	Mass Flux	Mlo 50m No Q-flux	330	660	5.4	C: 20 I: 10
CCM1	R15	12	No	Bucket	MCA	Mlo 50m No Q-flux	330	660	8.2	C: 20 I: 5
CCM1-OZ	R15	12	Yes	BATS	MCA	Mlo 50m Q-flux	330	660	2.7Θ	C: 20 I: 10
CCM1W	R15	12	No	Bucket	MCA	Mlo 50m Q-flux	354 (+ other trace gases)	539 (+ other trace gases)♣	4.2	C: 10 I: 10
CCM2	version R15♦	18	Yes	Specified wetness factor	Mass Flux	From climatology (Shea et al. 1990)	330	n/a	n/a	C: 10 I: n/a

♦The original resolution of CCM2 is T42

* All models use a σ coordinate and CCM2 uses a hybrid system (i.e. σ + pressure coordinate)

ΔT : Model sensitivity, i.e. $2 \times CO_2$ minus $1 \times CO_2$ change in the global annual average surface temperatures

Θ Model sensitivity is given for the surface temperatures from all models, and for the surface air temperatures with CCM1-OZ

♣ i.e. 660 'effective' $2 \times CO_2$

MCA: Moist convective adjustment

Mlo: Mixed layer ocean

16.3.1 Comparison of the simulations of vegetation distribution

From control simulations of these models (i.e. for the present day climate), monthly surface temperatures (and surface air temperatures for CCM1-OZ), precipitation and cloudiness are averaged over a period of 20 years for CCM1-OZ, CCM1, CCM0 and 10 years for CCM1W, BMRC, CCM2. The observed climate datasets used are Legates and Willmott (1990a, 1990b) for the precipitation and surface air temperature and ISCCP (Rossow & Schiffer 1991) for the total cloud fraction. These observed datasets were spatially aggregated to match the R15 resolution.

The vegetation maps predicted by the Holdridge and BIOME models using simulated climates and observed climate will be called GCM maps and reference maps, respectively, throughout the text. The vegetation distributions predicted using the BIOME model and the climates from the different GCMs and from the observed climate datasets are shown in Colour Plate 16.1.

The analysis of the results can be done first by simply comparing the percentages of land area occupied by each vegetation type. Differences in the geographical distribution of the vegetation will be examined next using the Kappa statistic. The objective of this comparison is to evaluate the sensitivity of each biome to the input climate. The focus is therefore on the individual biome, and not on the performance of particular climate models.

The results in Figures 16.3 and 16.4 show the frequency distribution of the biomes simulated by the BIOME and Holdridge models using observed climate data and simulated climate results from the six GCMs. These bar charts represent the proportions of land area (Antartica excluded) covered by the different biomes. If the results obtained with the simulated climates are considered together, the variation of area $Dif(i)$ relative to the total land area can be expressed by

$$Dif(i) = \left(\sum_{m=1}^{m=6} (area_m(i) - area_{obs}(i))/N \right) * 100$$

where $area_m(i)$ is the area covered by a biome i (expressed as a percentage of the total land area) obtained with simulated climate, $N = 6$, and $area_{obs}(i)$ is the 'reference' area (expressed as a percentage of the total land area) obtained with observed climate. Table 16.2 presents the results for both vegetation models.

With BIOME, two vegetation types are considerably underpredicted in all cases: the biomes 13 and 16 (cool grass and semidesert). The biome 11 (xerophytic woods/shrub) is overestimated by all GCMs, while the biome 15 (hot desert) is underestimated by all GCMs. The Student's t-test was applied to find if the variations of area were statistically significant. The null hypothesis (observed area equal simulated area) can be rejected at 5% significance level for the biome 15 and at 1% significance level for the biomes 11, 13, and 16 (cf. Table 16.2).

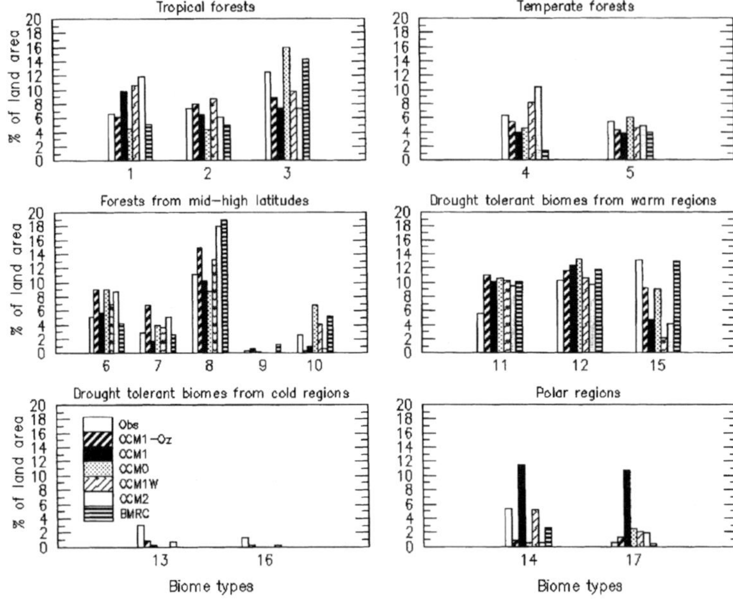

Figure 16.3 Bar chart showing the distribution of each vegetation type, expressed as a percentage of land area, predicted by BIOME using Legates and Willmott/ISCCP datasets and simulated climates from CCM1-OZ, CCM1, CCM0, CCM1 W, CCM2 and BMRC (aggregated to the R15 grid).

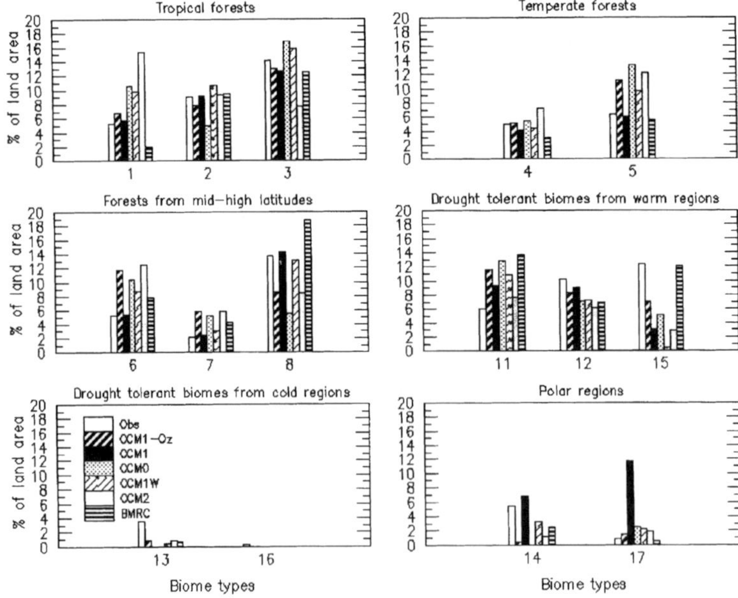

Figure 16.4 Bar chart showing the distribution of each vegetation type as in Figure 16.3 but with the Holdridge scheme.

Table 16.2 Variation in the areal coverage of vegetation predicted by the BIOME and Holdridge models when (i) observed climate is used and (ii) simulated climate from the six GCMs (see text for definition of Dif). The Student's t-test is given when the significance level is 1% or 5%

Biome types	BIOME		Holdridge	
	Dif (%)	t-test (%)	Dif (%)	t-test (%)
1. Tropical rainforest	+1.4		+3.0	
2. Tropical seasonal forest	−0.8		−0.4	
3. Tropical dry forest/savanna	−1.9		−1.1	
4. Warm-mixed forest	−0.8		−0.1	
5. Temperate deciduous forest	−0.9		+3.3	
6. Cool mixed forest	+2.1		+4.2	5
7. Cool conifer forest	+1.0		+2.3	5
8. Taiga	+2.9		−2.2	
9. Cold mixed forest	−0.1		n/a	
10. Cold deciduous forest	+0.4		n/a	
11. Xerophytic wood/scrub	+4.7	1	+4.9	1
12. Warm grass/shrub	+1.3		−2.8	1
13. Cool grass/shrub	−2.8	1	−3.0	1
14. Tundra	−1.8		−3.2	5
15. Hot desert	−6.1	5	−7.3	1
16. Cool semi-desert	−1.2	1	−0.2	1
17. Ice-polar desert	+2.6		+2.5	

Apart from these vegetation types, no other biome are statistically significantly underestimated or overestimated if the results from all the climate models are considered altogether. There is an overestimation of the biome 17 (ice-polar desert) when the CCM climates are used and an overall underestimation of the area of biome 14 (tundra). In the latter case, however, the variation amongst simulations is considerable (from an 89.0% absolute decrease with CCM0 and CCM2 to a 112.0% absolute increase with CCM1).

Biome 6 (cool mixed forest) is overestimated by all CCM models (the Student's t-test applied on the results of CCM models shows that the variation of cold mixed forest is significant at 5% significance level). Overall, the forests from the cold mid-high latitude regions (biomes 6, 7, 8) are overestimated.

With the Holdridge scheme, it appears that the signs of variation are similar to those obtained with the BIOME model, with the exception of the biome 12 (warm grass), the biome 8 (taiga) and the biome 5 (deciduous forest). Discrepancies between reference map and GCMs maps similar to those obtained with BIOME are found for the xerophytic woods, hot desert, ice-polar desert, cool grass, semidesert, and tundra.

The Student's t-test applied on the results obtained with the Holdridge model shows that the null hypothesis (observed area equal simulated area) can be rejected at 5% significance level for the biomes 6, 7 and 14 and at 1% significance level for the biomes 11, 12, 13, 15, and 16.

Table 16.3 Differences in vegetation distributions, expressed as a percentage of total land area (cosine weighted), predicted by the BIOME and Holdridge models when (i) observed climate and (ii) simulated climates are employed, and level of agreement between vegetation maps, expressed with the Kappa statistics (see text for explanations about the Kappa statistics)

	BIOME		Holdridge	
GCMs	Difference of area (%)	KAPPA	Difference of area (%)	KAPPA
CCM1-OZ	60.7	0.36	65.9	0.30
CCM1	73.1	0.23	73.9	0.22
CCM0	56.2	0.40	73.7	0.22
CCM1-W	63.0	0.33	66.8	0.27
CCM2 (R15)	59.4	0.37	67.5	0.27
BMRC	54.0	0.42	53.3	0.42

The main outcome of the comparison of global vegetated areas is the common tendency of the GCMs to induce the overprediction or underprediction of certain vegetation types (e.g. hot desert, xerophytic woods, cool grass, semidesert). It seems that common biases generated by climate models are more apparent with the Holdridge model than with the BIOME model. The biomes for which little bias is observed are the forests of the tropical regions.

The next step in the comparison of vegetation distributions is to evaluate the differences in the spatial distributions. These differences are expected to be greater than the differences found by comparing solely the proportions of area covered by the biomes. Indeed the climate models may be able to predict the 'right' amount of precipitation or the correct range of annual temperature that would lead to a correct prediction of, for instance, tropical rainforest or taiga. However, the climate models also need to be accurate in their prediction in order to obtain a spatial distribution of bioclimatic variables comparable to the observed one.

Table 16.3 shows the difference of distributions expressed as a percentage of the total land area (cosine weighted). For more than 53% of the total land area, the vegetation is 'incorrectly' predicted, if it is considered that the distribution of natural vegetation obtained with observed climate is the 'correct' one.

The comparison of maps can be made visually, but this method is often not sufficiently objective. The Kappa statistics, as described by Monserud and Leemans (1992), is an objective tool for intercomparison of vegetation maps. Kappa has a value of 1 when the agreement is perfect, and a value close to 0 when the observed agreement is approximately the same as would be expected by chance. If the classification of Kappa proposed by Monserud and Leemans (1992) (Table 16.4) is adopted, it follows that the degree of agreement between the reference map and the maps obtained using simulated climate is *poor* for all CCM models. The BMRC model appears to have the best agreement with both vegetation schemes (i.e. *fair* agreement).

Table 16.4 Kappa statistics and qualitative degrees of agreement given to the different ranges of Kappa (after Monserud & Leemans 1992)

Lower bound	Degree of agreement	Upper bound
0	no agreement	0.05
0.05	very poor	0.20
0.21	poor	0.40
0.41	fair	0.55
0.56	good	0.70
0.71	very good	0.85
0.86	excellent	

The agreement between the vegetation maps is worse overall with the Holdridge model, even though this scheme requires two climatic variables which do not account for seasonality of climate and would seem to be easier to predict.

The individual Kappa (Ki) was calculated for each biome and the results are displayed in Figures 16.5 and 16.6.

Following Table 16.4, the graphs can be divided into categories for which the degree of agreement is 'very poor', 'poor', 'fair' and 'good'. The vegetation type for which the agreement between maps is the best, in all cases, is the tropical rainforest (i.e. Ki is around 0.50). No other vegetation type of the BIOME model has such a uniformly 'fair' Kappa. These results mean that, even if the tropical rainforest is overestimated or underestimated (cf. Figure 16.3), the prediction of tropical rainforest is at least done in the correct location. If the biomes 1, 2, and 3 (tropical rainforest, seasonal tropical forest and dry forest/savanna, respectively) are considered together, it appears that the degree of agreement is very similar with all GCMs. There is only one case of 'very poor' agreement (i.e. less than 0.2). The vegetation types for which the agreement is the worst are, not surprisingly, the biomes 13 and 16 (cool grass and semidesert) which are absent, or strongly underpredicted when simulated climates are used (cf. Figure 16.3). The other biomes which are uniformly poorly predicted when the simulated climates are used are biome 9 (cold mixed forest) and biome 7 (cool conifer forest). Biome 9 does occupy a very small area in the reference map (0.38%) and its climatic requirements are very specific, so the likelihood of a poor prediction when a simulated climate is used is very high. The results for the biome 7 are, on the other hand, more significant. The cool conifer forest is clearly not predicted in the regions where it should occur, with all the climate models used here.

For the other biomes, the results are very heterogeneous and depend on the simulated climate employed. Biome 15 (hot desert) is remarkably well predicted when climates of CCM0 and BMRC are employed and the agreement is fair also when the climates from CCM1-OZ and CCM2 are evaluated. Although biome 15 is underpredicted by most GCMs, the Kappa statistic shows that the prediction is done in the correct geographical location.

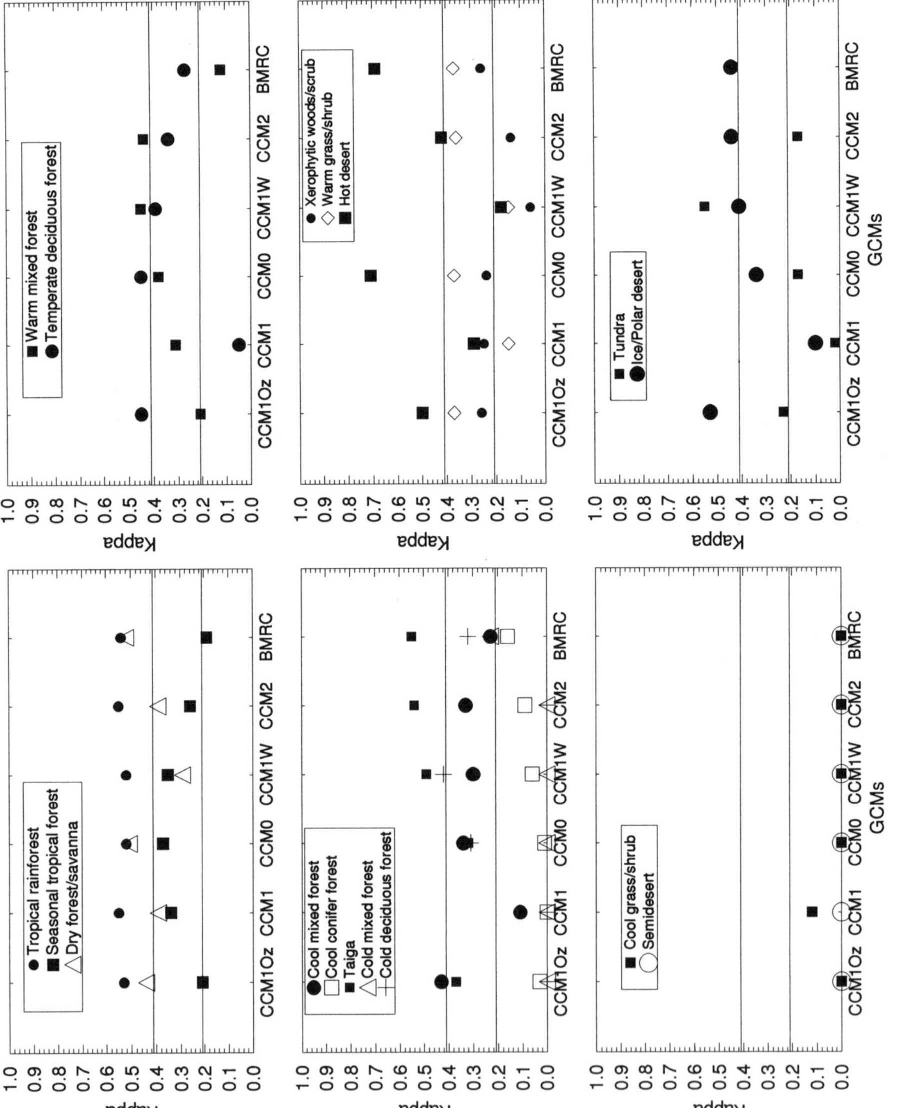

Figure 16.5 Individual Kappa statistics (BIOME model).

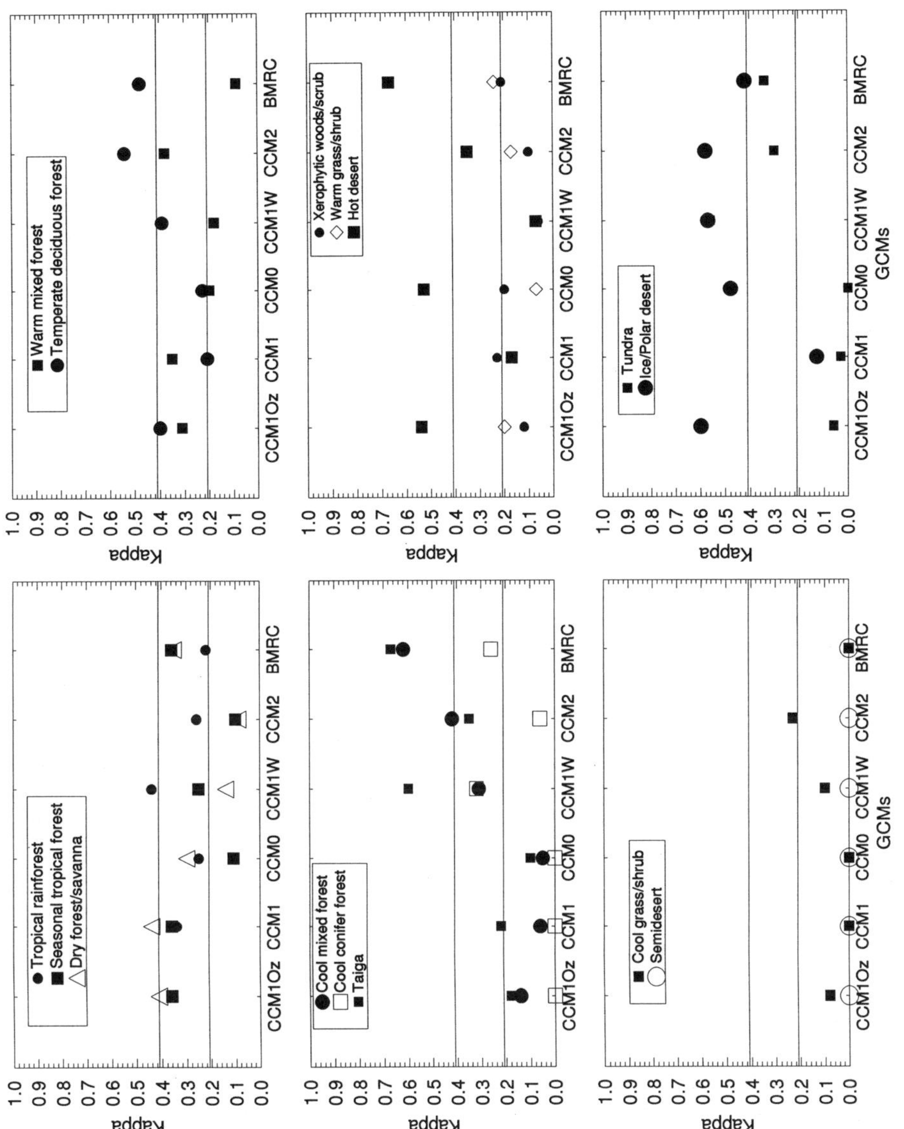

Figure 16.6 Individual Kappa statistics (Holdridge model).

Biome 11 (xerophytic woods) has a low Kappa with most models ('poor' or 'very poor' agreement) even if there is, in all cases, overprediction of the area occupied by this biome. That implies that the xerophytic woods area in the map obtained with the simulated climates does not encompass the xerophytic woods area in the reference map. The xerophytic woods in the reference map are located in many different regions (e.g. Mediterranean region, Australia, margin between desert and savanna in Africa). The likelihood of simulating the climate correctly in all these different regions is less than predicting the climate conditions required by the hot desert in North Africa, for instance.

The individual Kappa statistics obtained with the Holdridge scheme (Figure 16.6) have overall lower values than the ones obtained with BIOME. The results for biomes 15, 13 and 16 (hot desert, cool grass and semidesert, respectively) are comparable to the ones found with BIOME. However the values of Ki for the tropical rainforest are lower than 0.41 (i.e. 'poor' agreement) and it does not seem that there is any consensus between GCMs for this particular biome when it is predicted by the Holdridge scheme (that is when it is predicted solely as a function of total annual precipitation).

16.3.2   Impact of the GCM climates

The preceding section focused on the type of vegetation distribution predicted using climate models. This section will try to identify which are the bioclimatic variables responsible of the discrepancies found between the reference maps and the GCM vegetation maps.

With the Holdridge scheme, the identification of the causes of differences in the vegetation distributions is simple: if the vegetation types differ, it might be due to differences of precipitation values, to differences of biotemperature values, or to the combination of both, whereas with BIOME it is a great deal more complex. There are five bioclimatic variables, so the number of cases and combinations is much greater, even if each vegetation type has a maximum of three climatic predictants (cf. Figure 16.1). The different combination of variables responsible for the differences in vegetation are displayed in Figure 16.7 for BIOME and Figure 16.8 for the Holdridge scheme.

16.3.3   Total annual precipitation and soil moisture index

The soil moisture index (MI) and the total annual precipitation (P) are, by far, the principal causes of changes in the prediction of biomes with both vegetation models. The changes of vegetation attributable to these two variables represent a very important proportion (59% to 83%) of the area where differences of vegetation are found.

The overestimation of the total annual precipitation by the CCM climate models is clearly the main cause of discrepancies between the vegetation maps (Figure 16.9).

For all climate models, the underestimation of precipitation mainly occurs in the intertropical regions. The regions affected by the underprediction of rainfall are (i) India (with CCM1-OZ, CCM1, CCM0, CCM1W), (ii) Malaysia (CCM1, CCM1W, BMRC),

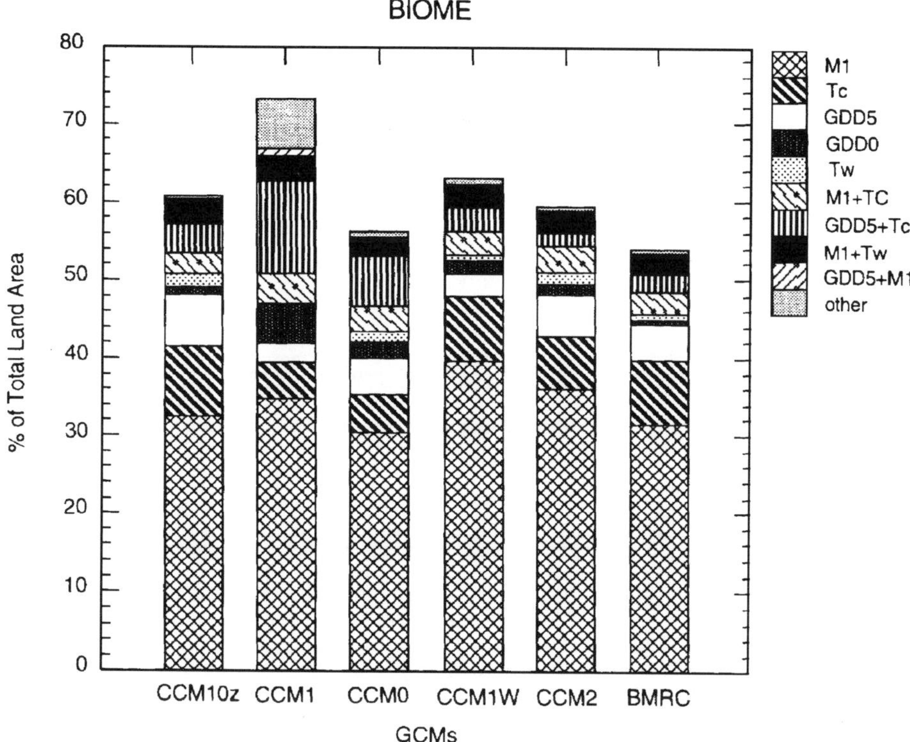

Figure 16.7 Bioclimatic variables responsible of the differences of vegetation between reference map and GCM maps with the BIOME model: moisture index MI, temperature coldest month Tc, growing degree day base 5°C and 0°C, GDD5 and GDD0, respectively, and temperature of the warmest month Tw.

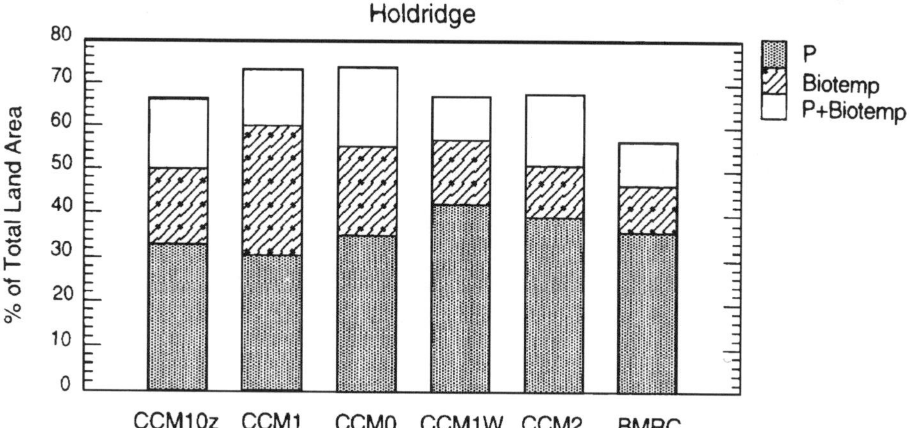

Figure 16.8 Bioclimatic variables as in Figure 16.7, except with Holdridge model; total annual precipitation, P and biotemperature, Biotemp.

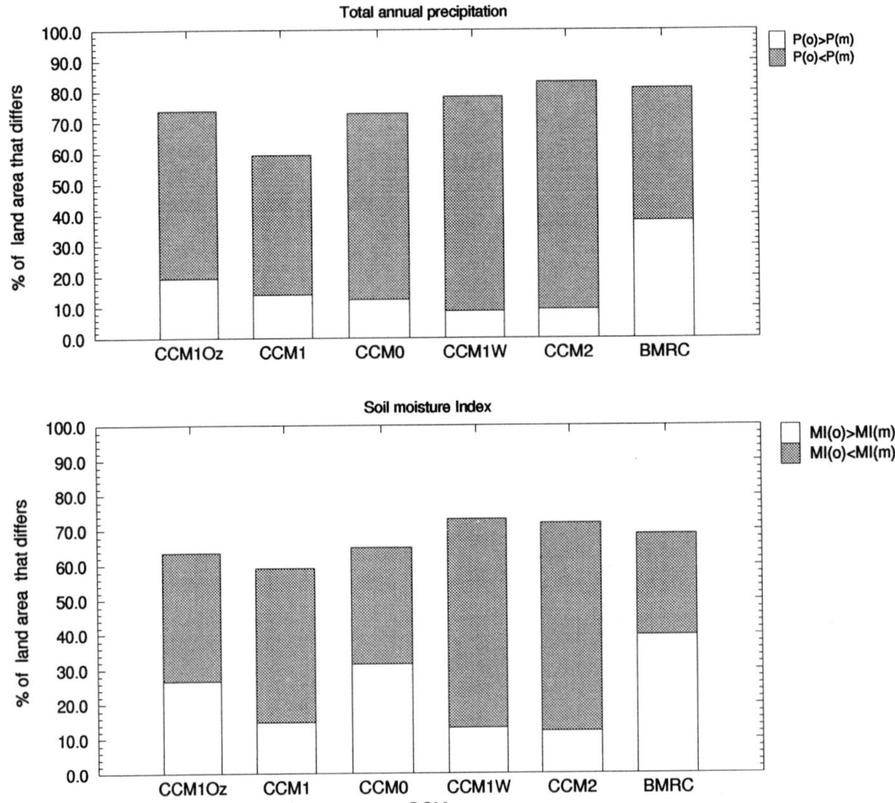

Figure 16.9 Top graph: proportion of the global land area where differences of vegetation are due to overprediction or underprediction of the total annual precipitation P; $P(o)$ is the observed total annual rainfall and $P(m)$ is the total annual rainfall obtained from simulated climates. Bottom graph: same as top graph for the soil moisture index MI.

(iii) part of Amazon Basin and western Africa (maps not shown here). In the mid-high latitudes (i.e. north of 45°N), discrepancies in the vegetation distributions are most often due to total annual precipitation which is *greater* than observed. In these regions, however, the causes of the variation of vegetation are due to a combination of biases in precipitation and biotemperature. The results obtained with MI (Figure 16.9) show a similar range and distribution to those found with P; the exception being with the model CCM0.

16.3.4 Growing season

Both vegetation models use as predictants a representation of the growing season (i.e. GDD5 and GDD0 for BIOME and the annual mean biotemperature for the Holdridge scheme). With three of the climate models (CCM1-OZ, CCM0, CCM2), the differences in vegetation

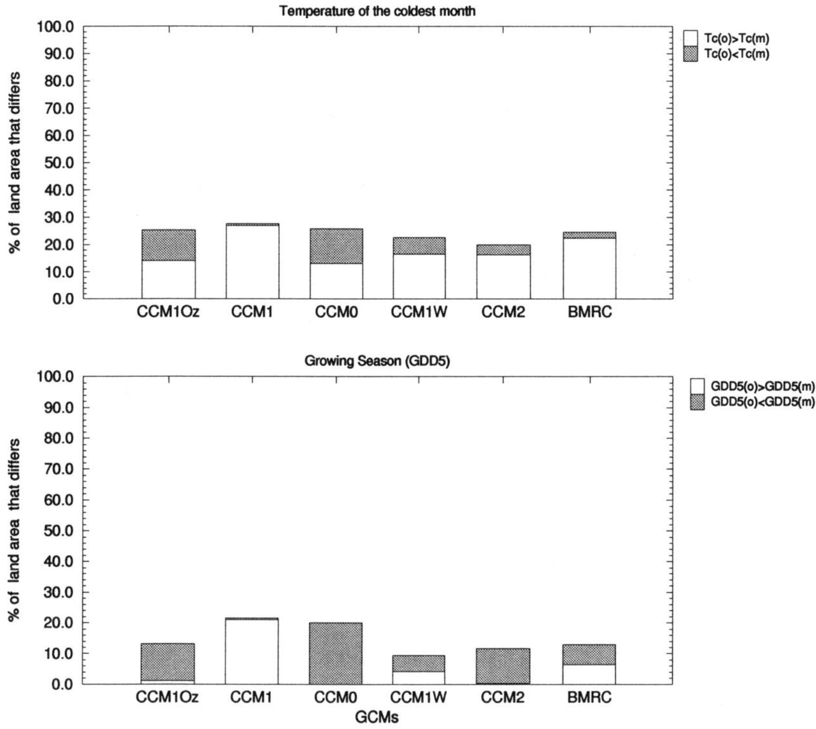

Figure 16.10 Proportion of the global land area as in Figure 16.9, except top graph: temperature of the coldest month Tc and bottom graph: growing degree day GDD5.

distributions due to GDD5 are almost entirely due to an *overestimation* of the length of the growing degree day base 5°C (Figure 16.10).

Yet paradoxically there is underestimation of the length of the growing degree day base 0°C for these three GCMs (map not shown here). The region where differences of GDD0 are responsible for differences in vegetation is Greenland. With CCM1, the region where GDD0 is overestimated encompass a much larger region than Greenland only. Indeed the climate simulated from this version of the model is abnormally cold in the northern high latitude regions. The regions where GDD5 is wrongly predicted depend on the simulated climate employed. These regions are located in the northern high latitudes (Canada, Siberia) but also in the region of Tibet-Himalayas. In this region the simulated climates are both too cold and too wet and it seems to be a common feature for all climate models.

16.3.5 Temperature of the coldest and warmest month

The temperature of the coldest month (Tc) is a predictant for all the forests and woody types of vegetation (i.e. 11 biomes). Tc, after MI, is the second most important cause of disagreement between vegetation maps (i.e. approximately 21% to 27% of the area where differences of vegetation occur). There is, with all climate models, an overall underestimation of the temperature of the coldest month even when GDD5 is overestimated (cf. CCM1-OZ, CCM0 and CCM2 in Figure 16.10). The fact that the growing degree days are too high and, at the same time, the temperature of the coldest month too low explains why there are discrepancies in the mid-high latitudes between the BIOME and Holdridge models when the climates from CCM1-OZ, CCM2 and CCM0 are used. In these regions, the winter temperatures simulated by the GCMs do not allow temperate forests to be predicted by the BIOME model, whereas the biotemperatures are sufficiently high to enable the prediction of temperate forests by the Holdridge scheme.

Discrepancies in the simulation of Tc have an impact on the prediction of warm-mixed forest, cool mixed forest, cold conifer forest and cool conifer forest.

The temperature of the warmest month (Tw) is a predictant for the grass/desert biomes. Differences in Tw do not have an important impact on the prediction of hot desert and warm grass biomes, but has a greater influence on the prediction of the cool grass and semidesert. The regions where Tw is underpredicted are located in Central Asia which is in agreement with the fact that temperatures are too cold in these regions. The regions where Tw is overestimated are located in the north of Central Asia, and in western United States.

16.4 Sensitivities of the vegetation schemes to climate change scenarios

The two basic prerequisites to undertake climate change assessment are: (1) information about the future climate state and, (2) transfer functions capable of transforming this information into impacts (biophysical or societal) (Fowler & de Freitas 1990). Since there is not yet a method, or a model, that can provide an acceptable prediction of the future climate at the regional level, many climate impacts studies have to rely on the use of climate change scenarios. A climate scenario can be defined as 'a description of a possible future climate, developed for some given purpose, and based on a number of assumptions' (Pittock 1993, p. 481), or as 'an internally consistent representation of future climate, that is constructed from methods based upon sound scientific principles, and that can be used to give an understanding of the response of environmental and social systems to future climate change' (Viner & Hulme 1992, p. 33).

A climate scenario may be constructed using different approaches, e.g. 'analogue' method based on palaeoclimatic or historical information and 'arbitrary' scenario (cf. Giorgi & Mearns 1991; Carter et al. 1992, for a review).

In most climate impacts analysis, however, the climate scenarios are based on climate models results. The technique the most commonly used consists of obtaining the difference or ratio between a $2 \times CO_2$ equilibrium experiment and the control run (i.e. $1 \times CO_2$) experiment.

Once the scenario anomaly field is created, the problem of spatial scale has then to be addressed since most impacts models require accurate information at a regional scale. Several methods can be employed to overcome the problem of the coarse GCM's spatial scale: (i) the scenario anomaly in each grid box is combined with the nearest station data, (ii) the scenario anomaly field is spatially interpolated, (iii) statistical relationship are established between observed climate at local scale and at the scale of the GCM (e.g. use of regression relations, Empirical Orthogonal analysis EOF, Perfect Prog cf. Wigley et al. 1990, and Giorgi & Mearns 1991) and, (iv) the scenario has already a fine resolution because a high resolution Limited Area Model was used (e.g. Giorgi et al. 1992).

In many studies that involve analysis of the impacts of the climate change at a global scale, the scenario anomaly field is simply interpolated to a higher spatial resolution and appended to a 'baseline' climate. The baseline climate is often a standard set of observed climate based on recent station, or satellite, data (Cohen 1990). The assumption is that the imperfections in the GCM are reduced by using a realistic climate baseline.

In the vegetation modelling community, this technique (i.e. difference/ratio between $2 \times CO_2$ and $1 \times CO_2$ interpolated to high resolution grid) was used by Emanuel et al. (1985a, 1985b), Leemans (1990), Smith et al. (1992), Monserud & Leemans (1992), Smith & Shugart (1993), Mellilo et al. (1993), Prentice et al. (1993) (in the latter study, the scenario was based on the difference between last glacial maximum and present climate), and Lenihan and Neilson (1995). It has also been used in regional climate impacts studies (e.g. Smith & Tirpak 1989, to study the impacts of climate change in United States).

Only few studies have assessed the effects of climate change by using the GCM output directly into an impact model. Claussen (1993) compared the differences in vegetation prediction using BIOME and three control and perturbed climate simulations.

The different methodologies applied to construct scenarios based on GCMs information may have some impacts on the simulation natural ecosystem. The response of the vegetation models to the climate forcing is not linear, so it is expected that some differences of prediction will occur when different approaches are used. The question is to know which biomes will be more affected by the changes in the methodology employed, for instance if the scenario anomaly field is spatially interpolated or not, if ratio or differences are used, if different baseline climate are used.

As noted earlier, most impacts studies employed climate scenarios, since it is assumed that the GCMs cannot realistically reproduce the current and perturbed climate at a regional level. Yet it is certainly interesting to evaluate what would be the prediction when the GCM climate integrations are used directly (i.e. $2 \times CO_2$ climate) or when the GCM climate is combined with a baseline climate. In this section, climate scenarios are created by using the difference between $2 \times CO_2$ and $1 \times CO_2$ for temperature T, and the ratio of $2 \times CO_2$ to $1 \times CO_2$ for precipitation P and cloudiness C. These scenario anomaly fields are then appended to the baseline climate, i.e. the spatially aggregated observed climate described in section 16.3. No spatial interpolation is undertaken at this stage.

Figures 16.11 and 16.12 shows the differences in vegetated areas between (i) '$2 \times CO_2$' and '$1 \times CO_2$' simulations and (ii) scenario and baseline. In both cases, the ΔT, ΔP and

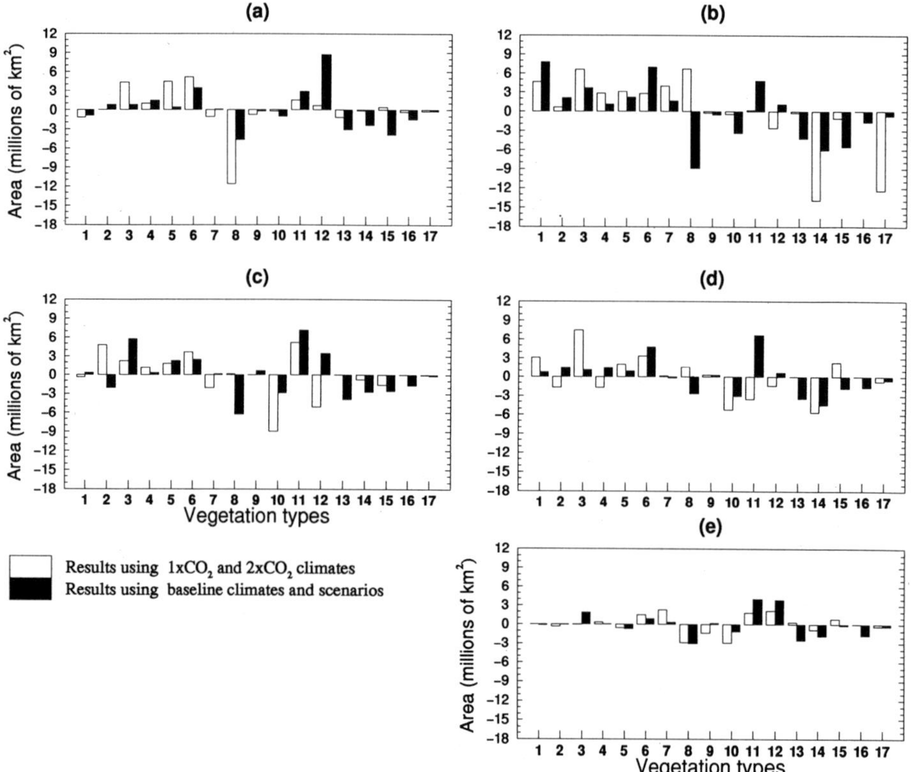

Figure 16.11 Bar charts showing the differences in biome areal coverage predicted by the BIOME model using method (i) (in white), that is comparing vegetation distribution using $1 \times CO_2$ and $2 \times CO_2$ climates and method (ii) (in black), that is comparing vegetation distribution using a baseline climate and climate change scenarios. The simulated climates are from: (a) CCM1-OZ, (b) CCM1, (c) CCM0, (d) CCM1W and (e) BMRC climate simulations. Positive values indicate that the biome areal coverage is predicted to increase under climate change conditions.

ΔC are the same. If the '$1 \times CO_2$' simulation were very close to the baseline simulation, then we would expect to see comparable trends in the variation of vegetated areas. From Figures 16.11 and 16.12 it can be seen that the response of the vegetation models (i.e. increase or decrease of a given vegetation type) differ dramatically whether (i) or (ii) are considered, at least for certain biomes (i.e. taiga, warm grass, xerophytic woods). When scenarios are used as in (ii), it appears that the responses of the different biomes to the predicted climate change are more uniform whereas when GCM climates are employed directly as in (i) the variation of area for a given biome can be of opposite sign (cf. biomes 7, 8, 11, 12, 15 in Figure 16.11). It is due to the fact that the vegetation maps generated with control climate are, in some cases, very different from the vegetation map generated with the baseline climate (cf. section 16.3).

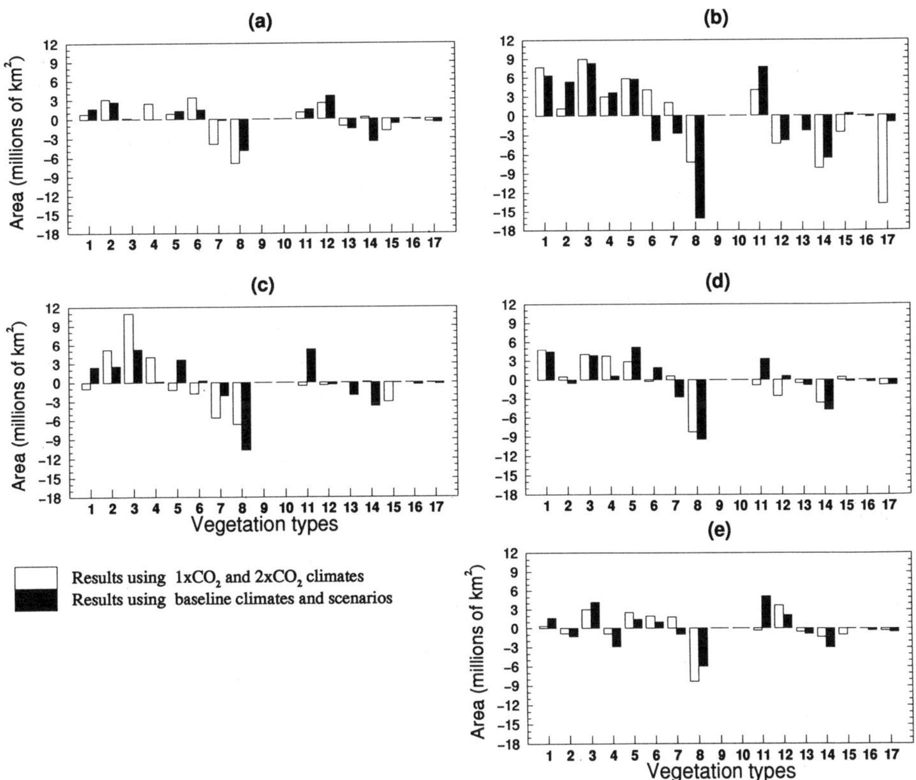

Figure 16.12 Bar charts showing the differences in biome areal coverage predicted by the Holdridge scheme using method (i) (in white), that is comparing vegetation distribution using $1 \times CO_2$ and $2 \times CO_2$ climates and method (ii) (in black), that is comparing vegetation distribution using a baseline climate and climate change scenarios. The simulated climates are from: (a) CCM1-OZ, (b) CCM1, (c) CCM0, (d) CCM1W and (e) BMRC climate simulations. Positive values indicate that the biome areal coverage is predicted to increase under climate change conditions.

For instance, the $2 \times CO_2$ climate simulated by CCM1 is considerably warmer (cf. ΔT in Table 16.1) than the control climate. Since the taiga is underpredicted and the tundra grossly overpredicted with the $1 \times CO_2$ climate conditions generated by CCM1, the taiga in the 'CCM1–$2 \times CO_2$ world' increases dramatically and replaces partly the tundra and ice-polar desert (cf. Figure 16.11(b), shaded). Yet the response of taiga to climate change using CCM1-based *scenario* (cf. Figure 16.11(b), clear) is more reasonable (i.e. decrease) and, overall, comparable to the response of taiga with the other GCM-based scenarios. It means that the use of scenarios generates more realistic (e.g. warmer climate leads to *decrease* of taiga) and more homogeneous responses from the exposure unit considered, but the flaws of the GCM climate predictions are certainly concealed if this method is used solely.

Another example is given by biomes 13 and 16 (cool grass and semi-desert): the response of these biomes to climate change *scenarios* is a decrease of area. However when $2 \times CO_2$ climates are direct inputs into the vegetation models, little changes occur because these biomes are almost absent from the $1 \times CO_2$ maps, since the GCMs used in this study do not simulate the climatic conditions favourable for cool grass and semi-desert. Similar conclusions can be drawn for biome 14 (tundra) (cf. Figure 16.11(a), (c), and (e)).

It can probably be argued that the more realistic approach to assess the impacts of climate change is to *build* a climate scenario, and not to use the GCM output directly. The type of comparison depicted in Figures 16.11 and 16.12 can be seen, however, as a way of providing an estimation of the uncertainties related to the use of GCM climate predictions. Perhaps a GCM simulation which generates dramatically different vegetation responses when method (i) and (ii) are employed should be treated very cautiously and, conversely, a GCM simulation (cf. Figures 16.11(e) and 16.12(e)) which generates similar vegetation responses with method (i) and (ii) can be considered with more confidence.

16.5 Summary

The comparison of the vegetation distributions predicted using simulated climates and observed climate shows that the performance of the climate models, with respect to the prediction of natural vegetation, has to be given an overall evaluation of 'poor'. Some climate models give better results than others (i.e. best results with BMRC and CCM1-OZ and very poor prediction of vegetation with CCM1). The climate models used in this study have different characteristics including the inclusion/exclusion of a diurnal cycle, the sophistication of the land surface scheme and the ocean model and the differing degree of sophistication of physical parameterisation (cf. Table 16.1). Some GCMs are more recent than others (e.g. CCM2 versus CCM0). Yet the predictions of vegetation obtained using these different climate model results do not seem to be simply or directly related to these climate model characteristics. With BIOME, for instance, the best results are obtained with CCM0 and BMRC. It must be noted however that the relatively poor simulation of bioclimatic variables obtained with CCM2 are due to the fact that the R15 version of the model was used, and not the original resolution of the model, i.e. T42. Indeed a sensitivity analysis (not presented here) shows that the more 'realistic' distribution of vegetation simulated by BIOME are obtained when the T42 version of CCM2 is used (Ciret, 1996).

The agreement between vegetation maps appears to be worst, overall, when the vegetation is predicted by the Holdridge model (cf. CCM0) even though the requirements of the Holdridge model (i.e. total annual precipitation and annual mean biotemperature) would seem to be easier to satisfy than the requirements of the BIOME model. The poor agreement between GCM maps and the reference maps does not mean that all biomes are equally poorly predicted. For some biomes, the agreement is fair, whether one simulated climate, or several simulated climates are used (cf. the case of tropical rainforest with BIOME, or hot desert and ice polar desert).

There are some common features in the prediction of natural vegetation using simulated climates. With both vegetation models, the biomes belonging to dry climates, that is hot desert, semidesert, cool grass (with BIOME), tundra, warm grass (with the Holdridge scheme) are underpredicted while the xerophytic woods are overpredicted. There is overprediction of forests from the temperate-cold regions with both the Holdridge and BIOME models. With BIOME, though, the overprediction is shifted toward the colder regions which implies that there is more taiga and less temperate forests. The difference in the prediction of temperate versus cool forest are particularly large with CCM1-OZ, CCM0, CCM2. These differences are due to the fact that, in the Holdridge scheme, there is no requirement for the temperature of the coldest month to be accurate for the prediction of boreal forests whereas with BIOME, the extreme temperature and the growing season length are both required.

This study has shown that the comparison of the vegetation model simulations using GCM *control* climates and *observed* climate is a first necessary step prior to climate change impacts assessment. The assessment of the sensitivity of the impact models to the manner the scenario is built should also be encouraged in order to explore the effect of uncertainties related to the use of GCM-based climate predictions.

Acknowledgments

This paper is CIC contribution number 95/42.

References

Carter, T.R., Parry, M.L., Nishioka, S. and Harasawa, H. (eds.) (1992) *Preliminary Guidelines for Assessing Impacts of Climate Change*. Environmental Change Unit (UK) and Center for Global Environmental Research (Japan), UK, Working Group II of the Intergovernmental Panel on Climate Change.

Ciret, C. (1996) Linking Global Vegetation models and global climate models, PhD thesis, Macquarie University, Sydney.

Claussen, M. (1993) Shift of biome patterns due to simulated climate variability and climate change, Technical Report No. 115, Max-Planck-Institut fur Meterologie, Hamburg.

Claussen, M. and Esch, M. (1994) Biomes computed from simulated climatologies. *Climate Dynamics*, **9**, 235–243.

Cohen, S.J. (1990) Bringing the global warming issue closer to home: The challenge of regional impacts studies. *Bulletin of the American Meteorological Society*, **71**, 520–526.

Colman, R.A. and McAvaney, B.J. (1995) Sensitivity of the climate response of an atmospheric general circulation model to changes in convective parameterization and horizontal resolution. *Journal of Geophysical Research*, **100**(D2), 3 155–3172.

Emanuel, W.R., Shugart, H.H. and Stevenson, M.P. (1985a) Climatic change and the broad-scale distribution of terrestrial ecosystem complexes. *Climatic Change*, **7**, 29–43.

Emanuel, W.R., Shugart, H.H. and Stevenson, M.P. (1985b) Response to comment: Climatic change and the broad-scale distribution of terrestrial ecosystem complexes. *Climatic Change*, **7**, 457–460.

Fowler, A.M. and de Freitas, C.R. (1990) Climate impact studies from scenarios: Help or hindrance? *Weather and Climate*, **10**, 3–9.

Gates, W. (1992) AMIP: The Atmospheric Model Intercomparison Project. *Bulletin of the American Meteorological Society*, **73**, 1962–1970.

Giorgi, F. and Mearns, L.O. (1991) Approaches to the simulation of regional climate change: A review. *Review of Geophysics*, **29**, 191–216.

Giorgi, F., Marinucci, M.R. and Visconti, G. (1992) A $2\times CO_2$ climate change scenario over Europe generated using a limited area model nested in a general circulation model 2. Climate change scenario. *Journal of Geophysical Research*, **97**(D9), 10011–10028.

Henderson-Sellers, A., Dickinson, R.E., Durbidge, T.B., Kennedy, P.J., McGuffie, K. and Pitman, A.J. (1993) Tropical deforestation: Modeling local- to regional-scale climate change. *Journal of Geophysical Research*, **98**(D4), 7289–7315.

Holdridge, L. (1947) Determination of world plant formations from simple climatic data. *Science*, **105**, 367–368.

Kates, R. (1985) The interactions of climate and society. In *Climate impacts assessment: Studies of the interactions of climate and society*, edited by R. Kates, J. Ausubel and M. Berberian, SCOPE 27. Chichester: John Wiley & Sons Ltd.

Köppen, W. (1936) Das geographische system der klimate. In *Handbuch der Klimatologie*, Vol. 1, edited by W. Köppen and R. Geiger. Berlin: Gebruder Borntraeger.

Leemans, R. (1990) Possible changes in natural vegetation patterns due to a global warming. Technical Report IIASA/WP-90-08, International Institute for Applied Systems Analysis, Laxenburg, Austria.

Legates, D. and Willmott, C. (1990b) Mean seasonal and spatial variability in gauge-corrected global precipitation. *International Journal of Climatology*, **10**, 111–127.

Legates, D. and Willmott, C. (1990a) Mean seasonal and spatial variability in global surface air temperature. *Theoretical Applied Climatology*, **41**, 11–21.

Lenihan, J.M. and Neilson, R.P. (1995) Canadian vegetation sensitivity to projected climatic change at three organizational levels. *Climatic Change*, **30**, 27–56.

Malanson, G.P. (1993) Comment on modeling ecological response to climate change. *Climatic Change*, **23**, 95–109.

Mellilo, J.M., McGuire, A.D., Kicklighter, D.W., III, B.M., Vorosmarty, C.J. and Schloss, A.L. (1993) Global change and terrestrial net primary production. *Nature*, **363**, 234–240.

Monserud, R.A. and Leemans, R. (1992) Comparing global vegetation maps with the Kappa statistic. *Ecological Modelling*, **62**, 275–293.

Oglesby, R.J. and Saltzman, B. (1992) Equilibrium climate statistics of a general circulation model as a function of atmospheric carbon dioxide. Part 1: Geographic distributions of primary variables. *Journal of Climate*, **5**, 66–92.

Pittock, A.B. (1993) Climate scenario development. In *Modelling Change in Environmental Systems*, edited by A.J. Jakeman, M.B. Beck and M.J. McAleer. Chichester: John Wiley and Sons Ltd.

Prentice, C., Cramer, W., Harrison, S., Leemans, R., Monserud, R. and Solomon, M. (1992) A global biome model based on plant physiology and dominance, soil properties and climate. *Journal of Biogeography*, **19**, 117–134.

Prentice, C. and Solomon, A. (1991) Vegetation models and global changes. In *Global Change of the Past*, edited by R.S. Bradley. Boulder: UCAR/Office for Interdisciplinary Earth Studies.

Prentice, C., Sykes, M.T., Lautenschlager, M., Harrison, S.P., Denissenko, O. and Bartlein, P.J. (1993) Modelling global vegetation patterns and terrestrial carbon storage at the last glacial maximum. *Global Ecology and Biogeography Letters*, **3**, 67–76.

Rossow, W. and Schiffer, R. (1991) ISCCP cloud data products. *Bulletin of the American Meteorological Society*, **72**, 2–20.

Shea, D.J., Trenberth, K.E. and Reynolds, R.W. (1990) A global monthly sea surface temperature climatology. Technical Report NCAR/TN-345+STR, National Center for Atmospheric Research, Boulder, Colorado.

Smith, J.B. and Tirpak, D.A. (1989) The potential effects of Global Climate Change on the United States. Technical Report, pp. 413. Washington, DC: Environmental Protection Agency (EPA).

Smith, T.M., Leemans, R. and Shugart, H.H. (1992) Sensitivity of terrestrial carbon storage to $2\times CO_2$-induced climate change: Comparison of four scenarios based on general circulations models. *Climatic Change*, **21**, 367–384.

Smith, T.M. and Shugart, H.H. (1993) The potential response of global terrestrial carbon storage to a climate change. *Water, Air and Soil Pollution*, **70**, 629–642.

Thornthwaite, C. (1948) An approach toward a rational classification of climate. *Geographical Reviews*, **38**, 55–94.

Viner, D. and Hulme, M. (1992) Climate change scenarios for impacts studies in the UK: General circulation models, scenario construction methods and applications for impacts assessment, Technical Report. UK: Climatic Research Unit.

Wang, W.-C., Dudek, M.P. and Liang, X.-Z. (1992) Inadequacy of effective $2\times CO_2$ as a proxy in assessing the regional climate change due to other radiatively active gases. *Geophysical Research Letters*, **19**(13), 1375–1378.

Washington, W. and Meehl, G. (1993) Greenhouse sensitivity experiments with penetrative cumulus convection and tropical cirrus albedo effects. *Climate Dynamics*, **8**, 211–223.

Wigley, T.M., Jones, P.D., Briffa, K.R. and Smith, G. (1990) Obtaining sub-grid-scale information from coarse resolution general circulation model output. *Journal of Geophysical Research*, **95**, 1943–1953.

Williamson, D.L., Kiehl, J.T. and Hack, J.J. (1995) Climate sensitivity of the NCAR Community Climate Model (CCM2) to horizontal resolution. *Climate Dynamics*, **11**, 377–397.

Woodward, F. (1987) *Climate and Plant Distribution*. Cambridge: Cambridge University Press.

CHAPTER 17

TROPICAL CYCLONES AND CLIMATE CHANGE:
A PRELIMINARY ASSESSMENT

Heather Tonkin, Gregory Holland, Christopher Landsea and Shuhua Li

17.1 Introduction: Why consider tropical cyclones and greenhouse?

Tropical cyclones, also called hurricanes and typhoons, are intense synoptic scale storms which form over warm tropical ocean regions and are the cause of major natural disasters. These storms can produce damage to coastal regions and islands through extreme wind, rain, storm surge and wave action. Increases in human settlement in tropical coastal regions is rapidly increasing global vulnerability to these tropical storms. The Pacific islands suffer extensive housing and coastal agriculture loss from tropical cyclone encounters (see, for example, Henderson-Sellers 1996). In addition tropical cyclones continue to cause considerable loss of life. A tropical cyclone in the Bay of Bengal in 1990 killed around 135 000 people. The largest damage costs come from the USA because of the high coastal infrastructure costs associated with several recent hurricanes, especially Hurricane Andrew which resulted in over US$30 billion in damage.

The characteristics of tropical cyclones vary among individual storms and between ocean regions. North Atlantic hurricanes have been linked to the El Niño-Southern Oscillation (ENSO) (Gray 1984). Although the existence and degree of an El Niño is only one of the factors affecting hurricane season statistics, the damage caused in the Florida Peninsula, the US East Coast and in the Caribbean is strongly correlated to El Niño events (cf. Chapter 6).

From hurricane data for the period 1950–1990 (Gray and Landsea 1992; Gray et al. 1993, 1994), during the 10 warmest ENSO years hurricanes have caused almost US$19 billion along the Florida peninsula and US East Coast versus US$1.5 billion during the 10 coldest ENSO years (or La Niñas). Gray et al. have also found a similar modulation in the number of intense hurricanes (those with sustained winds of at least 50 m/s): during El Niño years there were 15 intense hurricanes in the Atlantic basin whereas there were 32 during the La Niña years.

Although no detailed damage statistics are available, ENSO also effects tropical cyclone activity in the Pacific Ocean. Tropical cyclones tend to spread eastward into the central

Pacific during El Niño events, with devastating consequences for small island nations (Dong and Holland 1994).

Differences in current tropical cyclone regimes, links to complex atmosphere–ocean phenomena such as El Niño, and the difficulties of measuring the exact structure of tropical cyclones, means that the precise nature and formation process of tropical cyclones remain unclear and makes prediction of future changes in tropical cyclone characteristics very difficult (e.g. Lighthill et al. 1994). Nonetheless present vulnerability to tropical cyclones could be exacerbated by future changes in their occurrence, intensity or frequency.

The possibility of changed impacts in the future has prompted studies of greenhouse-induced changes in tropical cyclone characteristics (Emanuel 1987; Broccoli & Manabe 1990).

Concerns about possible future tropical cyclone changes can be summarised under: (i) the frequency, (ii) the area of occurrence, (iii) the mean intensity and (iv) the maximum intensity.

Two approaches have been employed to try to predict changes in tropical cyclone frequency: (1) to simply count the tropical cyclone-like vortices that appear in GCM simulations, and (2) to analyse the large-scale fields that best relate to tropical cyclone genesis (tropospheric wind shear, moist static stability, vorticity, mid-level relative humidity etc.) for changes between simulations of today's climate and in a greenhouse-warmed climate. The first approach has been attempted by Broccoli and Manabe (1990), Haarsma et al. (1993) and Bengtsson et al. (1994a, 1994b). The Haarsma et al. study suggests more tropical cyclones, the Bengtsson et al. studies suggest fewer tropical cyclones and the Broccoli and Manabe study suggests both more and fewer, depending on what type of cloudiness (either fixed or variable, respectively) is utilised.

The second approach was taken by Ryan et al. (1992) who found a substantial increase of tropical cyclones in a $2 \times CO_2$ scenario because of an increase in favourable thermodynamic parameters while the dynamic parameters changed little. However, Ryan et al. applied the Gray (1968, 1975) genesis parameter to GCM fields to obtain these findings. We have reservations about this approach, as the Gray parameter was highly tuned to fit current conditions and does not explicitly account for such thermodynamic changes as the upper-tropospheric warming in the tropics for $2 \times CO_2$. Bengtsson et al. (1994a, 1994b) found that this increased stabilisation leads to decreased frequency of cyclone type systems in a GCM. For changes in the area of occurrence, again the two methods have been utilised with mixed results. Ryan et al. (1992) suggest increases in the area of occurrence essentially because of the areal increase of oceans warmer than 26°C. We show in the following sections that the ocean temperature cut-off for cyclone formation is likely to increase in a $2 \times CO_2$ world. This is likely to result in little change in the region of cyclone development, as has been found by Bengtsson et al. (1994a, 1994b) and Haarsma et al. (1993).

Lastly, the mean intensity and maximum achievable intensity of tropical cyclones has been studied using GCM results directly by measuring the intensities of the simulated disturbances. This approach has a significant limitation in that the GCM with its coarse grid spacing is not able to capture the mesoscale structure of the inner core (and thus the true winds and central pressure) of the storms. However GCM studies have been able to show realistic outer core circulation and thermodynamic structure of these storms. Haarsma et al. (1993) find increases in both the mean and maximum intensities in their GCM-generated storms, while Bengtsson et al. (1994a, 1994b) find no change in either the mean or maximum intensities. Broccoli and Manabe (1990) do not report how the intensities of their storms changed.

All these GCM studies suffer from the difficulty that the spatial resolution (and also probably the parameterisations) of the atmospheric models employed are much too poor to permit simulation of realistic tropical cyclones. The MECCA Analysis Team tried to evaluate the potential of exploiting 'embedded' regional climate model results in regions of tropical cyclone incidence. Careful analysis of the only MECCA simulation in such an area (Hirakuchi & Giorgi 1995) revealed that the regional model produced a worse climatology, at least in terms of the meteorological factors important in generating and sustaining tropical cyclones, than the GCM used to 'drive' it. Whether the use of regional climate models embedded within a more correct GCM simulation can provide helpful information for greenhouse scenarios is an open question.

It is clear that other means of evaluating likely future tropical cyclone activity must be sought. The following two sections deal with an attempt to use a very simple relationship between ocean temperatures and tropical cyclones, followed by the examination of a new 'downscaling' model which uses GCM results directly.

17.2 'Predicting' tropical cyclones from sea surface temperatures

While Palmen (1948) demonstrated that sea surface temperatures greater than about 26.5°C were necessary for tropical cyclone genesis to occur, it is surprising the sea surface temperature itself has nearly no relationship to mean tropical cyclone intensity (Miller 1958; Merrill 1987; Evans 1993). Sea surface temperature does, however, provide an upper bound beyond which tropical cyclones cannot become more intense (Miller 1958; Emanuel 1987; Merrill 1987; Evans 1993).

This is illustrated here by comparing the maximum intensity of tropical cyclones occurring from 1967–1992 (National Climate Data Center 1994a) as a function of climatological sea surface temperature (SST) from the Comprehensive Oceanic Atmospheric Data Set (COADS) for 1957–79 (Figures 17.1–17.3). These figures show the sea surface temperature corresponding to the location where each tropical cyclone reached its maximum intensity. The 1967–1992 period is the modern satellite era, when application of quasi-objective analysis methods, such as those of Dvorak (1975) produced the first stable cyclone data set for the majority of regions without aircraft reconnaissance (e.g. Holland 1981).

In Figure 17.1(a), showing the south west Pacific region, there are two horizontal clusters of points near 25 m s^{-1} and 33 m s^{-1}. These arise from average values of maximum wind speed for a large number of cyclones where only the Saffir-Simpson category of the storm was provided. Tropical cyclones of intensity category 2 (wind speeds greater than 16 m s^{-1} and less than 33 m s^{-1}) were presumed to have an average wind speed of 25 m s^{-1}. Category 4 storms were set to 33 m s^{-1}.

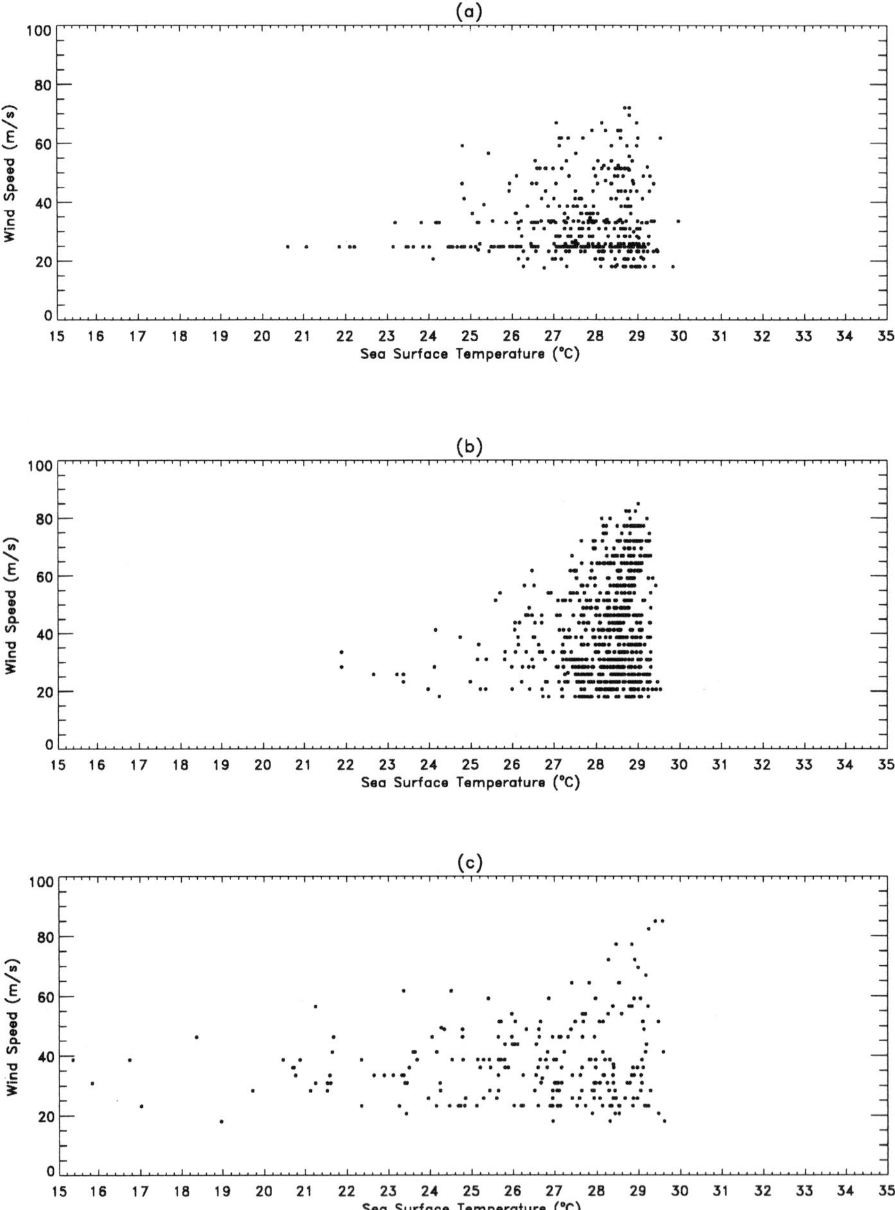

Figure 17.1 Observed maximum wind speed for individual tropical cyclones in the period 1967–1992 as a function of observed sea surface temperature for: (a) south west Pacific, (b) north west Pacific, and (c) north Atlantic.

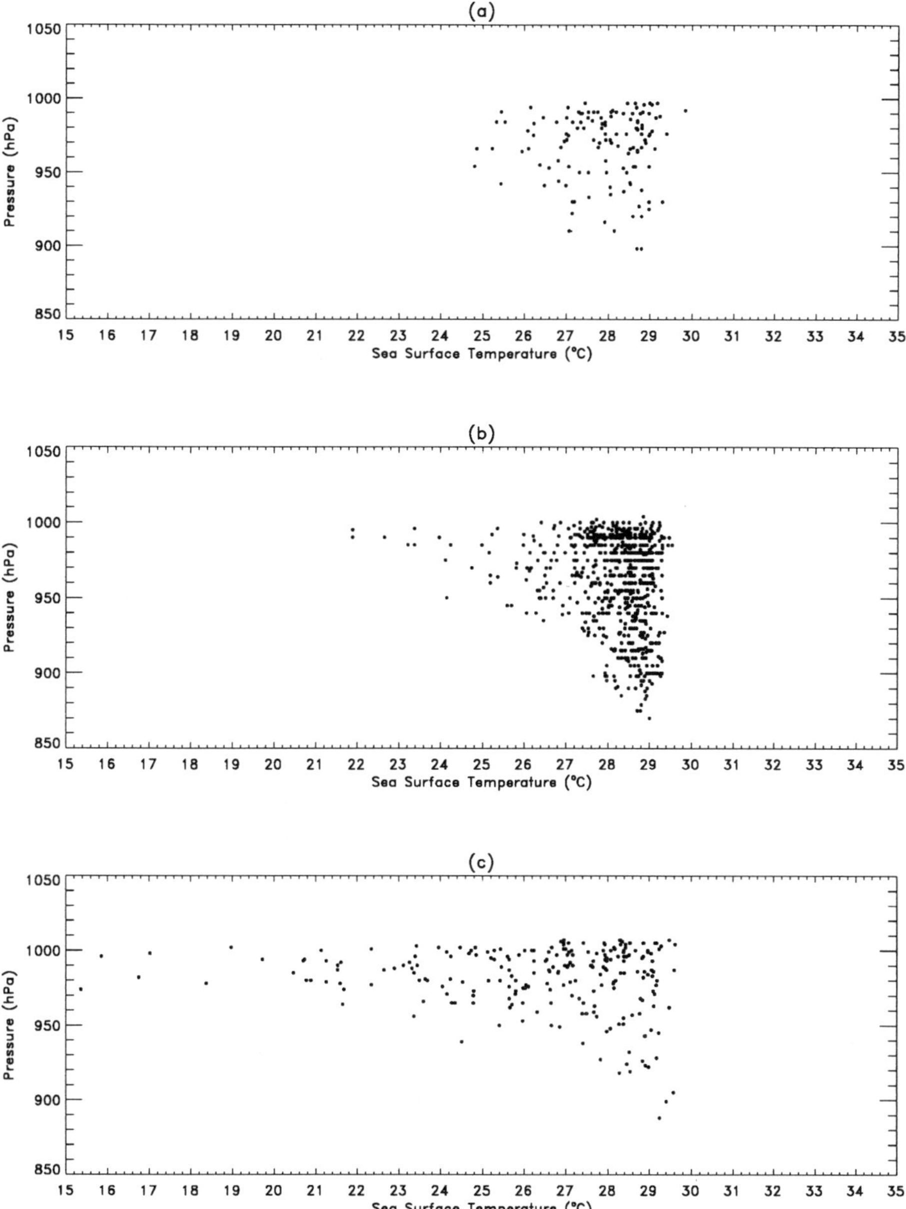

Figure 17.2 Observed minimum central pressure for individual tropical cyclones in the period 1967–1992 as a function of sea surface temperature for: (a) south west Pacific, (b) north west Pacific, and (c) north Atlantic.

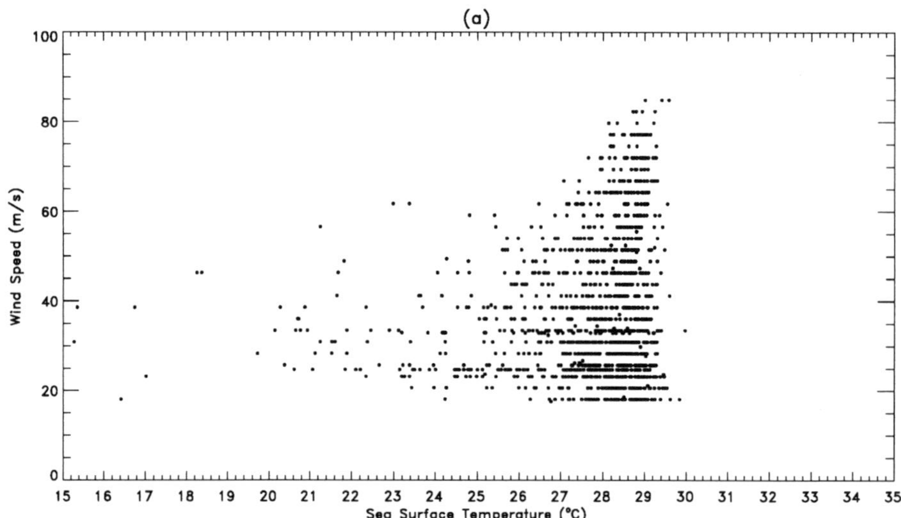

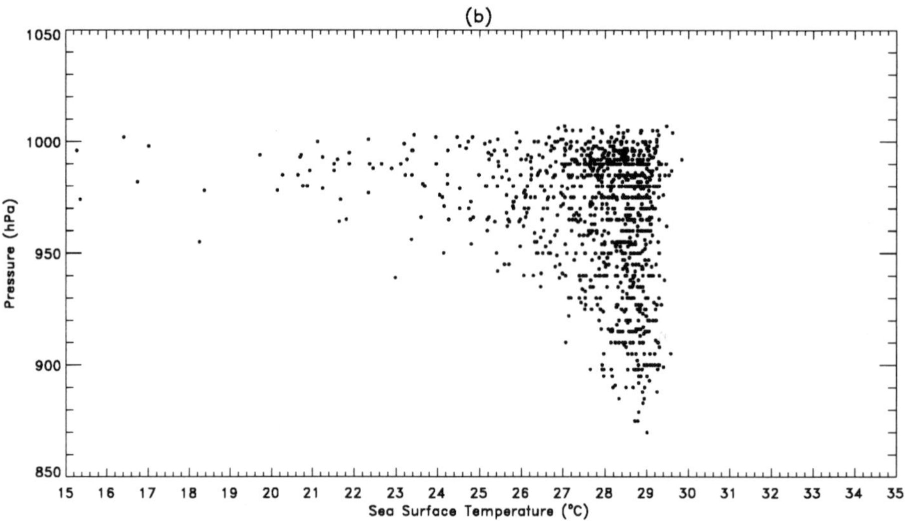

Figure 17.3 Observed maximum intensity for individual tropical cyclones in the period 1967–1992 as a function of sea surface temperature for all three ocean areas combined: (a) maximum wind speed and (b) minimum central pressure.

The results in Figures 17.1 and 17.2 illustrate an increase in maximum intensity with SST greater than 26°C (near the SST believed necessary for cyclones to form; Palmen 1948; Gray 1979). While the intensity of tropical cyclones for SST values above 26°C does not increase for the majority of storms, there appears to be a slight increase in

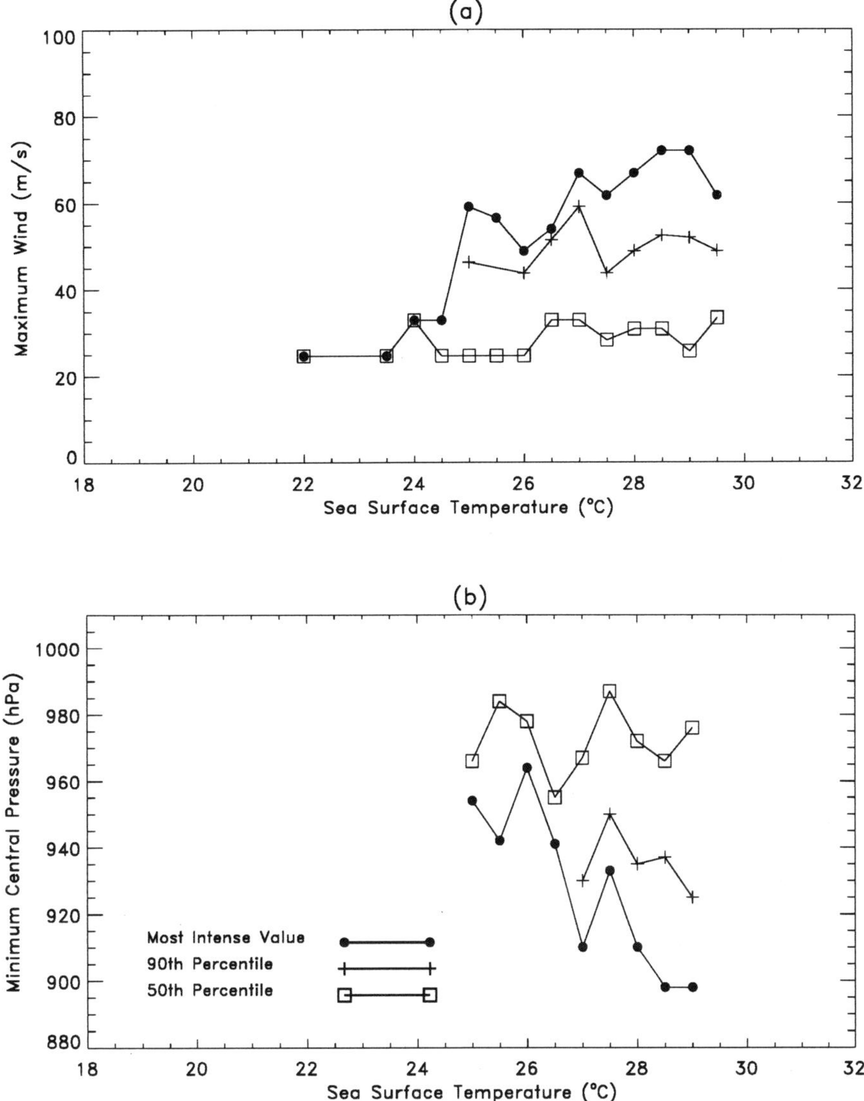

Figure 17.4 Observed most extreme, 90th and 50th percentile intensity for individual tropical cyclones in the period 1967–1992 as a function of each half degree sea surface temperature value for the south west Pacific: (a) maximum wind speed and (b) minimum central pressure.

the most intense tropical cyclones. However, above 28°C this trend is based upon a few individual cyclones. Furthermore, limitations to the accuracy of the observed data means that there is no clear trend in intensity above 28°C. Figure 17.3 combines all the available observations of tropical cyclone intensity from these three ocean regions.

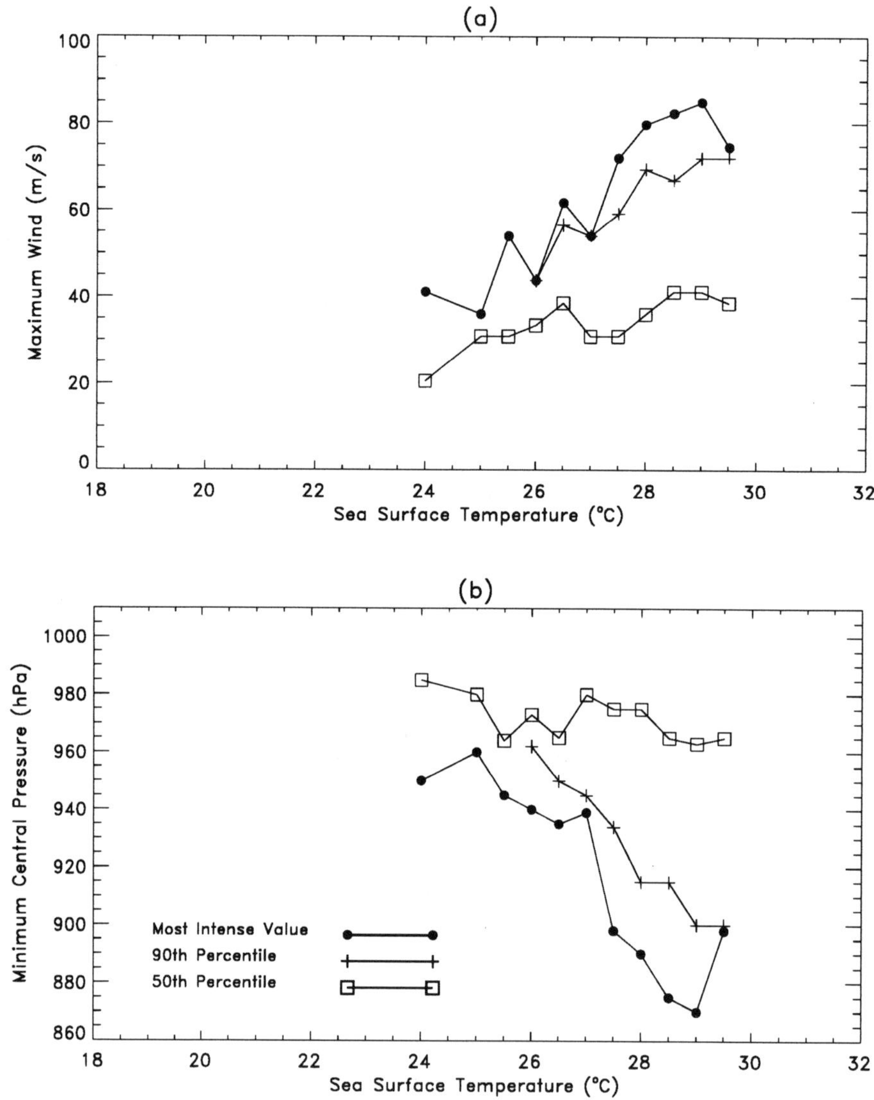

Figure 17.5 Observed most extreme, 90th and 50th percentile intensity for individual tropical cyclones in the period 1967–1992 as a function of each half degree sea surface temperature value for the north west Pacific: (a) maximum wind speed and (b) minimum central pressure.

Limitations on the accuracy of these graphs include: (i) the maximum intensity often occurs over a range of locations, as the tropical cyclone may have retained a maximum intensity level for a number of hours and so the location of maximum intensity are average values; (ii) measurements of tropical cyclone intensity are not very accurate — especially

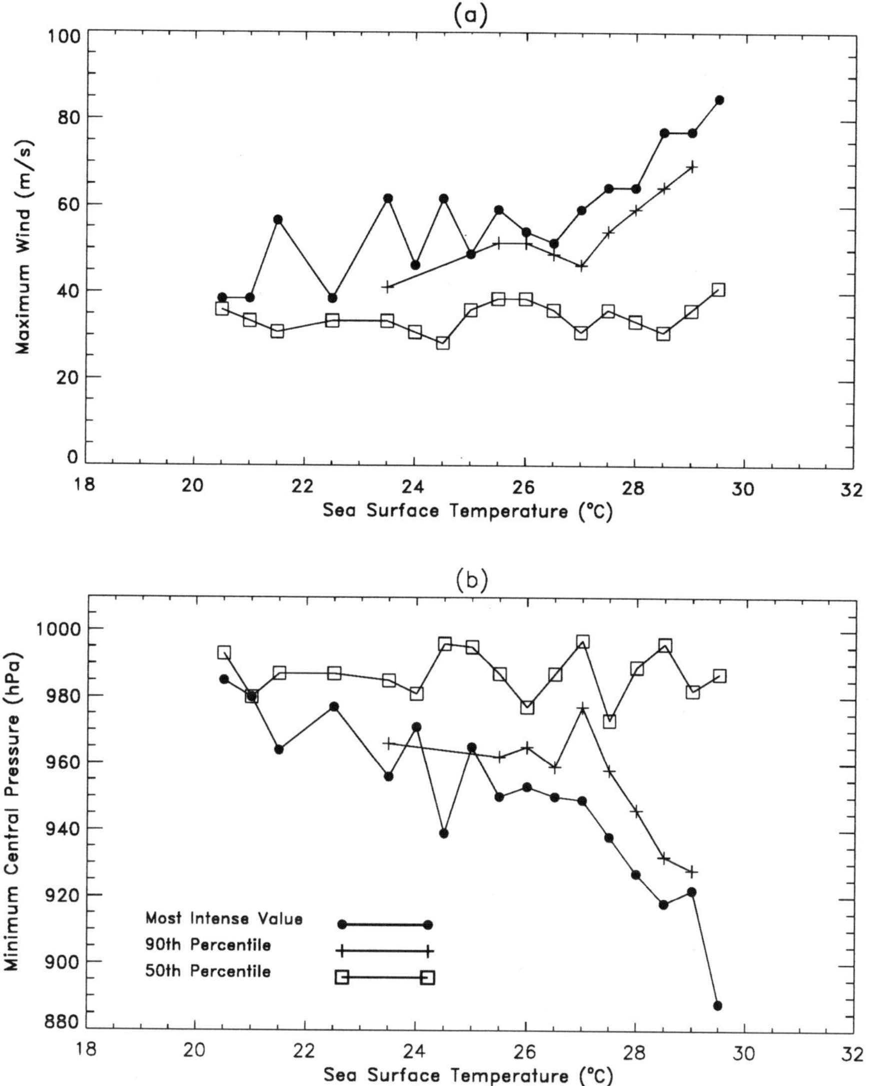

Figure 17.6 Observed most extreme 90th and 50th percentile for individual tropical cyclones in the period 1967–1992 as a function of each half degree sea surface temperature value for the north Atlantic: (a) maximum wind speed and (b) minimum central pressure.

if satellite pictures are used to infer (not measure directly) tropical cyclone intensity, as is done in the south west Pacific and in the north west Pacific since 1987; (iii) COADS is comprised of monthly averaged ship observations of SST within a 2×2 degree box, time and space scales smaller than this are smoothed out; and (iv) no account is taken of local

SST cooling arising from the direct interaction between the tropical cyclone and the ocean (Shay et al. 1992).

However these results compare quite well with Evans (1993) who produced similar scatter plots using data from 1967–1986 from the Constructed World Wide Tropical Cyclone Data Set.

Figures 17.4 to 17.6 are plots of observed maximum intensity against SST for the south west Pacific, north west Pacific and north Atlantic ocean regions. COADS data was used to find the SST value for the locations where each tropical cyclone reached maximum intensity. Maximum intensity data is taken from National Climate Data Center (NCDC) (1994a). These plots represent the most intense, 90th percentile and 50th percentile storm which occurred at each half-degree ocean temperature from $18°C$ to $32°C$. Each ocean region is represented by two sets of plots where maximum intensity values were taken as either the maximum wind speed or minimum central storm pressure. There are slight variations between the cyclones selected by each of the processes because cyclones of the same minimum central pressure will not necessarily have the same maximum wind speed.

In the Australian region values of minimum pressure were not generally provided in the NCDC data set until 1980, and this explains the small range of SST values corresponding to maximum tropical cyclone intensity in Figure 17.4(b) compared to Figure 17.4(a). Therefore, in the south west Pacific (Australian) region it is more useful to examine the trend in maximum intensity with SST by considering wind speed values. Figure 17.4(a) shows a clear increase in maximum intensity with SST up to $29°C$. However the increase in maximum intensity appears to reduce as SSTs increase. Potential reasons for this are presented later. Beyond $29°C$, the number of observations for each half degree SST value decreases, which means the recorded maximum intensity values are unlikely to be the most intense possible for these SST values.

Figure 17.4(a) might be divided into two or even three segments: SSTs below $25°C$ have low maximum intensity (less than 35 m s^{-1}); between $25°C$ and $26.5°C$ there is an intermediate intensity range where maximum intensity are from 50 m s^{-1} to 60 m s^{-1}. For SST values greater than $26.5°C$ cyclone intensity appears to be above 60 m s^{-1}. The cluster, however, may be partly due to the NCDC data set which frequently only supplied the cyclone intensity category and so many hurricane strength storms were given the intensity value of 33.0 m s^{-1} which is the minimum wind speed for hurricane strength.

Figures 17.5(a) and 17.5(b) show maximum intensity trends for the north west Pacific region where 604 cyclones occurred from 1967–92. There is a clear increase in tropical cyclone intensity with increasing SST. There appears to be a cluster of points in Figure 17.5(b), with SST values less than $27.5°C$ maintaining cyclones of greater than 940 hPa. For SSTs above $27°C$, the maximum intensity ranges from 870–900 hPa. This cluster of points is not as apparent in Figure 17.5(a), although low intensity storms may be said to exist at SST values below $25°C$, medium intensity storms between $25°C$ and $27°C$, while strong storms occur where SSTs exceed $27°C$.

North Atlantic cyclones reach maximum intensity over colder ocean regions than in the south or north west Pacific regions. In Figure 17.6(b) there is a cluster of maximum intensity points below SST values of 23°C where intensity values range from 960–990 hPa. From 22.5°C to 27°C the tropical cyclone intensity ranges from 940–970 hPa, and beyond 27°C maximum intensity increases to 890 hPa. The cluster of maximum intensity values exists for wind speed maximum intensity values, (Figure 17.6(a)) though they are less clearly defined. Least intense storms exist below 23.5°C. Medium intensity tropical cyclones range from 23.5°C–26.5°C. Above 26.5°C storm intensity increases, and appears to level out between 28.5°C and 29°C, before increasing again at 29.5°C. Figure 17.7 illustrates the trend in maximum intensity for increasing SSTs for all three ocean regions combined. Significant features of Figure 17.7(b) include the increase of maximum intensity with SST and clusters of points below SSTs of 23°C (above 960 hPa), between 23°C and 26.5°C (940–960 hPa) and from 26.5°C (870–910 hPa). Figure 17.7(a) may follow the same general clustering of points, however this is less clearly defined. Examining the scatter graphs from the three regions combined, there appears to be three main levels of maximum tropical cyclone intensity determined by a region's SST. These are least intense storms where central pressure is greater than 960 hPa, intermediate intensity where maximum intensity is ~940 hPa and most intense storms between 870–910 hPa.

The observed relationships between SST and maximum intensity might form the basis of a prediction method if there was: (a) a better correlation and (b) greater skill in modelling ocean surface temperatures (see, for example, Chapter 7). However such analysis assumes that the SST/maximum intensity relationship for current climate is maintained in a $2 \times CO_2$ climate. In the following section, we establish a predictive model for tropical cyclone maximum intensity which draws on a larger range of GCM fields.

17.3 Downscaling models of tropical cyclones

In common with most other areas of impact assessment, those concerned with tropical cyclone simulations have sought means by which coarse-resolution global circulation models can be utilised to estimate fine-resolution characteristics such as tropical cyclone intensity. One such approach is to develop a set of physical relationships between the thermodynamic state of the atmosphere and ocean, and the maximum potential intensity (MPI) of tropical cyclones.

The MPI achievable by a tropical cyclone was first assessed by Kleinschmidt (1951) and Miller (1958) who extended and quantified earlier work by Palman (1948), Riehl (1954) and Abdullah (1954). Miller derived a relationship between MPI and SST by assuming saturated moist ascent in the eye wall clouds to define a cloud-top from which air subsided in the eye. The subsiding air warms from dry adiabatic compression and Miller included a parameterised cooling from entrainment of eye-wall cloud into the eye. The method predicted increasing MPI with SST though the MPI/SST relationship devised was not uniquely defined. Miller did not include the important feedback between the ocean and tropical cyclone that occurs (Byers 1944; Riehl 1954; Malkus & Riehl 1960). This feedback arises through the dependence of equivalent potential temperature on pressure as well as temperature and humidity. As the surface pressure falls, whilst the temperature and humidity

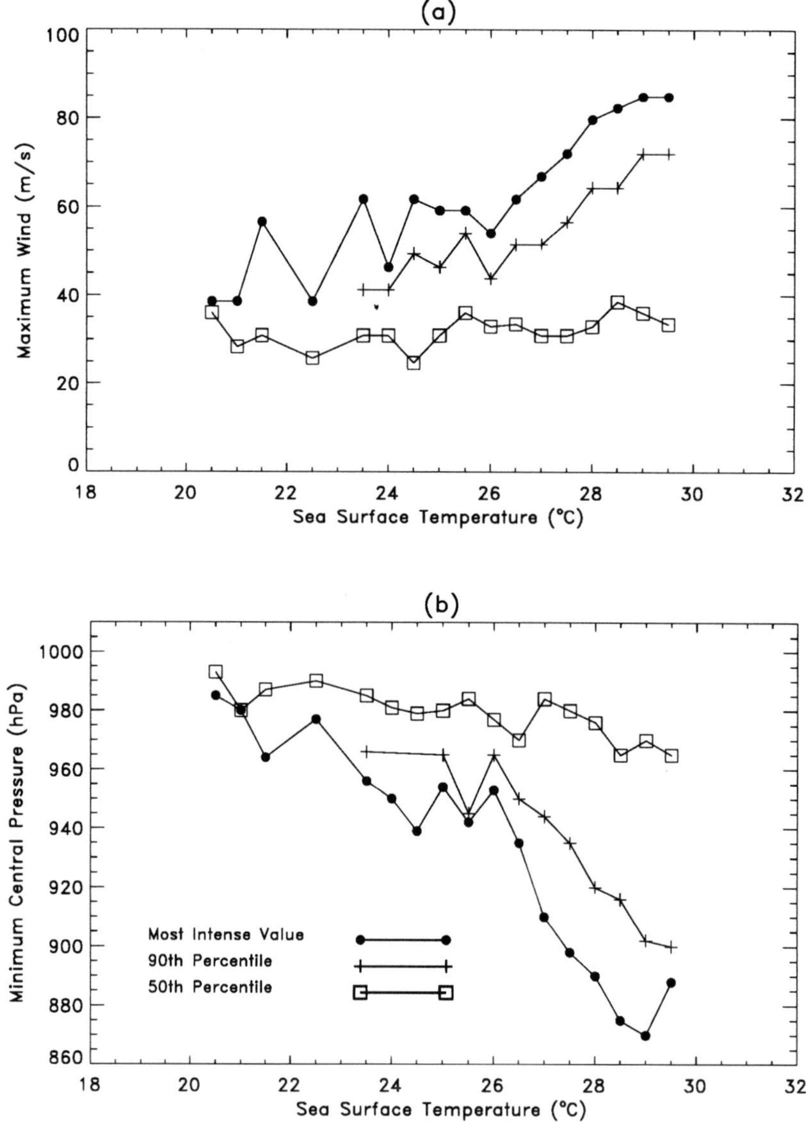

Figure 17.7 Observed most extreme, 90th and 50th percentile intensity for individual tropical cyclones in the period 1967–1992 as a function of each half degree sea surface temperature value for all three ocean areas combined: (a) maximum wind speed and (b) minimum central pressure.

remain constant, the equivalent potential temperature increases. Thus the ocean is able to release additional energy to the atmosphere with decreasing central pressure. A positive feedback cycle thus results to support the development of intense tropical cyclones (see Holland 1996 for a detailed discussion).

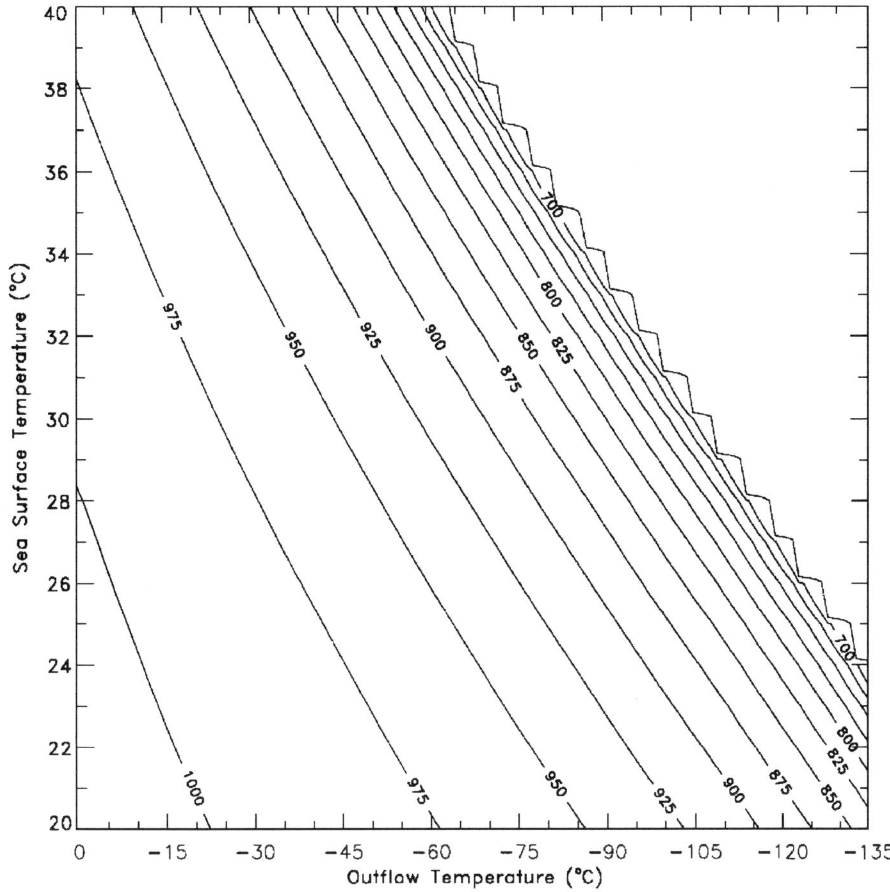

Figure 17.8 Relationship between sea surface temperature, tropical cyclone outflow layer temperature and estimated maximum potential tropical cyclone intensity (Emanuel Model). RH of 75%, ambient surface pressure of 1015 hPa, latitude 20°N, tropical cyclone radius of 500 km.

Emanuel (1987) applied one such thermodynamic model to the Goddard Institute for Space Studies (GISS) GCM and found that the MPI increased markedly in a $2 \times CO_2$ environment. Emanuel's model is based on a conceptual view of a tropical cyclone as a Carnot heat engine. The model assumes a steady state balance where angular momentum, total entropy and total water are conserved and the atmosphere is taken to be neutral to slantwise moist convection. The cyclone intensifies through a feedback mechanism: evaporation from the sea surface to the atmosphere increases as wind speeds increase in response to greater pressure gradient forces arising from falling central pressure values. Decreasing central pressure is due to increasing latent heat release in the cyclone core as converging wind speed increases. The model predicts maximum tropical cyclone intensity will rise with rising SST and with colder (higher) outflow conditions.

Emanuel used this method to derive a relationship between SST, outflow temperature and MPI (Figure 17.8). This relationship predicts an increase of MPI for SST increases, similar to the observations in Figures 17.4–17.7 however the predicted increase tends to be stronger than that which is observed, and an exponential increase at higher SST is not sustained by the observed intensity which shows a decrease in the rate of intensity change.

There are difficulties with relation to the highly asymmetric outflow of a real tropical cyclone to Emanuel's axisymmetric outflow temperature. Emanuel includes no eye processes and assumes a saturated cloud-filled core. Further, highly complex ocean interactions occurring under high wind conditions (Pudov & Holland 1994; Fairall et al. 1995) are neglected. As shown by Holland (1995) the major sensitivity is to the assumed values of relative humidity in the core region and the environment.

Holland (1995) has developed an approach to the thermodynamic prediction of MPI which contains both the ocean feedback and an explicit eye. In this approach the ocean feedback occurs under the eye wall where the relative humidity is assumed to be 90%. The eye is parameterised in a similar fashion to Miller (1958). This approach requires no estimate of the outflow temperature and explicitly derives the cloud detrainment level which defines the eye subsidence and the height of eye-wall warming. A direct relationship between SST and MPI is derived by using series of monthly-mean soundings in tropical cyclone ocean basins.

This approach agrees closely with observed maximum intensity and also explicitly predicts that intense tropical cyclones will not develop for SSTs below 26°C in the current climate. Holland also shows that for SSTs greater than 26°C the relationship is quite stable across ocean basins.

Limitations of Holland's approach include the assumption of a vertical eye-wall, lack of inclusion of spray effects in high wind conditions, lack of entrainment into the eye-wall and the assumption of 90% relative humidity (RH) at the surface under the eye wall. Holland argues that the lack of a sloping eye-wall is not a major limitation. The results are quite sensitive to the RH assumption, but less so than that in Emanuel's approach.

Figures 17.9(a) and 17.9(b) show MPI from the Holland and Emanuel 'downscaling' models for eleven radiosonde station locations in coastal north Australia (Figure 17.10).

For the Holland model, observational data include radiosonde retrievals of monthly averaged air temperatures from 850 hPa to 100 hPa and average monthly surface pressure data (Maher & Lee 1977). Seventeen vertical temperature and pressure levels were used to calculate the MPI, archived radiosonde data provided ten levels and the remainder were obtained by interpolation. Relative humidity in the tropical cyclone eye wall was taken to be 90%. Surface air temperatures were calculated as one degree less than SST values of the nearest 2 × 2 degree ocean grid to each radiosonde station. The SST data was taken from COADS (Slutz et al. 1985) consisting of monthly averaged values from 1951–1979. For the Emanuel model, local temperature profiles used to calculate the tropical cyclone outflow temperature

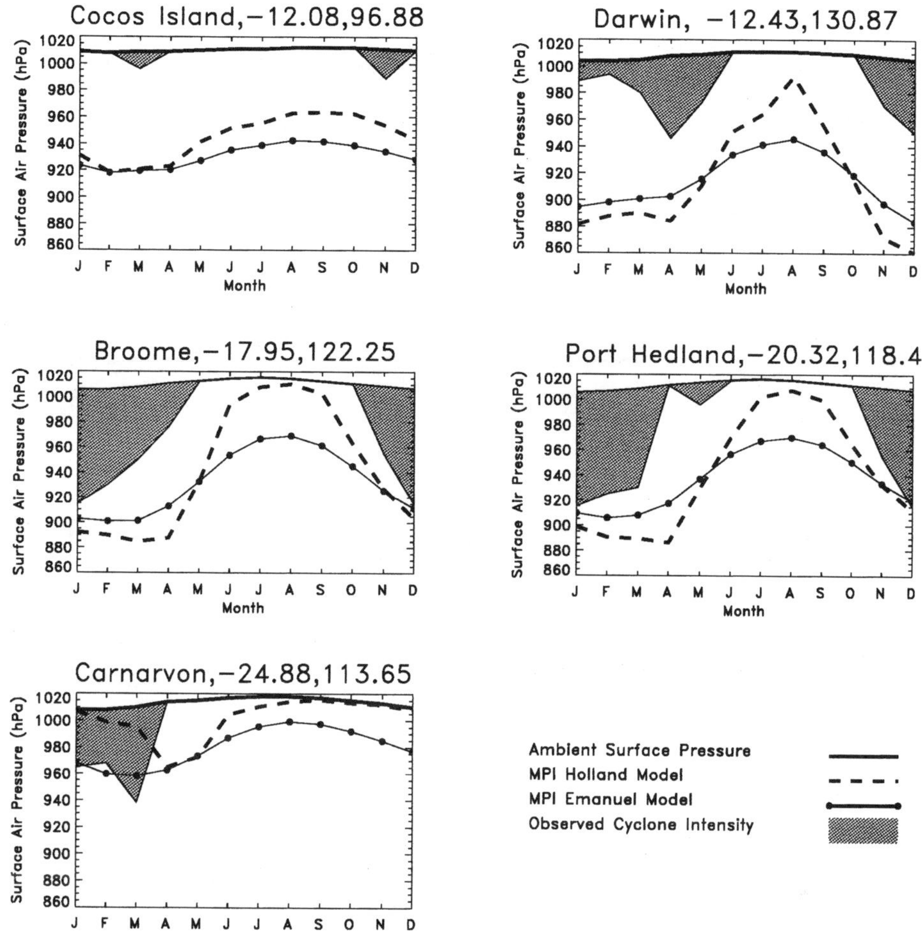

Figure 17.9(a) Simulated (Holland and Emanuel models) and observed maximum potential intensity for the western Australian region.

and ambient surface pressure were also taken from Maher and Lee (1977). Ambient relative humidity was set at 80% and surface relative humidity at the storm centre was set to 100%. Surface temperatures were taken as the SST value of the nearest 2×2 degree ocean grid to each radiosonde station (COADS). Observed cyclone intensity values were collected from the Tropical and Extra Tropical Cyclone Track CD (National Climate Data Center 1994a) and consist of the lowest central pressure within 5 degrees latitude of each station from 1967–92. Only storms of at least tropical cyclone intensity (wind speed greater than 17.5 m s^{-1}) were used.

Care needs to be taken with the use of soundings adjacent to a major landmass as representative of the oceanic regime in which tropical cyclones form. The soundings may

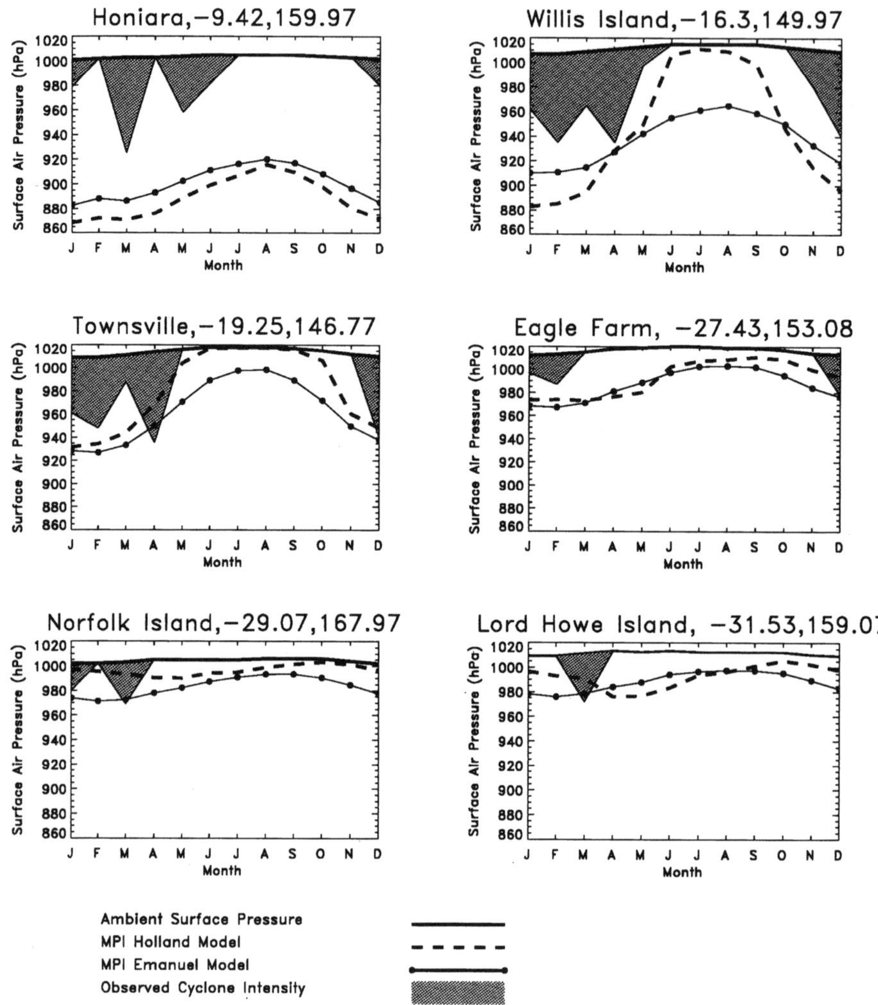

Figure 17.9(b) Simulated (Holland and Emanuel models) and observed maximum potential intensity for the eastern Australian region.

be defined more by radiative equilibrium over the land mass, which contains extensive areas of hot desert country. This is more of a problem for the Holland model, as the Emanuel approach assumes an 80% RH in the environment (which can be very dry near the land). As a general rule, use of such soundings in the Holland model will underestimate the capacity of the oceanic regime to generate intense tropical cyclones. At low latitude stations such as Broome, Port Hedland, Willis Island and Townsville, the Holland model predicts

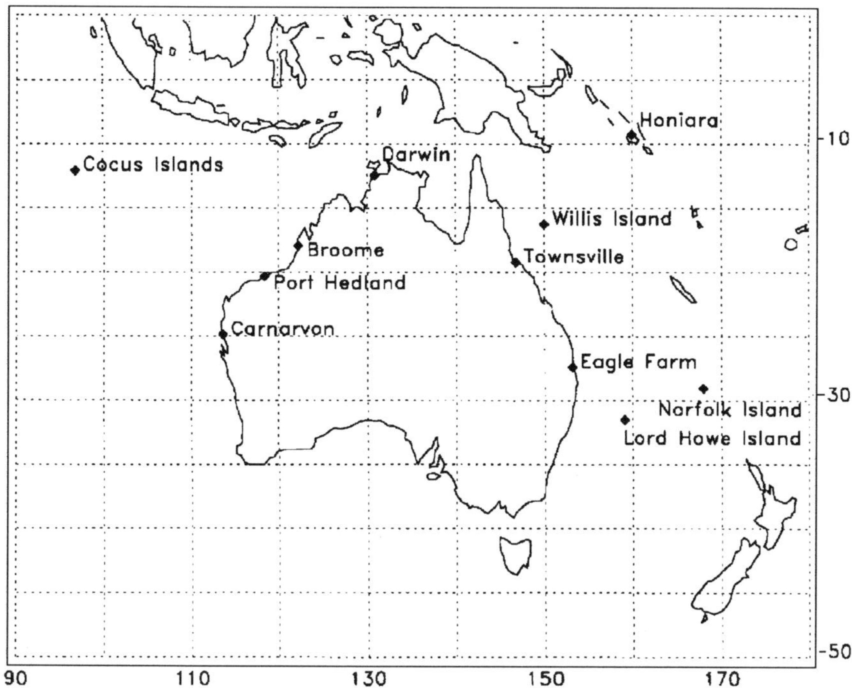

Figure 17.10   Australian region station locations for Figures 17.9(a) and 17.9(b).

definite MPI differences between summer and winter months, suggesting a well-defined tropical cyclone season exists at these locations. The exception to this is Honiara, Cocus Island and to a lesser extent Darwin, which show potential tropical cyclone activity through most of the year. At these latitudes, the Emanuel model predicts tropical cyclones throughout the year. As well, MPI estimated from the Emanuel model has less monthly variability of MPI than the Holland model.

At mid-latitude locations, both models show a potential for weak tropical cyclones within a short tropical cyclone season. At Carnarvon, Norfolk Island and Lord Howe Island, the Holland model appears to slightly underestimate tropical cyclone intensity. The observed intensities at these locations may be higher than the available thermodynamic energy would permit because tropical cyclones which move poleward may have intensities corresponding to slightly warmer conditions.

In general, the model results correspond reasonably well with observations in the south west Pacific. However, the short observational period means that there are few observed tropical cyclones at each location with which to provide an accurate comparison of model results.

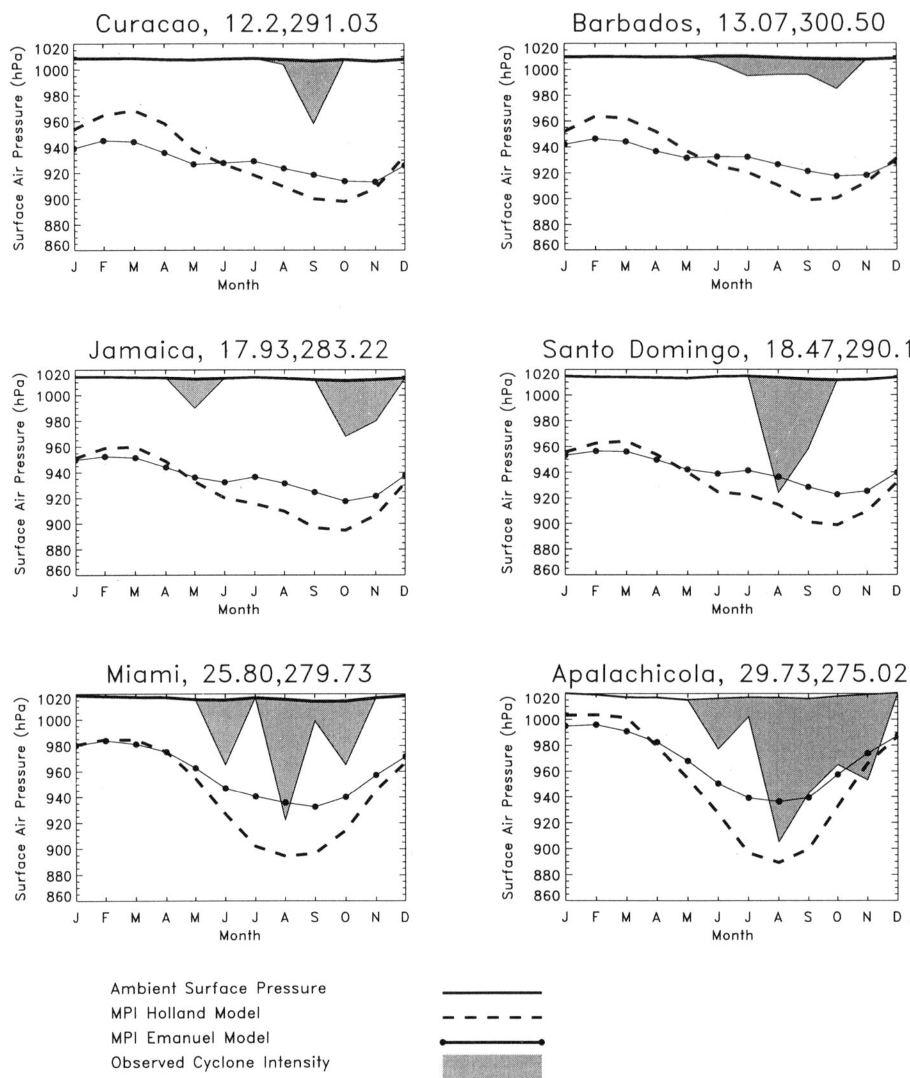

Figure 17.11(a) Simulated (Holland and Emanuel models) and observed maximum potential intensity for the north Atlantic region.

Figures 17.11(a) and 17.11(b) contain estimated MPI values for the Holland and Emanuel 'downscaling' models for nine radiosonde stations in the north Atlantic Ocean region (Figure 17.12). Data sources for these calculations were as for the Australian region except radiosonde data retrievals of monthly averaged air temperature and monthly averaged surface pressure were from the NCDC (1994b) and extended to 30 hPa. For the Emanuel model in this region MPI calculations used temperature values up to 30 hPa.

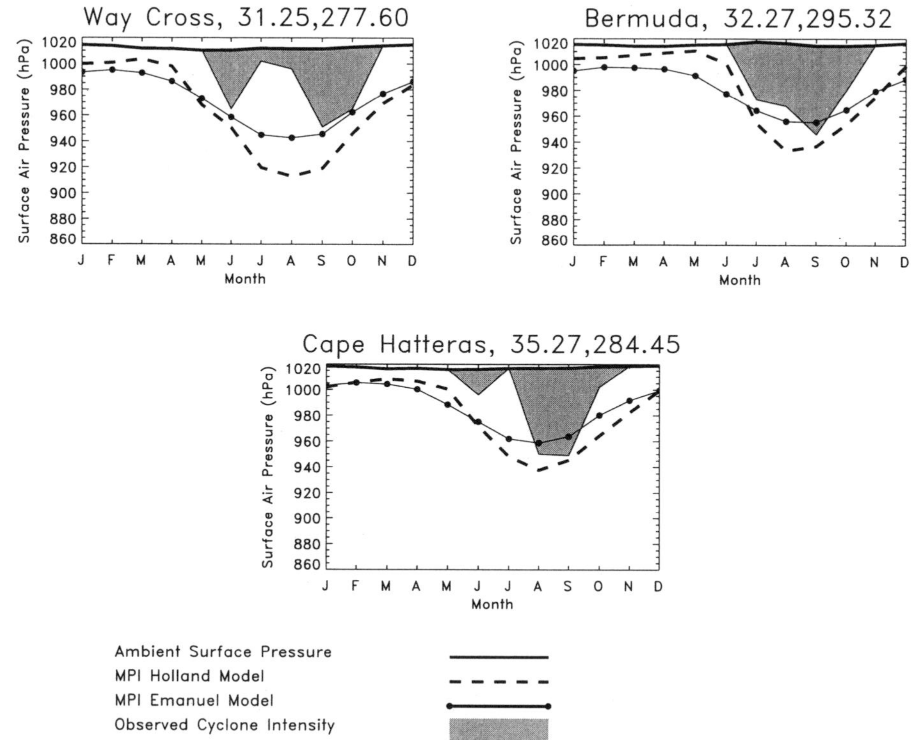

Figure 17.11(b) Simulated (Holland and Emanuel models) and observed maximum potential intensity for the north Atlantic region.

For low latitude stations, both models predict tropical cyclones all year round, although the Holland model predicts decreased intensity in colder months, and the Holland model predicts more intense tropical cyclones than the Emanuel model during the tropical cyclone season.

At low latitude locations the observed intensity appears significantly less than the MPI estimated from both models. Possible reasons for this may be the small number of observed cyclones near these stations over the last 30 years. A more likely explanation, however, is that the dynamics of cyclone development generally require time scales of days to move from formation to maximum intensity. During this time, cyclones typically move poleward at several meters per second, so that they do not generally remain in the deep tropics for sufficient time to develop to maximum intensity.

Figures 17.11(a) and 17.11(b) show considerable seasonality, at higher latitude locations such as Miami, Apalachicola, Way Cross, Bermuda and Cape Hatteras where cyclones are mostly predicted during the tropical cyclone season (between May and October) for both models. For these stations, the Holland model predictions correspond well with observations, and the Emanuel model estimates appear underestimated at some locations.

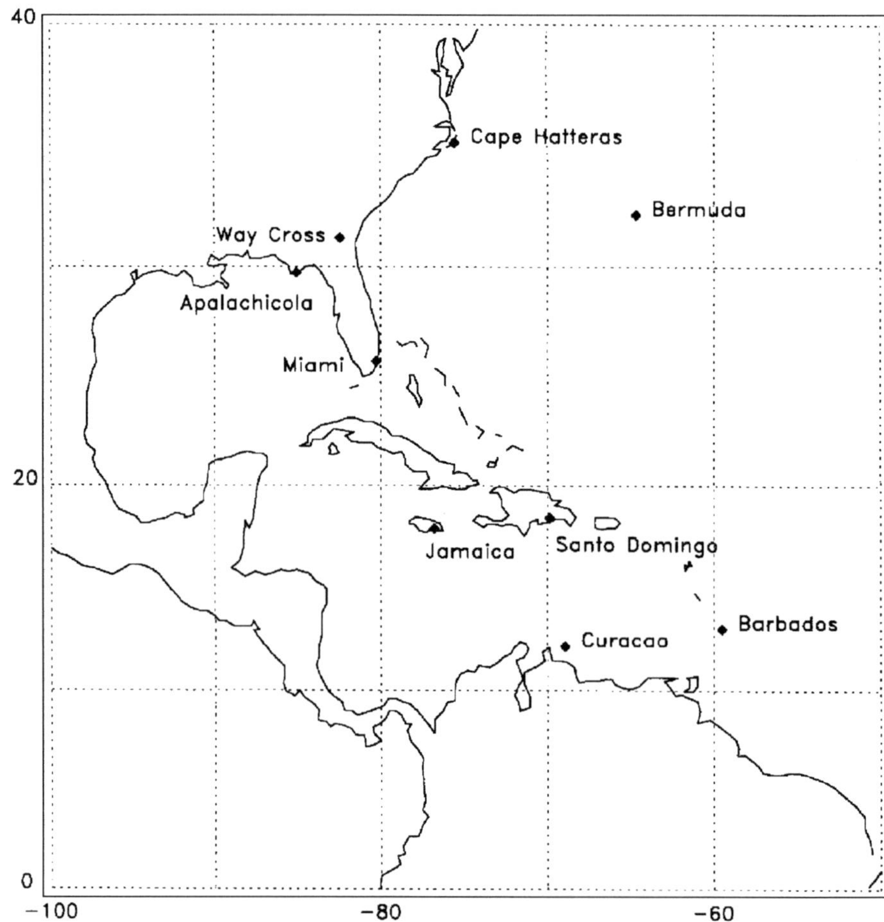

Figure 17.12 North Atlantic region station locations for Figures 17.11(a) and 17.11(b).

Figures 17.13(a) and 17.13(b) show MPI from both models for the north west Pacific region (Figure 17.14). Data sources and calculation methods are the same as those used for the north Atlantic region.

For low latitude stations of Palau, Kwajalein and Guam, tropical cyclones are predicted all year for both models. This appears reasonable at Guam, where tropical cyclones are observed throughout the year. Palau and Kwajalein are located close to the equator in a region of relatively low planetary vorticity, which is considered unfavourable for tropical cyclone development (Gray 1968). Tropical cyclones in this region are typically in the early stages of development, and tend to move towards the north west in the easterly air flow, reaching maximum intensity at higher latitudes. This may explain the lack of observed tropical cyclone at these stations.

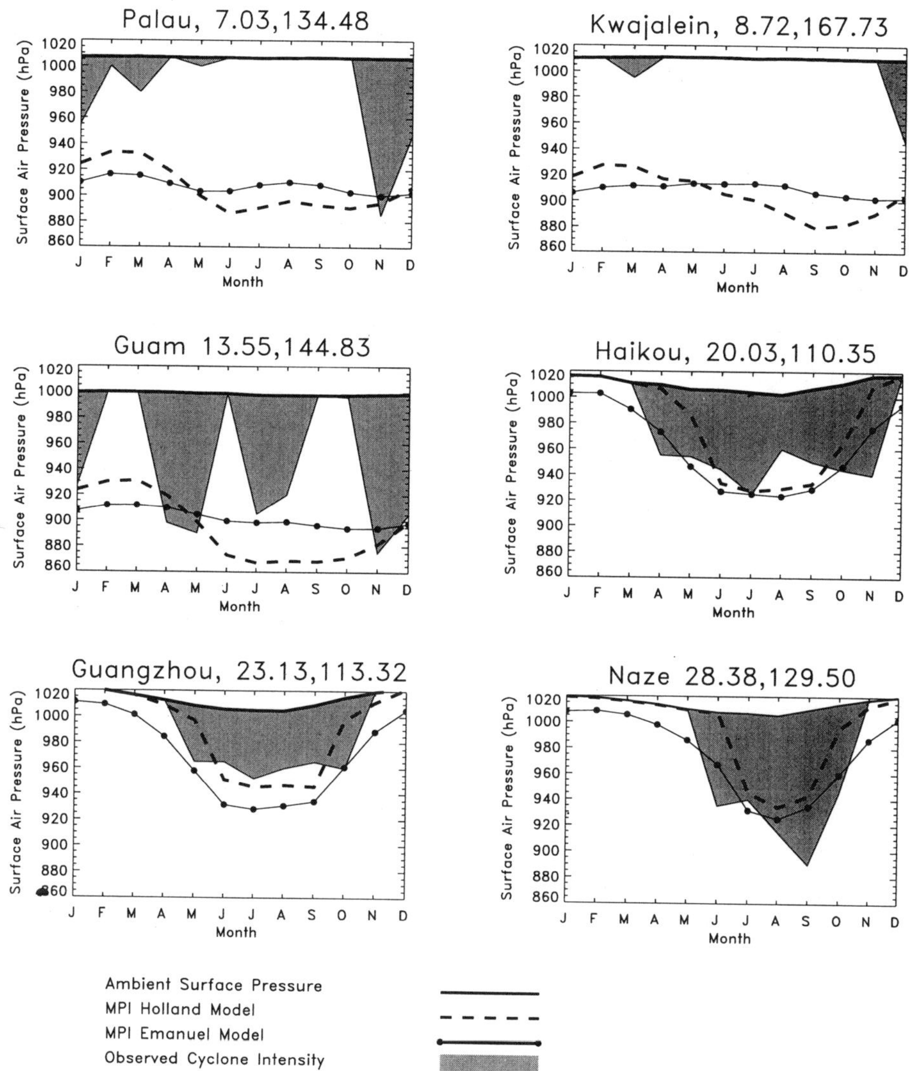

Figure 17.13(a) Simulated (Holland and Emanuel models) and observed maximum potential intensity for the north west Pacific region.

In this ocean region the Emanuel model is slightly more intense than the Holland model at latitudes above 20°N. For Haikou, Guangzhou and Naze, both models appear to underestimate MPI. For high latitude stations, predicted MPI from both models is much reduced and limited to a short cyclone season. At Tateno, the effect of advection of weakening storms from lower latitudes is clearly seen.

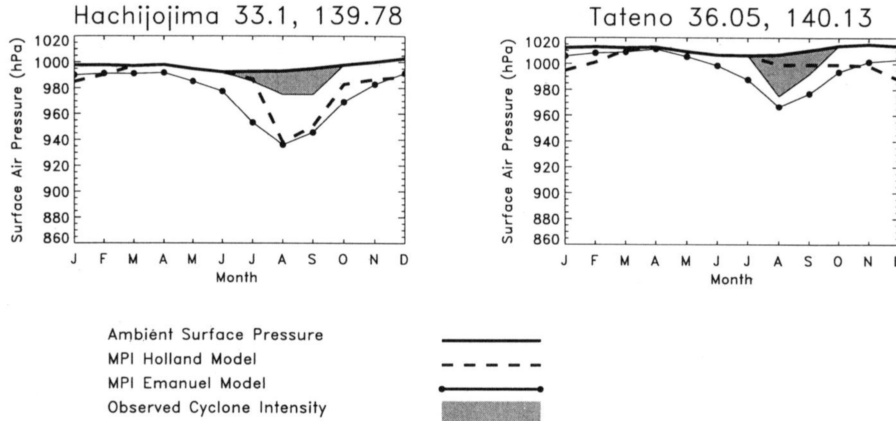

Figure 17.13(b) Simulated (Holland and Emanuel models) and observed maximum potential intensity for the north west Pacific region.

In general, the Holland and Emanuel models predict similar tropical cyclone intensity which corresponds reasonably with observations, the major differences being a larger seasonal variability of MPI from the Holland model generally, and in particular at low latitudes.

Figure 17.15 shows the relationship between observed and predicted maximum tropical cyclone intensity, together with the 50th percentile of observed intensities for comparison. Significant features include the near linear increase of maximum intensity for SSTs greater than 26°C for both models and observations. The Holland model captures actual intensities more accurately than does Emanuel's model, though both provide generally good indications of the maximum cyclone intensity/SST relationship. As discussed by Holland (1996), SST is a dominant factor in determining the radiative/convective equilibrium of the overlying atmosphere and thus provides a good, first order proxy to the full thermodynamics.

Results from the Holland and Emanuel models in the south west Pacific, north Atlantic and north west Pacific regions indicate the Holland model captures tropical cyclone seasonality quite well. MPI values from the models appear reasonable (within the range of intensities experienced in each region) although MPI values do not always closely match observed intensities.

While both the Emanuel (1991) and Holland (1996) thermodynamic methods of predicting MPI have significant limitations, they show agreement with observations and provide an objective method of interpreting GCM predictions of $2 \times CO_2$ changes to climate for potential tropical cyclone changes. It is especially important to note that the use of SSTs alone is not a valid approach and the models must be applied to the full soundings as SST changes arising from anthropogenic climate change are associated with different mechanisms and much different overlying atmospheric structures than are changes associated with seasonal or transitory changes in current climate.

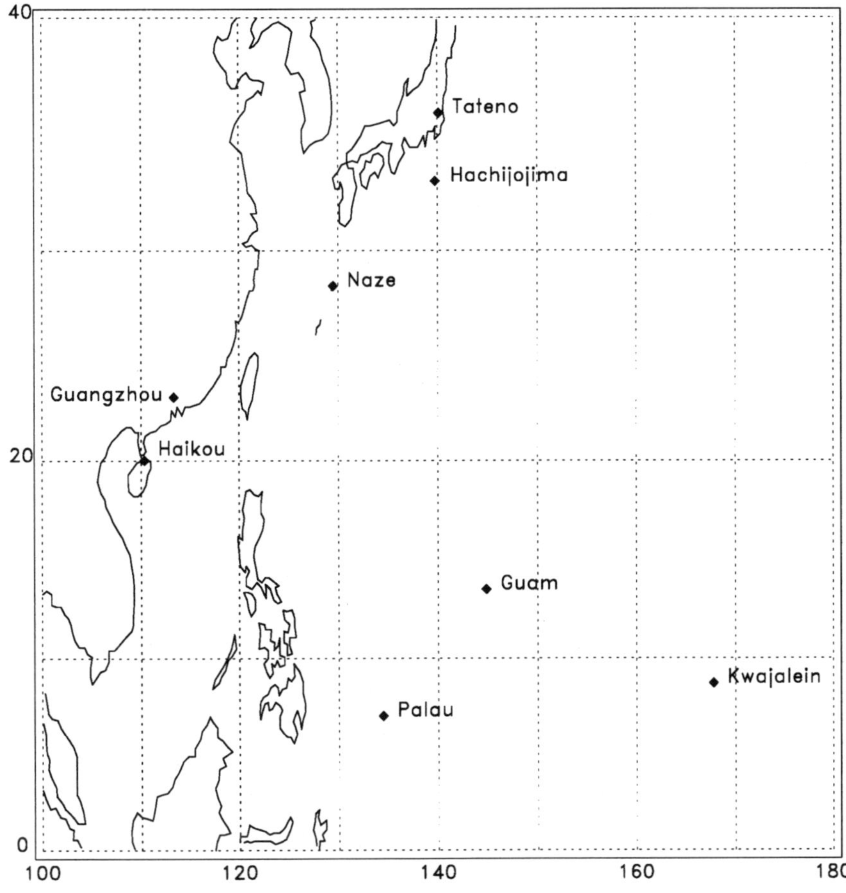

Figure 17.14 North west Pacific locations for Figures 17.13(a) and 17.13(b).

Both of these models can be used as downscaling tools since the problem with using GCM models to explicitly capture the structure of tropical cyclones is bypassed by these thermodynamic models. In addition using information averaged over GCM grid areas is less problematic than other applications, as parameters such as SST do not vary greatly in the tropics. Such tools have the potential to estimate information about future tropical cyclone intensity potential over GCM grid size areas.

Overall, however, the Holland model performs better than the Emanuel model in the three regions examined. The predictions are good enough to justify the next stage in this preliminary analysis.

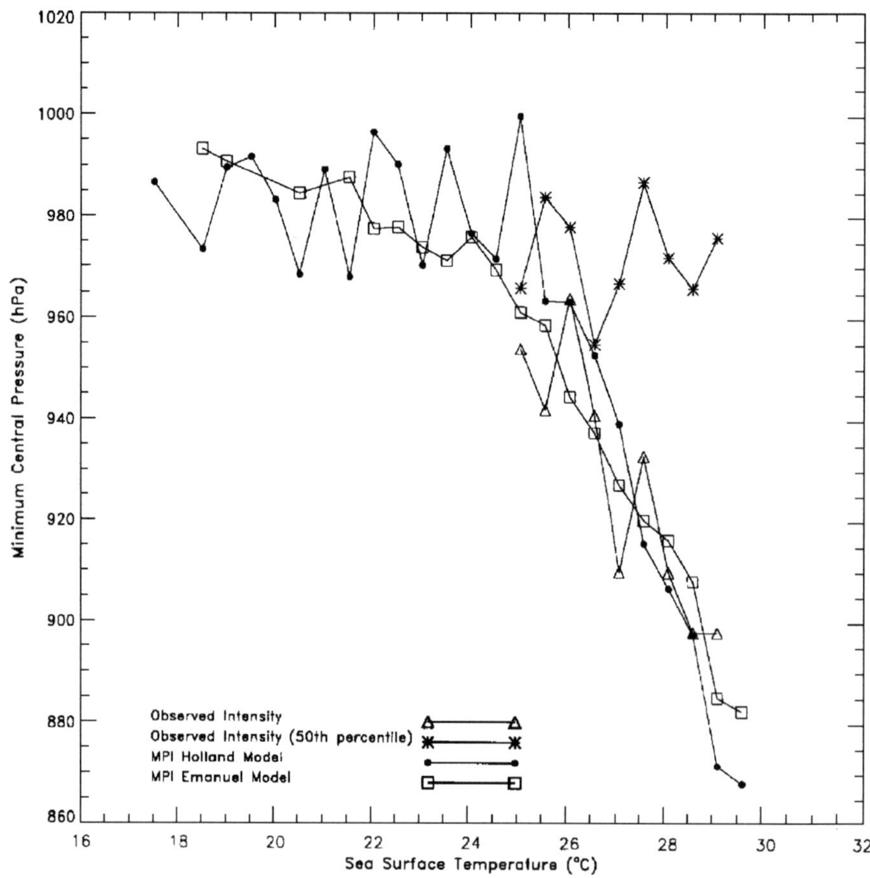

Figure 17.15 Maximum and 50th percentile observed and maximum predicted potential intensity (Holland and Emanuel models) for tropical cyclones as a function of sea surface temperature for the south west Pacific. (Only where three or more observed cyclones occurred at a given temperature were the points included in the profiles.)

17.4 Estimations of tropical cyclone changes from selected MECCA GCM simulations

The following analysis examines both the capacity for GCMs to reproduce current climatic conditions for tropical cyclones and the predicted changes that may occur. The analysis is focussed on application of the Holland model to six climate model simulations over the western North Pacific in August, the month and location of maximum global MPI. The data consist of mean fields from the following selection of the MECCA Phase 1 experiments (e.g. Henderson-Sellers et al. 1995).

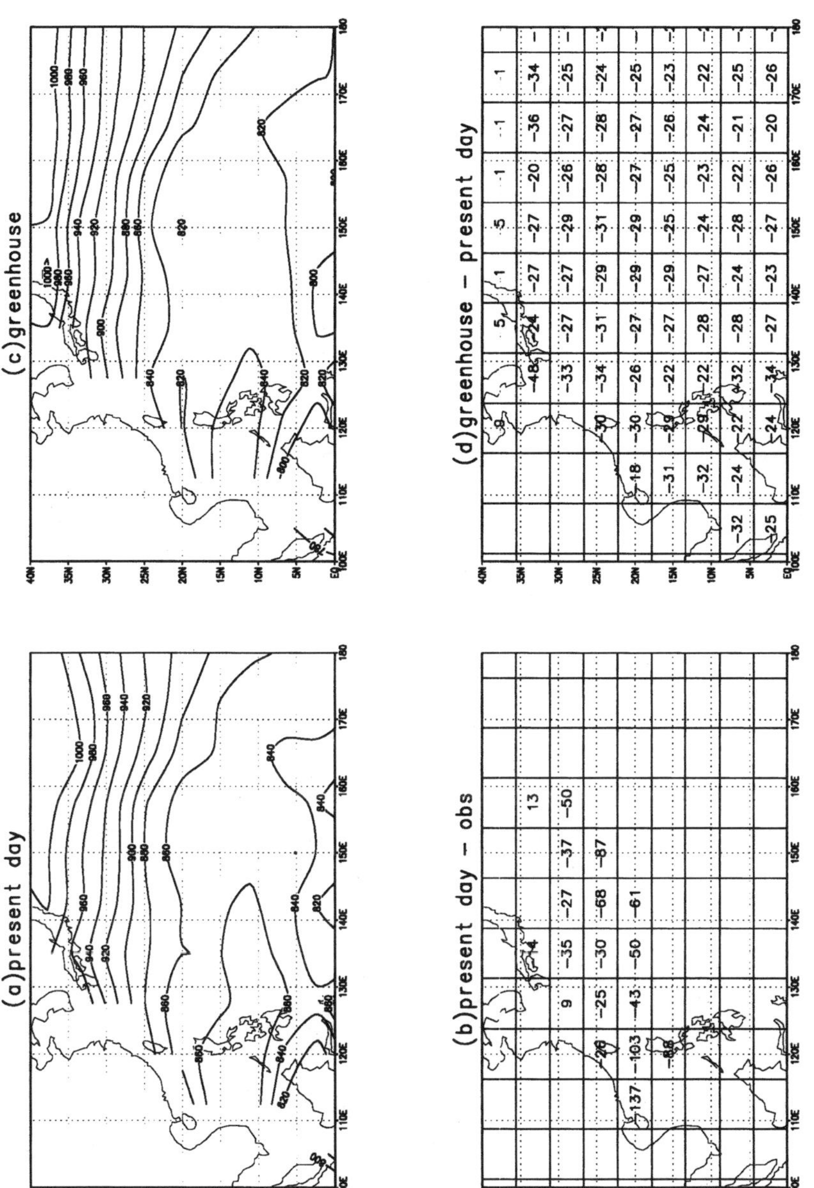

Figure 17.16 (a) Simulated maximum potential intensity (Holland model) for tropical cyclones in the north west Pacific from present-day climate simulations with the CCM1-OZ for August; (b) differences (GCM simulations — observed) at GCM grid locations; (c) simulations of maximum potential intensity for a greenhouse-warmed climate; (d) differences ($2 \times CO_2$ GCM simulations — current GCM simulations).

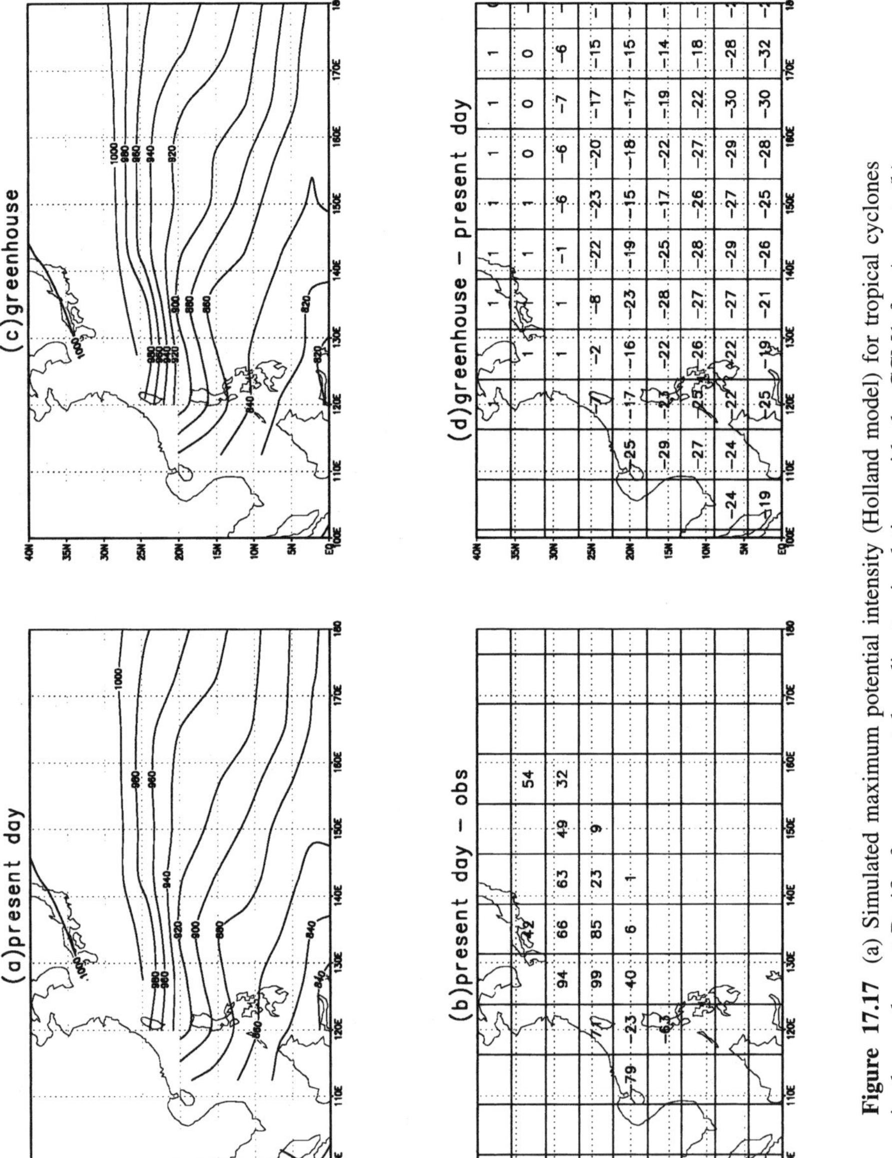

Figure 17.17 (a) Simulated maximum potential intensity (Holland model) for tropical cyclones in the north west Pacific from present-day climate simulations with the CCM1 for August; (b) differences (GCM simulations — observed) at GCM grid locations; (c) simulations of maximum potential intensity for a greenhouse-warmed climate; (d) differences ($2 \times CO_2$ GCM simulations — current GCM simulations).

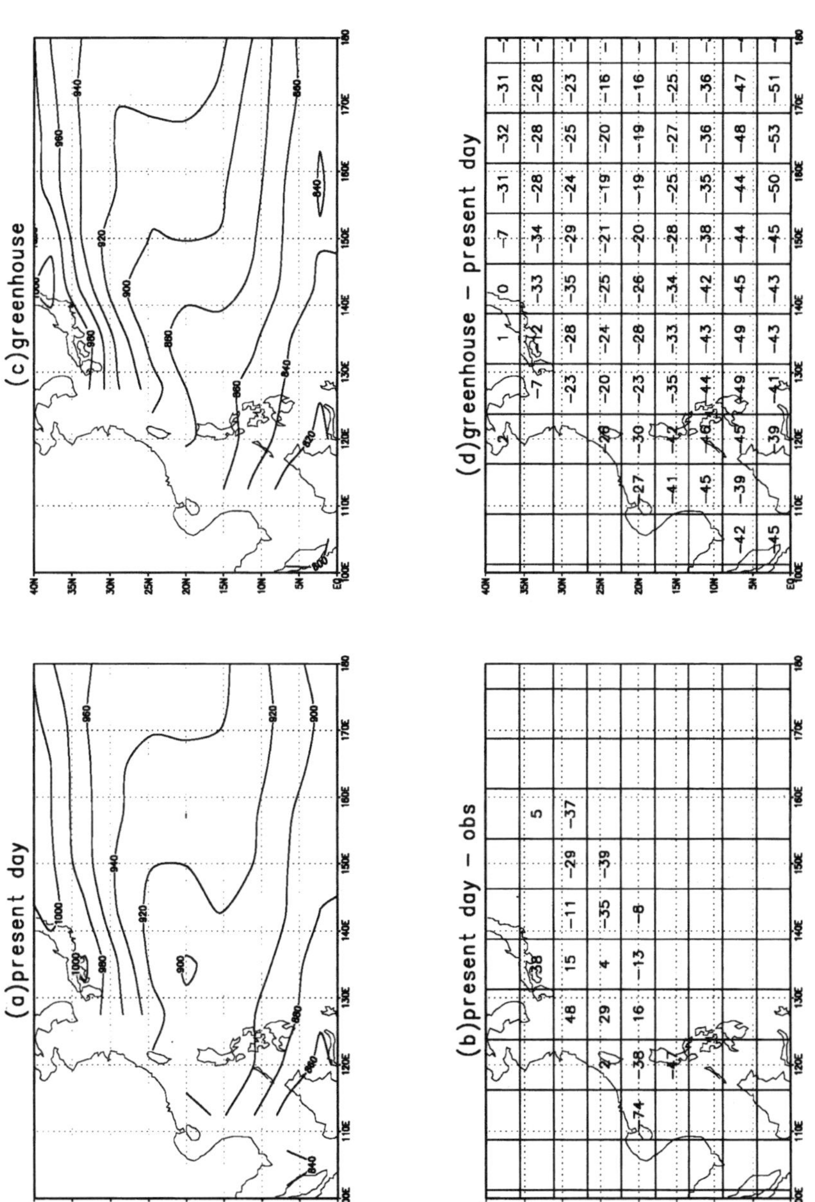

Figure 17.18 (a) Simulated maximum potential intensity (Holland model) for tropical cyclones in the north west Pacific from present-day climate simulations with the CCM1W for August; (b) differences (GCM simulations — observed) at GCM grid locations; (c) simulations of maximum potential intensity for a greenhouse-warmed climate; (d) differences ($2 \times CO_2$ GCM simulations — current GCM simulations).

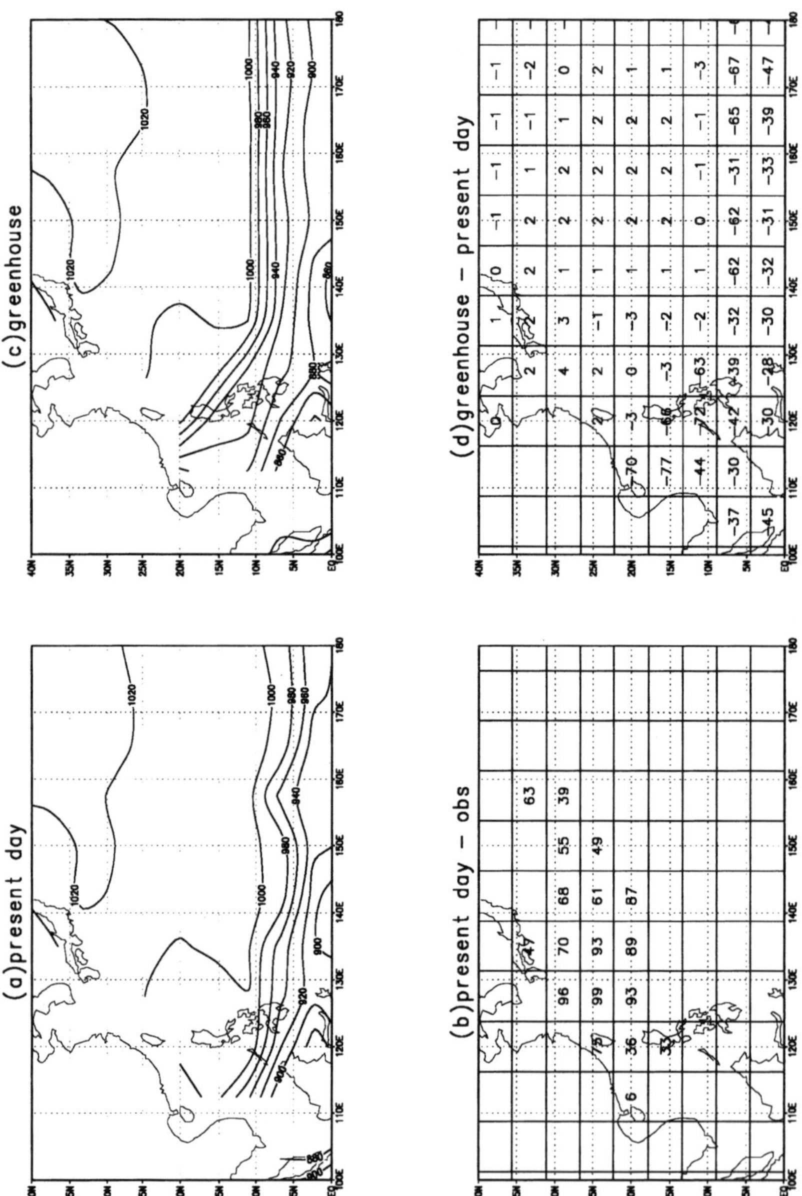

Figure 17.19 (a) Simulated maximum potential intensity (Holland model) for tropical cyclones in the north west Pacific from present-day climate simulations with the CCM0 for August; (b) differences (GCM simulations — observed) at GCM grid locations; (c) simulations of maximum potential intensity for a greenhouse-warmed climate; (d) differences ($2 \times CO_2$ GCM simulations — current GCM simulations).

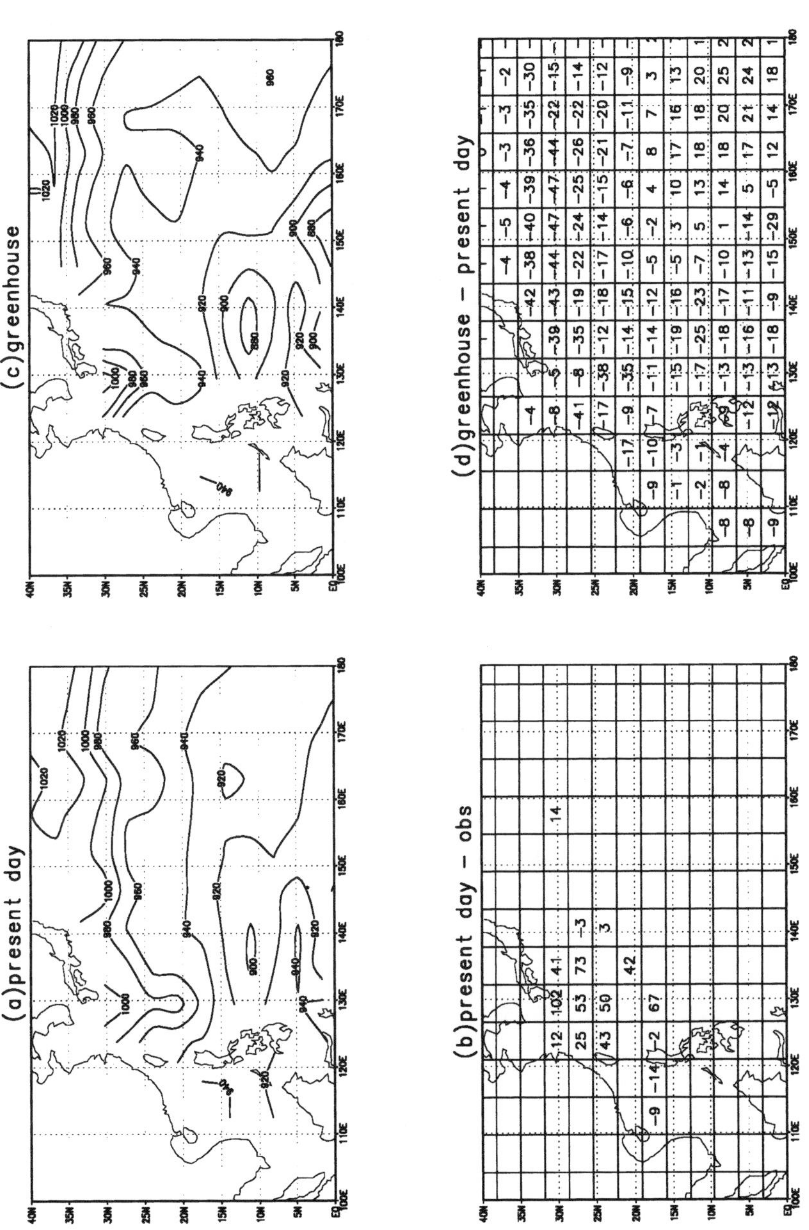

Figure 17.20 (a) Simulated maximum potential intensity (Holland model) for tropical cyclones in the north west Pacific from present-day climate simulations with the BMRC for August; (b) differences (GCM simulations — observed) at GCM grid locations; (c) simulations of maximum potential intensity for a greenhouse-warmed climate; (d) differences ($2 \times CO_2$ GCM simulations — current GCM simulations). (This GCM has a higher spatial resolution than the others so that more grid locations are displayed in part (b)).

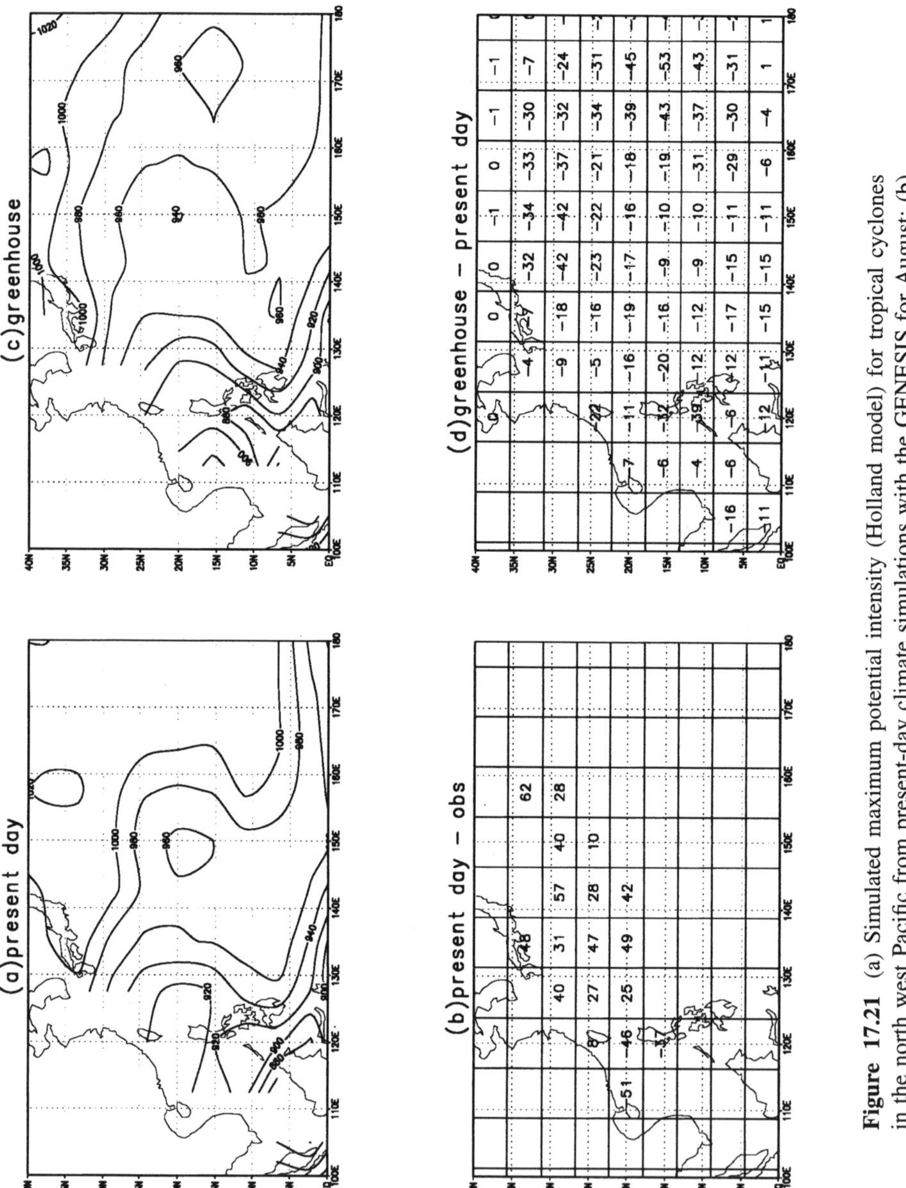

Figure 17.21 (a) Simulated maximum potential intensity (Holland model) for tropical cyclones in the north west Pacific from present-day climate simulations with the GENESIS for August; (b) differences (GCM simulations — observed) at GCM grid locations; (c) simulations of maximum potential intensity for a greenhouse-warmed climate; (d) differences ($2 \times CO_2$ GCM simulations — current GCM simulations).

The six MECCA models include five GCMs from NCAR and one GCM from BMRC. Each climate model includes a 50 m deep mixed ocean layer. The horizontal resolution of five of the GCM models is 7.5 degrees latitude by 4.5 degrees longitude (i.e. 48×40 grid points globally) while the BMRC model has a higher resolution of 64×56 grid points. Among the six models, CCM1-OZ, CCM1, CCM1W and GENESIS have 12 levels in the vertical, and CCM0 and BMRC have nine levels. The model simulations employed in the study are averaged over the last 10 years except for GENESIS where only five years of results were archived for analysis.

Figures 17.16–17.21 show the MPI results for each climate model. In each case, we indicate the MPI for current climatic conditions, the difference between simulated and observed MPI, obtained from the data sets described above, the predicted climatic MPIs, and the changes from current climate. Each grid square indicates the resolution of the climate model and only those grid squares with more than three cyclone observations were included in the analysis. We note that all thermodynamic models over estimate potential intensity in equatorial regions, where the dynamics are not able to support cyclone development (e.g. Gray 1968). For this reason, the region between the equator and 10°N is not analysed.

For CCM1-OZ (Figure 17.16), the model overpredicts MPI throughout most of the basin. We note that some overprediction is expected, because the maximum possible cyclone may not have been observed over the relatively short period of cyclone observations. Nevertheless, this model considerably overestimates MPI. The climatic changes in MPI are predicted to be a modest reduction of 20–30 hPa in central pressure, which is a fraction of the errors in the current climate assessment.

The MPI errors for CCM1 (Figure 17.17) vary from −79 hPa in the South China Sea to 99 hPa south of Japan. However, this model simulates generally weaker MPI for current climate than is observed, and substantially weaker than CCM1-OZ. Only modest MPI changes are predicted for a greenhouse climate.

The MPI distribution of CCM1W (Figure 17.18) contains a consistent error variation from negative in the south and east to positive in the north west. Errors range from −74 to 48 hPa, which is less than that for the previous two models. Predicted climate changes are modest, ranging from −46 to 1 hPa.

CCM0 produces very weak tropical cyclone MPI (Figure 17.19) for the current climate, with an error range of 6–99 hPa. The predicted climate changes are also modest, ranging from −77 to 4 hPa.

The BMRC model (Figure 17.20) has an error distribution ranging from −14 to 102 hPa. Interestingly, this model predicts weaker cyclones in some parts of the domain for a greenhouse climate.

The GENESIS model simulates generally weaker than observed MPI (Figure 17.21), except in the South China Sea. Predicted changes of MPI are similar to other models and range from −53 to 0 hPa.

17.5 Summary

This preliminary analysis indicates that thermodynamic models of tropical cyclone MPI are in relatively good agreement with available observations. The main differences are that the model MPIs are often more intense than observed, but this is at least partially explained by the relatively short period of observations. Application of the Holland model to MECCA climate models indicates that substantial errors occur in estimating current climatic conditions. These are mostly due to errors in SST. The models consistently predict a modest increase of around 20 hPa in MPI, with the exception of the BMRC model which predicts a weakening for part of the domain. Care needs to be taken with the application of this prediction, however, as the changes are much smaller than the errors in simulating the current climate.

Acknowledgments

This paper is CIC contribution number 95/41.

References

Abdullah, A.J. (1954) A proposed mechanism for the development of the eye of a hurricane. *Journal of Meteorology*, **11**, 189–195.

Bengtsson, L., Botzet, M. and Esch, M. (1994a) Hurricane-type vortices in a general circulation model, Part I. Max-Planck-Institut fur Meteorologie, Report No. 123.

Bengtsson, L., Botzet, M. and Esch, M. (1994b) Will greenhouse gas-induced warming over the next 50 years lead to higher frequency and greater intensity of hurricanes? Max-Planck-Institut fur Meteorologie, Report No. 139.

Broccoli, A.J. and Manabe, S. (1990) Can existing climate models be used to study anthropogenic changes in tropical cyclone climate? *Geophysical Research Letters*, **17**(11), 1917–1920.

Byers, R.H. (1944) *General Meteorology*, 2nd Ed., p. 645. New York: McGraw-Hill Book Company.

Dong, K. and Holland, G.J. (1994) A global view of the relationship between ENSO and tropical cyclone frequencies. *Acta Meteorology Sinica*, **8**, 19–29.

Demaria, M. and Kaplan, J. (1994) Sea surface temperature and the maximum intensity of Atlantic tropical cyclones. *Journal of Climate*, **7**, 1324–1334.

Dvorak, V.F. (1975) Tropical cyclone intensity analysis and forecasting from satellite imagery. *Monthly Weather Review*, **103**, 420–430.

Emanuel, K.A. (1987) The dependence of hurricane intensity on climate. *Nature*, **326**, 483–485.

Emanuel, K.A. (1988) The maximum intensity of hurricanes. *Journal of the Atmospheric Sciences*, **45**(7), 1143–1155.

Emanuel, K.A. (1991) The theory of hurricanes. *Annual Review in Fluid Mechanics*, **23**, 179–196.

Evans, J. (1993) Sensitivity of tropical cyclone intensity to sea surface temperature. *Journal of Climate*, **6**, 1133–1140.

Fairall, C.W., Kepert, J.D. and Holland, G.J. (1994) The effect of sea spray on surface energy transports over the ocean. *The Atmosphere Ocean System*, **2**, 121–142.

Gray, W.M. (1968) Global view of the origin of tropical disturbances and storms. *Monthly Weather Review*, **96**, 669–700.

Gray, W.M. (1975) Tropical cyclone genesis. Department of Atmospheric Science, Paper No. 232, Colorado State University, Ft. Collins, CO, p. 121.

Gray, W.M. (1979) Hurricanes: Their formation, structure and likely role in the tropical circulation. In *Meteorology over the Tropical Oceans*, edited by D.B. Shaw, pp. 278. Bracknell: Royal Meteorological Society.

Gray, W.M. (1984) Atlantic seasonal hurricane frequency: Part I: El Niño and 30mb quasi-biennial oscillation influences. *Monthly Weather Review*, **112**, 1649–1668.

Gray, W.M. and Landsea, C.W. (1992) African rainfall as a precursor of hurricane-related destruction on the US East Coast. *Bulletin of the American Meteorological Society*, **73**, 1352–1364.

Gray, W.M., Landsea, C.W., Mielke, P.W. Jnr. and Berry, K.J. (1993) Predicting Atlantic seasonal tropical cyclone activity by 1 August. *Weather Forecasting*, **8**, 73–86.

Gray, W.M., Landsea, C.W., Mielke, P.W. Jnr. and Berry, K.J. (1994) Predicting Atlantic seasonal tropical cyclone activity by 1 June. *Weather Forecasting*, **9**, 103–115.

Haarsma, R.J., Mitchell, J.F.B. and Senior, C.A. (1993) Tropical disturbance in a GCM. *Climate Dynamics*, **8**, 247–257.

Hirakuchi, H. and Giorgi, F. (unpub.) Multi-year present day and $2 \times CO_2$ simulations of monsoon climate over eastern Asia and Japan with a regional climate model nested in a GCM.

Henderson-Sellers, A., Howe, W. and McGuffie, K. (1995) The MECCA analysis project. Special issue of *Global and Planetary Change*, **10**(1–4), 3–21.

Henderson-Sellers, A. (1996) Adaptation to climatic change: Its future role in Oceania. In *Greenhouse, coping with climate change*, edited by W.J. Bouma, G.I. Pearman and M. Manning, p. 349–376. Collingwood: CSIRO Publications.

Holland, G.J. (1981) On the quality of the Australian tropical cyclone data base. *Australian Meteorological Magazine*, **29**, 169–181.

Holland, G.J. (forthcoming) The maximum potential intensity of tropical cyclones. *Journal of Atmospheric Sciences*.

Kleinschmidt, E. Jnr. (1951) Principals of the theory of tropical cyclones. *Archives of the Materials of Geophysical Bicklimatology*, **4a**, 53–72.

Lighthill, J., Holland, G.J., Gray, W.M., Landsea, C.W., Craig, G., Evans, J., Kurihara, Y. and Guard, C. (1994) Global climate change and tropical cyclones. *Bulletin of the American Meteorological Society*, **75**(11), 2147–2157.

Maher, J.V. and Lee, D.M. (1977) Upper air statistics, Australia: Surface to 5 hPa, 1957 to 1975, pp. 202. Canberra: Australian Government Publishing Service.

Malkus, J.S. and Riehl, H. (1960) On the dynamics and energy transformations in steady-state Hurricanes. *Tellus*, **12**(1), 1–20.

Merrill, R.T. (1987) An experiment in statistical prediction of tropical cyclone intensity change. NOAA Technical Memo, NWS NHS-34, pp. 34.

Miller, B.I. (1958) On the maximum intensity of hurricanes. *Journal of Meteorology*, **15**, 184–195.

National Climate Data Center (1994a) Global Tropical and Extratropical Cyclone Climatic Atlas (GTECCA), Version 1, March 1994.

National Climate Data Center (1994b) Monthly Climatic Data for the World, April 1993, Vol. 46, No. 4.

Palmen, E. (1948) On the formation and structure of tropical hurricanes. *Geophysica*, **3**, 26–38.

Pudov, V.D. and Holland, G.J. (1994) Typhoons and oceans: Results of experimental investigations. BMRC Research Report No. 45. Melbourne: Bureau of Meteorology.

Riehl, H. (1954) *Tropical Meteorology*. New York: McGraw-Hill Book Company.

Ryan, B.F., Watterson, I.G. and Evans, J.L. (1992) Tropical cyclone frequencies inferred from Gray's yearly genesis parameter: Validation of GCM tropical climates. *Geophysical Research Letters*, **19**(18), 1831–1834.

Shay, J.K., Black, P.G., Mariano, A.J., Hawkins, J.D. and Elsberry, R.L. (1992) Upper ocean response to Hurricane Gilbert. *Journal of Geophysical Research*, **97**, 20227–20248.

Slutz, R.J., Lubker, S.J., Hiscox, J.D., Woodruff, S.D., Jenne, R.L., Joseph, D.H., Steurer, P.M. and Elms, J.D. (1985) COADS Comprehensive Ocean-Atmosphere Data Set. Release 1. Boulder, CO: Climate Research Program, Environmental Research Laboratories.

PART 6

LESSONS LEARNED

CHAPTER 18

MECCA ACHIEVEMENTS AND LESSONS LEARNED

Ann Henderson-Sellers and Wendy Howe

18.1 Introduction: Uncertainty revealed

Climatic assessments will, increasingly, provide part of the basis for negotiating global and transnational protocols addressing climatic change issues which lie outside the jurisdiction of individual policymakers. In particular, the Framework Convention on Climate Change, signed by 155 nations at and since UNCED in June 1992, came into force on 21 March 1994 after its ratification by the 50th signatory nation. The Convention's ultimate objective as stated in Article 2 is to achieve 'stabilisation of greenhouse gas concentrations in the atmosphere at a level that would prevent dangerous anthropogenic interference with the climate system. Such a level should be achieved within a timeframe sufficient to allow ecosystems to adapt naturally to climate change, to ensure that food production is not threatened and to enable economic development to proceed in a sustainable manner' (refer DEST 1995, p. 14).

MECCA was created during the preliminary stages of development of national perspectives on the international issue of greenhouse warming. The uncertainties in concentrations of greenhouse gases, in timeframes needed for stabilisation and in the impacts of the projected changes underpinned the development of MECCA. For these reasons MECCA's goal was to assess (and reduce) uncertainty in projections of future climate change. Despite the need for more confidence and the efforts of MECCA and many other organisations, it is probably the case that uncertainty about climate change resulting from greenhouse gas increases in the atmosphere has not been reduced during the four-year period since MECCA was created.

This is not the 'fault' of MECCA. It is generally agreed that, although uncertainty seems to be larger now than in 1990 when the First IPCC Scientific Assessment was published, the causes of (at least) some of this uncertainty are now known and better understood (e.g. the role of sulphate aerosols). This chapter tries to answer the question — was the MECCA experiment worthwhile? — from the point of view of the two people responsible for the day-to-day activities of MAT. Specifically, this chapter seeks to evaluate the important

Table 18.1 Desirable characteristics of climate change prediction (EPA 1989, p. 57)

1.	The scenarios should be internally consistent with global warming caused by increases in greenhouse gas emissions. A doubling of the CO_2 concentration in the atmosphere is thought to increase global temperatures by approximately 1.5 to 4.5°C (3 to 8°F). The regional temperature changes and seasonal distributions may be higher or lower, as long as they are internally consistent with the global range.
2.	The scenarios must include a sufficient number of meteorological variables to meet the requirements for using effects models. These effects models include models of crop growth, forest succession, runoff, and other systems. Some models of the relationship between climate and a system use only temperature and precipitation as climate variables, while others also need solar radiation, humidity, winds, and other variables.
3.	The meteorological variables should be internally consistent. While a scenario is not a prediction, it should at least be plausible. The laws of physics limit how meteorological variables may change in relationship to each other. For example, if global temperatures increase, global precipitation must also rise. Regional changes should be internally consistent with these large-scale changes.
4.	The scenarios should provide meteorological variables on a daily basis. Many of the effects models used in this study, such as crop yield and hydrology models, need daily meteorological inputs.
5.	Finally, the scenarios should illustrate what climate would look like on a spatial scale fine enough for effects analysis. Many effects models consider changes in individual stands of trees or farm fields. To run them, scenarios must illustrate how climate may change locally.

question: to what extent, if any, can MECCA claim to have assisted in *revealing* the causes of this, increased, uncertainty and in understanding them?

The overall aim of MECCA was to help quantify the reliability of climate change forecasts that are based on various scenarios of increasing emissions of greenhouse gases and the resulting increased concentration of these radiatively active gases in the atmosphere. To achieve this goal, MECCA's research plan included two key elements:

1. a set of carefully selected model experiments, designed to supplement ongoing work by other modelling groups; and
2. an analysis strategy that focused on impacts and evaluation.

In order to use climate information for impacts assessment and, ultimately, as one component of the input to policy advice, the climate predictions provided must be compatible with the impacts and policy users' requirements. Delivering this compatibility is not straightforward. It presumes a sound understanding of the social, economic and cultural needs of nations and communities which rarely exists in fora discussing climate and global change. Indeed, compatibility between supplied and needed information is difficult to attain even within a developed nation in the northern hemisphere. For example, the USA EPA Report to Congress titled 'The Potential Effects of Global Climate Change on the United States' summarised the desirable characteristics of a climatic change prediction (Table 18.1) (EPA 1989). It is doubtful if these requirements can be met

Table 18.2 Required climatic variables (adapted from McKenney & Rosenberg 1991, Table 18.1)

- Temperature
  - mean, minimum, maximum
  - variance
  - temporal autocorrelation
  - frequency, magnitude and duration of extreme values/values which exceed threshold values (e.g. heat waves, severe frost)
  - frequency, timing and intensity of climate-triggered events like snowmelt
- Precipitation
  - amounts
  - variability
  - frequency
  - persistence (temporal autocorrelation)
  - intensity
  - form (rain, snow, hail, etc.)
  - frequency of extreme events (e.g. drought, floods)
  - monsoonality (timing, duration, intensity, location)
- Radiation (solar, net)
  - intensity
  - duration
  - cloud cover
  - quality (wavelength)
- Humidity
  - means
  - diurnal and seasonal variability
- Wind speed and direction
  - variability
  - frequency, magnitude and duration of extreme events
  - (storms, hurricanes)
  - turbulent intensity
- Evapotranspiration
  - mean, variability, extremes
- Soil moisture profile
  - daily, seasonal, long-term
- Surface runoff
  - mean, variability, extremes, time-series
  - frequency, magnitude of extreme events
- Thermal inversions
  - frequency, intensity, duration
- Growing degree days
- Ocean climate
  - sea level
  - sea-surface temperature
  - salinity
  - ocean circulation
  - extent of sea ice

even now, six years later. At present, lists of requested model output variables (e.g. Table 18.2) are not especially valuable guides for future evaluations as they, often, cannot be produced from climate models and even if they are generated there are very few assessment groups with the necessary interpretative skills. MAT set itself the goals of trying to help bridge the gap between users of climatic information and providers.

The creation of MAT was an innovative and unique contribution to the climate assessment process. Their tasks, turning out to be much greater than anyone anticipated, included:

- establishing and describing the experiments conducted, model versions employed and output archived;
- the quality control of climate model simulation results;
- developing an homogeneous archive of simulation results in a standard and readily accessible format and on various media; and
- using subsets of this archive to produce displays (static and movies) that permit comparison between climate model simulations.

These tasks required continuing and detailed discussions with three different international groups:

1. the MECCA Consortium members and associates;
2. the MECCA-sponsored scientific investigators; and
3. the international science, impacts and assessment communities, especially but not exclusively involved with the Intergovernmental Panel on Climate Change (IPCC), World Climate Research Programme (WCRP), International Geosphere–Biosphere Programme (IGBP) and Human Dimensions Programme (HDP).

Developing successful dialogues with these three groups was time consuming and demanded different terminology, formats, frequency and styles. The level of success was probably greatest with the individual investigators, fortunately, as without effective communication with these people little reliance could be placed on the archive of model results or any other MAT products. The (unexpected) need for quality assurance resulted in considerable MAT effort being focussed on straightforward but time-consuming evaluations of the MECCA-sponsored GCM results.

In trying to evaluate MECCA's outcomes the following questions are reviewed: (1) Were the experiments and analysis well planned? (2) Were the right questions asked? (3) Were appropriate resources available? (4) Are the scientific outcomes recognised and are they appreciated? In attempting to answer these questions it must be recognised that MECCA was a remarkable and novel concept. The traditional mode of operation of many of the consortium members was (and is) to engage consultants. There is a considerable difference between curiosity- and application-driven research: one is open-ended, externally reviewed, curiosity-prompted; the other is task-oriented, product, outcome-specific and timetable driven. MECCA's 'greatest mistake' might have been its failure to fully disentangle these two modes of research sponsorship. On the other hand, MECCA's 'greatest achievement' may be this overlapping of motivation and sponsorship.

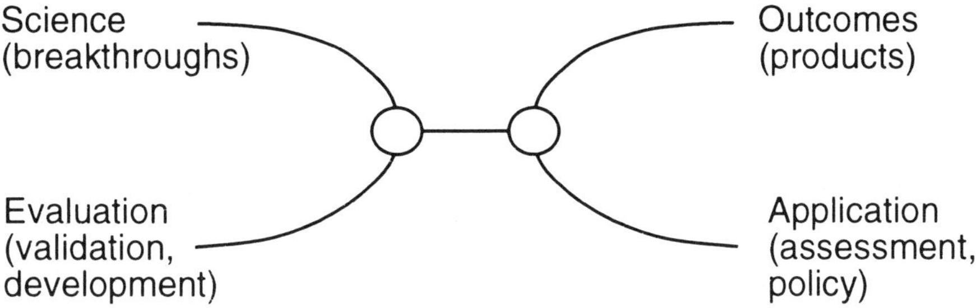

Figure 18.1 'Creative tensions' among MECCA participants and the wider community.

Creative tension between the 'scientific goals' of MAT (especially the team leader) and need for 'products' for display on the part of the EPRI management (which was responsible for the MAT contract and the month-to-month project control) operated throughout the lifetime of MAT and led to both good and less good outcomes (Figure 18.1). These tensions were exacerbated by concerns in the wider scientific community about the motivation of some MECCA Consortium members who represent major fossil fuel users and, hence, greenhouse gas emitters. For example, in 1993, MECCA sponsorship was compared (in a discussion with one of us: Ann Henderson-Sellers) with lung cancer research being supported by funds from major tobacco companies. Even some of the sponsored investigators expressed concerns about the extent and nature of Consortium members' potential involvement in research projects.

MECCA's successes include the fact that it sponsored a wide variety of climate change activities (e.g. AGCMs, OGCMs, RegCMs, OAGCMs, impact and assessment studies; see, for example, Colour Plate 18.1). Moreover, this sponsorship was established very rapidly and effectively so that experiments identified as important were able to be conducted in a timely manner, peer-reviewed and published.

This chapter is aimed at identifying and placing the MECCA achievements and contributions into a wider context as seen from the viewpoint of the MAT Leader and MAT Project Manager. The next section deals with the role, successes and failures of the MAT. In section 18.3 the achievements of MECCA are viewed from the point of view of international science. Chapter 19 deals with similar overall issues but from the point of view of some of the members of the MECCA Consortium. These chapters are aimed to increasing understanding of the MECCA 'experiment' of industry-linked research in the context of the uncertainties surrounding future climate change.

Table 18.3 MECCA Analysis Team resources (full-time equivalent persons p.a. over 3.5 years)

Leader:	Henderson-Sellers	0.1
Manager:	Howe	0.6
Computing:	Brescianini, Gross	1.0
Visualisation:	Durbidge, Horne, Scheitlin (UCAR)	0.5
Postdoctoral researchers:	Larson, Landsea	1.0
Graduate students:	Ciret, Kothavala, Balachova, Yuan, Nakken	2.0*
Consortium:	Hoekstra (N.V. KEMA)	0.3**
	and Scheitlin (UCAR)	0.2
Consultant:	McGuffie	0.05

* The Climatic Impacts Centre at Macquarie University also supported Zhang, Zhong, Tonkin, Li and Hansen
** MECCA 'promised' 6–8 co-opted researchers (Jan 1993), Mearns; Oglesby; Mokhov
* and ** Southern California Edison supporting Nakken.

18.2 MECCA: The Analysis Team viewpoint

18.2.1 Analysis strategy and resources

The MECCA Analysis Team plan was established in June 1992 and was offered resources that the Consortium felt sufficient for the tasks planned (Table 18.3). There were strong suggestions early in the life of the MAT that there would be significant contributions both from Consortium members and from other groups (for example the USA's National Oceanographic and Atmospheric Administration (NOAA)) but these expectations did not eventuate. On the other hand, there were no attempts to control either the individual experiments or the activities of MAT by the Consortium as a whole or by individual members. Indeed, this lack of guidance might be viewed as one of the reasons for the mismatches between some expectations of, and outcomes delivered by, MAT.

The MECCA Analysis Team plan (Table 18.4), also developed in mid- to late-1992, was designed to address the MECCA goals by delivering:

(a) links to international science (e.g. IPCC, WCRP, IGBP);
(b) a 'mandate' for enhanced exchange of climate information (since adopted by US DOE; and used in the development of AMIP);
(c) an archive of GCM-derived results for impacts (MADD); and
(d) visualisation and communication examples for policy 'education' (CD and two atlases). See, for example, Colour Plates 18.1 and 18.2.

MAT tried to use either existing standards, or where none existed, the most likely forthcoming standards. For example, there is as yet no 'standard' set of GCM diagnostics

Table 18.4 MECCA Analysis Team timetable and foci

		Delivery	Foci
Modellers			
(i)	homogenised results	8/93	– same statistics
(ii)	uncertainty measures	8/94	– same sub-components
(iii)	MECCA comparisons	8/95	– same experiments
(iv)	Phase 2 models included	8/96	– atlas of $1 \times CO_2$
			– atlas of $2 \times CO_2$
			– electronic atlas
Impacts			
(i)	MECCA mandate	8/93	– impacts assessment
(ii)	agreement on mandate	8/94	– downscaling
(iii)	data on CD-ROM	8/95	– regional evaluation
(iv)	Regional Atlas (W. USA)	8/96	– land-use change
			– variability and extremes
			– tropical storms
Policy			
(i)	MECCA videos	8/93	– uncertainty propagation
(ii)	MECCA atlases	8/94	
(iii)	MECCA CD-ROM (and WWW)	8/95	– communication
(iv)	MECCA book	8/96	

although those requested as part of the AMIP (Gates 1992) may become a *de facto* standard. MAT developed its archive format and contents in conjunction with the Program for Climate Model Diagnostics and Intercomparison (PCMDI) team responsible for AMIP.

In part because of the late initiation of the MAT (compared with the starting dates of MECCA and Phase 1 experiments) and in part because of the failure to identify delivery of results as obligations in MECCA contracts, there was always a mismatch between Phase 1 experiments and MAT goals and tasks. Although MECCA contracts mentioned MAT together with its role and anticipated participation in analysis and impacts assessment and in results dissemination, after selection of the MAT Leader there was very little guidance from the Consortium to individual investigators about MAT's role and tasks. The 'standard' EPRI contract with the Phase 1 PIs state that:

Task 2 Analysis and Interpretation of Experiment Result: The contractor will, in accordance with the general outline of the MECCA Experiment and Analysis Plan, coordinate his analysis and interpretation of the model experiment results with, and will be a contributing member of, the MECCA analysis support team. A detailed plan of action for the MECCA analysis support team is being developed by the EPRI Project Manager through directed activities of other MECCA contractors and the MECCA Technical Committee.

MAT, however, is not aware of such a 'detailed plan of action' from the MECCA Consortium or the MECCA Technical Committee. In addition some 'MECCA' (strictly EPRI) contracts were never fully negotiated with sponsored investigators so no obligations could be insisted upon.

When Phase 2 was developed by the Consortium there was little additional effort made to match the call for proposals or the selection of experiments to the MAT goals or needs. Specifically the Phase 2 call and selection failed (despite recommendations by MAT) to request or note as highly desirable:

- quantitative verification as part of the experimental design;
- uncertainty assessment statistics as part of the experimental design;
- tools for evaluating uncertainty used or developed in the experiment; and
- output designed for impact and/or reduced form models.

The result of this 'on the run' design of MECCA was that there were few means by which MAT could co-ordinate its goals with those of individual experimenters and MAT had no means (except repeated petition) of gaining access to results, model descriptions and simulation design. Specifically, the MECCA Consortium never followed through with specific tasks (beyond their original proposal) for individual investigators.

An unexpected drain on MAT resources appeared in the form of communication with individual PIs and (particularly) quality control of GCM results obtained. As has also been found in AMIP (Gates 1992) quality assurance is *slow* but very necessary. It consumes resources (people, time, funds) for checking and processing GCM results and community (investigator) enthusiasm and involvement is *vital*. MECCA's planning considerably underestimated this requirement and consequently allocated very much less overall resource to it than did AMIP (Table 18.5).

The Consortium members were briefed on the plans and progress of the MAT but rarely made suggestions about directions. However, there were a few instances of specific directing

Table 18.5 Comparison of two recent international intercomparisons: AMIP and MECCA

Features	AMIP	MECCA
Date established	1990	1991
No. of experiments	30	27
Sponsors	US DOE	EPRI, UCAR etc
QA personnel	16	1–2
Max. resoln. of results	5 out of 30: 6 hourly	7 out of 27: daily
International cooperation	IPCC/WCRP	IPCC/WCRP/IGBP
Analysis team(s)	26	1
Analysis personnel	50–100	10
Products	none (12/95)	3 (–5)
Follow-on project	AMIP II	ACACIA

or, perhaps, attempted control of MAT activities. When these occurred, it was difficult to determine whether the directions came from the Consortium as a whole (i.e. having been debated and agreed to by the MECCA Policy Committee); from an individual member of the Consortium (e.g. UCAR, N.V. KEMA); or from the MECCA Project Manager, an EPRI employee. Examples include:

- editorial revisions of the *Climate Change Atlas: Greenhouse Simulations from the Model Evaluation Consortium of Climate Assessment* (Henderson-Sellers & Hansen 1995);
- instructions regarding evaluation of the MECCA-sponsored regional model simulations; and
- directions regarding a proposal to seek feedback from individual investigators on the MECCA project.

The *Climate Change Atlas: Greenhouse Simulations from the Model Evaluation Consortium for Climate Assessment* was first presented to the MECCA Policy Committee in February 1994. It was approved in principle. The Policy Committee expressed the view that regional displays should be included in the atlas and one member of the Policy Committee offered detailed editorial suggestions on the text. After the atlas displays were completed and it was prepared for publication, one Policy Committee member conducted discussions with a number of publishing houses and, having identified Kluwer Academic Publishers as a suitable publisher, entered into negotiations about subsidising the publication costs. The result of these discussions were that N.V. KEMA sponsored the publication. Feedback on the Atlas was received from the publisher-selected reviewers, a Policy Committee member and the EPRI Project Manager. Each review dealt only with the scientific content and the literary style; no one attempted to include a 'MECCA view'. However, after publication, the Policy Committee complained about the lack of specific mention of MECCA or any honorific listing of the Consortium members.

MAT conducted selected evaluations of some of the MECCA-sponsored simulations. These evaluations were detailed in quarterly reports to EPRI. In April 1995, the report included an evaluation of the regional simulation over East Asia which showed a poorer overall climatology in the nested model than in the forcing GCM, itself fairly poor. The MECCA Policy Committee reacted more strongly to this evaluation than to any others presented. Some members were interested in the underlying science (that intensification of the hydrologic cycle in the embedded model increased upper atmospheric heating and hence decreased the verity of the climatology); some members were more interested in whether similar negative results could occur in other regions; one or two members expressed considerable concern about the suggestion that any MECCA simulation was regarded as 'poor' by MAT. MAT was criticised for an apparent lack of communication with the PIs even though this had been attempted but no reply received. An important, and rather disturbing, result of this discussion was the specific direction to omit reference to regional evaluations in the final MAT report and this book.

MAT conducted a survey of the PIs early in its activities (Henderson-Sellers et al. 1995) and planned a follow-up questionnaire towards the end of MECCA. The initial draft of this

Table 18.6a Set of 'preferred' variables*

1. Surface type (flag for ocean, land, sea ice) (flag)
2. Mean sea-level pressure (hPa)
3. Ground temperature (K)
4. Lowest air (surface air) temperature (preferably at 1.5 m) (K)
5. Max. ground temperature (over diurnal cycle) (K)
6. Min. ground temperature (over diurnal cycle) (K)
7. Total precipitation (mm)
8. Incident solar radiation at the surface (W m^{-2})
9. Absorbed solar radiation at the surface (W m^{-2})
10. Upward longwave radiation from the surface (W m^{-2})
11. Downward longwave radiation at the surface (W m^{-2})
12. Evaporation from the surface (or latent heat flux) (W m^{-2})
13. Sensible heat flux from the surface (W m^{-2})
14. Total runoff (mm)
15. Surface runoff (mm)
16. Total soil moisture (mm or ratio)
17. Amount of snow (in depth in water equiv.) (mm)
18. Sea-ice thickness (m)
19. Total cloud amount (from random overlap in layers) (ratio)
20. Top-of-the-atmosphere downward solar radiation (W m^{-2})
21. Top-of-the-atmosphere absorbed solar radiation (W m^{-2})
22. Top-of-the-atmosphere outgoing longwave radiation (W m^{-2})
23. Total precipitable water in atmosphere (m)
    (Multiple layer values: (archived at 200, 500 and 850 hPa)
24. Moisture (kg/kg)
25. Horizontal wind (u and v) (m s^{-1})
26. Temperature (K)
27. Geopotential height (m)

* As defined by Users of Climatic Change Results in response to a MAT survey.
Note: Data parameters: time scales (for every simulated year); daily data and monthly means for all fields; space scales: GCM model resolution (no downscaling)

survey was sent to the EPRI Project Manager in August 1995 prompting an unexpected reaction. He expressed the view that it was 'inappropriate' for MAT to seek out the views of PIs and went on to comment (very) negatively on surveying by 'faceless e-mail'. This was a particularly odd reaction as the MAT Project Manager was, by then, personally known to all the PIs and was probably the person with whom most regular contact had been maintained. The concern seemed to arise from a disinclination (on the part of the EPRI Manager at least) to have a survey of 'views on MECCA' conducted outside the control of MECCA.

In order to satisfy its goals of intercomparison and analysis, MAT initiated the development of parameter lists of variables which were of value to users and able to be produced in a comparable format from at least some of the MECCA models. MAT created

Table 18.6b Daily data available from MADD

- surface (ground/ocean/sea-ice) temperature (K)
- total precipitation (mm)
- mean sea-level pressure (Pa)
- snow depth (mm H_2O equiv.)
- evaporation (W m^{-2})
- lowest-level air temperature (K)
- total cloud amount (fraction)
- sea-ice extent (flag)

Note: MAT has selected only those models which include an ocean sub-scheme and which offer daily data as these are believed to be a unique contribution to climate assessment and of considerable value to impact modellers and climate policy advisers. A significant number of the archive data are from experiments that ran for over 100 years.

a 27-element 'of use by impact assessors' list of results (Table 18.6a) after a four month question and answer process. To date, only the eight-element set given in Table 18.6b has been generated from available MECCA models because of the lack of availability and consistency among the archived GCM fields (Henderson-Sellers et al. 1995). Nonetheless, the MECCA Archive of Daily Data (MADD) is seen as one of the most important 'bridging' products derived from the overall MECCA effort. The data has been used in a wide variety of applications ranging from crop hail damage in Australia to agricultural land price projections in America and water resource simulations in California.

The IPCC Regional Assessment which involved only four OAGCMs encountered similar difficulties in providing consistent and useful information. For example, some fundamental fields were not made available (e.g. precipitation from the USA's Geophysical Fluid Dynamics Laboratory (GFDL)) and the resulting data source, although somewhat more up-to-date than MADD, is very much less homogeneous (see, for example, Table 2.5).

18.2.2 Products

The MECCA Analysis Team generated a series of 'products' including:

- the MECCA Archive of Daily Data (MADD) (Table 18.6b), 1993–1996;
- the MECCA Protocol (Figure 2.2), mid-1993;
- a compact disc (Figure 2.3), early-1994;
- the *Climate Change Atlas: Greenhouse Simulations from the Model Evaluation Consortium for Climate Assessment* (Henderson-Sellers & Hansen), mid-1995;
- a WWW site (CD duplicate), mid-1995; and
- the *Regional Atlas for Western USA*, late-1995.

These products were selected in consultation with potential users (especially for the MADD archive), with the aim of fulfilling the four goals outlined in section 18.2.1.

All of these products have been used or reviewed by scientists, policy developers and other interested parties external to MECCA. (Specific examples of the type of use made and its impact on international climate change evaluation are given in section 18.3.) In addition to these products, and sometimes as a result of their creation, MAT has undertaken a wide-ranging series of assessments and evaluations of the results of the MECCA-sponsored climate simulations. These include global evaluations of the cryosphere (frozen water), the hydrosphere (atmospheric and surface moisture), cloudiness, precipitation and temperature. MAT also supported a smaller set of impact assessments including an evaluation of vegetation and crop modelling, an assessment of the potential for future climate-induced changes in tropical cyclone frequency, intensity or location and an assessment of the potential for changes in drought and flood frequency in greenhouse climates.

In retrospect, the MECCA project can be illustrated as a four-way tug-of-war between *science* (mostly the MECCA investigators and the wider community), *evaluation* (MAT, peer reviewers and others), *outcomes* (mostly the Consortium members and their associates), and *application* (MAT again, governments and policy advisers) (Figure 18.1). MAT's ambiguous, but vital, role is clear from its positions on both 'sides' of this tug-of-war. MAT's responsibilities ranged from science, through evaluation and application into outcomes. It was also the only element of MECCA able to work towards retaining the (dynamic) equilibrium between science and outcomes and evaluation and assessment. This tension was a vital component of the MECCA experiment since, without it, the overall effort could easily have overbalanced. Learning how to maintain both the balance and tension required to sustain this tug-of-war was an important and challenging lesson for MAT and one that can valuably be shared with future participants in climate change assessment.

18.3 MECCA: An international perspective

18.3.1 The international context for MECCA

Inclusion of the precautionary principle as stated in Article 3.3 of the Framework Convention on Climate Change is significant in that it is an important environmental management guide in a situation of scientific (un)certainty. The Article states:

> The Parties should take precautionary measures to anticipate, prevent or minimise the causes of climate change and mitigate its adverse effects. Where there are threats of serious or irreversible damage, lack of full scientific certainty should not be used as a reason for postponing such measures, taking into account that policies and measures to deal with climate change should be cost-effective so as to ensure global benefits at the lowest possible cost...

Thus, information on climate and possible climatic impacts is required by policy makers now and scientific (un)certainty is no longer deemed an adequate reason for withholding such advice (cf. Taplin 1995). Indeed, it has not been viewed so for some time, although this is still not widely recognised in the climate assessment community.

As negotiations begin on emission reductions of greenhouse gases, assessments of changes in (global mean) temperature and precipitation will not suffice. Society, and therefore policy makers, must be informed about whether changes in climate might inflict serious effects and, especially, on whom and to what extent. Assessing climate change in a relative risk context is difficult, but extremely important, since the public tends to respond urgently to perceived crises. Assigning relative risks could assist decision makers in distinguishing between verifiably serious threats and possibly misplaced public concern.

The overall aim of MECCA was to help quantify the reliability of climate change forecasts that are based on various scenarios of increasing emissions of greenhouse gases and the resulting increased concentration of these radiatively active gases in the atmosphere. These aims were being pursued in the context of the negotiations which developed the Framework Convention on Climate Change for UNCED. The aims of MECCA were, necessarily, very similar to those of the IPCC. Indeed, the MECCA-funded experiments overlapped to some extent with experiments being used in the IPCC 1992 Science Update (Houghton et al. 1992) and the anticipated requirements for the Second IPCC Scientific Assessment developed during 1995.

Moreover, IPCC and MECCA were not the only intercomparisons that were being undertaken in the first half of the 1990s. The World Climate Research Programme had initiated at least three others: AMIP (Atmospheric Model Intercomparison Program), PILPS (the Project for Intercomparison of Land-surface Parameterization Schemes) and FANGIO (Feedback Analysis of GCMs and Intercomparison with Observations).

Recalling that to achieve its stated goals in MECCA's research plan of established key elements it may be useful to pose the following questions:

1. Were the MECCA model experiments 'carefully selected'?
2. Were they selected to 'supplement ongoing work' elsewhere?
3. Was the 'impact-oriented' strategy successful?

Indeed, it is worthwhile to ask which, if any, of the current international comparisons have even posed questions to be answered as a result of the intercomparisons. IPCC has a mandate to update the scientific understanding of climate and climate change, but poses no specific 'questions' beyond the need for improved assessment. AMIP, PILPS and FANGIO are all intercomparisons that are designed to achieve two (interlinked) goals: (1) overall (i.e. group) improvement, and (2) common agreement on what is (and what is not) acceptable performance (e.g. Cess et al. 1991, 1995; Gates 1992; Henderson-Sellers et al. 1993, 1995). Specific questions or tasks are more difficult to identify.

The idea of 'group advancement' does not seem to have been part of the goals of the MECCA Consortium. Rather, MECCA seems to have wished for a better understanding (mostly but not exclusively) for its members (the Consortium) and their improved visibility (in climate arenas).

For the other, concurrent intercomparisons community (i.e. many investigators) enthusiasm and involvement is *vital*. MECCA did not have this, but maybe it did not wish, or seek, it. Indeed, MECCA was not originally planned as an intercomparison project.

Table 18.7 'Usefulness' test for GCMs

Test Types:
I   Balance/conservation (good housekeeping)
  – to a radiative balance
  – continental surface balance
  – ocean energy input (flux correction)
  – atmospheric energy and water balance
II   Integrative quantities (quick look)
  – length of day (angular momentum)
  – Köppen classification of surface climate
  – SOI or Asian monsoon
  – jet streams, position and strength
III   Community consensus (current wisdom)
  – minimum horizontal and vertical resolution
  – spin-up requirements (cf. cold start)
  – AMIP-type exp. (successfully) completed
  – ensemble simulation
IV   Assessment specific (adequate for job)
  – $2 \times CO_2$ sensitivity (for climate)
  – surface climate 'predicts' biome

Note: In general: (i) no one complete test; (ii) no GCM performs well in all tests; and (iii) performance is a function of application.

The lack (or apparent lack) of focussed goals and the contract nature of the experiments (as compared with AMIP, PILPS and FANGIO which are all open and not directly financially sponsored) placed MECCA in a different situation from other concurrent intercomparisons. This difference was also clear in the different expectations of the Consortium members, the Analysis Team and the investigators themselves.

For example, eighteen months after the demonstration CD was produced by MAT it was placed on the World Wide Web by EPRI staff. The EPRI Project Manager stated in August 1995: 'Far more people can be exposed to this science on the World Wide Web page ... than will ever see it on the original CD ... several thousand individuals join the World Wide Web every month ... the MECCA web pages are being visited by more than 200 people per month, not including those within EPRI.' He went on to say: 'I would like to see more changes made—some of the figures in the analysis papers have very obscure legends and titles. I would like to see these modified.' This request underlines the conflicting goals of different parts of MECCA as the main bulk of the content of the CD (including the obscure legends!) had already been peer-reviewed and published, and (more importantly for most researchers) it was all at least eighteen months out of date. Clearly the goal of this WWW display was *not* novel research results' dissemination but rather a 'demonstration' or information distribution to a wider community. If so, one is forced to wonder whether the contents of the CD (produced for other reasons) are most valuable or appropriate to the task of disseminating MECCA project results (cf. the discussion of the reactions to the MECCA Video detailed in Chapter 2).

MAT pursued international intercomparisons and evaluations, especially during 1995 when the Second IPCC Scientific Assessment was conducted and AMIP was reviewing the results of its first intercomparison. An important result of meetings conducted during May 1995 was an overview of evaluation of model results and simulations and an agreed list of tests for GCMs (Table 18.7).

The outcomes from MAT have contributed to the overall understanding of climate change research by highlighting the following points.

Among the model simulations which are 'credible':

- There is *no* one model that performs best in all respects.
- There is *no* one measure that fully evaluates performances.
- Dispersion in simulations is *not* trivially related to complexity, season, latitude etc.
- Differences between models are *much* greater than among ensemble members from one model.

Among the model results which are 'credible':

- even those from 'good' models may *not* be 'useful' for your application;
- they have a shelf life of ≤ 5 years; and
- they are 'like children': need lots of attention, discipline and continual care.

These evaluation statements may seem somewhat 'weak' to someone from outside the climate modelling community, but they represent a rather significant breakthrough in terms of widespread agreement on what constitutes a good model assessment process.

18.3.2 MECCA's contribution to international understanding

MECCA was perceptive in anticipating some of the most critical aspects of climate science in the early 1990s. In particular it was an early sponsor of experiments on (see also Table 2.2):

1. greenhouse gases other than CO_2 (Wang et al.);
2. long integrations (Schneider & Kinter);
3. sulphate aerosols (Penner & Taylor);
4. regional simulations (Giorgi et al.); and
5. high resolution AGCMs (Williamson).

Although MECCA exhibited considerable foresight in identifying important topics for study it was only partially successful in attracting the leading modelling groups into its activities, for example there was little or no involvement with GFDL, the Hadley Centre or German and French modelling groups. In addition MECCA's foresight was a handicap in some senses because the experiments which it sponsored occurred a little too early to be 'state-of-the-art' for the Second IPCC Scientific Assessment: the long timescale variability is being taken from the GFDL and the Max Planck Institute integrations and the impact of sulphate aerosols from the Hadley Centre and the Max Planck Institute. Results from the

Wang et al. 'MECCA' integrations were included in the 1992 IPCC Update and the Giorgi et al. simulations were among the only regional simulations in the Second IPCC Scientific Assessment.

Major sources of uncertainty in climate assessment today (the end of 1995 as Second Scientific Assessment of IPCC goes to press) in the context of its three Working Groups (WGs) are:

1.   'Validity' of climate model predictions (WG I)
      – How well do they compare to observations?
      – How big are the ocean flux adjustments used in some coupled models?
      – Do they matter if used?
      – Do models with flux adjustment give the same predictions as same/other models without?
      – Is the assumption of linearity of change (and hence unimportance of the large fluxes that have to be used) valid?
2.   Regional climate prediction and hence impacts (WG II)
      – Do all methods e.g. embedding cf. downscaling give same results?
      – Do any methods give 'valid' results?
      – Are GCM 'good enough' to use as basis for regional prediction?
      – What impact do sulphate aerosols have on regional predictions?
3.   Tools and processes used in policy evaluation (WG III)
      – Can results from GCMs be 'simplified' for policy use?
      – Can ecology, economics, energy projections be (simply) linked?

In trying to assess the success of MECCA it is useful to ask to what extent were these sources of uncertainty foreseeable when the MECCA experiments were selected and MAT tasks undertaken; and how has MECCA contributed to international understanding of these and other issues.

Validating climate models

The issue of validation or evaluation of the performance and predictions of climate models has, rather surprisingly, only recently come to the fore. The IPCC failed even to recognise the potential sensitivity to the term 'validation' in naming its chapters. Validation is not an implication of truth nor a 'verification', since to be completely consistent with observations does not (yet) imply correctness. Validation is an estimation of the degree of correspondence between simulations and the real world (Oreskes et al. 1994).

MECCA shares this lack of foresight with the IPCC having also failed to emphasise validation as part of its programme. Certainly, the individual MECCA investigators did not undertake fewer evaluations of their models and results than the rest of the international community, but neither did they undertake more.

As most of the MECCA simulations were undertaken with mixed-layer ocean sub-models, the issue of flux adjustment was not as critical to the results as it has since become for the

IPCC. Indeed, MECCA should be commended for its foresight in sponsoring what was then the only coupled ocean-atmosphere GCM which does not employ flux adjustment to force the present-day climate to approximate reality (i.e. the Washington and Meehl experiment).

Regional climate prediction and impacts

MECCA's interest in regional climates and their assessment was underlined by its co-sponsorship (with the IPCC) of a process of 'evaluation of regional climate simulations' (Cubasch et al. 1994). The aims of this project as set out in the IPCC terms of reference were to:

1. assess the performance of current GCMs in simulating contemporary climate and to intercompare the response of the different GCMs to increasing CO_2 on regional scales;
2. establish a methodology for the conduct of similar comparisons in the future;
3. encourage the wider analysis and use of data from GCMs in countries and by suitable organisations which are separate from the modelling centres themselves; and
4. through the assessment, suggest possible improvements to improve the quality of model simulations of regional climates and to reduce uncertainties.

MAT supported this regional evaluation and, in conjunction with the development of the MECCA Protocol, made the MADD and IPCC archive (Table 2.5) available to a very wide range of international scientists and climate assessors.

One of the major problems faced in applying GCM projections to regional impact assessments is the coarse spatial scale of the numerical model results. 'Downscaling', as the production of increased temporal and/or spatial resolution climates from GCM results has come to be termed, has two basic forms: statistical and modelling. In the model-based downscaling, a high resolution, limited-area model is run off line using boundary forcing from the GCM (e.g. Giorgi & Mearns 1991). Statistical downscaling can take many forms ranging from a standard interpolation (e.g. Gaussian filtering, kriging) or the application of Model Output Statistics (MOS). These methods generate higher spatial resolutions. Although GCMs typically have high temporal resolutions, downscaling has been applied in the time domain also because GCM results are found to have rather poor temporal characteristics when examined at timescales less than about a month (e.g. Reed 1986).

Several methods have been adopted for developing regional GCM-based scenarios at sub-grid-scale, including assignment of the nearest grid box estimate (e.g. Croley 1990); objective interpolation (e.g. Parry & Carter 1988; Cohen 1991); statistical analysis of local climatic fields (e.g. Wilks 1992; Barros & Lettenmaier 1993); the merging of several scenarios based on expert judgement (e.g. Pearman 1988; CIG 1992); and the use of mesoscale models embedded within GCMs (e.g. Giorgi & Mearns 1991; Mearns et al. 1994). Root and Schneider (1995, p. 337) discuss some of the difficulties of embedding: 'If, however, the GCM's wind pattern is not very accurate in the region of embedding, then the mesoscale model will translate the error in its boundary conditions into a higher resolution erroneous pattern.' Tanaka and Hasegawa (1996) show that the results from a regional model embedded in a coarse resolution GCM (the MECCA technique) will have a fundamentally different form from results derived using a higher resolution global model.

MECCA can be applauded for two aspects of its consideration of regional climate prediction. Several MECCA members stipulated that a portion of the experiments cover one or more geographic domains pertinent to their region of the world. This resulted in a series of embedded mesoscale modelling exercises which have achieved considerable publicity and have been included in the Second IPCC Scientific Assessment.

The negative aspect of the Consortium-demanded regional emphasis was highlighted at the May 1995 Policy Committee meeting. During the course of this meeting, it was made clear that the evaluation by MAT of one regional simulation was unwelcome. This was a surprising outcome because the analysis offered to MECCA specifically addressed one aspect of the necessary regional evaluation: whether poor GCM forcing can negatively influence the results from embedded mesoscale models. Landsea (1995) shows that the Asian regional simulations by Hirakuchi and Giorgi (1995) contrived to render a poor GCM-based regional climate into a worse, but higher resolution, local climate. It is interesting to note also that at the Third International Conference on Modelling of Global Climate Change and Variability, hosted by the Max Planck Institute in September 1995, several speakers commented with regard to an EC-sponsored comparison of regional embedded model simulations from the Hadley Centre (RegCM), Max Planck Institute (HIRHAM), and Meteo France's variable resolution GCM with high resolution over Europe, that: 'regional climate prediction requires first and for all good global prediction by coupled climate models—without it regional simulations are useless' (Bengtsson 1995, pub. comm.). The summary statement made was: 'therefore, at the present state of development it seems that the high resolution climate change scenarios that can be produced by the dynamical regional techniques tested here are not yet accurate enough to be used for meaningful impact studies' (Machenhauer 1995, pub. comm.).

Tools for use in policy advice

Although the impact modellers initially made rather modest requests of their climate modelling colleagues (e.g. Robinson & Finkelstein 1991), they are becoming more demanding. However, they see themselves, and are seen as, being prepared to use 'old' information (e.g. IPCC I cf. IPCC II in 1990) and 'making do' with either out-of-date model results, or less than complete explanations of output offered (e.g. Tegart & Sheldon 1993).

It is probably true to say that the majority of numerical climate modellers remain more interested in understanding the climate system and how it works than in providing results to the climate impacts assessment community. Some, though by no means all, are quite hostile to the suggestion that they become 'service providers', which is part of the reason why many of the GCM experiments were (and still are) undertaken with little or no thought that the results might be used in this way. For example, most (but not all) modelling groups produce daily 'data drops' during their model runs which usually include a mixture of instantaneous, accumulated and maximum and minimum information for a large number of variables of interest to the modellers. Some GCM groups, on the other hand, only produce monthly averages of variables especially when computer storage is a major issue. Usually GCM data are presented in the model's own unique coordinate system and are then interpolated to pressure coordinates in later post-processing. Any potential user of 'raw' model history data must work together with the GCM group to do these interpolations before they can obtain the variables sought. In summary, model output is not designed to be readily accessible. Moreover, there is typically a very large amount of such data, far exceeding the requirements listed by potential impact modellers. In addition, many GCM modelling groups have developed their own 'in-house' postprocessors that are 'non transferable' so that even if the impacts community had access to the supercomputer archive they might not be able to use the necessary tools.

MAT identified policy advice as a major focus of its activities (Table 18.4). The initiation of the MECCA Protocol was an important achievement in response to this goal. In addition, the MECCA Consortium specifically asked MAT to work with members of a number of EPRI-funded 'integrated assessment' teams in order to facilitate the generation of 'reduced form' models for policy use. The extent to which MECCA has been successful in inputting ideas and results to the arena of integrated assessment cannot yet be fully evaluated. There have been some highly praised successes: Mendelsohn, Nordhaus and Shaw (1994) used the MAT archive results to develop a new type of economic tool for integrated assessment. On the other hand, the whole philosophy underpinning integrated assessment modelling is also being questioned (e.g. Kandlikar et al. 1995).

18.4 Summary

MECCA was ambitious in that it tried not only to foster interdisciplinary research, communication and understanding, but that it also involved multinational industrial participation in decisions concerning and sponsorship of open-ended climate research projects. The ideas, processes, research accomplishments and long-term collaborations that have grown out of the MECCA project have, in some ways, proved to be a commendation of the MECCA aims and goals as originally stated in 1991. In many ways MECCA was a considerable success.

MECCA's weaknesses and failures are no greater than those of, for example, IPCC and AMIP. In retrospect, the planning seemed to be too weak; good questions were posed but proved impossible to answer from the experiments and analysis conducted. There seemed to be too little discussion between Consortium members; among investigators; between these groups and with the Analysis Team.

On the other hand, MECCA's gains and achievements are considerable. The MECCA project represented an opportunity for the Consortium members to become partners and participants in state-of-the-art global climate change research at an international level, a task that would be difficult to accomplish on their own.

MECCA was recognised as a non-governmental organisation and, as such, participated in UNCED in 1992 and in the early stages of lead author selection and chapter content discussion for IPCC WG I in 1993. In addition, MECCA co-sponsored (with IPCC WGs I and II) the first international regional climate intercomparison in 1993/4 and the workshop on climate scenarios in 1995. Benefits were two-way: MECCA members gained visibility and exposure to the international science and policy processes while other agencies obtained some insight into the goals and aims of members and the consortium as a whole.

MECCA was in the vanguard of model intercomparison, evaluation of uncertainty, integrated assessment and policy development. It must be applauded for establishing and maintaining a four-cornered tension (illustrated in Figure 18.1) for over four-and-a-half years. This balance between creative tension and continuing dialogue must be a model for future climate change assessment.

That MECCA was unique and worthwhile is unchallengeable. This chapter, written by two leaders of the Analysis Team, has been almost as difficult to develop as the analysis effort itself. We have felt and still feel, at one and the same time, frustrated and elated by the opportunities and challenges of MECCA. We feared intervention and direction by the Consortium members and, when it did not arise, we criticised the lack of leadership and have even, sometimes, doubted their interest. We have challenged individual scientists and have been chastised for our efforts; we have begged for replies to emails and persisted where others might have given up; we have made enemies but we have also made many friends. The members of MAT have been mutually supporting and have worked exceedingly well as a group. Overall, we have all learned a great deal from the 'experience of MECCA'.

Acknowledgments

This paper is CIC contribution number 95/34.

References

Barros, A. and Lettenmaier, D.P. (1993) Dynamic modeling of the spatial distribution of precipitation in remote mountainous areas. *Monthly Weather Review*, **121**, 1195–1213.

Cess, R.D., Potter, G.L., Zhang, M.-H., Blanchet, J.P., Chalita, S., Colman, R., Dazlich, D.A., Del Genio, A.D., Dymnikov, V., Galin, V., Jerrett, D., Keup, E., Lacis, A.A., Le Treut, H., Liang, X.-Z., Mahfouf, J.-F., McAvaney, B.J., Meleshko, V.P., Mitchell, J.F.B., Morcrette, J.-J., Norris, P.M., Randall, D.A., Rikus, L., Roeckner, E., Royer, J.-F., Schlese, U., Sheinin, D.A., Slingo, J.M., Sokolov, A.P., Taylor, K.E., Washington, W.M., Wetherald, R.T. and Yagai, I. (1991) Interpretation of snow-climate feedback as produced by 17 general circulation models. *Science*, **253**, 888–892.

Cess, R.D., Zhang, M.H., Potter, G.L., Alekseev, V., Barker, H.W., Cohen-Solai, E., Colman, R.A., Dazich, D.A., Del Genio, A.D., Dix, M.D., Dymnikov, V., Esch, M., Fraser, J.R., Galin, V., Gates, W.L., Hack, J.J., Ingram, W.J., Kiehl, J.T., Le Treut, H., Lo, K.-K., McAvaney, B.J., Meleshko, V.P., Morcrette, J.-J., Randall, D.A., Roeckner, E., Royer, J.-F., Schlesinger, M.E., Sporyshev, P.V., Timbal, B., Volodin, E.M., Taylor, K.E., Wang, W. and Wetherald, R.T. (1996) Cloud-Climate Feedback as produced by Atmospheric General Circulation Models: An Update. *Journal of Geophysical Research*, **101**, 12791–12794.

Climate Impacts Group (CIG) (1992) *Climate change scenarios for the Australian region*, issued Nov. Victoria: CSIRO Division of Atmospheric Research.

Cohen, S.J. (1991) Possible impacts of climatic warming scenarios on water resources in the Saskatchewan River sub-basin, Canada. *Climatic Change*, **19**, 291–317.

Croley II, T.E. (1990) Laurential Great Lakes double-CO2 climate change hydrological impacts. *Climatic Change*, **17**, 27–47.

Cubasch, U., Meehl, G. and Zhao, Z.C. (1994) IPCC WG I Initiative on Evaluation of regional climate simulations, summary report, prepared for IPCC and MECCA, Aug.

DEST (1995) *A Guide to the UN Framework Convention on Climate Change.* Canberra: Department of Environment, Sport and Territories.

Dumenil, L. (ed.) (1996) Proceedings of the Third International Conference on Modelling of Global Climate Change and Variability. Max Planck Institute, September 1995.

EPA (1989) *The potential effects of global climate change on the United States, Report to Congress*, edited by J.B. Smith and D. Tirpak, USEPA, Dec.

Gates, W.L. (1992) AMIP: The Atmospheric Model Intercomparison Project. *Bulletin of the American Meteorological Society*, **73**(12), 1962–1970.

Giorgi, F. and Mearns, L.O. (1991) Approaches to the simulation of regional climate change: A review. *Review of Geophysics*, **29**(2), 191–216.

Henderson-Sellers, A., Yang, Z.-L. and Dickinson, R.E. (1993) The Project for Intercomparison of Land-surface Parameterization Schemes (PILPS). *Bulletin of the American Meteorological Society*, **74**(7), 1335–1349.

Henderson-Sellers, A., Howe, W. and McGuffie, K. (1995) The MECCA Analysis Project. *Global and Planetary Change*, **10**, 3–21.

Hirakuchi, H. and Giorgi, F. (1995) Multi-year present day and $2 \times CO_2$ simulations of monsoon climate over eastern Asia and Japan with a regional climate model nested in a GCM. *Journal of Geophysical Research*, **100**, 21105–21125.

Houghton, J.T., Jenkins, G.J. and Ephraums, J.J. (eds.) (1990) *Climate Change. The IPCC Scientific Assessment.* Cambridge University Press.

Houghton, J.T., Callander, B. and Varney, S.K. (1992) *Climate Change 1992: The Supplementary Report to The IPCC Scientific Assessment.* Cambridge University Press.

Kandlikar, M., Risbey, J. and Patwardhan, A. (1995) Assessing integrated assessments: the crisis of quality, discussion paper presented at the Snowmass Meeting on Integrated Assessments, Snowmass, CO, 16–23 August.

Landsea, C.W. (unpub.) Comments on 'Multi-year present day and $2 \times CO_2$ simulations of monsoon climate over eastern Asia and Japan with a regional climate model nested in a GCM'.

Mearns, L.O., Giorgi, F., McDaniel, L. and Brodeur, C. (1995) Analysis of the variability and diurnal range of daily temperature in a nested regional climate model: Comparison with observations and doubled CO_2 results. *Climate Dynamics*, **11**, 193–209.

Mendelsohn, R., Nordhaus, W.D. and Shaw, D. (1994) The impact of global warming on agriculture: A Ricardian analysis. *The American Economic Review*, **84**(4), 753–771.

McKenney, M.S. and Rosenberg, N.J. (1991) Climate data needs from GCM experiments for use in assessing the potential impacts of climate change on natural resource systems. Discussion Paper ENR91-15, Report to the U.S. Department of the Interior Task 3 under Cooperative Agreement No. 4-01-001-88-CA-37, Resources for the Future.

Oreskes, N., Shrader-Frechette, K. and Belitz, K. (1994) Verification, Validation, and Confirmation of Numerical Models in the Earth Sciences. *Science*, **263**, 641–646.

Parry, M.L. and Carter, T.R. (1988) The assessment of effects of climatic variations on agriculture: aims, methods and summary of results. In *The Impact of Climatic Variations on Agriculture, Volume 1: Assessments in Cool Temperate and Cold Regions*, edited by M.L. Parry, T.R. Carter and N.T. Konijn, pp. 11–95. Dordrecht: Kluwer Academic Publishers.

Pearman, G. (ed.) (1988) *Greenhouse: Planning for Climate Change.* Melbourne: CSIRO.

Reed, D.N. (1986) Simulation of time series of temperature and precipitation over eastern England by an atmospheric circulation model. *Journal of Climatology*, **6**, 233–253.

Robinson, P.J. and Finkelstein, P.L. (1991) The development of impact-oriented climate scenarios. *Bulletin of the American Meteorological Society*, **72**(4), 119–122.

Root, T. and Schneider, S.H. (1995) Ecology and climate: Research strategies and implications. *Science*, **269**, 334–340.

Taplin, R. (1995) Climate science and politics: The road to Rio and beyond. In *Climate Change, People and Policy: Developing Southern Hemisphere Perspectives*, edited by A. Henderson-Sellers and T. Giambelluca. Chichester: Wiley & Sons.
Tegart, W.J.McG. and Sheldon, G.W. (eds.) (1993) *Climate Change 1992. The Supplementary Report to the IPCC Impacts Assessment*. Canberra: Australian Government Publishing Service.
Wilks, D.S. (1992) Adopting stochastic weather generation algorithms for climate change studies. *Climatic Change*, **22**, 67–84.

CHAPTER 19

MECCA CONSORTIUM OVERVIEW AND MEMBERS' VIEWS

Wendy Howe and MECCA Consortium members

19.1 MECCA Consortium members

As previously mentioned, the MECCA Consortium comprised representatives of electric utility companies from France, Italy, Japan, The Netherlands and the USA, as well as academic institutions and government agencies. However, not all consortium members have been actively involved in the MECCA project; for example no EPA representative has ever attended a MECCA Policy Committee meeting nor have they had (or taken) any opportunity to speak with or contact either the Principal Investigators or the Analysis Team members.

19.1.1 Expectations

During the lifetime of the MECCA project, there were occasions when some members of MAT and the Consortium were able to discuss matters concerning the project. One example of these discussions focuses on the reasons the various sponsors became members of the MECCA Consortium. The questions asked of consortium representatives were:

1. Why did their organisation become a member of MECCA?
2. What did they hope to achieve by joining MECCA?
3. How do they feel now (1995/6) after MECCA's conclusion?

Not all the Consortium members chose to answer these questions, some answered only a few questions and some all three. Overall, however, the responses underline the valuable international cooperation and understanding links forged through this project.

19.1.2 Overview of responses

The responses to the first question may be summarised as follows:

- All of the responders said that they were aware of the global climate change problem.
- All expressed a desire to share in state-of-the-art research at an international level.
- All mentioned an interest in regional impacts of climate change.

The responses to the second question indicated that all of the members valued an awareness of current research directions offered by the MECCA Consortium. Several reiterated their desire for regional impact work while another member felt that they had an equitable role in decision-making processes even though they may contribute less cash than other Consortium members.

Responses to Question 3 were generally encouraging with most sponsors indicating that they valued their Consortium membership, the results of the project and offered suggestions for future directions. Details of the full responses may be found in the following sections.

19.2 Responses

19.2.1 Jean-Yves Caneill, EdF

Question 1. Why did their organisation become a member of MECCA?
EdF became involved in climate modelling and analysis 4–5 years ago for two reasons:

1. they were interested in the global climate change problem; and
2. they were interested in climate impacts even if the climate is not changing, for management plans.

EdF already had contacts with EPRI on other research projects and so it was EPRI that approached EdF to join MECCA.

EdF felt that 'MECCA represented an opportunity to share research on an international level', especially with different partners concerned by research in the electricity sector, for example CRIEPI and EPRI.

Question 2. What did they hope to achieve by joining MECCA?
EdF felt that an awareness of research by MECCA and the use of MECCA results (like those of Filippo Giorgi), with their own impact studies were important benefits of joining MECCA.

The Climate Group at EdF were working on deriving scenarios for hydrological models (river flow) and would like to use some of the MECCA project results in hydrological models developed by another group inside the Research Division, especially for impacts of global warming on surface temperature of rivers and riverflow (EdF use rivers as a source of cooling water for power plants).

[Jean-Yves sees MAT activities as a link between GCMs and impact models (assessing the propagation of uncertainties as in Catherine Ciret's work detailed in Chapter 16). Uncertainty assessment in climate model results is considered as an important issue.]

Question 3. How do they feel now (1995/6) after MECCA's conclusion?
It is very clear that the way the MECCA Consortium was created was quite original and somewhat unusual, at least in this scientific field. Historically the MECCA Consortium was

proposed by EPRI and UCAR by the end of 1990 to develop a strategy for understanding the origin of uncertainties in climate modelling questions and their consequences on the assessment of future climate constrained by greenhouse gas emissions. The formation of the Consortium, composed of members of the industrial sector concerned with energy production and of the academic research sector, was a response to accelerate the knowledge in the scientific field of climate with the idea that policy makers needed to get a better idea of the uncertainties present in the impact studies related to global warming. This process took place in the context of an increasing concern of global environmental questions on the international scene. The idea was to provide scientists with a dedicated computer to perform relevant climate simulation projects decided on a peer review basis with two planned phases. I think that the fact that MECCA formed and existed during five years is proof that this way of doing science can work. The industrial and academic research sectors can benefit from each other on clearly stated problems from an international point of view (in the future many issues could come from the international scene). This is not to say that the task was easy and we had all to face many problems during the life of the program, for example the bottom-up versus the top-down approach. However, the way we solved them showed that it was not impossible to do so and MECCA should be recognised as an example of an applied research action oriented to practical goals needing the support of fundamental research.

Additional comments

The achievements of MECCA: It is not my intention here to review in much detail all the MECCA projects but rather highlight some items and give my own view and judgment on the MECCA results. It is clear that when one reviews the different projects of Phases 1 and 2 of MECCA, the number of scientific achievements is quite impressive for such a short time. Many scientific studies were performed with different climate models, addressing key physical processes of the climate system. The ocean modelling was considered of particular importance and the projects chosen by MECCA obtained particularly interesting results. As the program was, however, oriented to impact assessment a specific concern has been given to that area in order to analyse the results of the projects along these lines. A technique of regionalisation (often used in weather forecasting) has been used and assessed for climate runs by one group, choosing different areas in the world for experimentation. MECCA hosted important projects on tropospheric chemistry and aerosols, subjects which started to become key issues during the course of the program. Lastly, it is important to mention the analysis phase of MECCA which has been undertaken at Macquarie University. It was recognised that it was a key issue in the MECCA process, and this activity permitted the achievement of a good synthesis of the whole work, given the fact that the program was run on an essentially bottom-up approach. The produced *Climate Change Atlas* is one of the products we can be proud to have. It is certain that it would have been beneficial if this analysis phase had been launched right at the beginning of the program. This would have solved the data availability for analysis before the runs were done, for instance, in order to be more efficient.

Lessons: bottom-up or top-down approach? Undertaking a project like MECCA needs some significant amount of 'top-down approach'. I feel that this is something that we have learned during the program. If one has only the will that scientific knowledge increases, a bottom-up approach can work very well. However, if one has some requirements or specific needs of the products of this research it is necessary to have the research done in an accepted framework. The difficult thing is to fulfil the conditions of this *accepted by both sides* framework. The scientists must accept that their results will be used in the context of impact studies and industry has to be aware of the difficulties of the research that cannot provide, in some cases, definitive assessments. One can see that when only the scientific community is involved in a top-down approach process, it can work very well (e.g. AMIP, FANGIO, etc.). In the present case it was more difficult, because on one side there were the industry needs and on the other side the scientists projects. One of the lessons of MECCA is that to guarantee the success of such an action, a clear compromise has to be made at the beginning of the program. The industrial sector has to express its needs and the scientists, who accept to play the game, must express where they do not want to go. From that the accepted framework can be established in order that the integrated analysis of the whole results is optimal to answering the questions raised.

Has MECCA been reactive? I would not like to give the impression that MECCA failed. I consider its achievements mainly as a success which is due to many things: the quality of the science work that has been accomplished, the originality of the framework (academic and industry) and its reactivity. Two examples can be given to illustrate this point. Initially it was agreed that the Phase 2 of MECCA would concern transient climate experiments. When we came to the point of deciding to launch this phase a quite different decision was eventually made. Projects were chosen with a special emphasis on the use of their results for impact assessment studies. We recognised that the final goal was important to accept the scientific projects together giving the PI's specific recommendations. This decision has been a turning point in the program. Something similar, but different in essence, happened when during the first phase the aerosol questions seemed to increase in interest in the scientific community. Although it had been agreed right at the beginning of the program, in a framework plan, that we would not study this area, it was decided that a specific 'call for proposals' was worthwhile to be launched on this topic. A project was funded which produced important inputs to the IPCC process. These two examples show that the program has reacted in the right way to external solicitations and that we have been able to achieve important contributions during these five years of the life of MECCA.

MECCA and the climate scientific community: I have the feeling that this point has been a hard one, not for the scientists participating in MECCA but for the scientific community in general and I think it is worthwhile commenting on this point. The counterpart of the originality of MECCA has been the fact that many scientists asked themselves questions about the interest of industry in conducting these kind of studies and why was it necessary to set up this consortium rather than to invest directly in the governmental agencies which usually provide funds for research in the climate area. One major reason was that the integration of research from climate modelling science (e.g. the uncertainties in climate modelling) towards the impact assessment studies (uncertainties in assessing the consequences of possible global warming) was somewhat missing when MECCA started. One of MECCA's goals was to partially fill this gap thereby bringing an original contribution to the whole exercise. Taking into account the fact that most of the MECCA projects permitted scientists' contributions to the IPCC process, and that a specific action has been organised jointly with IPCC (a workshop on regional assessment of climate and a mandate to provide climate data to developing countries) I have the feeling that we have succeeded and that MECCA has played a significant part in the whole process.

In conclusion, I would like to make some final comments that could help give some guidelines for the future:

1. I think that taking into account the innovative framework we have set up with MECCA, we have accomplished a very interesting work. The best proof of this is the different chapters of this book describing the excellent scientific work accomplished under the auspices of MECCA and often jointly with other academic organisations.

2. Industry and scientific community can work together in an innovative way if they are both motivated to get answers to complex problems, but with clear questions. An important requirement is to state clearly what can be done realistically on the science side, and what are the expectations from the industry side.

3. On this subject, reactivity has an important role. The advancement of scientific knowledge gives surprises (good or bad!) and one can sometimes have the impression that things go slowly. This is inherent to the scientific process. When new opportunities or questions arise one has to take them into account in the strategy in order to be more efficient.

4. And lastly, another program which would address similar goals in the future will have to act more 'top-down' than MECCA did, especially to make the integrated analysis of the results easier. It is necessary to ensure right at the beginning of the program that all ingredients are present to pursue the defined goal.

My personal view of MECCA is that I have always been enthusiastic about the MECCA program and about this way of doing science. Five years after the beginning of the program I keep a positive view of the different accomplishments. The MECCA program has given the opportunity to two different communities to learn to discuss together and converge towards common goals. This published book will testify to the MECCA experience and the quality of the scientific results described in this book will show, more than a lengthy speech, that MECCA was necessary and useful.

19.2.2 Maurizio Sciortino, ENEA

Question 1. Why did their organisation become a member of MECCA?
The Italian Agency for New Technologies, Energy and Environment (ENEA) is a

governmental institution responsible for research and support to decision makers and local authorities on issues of public interest that require specific scientific knowledge. The main focus of ENEA's activities is on issues related to energy so, because of the possible implications of global climate change to the energy sector, ENEA recognised the relevance of the global change problem. The delayed response of the Italian scientific and policy community on the climate issue moved ENEA, in 1990, to develop an activity that could provide a significant contribution to the improvement of the level of understanding of the problem. The opportunity to join MECCA came in the framework of a scientific collaboration between ENEA and NCAR. It was immediately clear that to reach significant results, synergistic efforts were necessary. Initially MECCA had been perceived by ENEA as an opportunity to make more productive use of an already existing scientific collaboration. However, it became immediately evident that the value of sharing in this ambitious project was in improving the state of knowledge with other international partners. ENEA felt that MECCA was a unique opportunity to promote regional climate studies on the Mediterranean basin and more focused research on issues of practical interest.

Question 2. What did they hope to achieve by joining MECCA?
The principal initial motivation for ENEA's participation in MECCA can be identified in the improvement of the understanding of the climate problem through research projects. This motivation stimulated ENEA to participate actively with MECCA on scientific projects, focusing on issues addressing questions of particular relevance to Italy . The ambitious goals of narrowing uncertainties would probably have needed a scientific plan with a top down approach rather than a selection of experiments according to a bottom up approach. ENEA and the other Consortium members learned from experience which projects could meet both scientific and sponsors' priorities. The work of translating scientific results into useable products started with MECCA's Analysis Team. This has been a valuable contribution because it helped to reach a larger community of users and gave a much bigger visibility to the activity of the consortium.

In conclusion, ENEA feels that it is getting value for money from MECCA membership and thinks that a new strategy is needed to overcome the budget constraints that may limit the fruitful collaboration started with MECCA.

Question 3. How do they feel now (1995/6) after MECCA's conclusion?
The perception of the climate change issue has progressively risen in the public opinion in Italy. Nevertheless, the level of attention given to climate change issues by the media is still far from that reached, for example in the United States where the *New York Times* published 28, 33, 24, 69 and 36 articles related to the climate change in the years 1990, 91, 92, 93 and 94 respectively while the *Washington Post* published 61, 35, 39, 27 and 7 articles. The newspaper *Corriere della Sera* in Italy devoted more attention to the climate change issue published in the same years 18, 12, 52, 10 and 21 articles. In Italy the awareness of the problem and of its implications is probably still insufficient in the public and in the policy

maker community. The climate change issue is still perceived only as something that perhaps will happen in the future and attracts the attention of the public only in correspondence of international events like the United Nations Conference on Environment and Development (UNCED) in Rio de Janiero in 1992. The opportunity to use the environment as a new view to set up a new consciousness and economically sustainable development is not debated enough and the link between scientific results and social and economic implications still needs to be explored and understood both by the scientific and policy communities.

Probably one of the reasons for the misperception of the relevance of the climate change issue is that until now this issue has been addressed only by specialists in highly technical language with insufficient occasions to translate scientific results into statements understandable to a larger audience.

The commitments consequent to the ratification of international agreements do not seem to modify any order of priorities still. The context briefly and perhaps pessimistically/realistically sketched for the Italian situation, can help one to understand the interest that a project like MECCA has immediately obtained from ENEA.

ENEA is, in fact, a governmental institution in Italy whose main role is to assist public administration and to promote the application of the results of scientific research in the fields of new technologies, energy and environment. Incidentally, it is worthwhile mentioning that all these three fields of activity could potentially be involved in the climate change issue. ENEA saw in MECCA the opportunity of filling a gap between science and applied research that could help to plan better management of global problems. ENEA devotes special attention to the Italian territory and all the Mediterranean environment of which Italy, because of its position, is probably one of the most sensitive and vulnerable countries.

The MECCA program addressed some of the topics that have the highest priority on the ENEA agenda like the quantification of uncertainties and the improvement of scenarios at a regional level. The MECCA project has really been a pioneering initiative and tried to address some of the most pressing questions about climate change, connecting the climate research community with the impact assessment community that should potentially produce the answers to the questions of policy makers. ENEA recognised the importance of the priorities expressed by several bodies like IPCC and strongly recommended that the resources of MECCA should be oriented to answer those questions.

There have been some additional specific goals that ENEA wanted to pursue through MECCA related to environmental modifications that will potentially affect social and industrial sectors of the Italian society. One of these goals, still very far from being reached, is worthwhile mentioning. Italy ascribes, respective to many other countries, a very high priority to the conservation of its cultural and historical heritage because it is conscious of its mission to preserve for future generations the memory of what humanity has been in the past and the testimony of its ingenious work. Perhaps the most outstanding example of threat by environmental changes is the city of Venice which would be damaged or even destroyed by an increase in sea level. Saving Venice from recursive flooding requires the adoption of prevention and adaptation measures that need to be planned and evaluated from both economical and environmental points of view. The special attention devoted by ENEA to the Mediterranean area is justified by the inadequacy of the climate evaluations at this scale. In fact, global atmospheric and ocean climate models are not still able to describe the features of the Mediterranean climate realistically. The main effort that ENEA pursued through its participation in MECCA has been to set up new scientific collaboration and new resources to improve the state of knowledge of the Mediterranean climate and of its possible evolution. Even from this point of view MECCA fulfilled the initial expectations producing results that wouldn't have been available at this time for the assessment of impacts.

ENEA, having been involved in the elaboration of the 'first Italian national communication to the framework convention on climate change' realised the importance of impact studies for the elaboration of prevention, mitigation and adaptation strategies. ENEA, for its institutional role in support of policy makers, strongly supports the further development of approaches to climate change issues that combines climate modeling with impact studies in order to improve the quality of information needed to elaborate on more comprehensive strategies for a sustainable development.

19.2.3 James Young, SCE

Question 1. Why did their organisation become a member of MECCA?
It is a company policy to recognise the potential significance of greenhouse gas warming and belonging to MECCA was seen as an opportunity of involvement with global climate change research.

SCE has an environmental research group working on about 20 research projects. These projects fall into three categories:

1. land-air, water quality and associated pollution;
2. the resources group: ocean water cooling, endangered species and habitat preservation; and
3. environmental quality strategies, e.g. risk assessment, cost/benefit analysis, climate change research (started about 5 years ago).

The research program is to help SCE formulate environmental plans and ongoing research, especially impacts research on climate change for industry is needed. Also SCE needs to demonstrate the value of the research program to its ratepayers.

Question 2. What did they hope to achieve by joining MECCA?
Most American electricity companies belong to EPRI but SCE didn't at the beginning of the MECCA program because they didn't feel that they would get value for money, i.e. the research would not be appropriate to SCE. In the 1980s EPRI spent a lot of dollars on acid rain problems and SCE are more interested in current projects rather than long term strategies. (It costs about one-third of SCE's research budget to belong to EPRI (US$15–18 million). So when SCE joined MECCA they weren't part of EPRI.)

SCE's current role in MECCA is satisfactory given that they have equitable voices in decision-making processes, even though they may contribute less dollars than other Consortium members. SCE also felt they were getting value for money because they are able to direct research to appropriate areas of interest.

Question 3. How do they feel now (1995/6) after MECCA's conclusion?
The Southern California Edison Company (Edison) recognised in the late 1980s the potential for serious regional climatic impacts due to a change in the global climate system. Of specific interest were changes in precipitation amounts, runoff patterns and temperature regimes that could significantly alter the way already tight water supplies are distributed in the western United States. Increasing regional temperatures would lead to an increased demand for electricity during summer months when regional electrical demand is already at its highest levels, potentially requiring new generation sources to be found. Increased variability in annual precipitation and uncertainty about the timing of the spring runoff could significantly affect the reliability of the regions most cost-effective and environmentally benign form of generation, hydroelectric power.

These concerns led Edison to develop a two-pronged approach to prepare for the possibilities of climate change. On the technical side, Edison embarked on a research program to explore variability in the regional climate of the western United States under various global climate change scenarios and to understand the uncertainties inherent in global climate models (GCMs). This technical effort was paired with a public policy position that supported an immediate effort to implement so-called 'no regrets' programs that would, among other economic and environmental benefits, reduce emissions of greenhouse gases at little or no incremental cost. Among the most effective examples of these types of programs are energy efficiency improvement programs and utility demand side management programs.

MECCA provided an opportunity for Edison to augment its existing climate change research in several important ways. First, MECCA was focused on understanding and quantifying the variability and uncertainties in GCMs. This objective is one few other research groups were addressing in a systematic manner. Furthermore, the results of the Consortium's work are to be presented to policy makers involved in the debate on how best to address the climate change issue. Second, a number of MECCA experiments investigated regional effects of climate change. These experiments provided a context in which to evaluate the seriousness of regional impacts in the western US that Edison's climate change research is investigating.

MECCA's ultimate value to climate change research may well be in its demonstration that barriers exist between the climate and impact modeling communities that, in some cases, prevent the two groups from working together effectively. The MECCA Protocol, or some similar means for dialogue between these two groups, is essential to ensure the scientific community provides the best estimates possible future climate impacts. Finally, in these days of ever shrinking research budgets both in the public and private sector, MECCA has shown that large research projects can be successfully undertaken, even when funded by a consortium of government agencies, non-government organisations and private industry each leveraging their own research funds. In fact, this paradigm may be the most suitable approach to tackle large multidisciplinary research programs whose objectives are shared by multiple organisations.

19.2.4 Shaw Nishinomiya, CRIEPI

Question 1. Why did their organisation become a member of MECCA?
CRIEPI participated in the MECCA studies because it felt that collaboration in world-wide networks are indispensable to climate change studies. It was also felt that to understand in detail the degree of influence of climate change, it is important to evaluate the uncertainties of the present global climate models and to encourage the improvement and reliability of climate prediction.

Question 2. What did they hope to achieve by joining MECCA?
In order to understand the potential impact of climatic change on Japan's electric utility industry, CRIEPI started studies on the prediction of climate change on a regional scale in East Asia. This was judged to be the best way to obtain appropriate and useful information about data and technique.

Question 3. How do they feel now (1995/6) after MECCA's conclusion?
There is still a high degree of uncertainty with regard to forecasting climate change related to global warming. There is a possibility that many irreversible impacts may occur. Thus, despite the uncertainty in climate change forecasts, there are many strong calls for implementing control measures as soon as possible. Measures to prevent global warming are unlike military preparedness to prevent war in that the required measures such as energy conservation and preventing depletion of resources are not minuses even if global warming does not occur in the future. Therefore, the implementation of control measures for global warming does not have any problems from this standpoint despite the uncertainty of climate change forecasts for this group.

However, if measures for preventing global warming beyond those actually required are implemented, this may prevent the economically developing nations from rising out of poverty. If the relationship between poverty and the population explosion are considered, implementing excessive measures for preventing global warming may have a double edged impact. Having said this, appropriate control measures for global warming should be implemented as soon as feasible.

Given this situation, climate change forecasts with high precision are extremely important for developing appropriate global warming control measures.

MECCA was the first international project on climate models. It was also the first primarily private sector project which involved joint private sector/government co-operation and the project had a number of successes.

Frankly speaking, MECCA's attempt was successful and there were no problems with MECCA's basic conception and objectives. But it ran into limits too. There were problems with implementation, that is in relation to funding, method, and management. This said, the issue is how to improve these points.

Starting with the lessons from MECCA, when considering the proposals for the new MECCA, large plans may be appropriate, it will be important to focus on the key basic points in the implementation plan. Starting with the feasible objectives, the project should be aggressively implemented by stages, desired results achieved, reported to the community, and expanded by stages. In other words, start small and expand.

We expect much from the next new MECCA.

19.2.5 Walter Ruijgrok, N.V. KEMA

Question 1. Why did their organisation become a member of MECCA?

The Laboratory for Environmental Research of KEMA is one of the principal environmental research centres in The Netherlands. Its mission is to provide state-of-the-art assessments of the impact on the environment, in the broadest sense, of energy use and supply. During the 1980s our research program slowly moved from more 'traditional' environmental sciences to fields which were more related to climate change issues. For two key areas (atmosphere/surface exchange processes and biological impacts on terrestrial vegetation) this slowly progressing redirection in research focus could be incorporated within ongoing efforts. However, such a redirection was virtually impossible within our budget and skills for our meteorologically oriented research projects. Yet, climate models form the only available tools for assessing the consequences on the Earth's climate of human activities. We considered it crucial to establish active links with the climate modelling community in order to get a better understanding of issues involved and to provide a background for our own research work. MECCA served as the vehicle to achieve this by offering an opening to climate modelling.

Question 2. What did they hope to achieve by joining MECCA?

What we hoped to achieve comprised essentially three reasons. First and most importantly, we hoped that MECCA's initiative might make a contribution to the scientific advancement of climate modelling research, in particular with respect to a better understanding of the nature of uncertainties in GCM results. The feeling was that getting such a better understanding would make a difference in the confused policy debate which still struggles with uncertainty. We realised that, although MECCA was a relatively small effort in this area and our group is only a 'beginner', the effort would be worthwhile, at least scientifically. The second expectation was to get from the MECCA experiments and MECCA's analysis efforts, up-to-date information which could be used in our work on assessments besides other available sources. The third aim was to establish links and co-operation with some of the climate modelling groups around the world. This process should serve to provide a better basis and background for our own research efforts we had started. To establish such links and co-operation, one member of our research staff joined the analysis team for nearly a year in Sydney and also worked earlier with a group in France.

Question 3. How do they feel now (1995/6) after MECCA's conclusion?
Here we will focus mainly on the achievement of MECCA and lessons to be learned from this program. When MECCA started in 1990, ambitious goals were set to achieve by its research. The founders hoped to accomplish a better understanding and quantification of uncertainties in climate modelling as well as an assessment of climate change induced by human activities. These ambitious goals have proven to be too large for MECCA to handle for a number of reasons. The most important reason was the inability of MECCA to attract the full support of the climate modelling community. Most of the world's leading groups did not show interest in participating, with the exception of NCAR. However, although NCAR's management and its parent organisation UCAR endorsed and supported the MECCA concept officially, most of NCAR's scientists regarded MECCA only as a welcome, additional source of funding and computer resources. A more or less similar attitude was present among many of the investigators who put in proposals to MECCA and got projects funded. As a result of the inability to attract leading groups and to carry out its original research plan, MECCA's research was defined in an ad hoc, incoherent and uncoordinated way, having some interesting and sometimes even exciting new pieces of research. Overall, however, more could have been gained.

A second important reason was the late start of the analysis team, which in itself was a good and very necessary activity to have within the program. On top of the late start, the analysis team had the difficult task of providing an integration and synthesis of results of individual projects which were not defined and carried out with such a purpose in mind. Therefore, this has resembled asking someone to make a nice jigsaw puzzle with pieces known not to fit. The group given this task was small and also relatively inexperienced. Yet, after struggling, they managed to come up with a few, but essential, achievements. One of the most important (but not very visible and usually unappreciated) accomplishment is the archive of daily data of climate model results. Putting such a set together was a painstaking effort but it may be a very valuable source for research projects focussing on impact assessments. In its present state (eight variables on a daily basis which are consistent for results from six models) it forms a unique source of data. On the other hand, the synthesis of results reached by the analysis team is still rather weak (although, in part, for understandable reasons).

A third reason involves the differing reasons with which organisations joined MECCA and their thoughts on guiding the program and setting goals to achieve. Some joined MECCA to establish an additional source of funding for research with little guidance only, while others wanted a more coherent program with a stronger selection of projects and a clear synthesis of results. These differing views were difficult to reconcile and many matters were left undecided. As a result MECCA turned out to be in many aspects a similar funding organisation of research to many others. Some of the reasons are understandable (such as the initial lack of interest for the program which prompted ad hoc decisions to be made). However, at later stages, chances were missed to redirect the program, research proposals and tasks for the analysis team, when the opportunities were there. At some points it seemed as if partners in MECCA had little interest in the program itself. An example is the lack of direct involvement in the analysis activities. Especially the larger research organisations which were members (NCAR, ENEA, CRIEPI) could have made essential contributions to this part of MECCA.

These notes give the impression that MECCA has failed as a concept (of industry research and academic institutions as partners in research) and in its results. However, most of MECCA was a success: its principal legacy is in the science it has fostered and accelerated. Some of the studies would not have been performed otherwise or at a much later stage in time. In addition to the mainstream results of MECCA there are some unique or early achievements. For the good of MECCA, it has also to be said that some of the points of criticism are not unique to MECCA. Establishing effective, well co-ordinated research programs is a tremendous effort, especially when 'integration', 'synthesis' and 'assessment' are part of it. Most of these programs show similar shortcomings as MECCA did, even at national levels where cultural differences count less.

There may be a future for a MECCA-type of program as originally envisioned. Within the issue of climate change there are still many issues waiting to be answered by the scientific community. On the other side, there is a great interest of society at large to have these questions answered. For instance, one point remains still very weak in climate change assessments and research: what will be the impacts of human-induced climate change? An 'answer' to this issue may be crucial to successful policy making in the near term. As measures to reduce emissions are seen by some parties involved to be costly, the willingness to act and pay will become increasingly difficult to maintain if risks are only quantified as 'warming', which is not perceived and appreciated by many of the public as a good indicator of climate risks. A co-ordinated research program (including state-of-the-art climate modelling in combination with impact modelling) with clear goals to achieve may make a contribution to the need of well informed assessments. Some of the lessons learned in MECCA and of its scientific legacies can be of use for such a program.

19.2.6 Bahram Nassersharif, NSCEE, ULV

The National Supercomputing Center for Energy and the Environment (NSCEE) joined the MECCA Consortium in 1992 under leadership of Dr David McNelis, Associate Vice President for Research at University of Nevada Las Vegas and Dr Bahram Nassersharif, NSCEE Director. NSCEE is funded by the US Department of Energy/Fossil Energy Research (DOE/FE) and the US Naval Research Laboratory (NRL) under the Strategic Environmental Research and Development Program (SERDP). Several of the MECCA projects (including Wang et al. and Penner et al.) were supported by the NSCEE supercomputing facilities and co sponsored by DOE/FE and SERDP.

Question 1. Why did their organisation become a member of MECCA?
NSCEE's primary interest in joining MECCA was to fulfil its mission in providing supercomputing resources and assistance to energy and environmental projects of US national significance. The close relationship between the mission and origins of NSCEE and MECCA's goals brought NSCEE into MECCA officially in 1992.

Question 2. What did they hope to achieve by joining MECCA?
NSCEE's role in MECCA has been to complement the MECCA dedicated numerical laboratory hosted at NCAR. NSCEE was able to provide supercomputing resources to several MECCA investigators through its existing agreements with the DoE/FE and the NRL.

Question 3. How do they feel now (1995/6) after MECCA's conclusion?
The relationship between NSCEE and MECCA has been mutually beneficial and has allowed both entities to advance their goals and missions. The high calibre of the scientific results achieved by the MECCA investigators and open publication of those results is a clear testimony to the success of the MECCA program.

19.2.7 Robert Serafin, NCAR

Question 1. Why did their organisation become a member of MECCA?
NCAR was very interested in the problem of climate variability and the uncertainties in climate models. We also welcomed the opportunity to make more computing available to conduct the simulations needed to address these issues and we were excited about the possibility of attracting scientists from around the world to launch a concentrated effort on these problems. Because of the above, NCAR strongly supported NSF's and NCAR's participation in and co-sponsorship of MECCA.

Question 2. What did they hope to achieve by joining MECCA?
It was NCAR's hope that, through MECCA, significant progress would be made in understanding better the strengths and weaknesses of global climate models and how regional climates might be affected within the framework of global climate change.

Question 3. How do they feel now (1995/6) after MECCA's conclusion?
I feel strongly that MECCA has been successful for several reasons:

1. It has been demonstrated that an international coalition of diverse organisations can join together for the pursuit of common scientific goals.
2. MECCA showed that world class scientists could be attracted to participate in the program and to leverage MECCA's computer contributions, with support from other sources. This outstanding document provides ample testimony to the scientific success of MECCA.
3. Some of the research undertaken within MECCA has helped to clarify the role of aerosols in climate change. This research has added credibility to climate simulations and has shown how aerosols can affect climate, globally and regionally.
4. Through MECCA, NCAR has established new and stronger scientific collaborations with scientists and organisations in Europe, Japan and Australia. Collaborations that continue beyond the formal execution and analysis phases of MECCA.

5. Beyond the scientific results, we demonstrated that NCAR could operate and make available a dedicated supercomputer for a diverse collection of researchers around the world, and that the machine could be utilised highly efficiently. About six terabytes of data, or roughly one-seventh of the data now resident in our mass storage system, were the result of MECCA simulations.

NCAR is happy to have had the opportunity to be a part of MECCA. International scientific collaborations on the many remaining problems related to climate will be absolutely essential for effective and timely progress to be made. MECCA has shown that such collaboration is indeed possible and that the best scientists are ready to take part.

Now, it is time to take the next steps. While the fundamental, grass roots science must continue, it is also appropriate to focus more on the use of scientific information to answer specific questions related to natural climate variability and human-induced climate change. Since these questions carry with them a global universality, it is appropriate and desirable that these more focused efforts also involve strong international collaboration. The past decade of global change research has yielded outstanding results. I am confident that the next decade will produce even more exciting and useful findings as humankind charts its way through the many climate-related challenges that it faces.

19.3 Comment on the MECCA concept by W.L. Gates

It is generally agreed that MECCA has made a useful contribution to the modeling of the climatic effects of increased atmospheric CO_2, and that it represented a timely and unique collaboration of the industrial, governmental and university research communities. As described in some detail by Anthes in Chapter 1, it is fair to say that the interest of UCAR in the formation of MECCA focused more on the possibility it offered to promote climate system modeling at NCAR through the provision of a dedicated supercomputer than it did on the stated goals of MECCA itself. UCAR/NCAR had been unsuccessful in achieving a consensus among its staff, its member universities and the supporting governmental agencies as to the need, priority and organisation of a nationally-coordinated program of climate system modeling, and had seen its efforts restricted to the sponsorship of conferences, reports and post-doctoral studies. Aside from having provided needed computer time to many NCAR scientists during the period 1991–1994, the most important legacy of MECCA to NCAR/UCAR is the justification it provided for the establishment of the Climate Simulation Laboratory in support of the NSF Climate Modeling and Analysis Program in 1995. While this facility will not in itself ensure the success (or even the existence) of a climate system modeling program, it clearly provides one of the prerequisites.

The stated goals of MECCA were compatible with those of the climate modeling community, but were in fact slightly different. Most climate modelers are primarily interested in improving (their) climate models, but have been less inclined to undertake studies leading to the quantification of the uncertainties involved in the modeled changes due to increasing greenhouse gases and even less interested in studies of climate impacts and their relationship to policy questions. MECCA's initial desire to focus on these latter issues was therefore met with a certain amount of skepticism, and the principal modeling groups capable of such studies did not show the expected interest in participating in MECCA. Rather than exclusively supporting long-term integrations of state-of-the-art coupled global models with realistic transient increases of greenhouse gases (and aerosols) as initially envisaged, the experimental plan for the use of the MECCA computational resources was therefore modified to include support for a wide range of climate modeling and analysis. In response to this broadening, numerous proposals were received in the first phase of MECCA for studies of the sensitivity of simulated climate to surface boundary conditions, resolution and physical parameterizations, and for studies of regional climate, and (relatively few) studies of equilibrium CO_2-induced climate change. Although a more coordinated set of experiments was hoped for in the second phase of MECCA, much the same mix of studies was supported. This unfocused nature of the MECCA experiments made their synthesis relatively difficult, and the goal of quantifying the uncertainty of climate change cannot be said to have been achieved in a completely satisfactory manner. The principal scientific legacy of MECCA is therefore the value of the individual scientific investigations.

Without the support of a group dedicated to the systematic analysis of the MECCA experiments, even the evaluation and intercomparison of those experiments with common elements would not have been accomplished. The establishment of the MECCA Analysis Team (MAT) was therefore a key decision of the MECCA programme. Although the MECCA simulations were unfortunately not designed with their comparative analysis in mind, the MAT succeeded in assembling results from six experiments with doubled CO_2 into an electronic data archive and a published atlas. These are valuable comparative documentations of the performance of the models (all but one of which were versions of the NCAR CCM). MAT is also to be congratulated for tackling the translation of model results into estimates of model uncertainty and hence into formats relevant to impacts and policy. The various questionnaires and the MECCA Protocol presented at length in Chapter 2 by Howe and Henderson-Sellers are valiant attempts to address these critical questions, which would otherwise have received scant attention in MECCA. These efforts constitute a diagnostic legacy of MECCA which I believe is as important as the scientific results themselves.

BIBLIOGRAPHY

Bates, G.T., Hostetler, S.W. and Giorgi, F. (1994) 'Multi-year simulation of the Great Lakes region with a coupled modeling system, *Preprints of the Sixth Conference on Climate Variations*, Nashville, Tenn., 23–28 January 1994, American Meteorological Society, Boston, Mass., pp. 399–407.

Campbell, G.C., Kittel, T.G.F., Meehl, G.A. and Washington, W.M. (1995) Low-frequency variability and CO_2 transient climate change. Part 2: EOF analysis of CO_2 and model-configuration sensitivity. *Global and Planetary Change*, **10(1–4)**, 201–216.

Dudek M.A., Wang, W.-C. and Liang, X.-Z. (unpub.) A General Circulation Model Study of Observed Stratospheric Ozone Depletion Between 1980 and 1990.

Erickson III, D.J., Oglesby, R.J. and Marshall, S. (1995) Climate response to indirect anthropogenic sulfate forcing. *Geophysical Research Letters*, **22**, 2017–2020.

Fan, Z. and Oglesby, R.J. (forthcoming) A 100–year CCM1 simulation of North China's hydrologic cycle. *Journal of Climate*.

Giorgi, F. (1995) Perspectives for regional earth system modeling. *Global and Planetary Change*, **10(1–4)**, 23–42.

Giorgi, F., Brodeur, C.S. and Bates, G.T. (1994) Regional climate change scenarios over the United States produced with a nested regional climate model. *Journal of Climate*, **7**, 375–399.

Giorgi, F., Hostetler, S.W. and Brodeur, C.S. (1994) Analysis of the surface hydrology in a regional climate model. *Quarterly Journal of the Royal Meteorological Society*, **120**, 161–184.

Giorgi, F., Bates, G.T., Hostetler, S.W., Brodeur, C.S., Marinucci, R. and Hirakuchi, H. (1993) 'Recent nested modeling experiments with the NCAR Regional Climate Model (RegCM), *Preprints of American Geophysical Union Fall Meeting*, San Francisco, Calif., 6–10 December 1993, American Geophysical Union, Washington, D.C., pp. 171.

Hack, J.J. (1993) Climate system simulation: basic numerical and computational concepts. In *Climate System Modeling*, edited by K. Trenberth, pp. 283–318. Cambridge University Press.

Hahmann, A.N., Ward, D.M. and Dickinson, R.E. (submitted) Land surface temperature and radiative fluxes response of the NCAR CCM2/BATS modifications in the optical properties of clouds. *Journal of Geophysical Research*.

Henderson–Sellers, A. (1993) A factorial assessment of the sensitivity of the BATS land-surface parameterization scheme. *Journal of Climate*, **6 (2)**, 227–247.

Henderson-Sellers, A. (1993) Continental vegetation as a dynamic component of a global climate model: A preliminary assessment. *Climatic Change*, **23**, 337–377.

Henderson-Sellers, A. (1993c) 'Tropical vegetation and climate change'. Preprint volume of the Fourth International Conference on Southern Hemisphere Meteorology, American Meteorological Society, pp. 328–329.

Henderson-Sellers, A. (1995) Adaptation to climate change: its future role in Oceania. In *Greenhouse '94*, edited by W.J. Bouman, G. Pearman and M. Manning, 349–376, Canberra: CSIRO.

Henderson-Sellers, A. (1996) Human effects on climate through land-use change. In *Future Climates of the World*, edited by A. Henderson–Sellers, pp. 433–475. Amsterdam: Elsevier.

Henderson-Sellers, A. (in press) Land–use change and climate prediction, invited paper for Joint Academies Symposium, 20–22 October 1993. In *Land Use and Land Cover in Australia*.

Henderson-Sellers, A. (1996) Climate modelling, uncertainty and responses to prediction of change. In *Mitigation and Adaptation Strategies for Global Change*, **1(1)**, 1–2.

Henderson-Sellers, A., Dickinson, R.E., Durbidge, T.B., Kennedy, P.J., McGuffie, K. and Pitman, A.J. (1993) Tropical deforestation: modeling local- to regional-scale climate change. *Journal of Geophysical Research*, **98 (D4)**, 7289–7315.

Henderson-Sellers, A., Dickinson R.E. and Pitman, A.J. (1993c) 'Atmosphere-landsurface modelling', *Proceedings of the International Congress on Modelling and Simulation*, edited by M. McAleer and A. Jakeman, MSSA/IMACS, pp. 31–36.

Henderson-Sellers, A. and Durbidge, T.B. (1993) 'Large-scale hydrological responses to tropical deforestation', *Proceedings of Sixth Scientific Assembly of IAMAP and Fourth Scientific Assembly of IAHS*: Symposium H1: Macroscale Modelling of the Hydrosphere.

Henderson-Sellers, A. and Hansen, A.-M. (1995) *Climate Change Atlas. Greenhouse Simulations from the Model Evaluation Consortium for Climate Assessment* Dordrecht: Kluwer Academic Publishers.

Henderson-Sellers, A., Holland, G.J., McGuffie, K., Tonkin, H. and Li, S. (1995). 'Implications of anthropogenic climate change for tropical cyclone intensity', *Proceedings AMS 21st Conference on Hurricanes and Tropical Meteorology*, American Meteorological Society, Boston, MA, pp. 234–237.

Henderson-Sellers, A., Howe, W. and McGuffie, K. (1995) The MECCA analysis project. *Global and Planetary Change*, **10(1–4)**, 3–22.

Henderson-Sellers, A., Howe, W., Nakken, M. and Yuan, X. (submitted) Global modelling to crop impacts: a case study in the climatic change information process. *Mitigation and Adaptation Strategies for Global Change.*

Henderson-Sellers, A. and McGuffie, K. (1994) Land–surface characterization in greenhouse climate simulations. *International Journal of Climatology*, **14**, 1065–1094

Henderson-Sellers, A. and McGuffie, K. (1995) Global climate models and 'dynamic' vegetation changes. *Global Change Biology*, **1(1)**, 63–75.

Henderson–Sellers, A., Zhang H. and Howe, W. (1996) Human and physical impacts of tropical deforestation. In *Climate Change: Developing Southern Hemisphere Perspectives*, edited by T. Giambelluca and A. Henderson-Sellers, pp. 259–292. Chichester: J. Wiley and Sons.

Hirakuchi, H. and Giorgi, F. (1995) Multiyear present–day and $2\times CO_2$ simulations of monsoonal climate over eastern Asia and Japan with a regional climate model nested in a general circulation model. *Journal of Geophysical Resources*, **100 (D10)**, 21105–21125.

Hirakuchi, H., Maruyama, K. Tsutsui, J. Kato, H. and Nishizawa, K. (1994) Prediction of $2\times CO_2$ climate over Eastern Asia — Regional Climate model (RegCM) nested in a GCM with mixed-layer ocean. CRIEPI Report, U94023 (in Japanese).

Hirakuchi, H., Maruyama, K., Tsutsui, J., Kato, H. and Nishizawa, K. (1995) Climate change over East Asia simulated by regional climate model nested in a GCM. In *Preprint of the 3rd Symposium on Global Environment*, Japan Society of Civil Engineers, pp. 211–217.

Holland, D.M., Ingram, R.G., Mysak, L.A. and Oberhuber, J.M. (1994) A numerical simulation of the sea ice cover in northern Greenland Sea. *Journal of Geophysical Research*, **100**, 4751–4760.

Holland, D.M., Mysak, L.A. and Oberhuber, J.M. (forthcoming) An investigation of the general circulation of the Arctic Ocean using an isopycnal model. *Tellus.*

Holland, D.M. Mysak, L.A. and Oberhuber, J.M. (forthcoming) Simulation of the mixed-layer circulation in the Arctic Ocean. *Journal Geophysical Research.*

Hostetler, S.W. and Giorgi, F. (1995) Effects of a $2\times CO_2$ climate on two large lake systems: Pyramid Lake, Nevada and Yellowstone Lake, Wyoming. *Global and Planetary Change*, **10(1–4)**, 43–54.

Krishnamurti, T.N., Bedi, H.S., Rohaly, G., Fulakeza, M., Oosterhof, D. and Ingles, K. (1995) Seasonal monsoon forecast for the years 1987 and 1988. *Global and Planetary Change*, **10(1–4)**, 79–96.

Liang, X.–Z., Wang, W.–C. and Dudek, M.P. (1995) Interannual variability of regional climate and its change due to the greenhouse effect. *Global and Planetary Change*, **10(1–4)**, 217–238.

Liu, Y., Giorgi, F. and Washington, W.M. (1994) 'Simulation of summer monsoon climate over east Asia with an NCAR regional climate model', *Preprints of the Sixth Conference on Climate Variations*, Nashville, Tenn., 23–28 January 1994, American Meteorological Society, Boston, Mass., pp. 393–394.

McAvaney, B.J., Dahni, R.R., Colman, R.A. Fraser, J.R., and Power, S.B. (1995) The dependence of the climate sensitivity on convective parameterization: statistical evaluation. *Global and Planetary Change*, **10(1–4)**, 181–200.

McCann, M.P., Semtner, A.J. and Chervin, R.M. (1994) Transports and budgets of volume, heat and salt from a global eddy-resolving ocean model. *Climate Dynamics*, **10**, 59–80.

McGuffie, K., Henderson-Sellers, A. and Zhang, H. (in press) Modelling climatic impacts of future rainforest destruction. In *Development and Destruction of the Tropical Rainforest*, edited by B.K. Maloney. Dordrecht: Kluwer Academic Publishers.

McGuffie, K., Henderson-Sellers, A., Zhang, H., Durbidge, T.B. and Pitman, A.J. (1995) Global climate sensitivity to tropical deforestation. *Global and Planetary Change*, **10(1–4)**, 97–128.

Marshall, S., Mann, M.E., Oglesby, R.J. and Saltzman, B. (1995) A comparison of the CCM1-simulated climates for pre-industrial and present-day CO_2 levels. *Global and Planetary Change*, **10(1–4)**, 163–180.

Marshall, S. and Oglesby, R.J. (1994) An improved snow hydrology for GCM's. Part I: snow cover fraction, albedo, grain size, and age. *Climate Dynamics*, **10**, 21–37.

Marshall, S., Oglesby, R.J., Larson J.W. and Saltzman, B. (1994) A comparison of GCM sensitivity to changes in CO_2 and solar luminosity. *Geophysical Research Letters*, **21**, 2487–2490.

Massaquoi, J. (1993) An Intercomparison of General Circulation Models over Africa: Implications for Future Climatic Change, report for the MECCA Analysis Team, Oct.

Mearns, L.O. (1995) Methodological issues concerning the determination of the effects of changing climatic variability on crop yields. In *Climate Change and International Impacts,* American Society of Agronomy Special Publication No. 59. ASA, Madison, Wi., pp. 123–146.

Mearns, L.O., Giorgi, F., McDaniel, L. and Shields, C. (1995) Analysis of daily variability of precipitation in a nested regional climate model: comparison with observations and doubled CO_2 results. *Global and Planetary Change*, **10(1–4)**, 55–78.

Mearns, L.O. and Rosenzweig, C. (1994) 'Use of a nested regional climate model with changed daily variability of precipitation and temperature to test related sensitivity of dynamic crop models', *Preprints of Fifth Symposium on Global Change Studies,* Nashville, Tenn., 23–28 January 1994, American Meteorological Society, Boston, Mass., pp. 142–155.

Mearns, L.O., Rosenzweig, C. and Goldberg, R. (forthcoming) The effect of changes in daily and interannual climatic variability on CERES-wheat: A sensitivity study. *Climatic Change.*

Meehl, G.A. (1994) Influence of the land surface in the Asian summer monsoon: External conditions versus internal feedbacks. *Journal of Climate, 7,* 1033–1049.

Meehl, G.A., Wheeler, M. and Washington, W.M. (1994) Low-frequency variability and CO_2 transient climate change. Part 3: intermonthly and interannual variability. *Climate Dynamics*, **10**, 277–303.

Meleshko, V.P. (1993) Cloud Parameterization Sensitivities in Climate Simulations, final report to the Model Evaluation Consoritum for Climate Assessment, Mar.

Mueller, P. K. and Hakkarinen, C. (1993) Confidence in Greenhouse Gas Forced Predictions: The Story of the Model Evaluation Consortium for Climate Assessment, paper presented at the Annual Meeting of the New Zealand Coal Research Association, Oct.

Oglesby, R.J. and Saltzman, B. (1992) Equilibrium climate statistics of a general circulation model as a function of atmospheric carbon dioxide. Part 1: Geographic distributions of primary variables. *Journal of Climate, 5,* 66–92.

Ojima, D.S., Parton, W.J., Chimel, D.S., Scurlock, J.M.O. and Kittel, T.G.F. (1993) Modeling the effects of climatic and CO_2 changes on grassland storage of soil C. *Water, Air and Soil Pollution*, **70**, 643–657.

Peck, S.C. and Teisberg, T.J. (1993) Global warming uncertainties and the value of information: an analysis using CETA. *Resource and Energy Economics*, **15**, 71–97.

Pitman, A.J., Durbidge, T.B., Henderson-Sellers, A. and McGuffie, K. (1993) Assessing climate model sensitivity to prescribed deforested landscapes. *International Journal of Climatology*, **13**, 879–898.

Pollard, D. and Thompson, S.L. (1994) Sea-ice dynamics and CO_2 sensitivity in a global climate model. *Atmosphere-Ocean*, **32**, 449–467.

Pollard, D. and Thompson, S.L. (1995) Use of a land-surface-transfer scheme (LSX) in a global climate model: the response to doubling stomatal resistance. *Global and Planetary Change*, **10(1–4)**, 129–162.

Roads, J. Marshall, S., Oglesby, R. and Chen, S. (submitted) Sensitivity of CCM1's hydrologic cycle to CO_2. *Journal of Geophysical Research.*

Schimel, D.S., Kittel, T.G.F, Ojima, D.S., Giorgi, F., Meterell, A., Pielke, R.A., Cole, C.V. and Bromberg, J.G. (1994) 'Models, methods and tools for regional models of the response of ecosystems to global climate change', *Proceedings of the International Workshop on Sustainable Land Management for the 21st Century,* edited by R.C. Wood and J. Dumannski, Vol. 2. Ottawa: Agricultural Institute of Canada.

Schneider, E.K. and Kinter III, J.L. (1994) An examination of internally generated variability in long climate simulations. *Journal of Climate*, **10**, 181–204.

Taylor, K.E. and Penner, J.E. (1994) Response of the climate system to atmospheric aerosols and greenhouse gases. *Nature*, **369**, 734–737.

Verdecchia, M., Visconti, G., Giorgi, F. and Marinucci, M.R. (1994) Diurnal temperature range for a doubled carbon dioxide concentration experiment: analysis of possible physical mechanisms. *Geophysical Research Letters*, **21**, 1527–1530.

Wang, W.-C., Dudek, M.P. and Liang, X.-Z. (1992) Inadequacy of effective CO_2 as a proxy to assess the greenhouse effect of other radiatively active gases. *Geophysical Research Letters*, **19(13)**, 1375–1378.

Wang, W.-C., Dudek, M.P., and Liang, X.–Z. (in press) The greenhouse effect of trace gases. In *Future Climates of the World*, edited by A. Henderson–Sellers, Chapter 16. Number 16 of the World Survey of Climatology. Elsevier Science Publishers.

Wang, W.-C., Liang, X.-Z., Dudek, M.P., Pollard, D. and Thompson, S.L. (1995) Atmospheric ozone as a climate gas. *Atmospheric Research*, **37**, 247–256.

Washington, W.M. and Meehl, G.A. (1993) Greenhouse sensitivity experiments with penetrative cumulus convection and tropical cirrus albedo effects. *Climate Dynamics*, **8**, 211–223.

Washington, W.M., Meehl, G.A., VerPlank, L. and Bettge, T. (1993) A world ocean model for greenhouse sensitivity studies: resolution intercomparison and the role of diagnostic forcing. *Climate Dynamics*, **8**, 321–344.

Williamson, D.L., Kiehl, J.T. and Hack, J.J. (1995) Climate sensitivity of the NCAR Community Climate Model (CCM2) to horizontal resolution. *Climate Dynamics*, **Vol.II**, 377–397.

Zhang, H., Henderson-Sellers, A., McAvaney, B.J. and Pitman, A.J. (1996) (submitted) Uncertainties in GCM simulations of tropical deforestation. *Climate Dynamics*.

Zhang, H., Henderson-Sellers, A. and McGuffie, K. (1996) Impacts of tropical deforestation I: Process analysis of local climatic change. *Journal of Climate.*, **9(7)**, 1497–1517.

Zhang, H., Henderson-Sellers, A. and McGuffie, K. (1996) Impacts of tropical deforestation II: The role of large-scale dynamics. *Journal of Climate*, **9(10)**, 2498–2521.

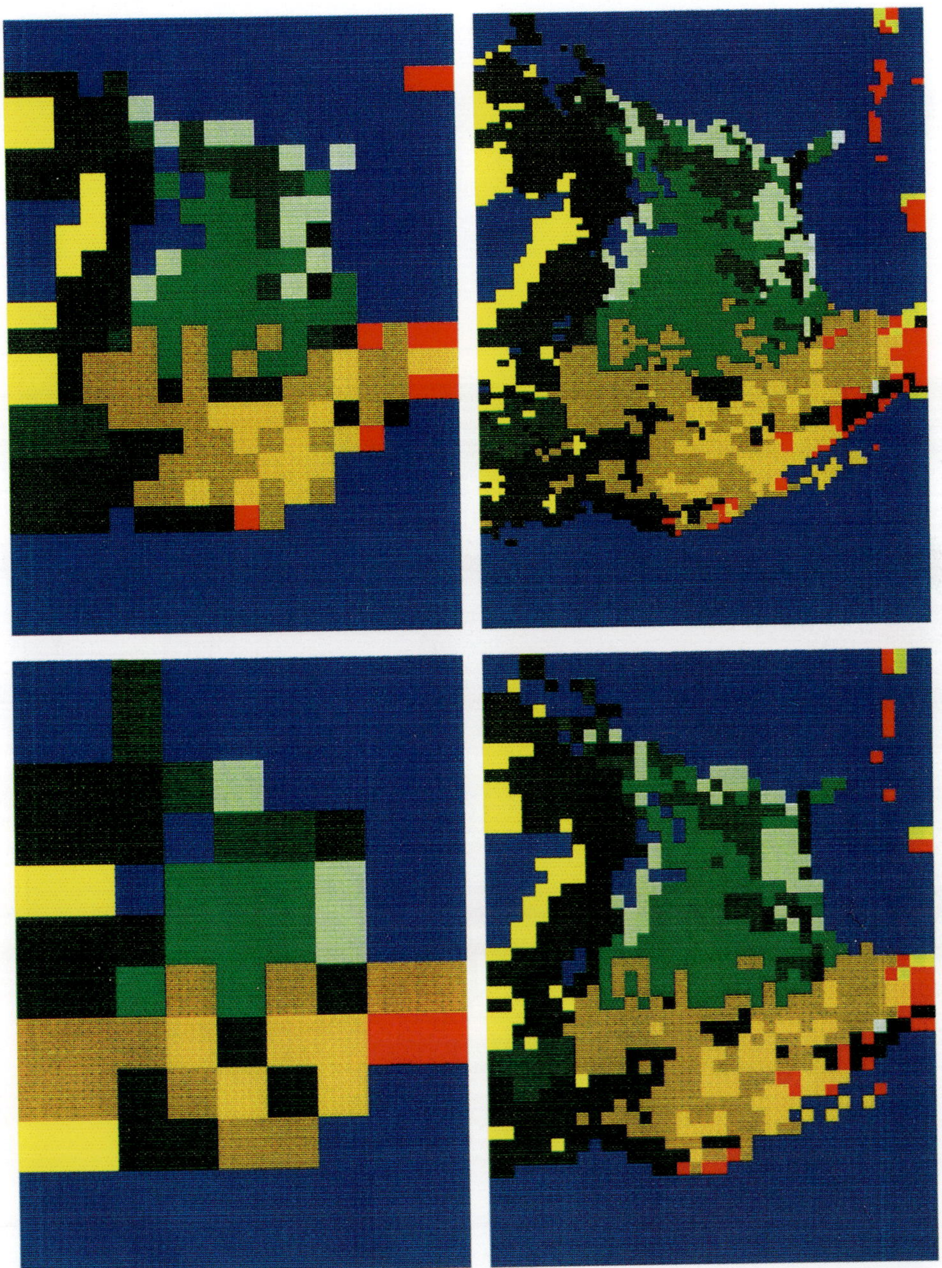

Colour Plate 1.1 Land use categories for North America at four horizontal resolutions (480km, 240km, 120km and 60km). This was one of the colour figures used in the First CSMI Workshop.

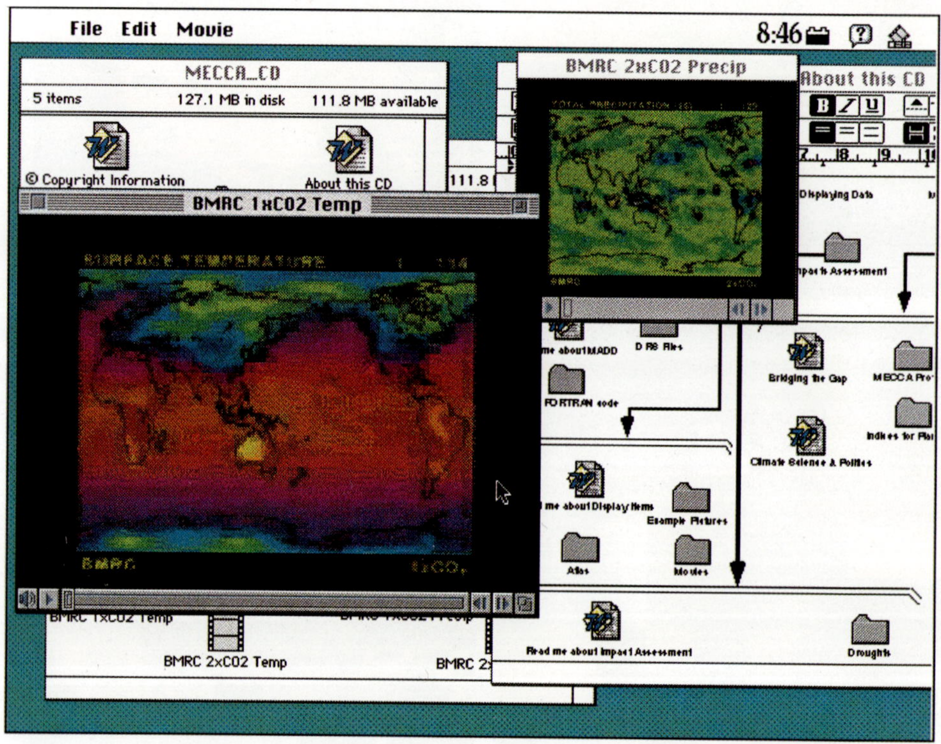

Colour Plate 2.1   Example display from the MECCA CD showing movie sequences of surface temperature and precipitation predicted for greenhouse conditions by the BMRC GCM and (behind) the CD file structure and contents.

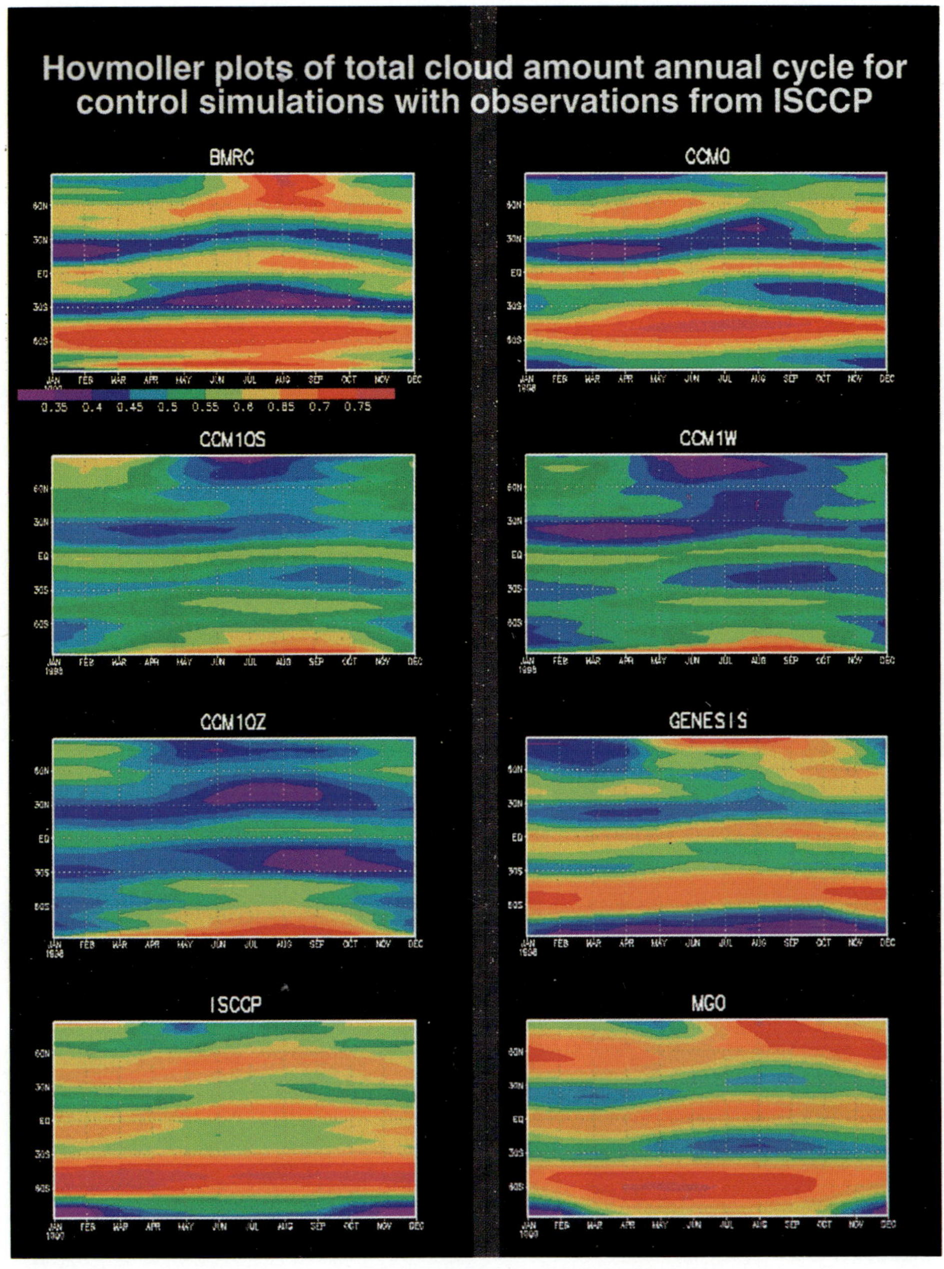

Colour Plate 3.1   Hovmöller plots of total cloud amount derived from a selection of MECCA models compared with observations from ISCCP.

ANNUAL MEAN ΔT (WITH GLOBAL-MEAN)

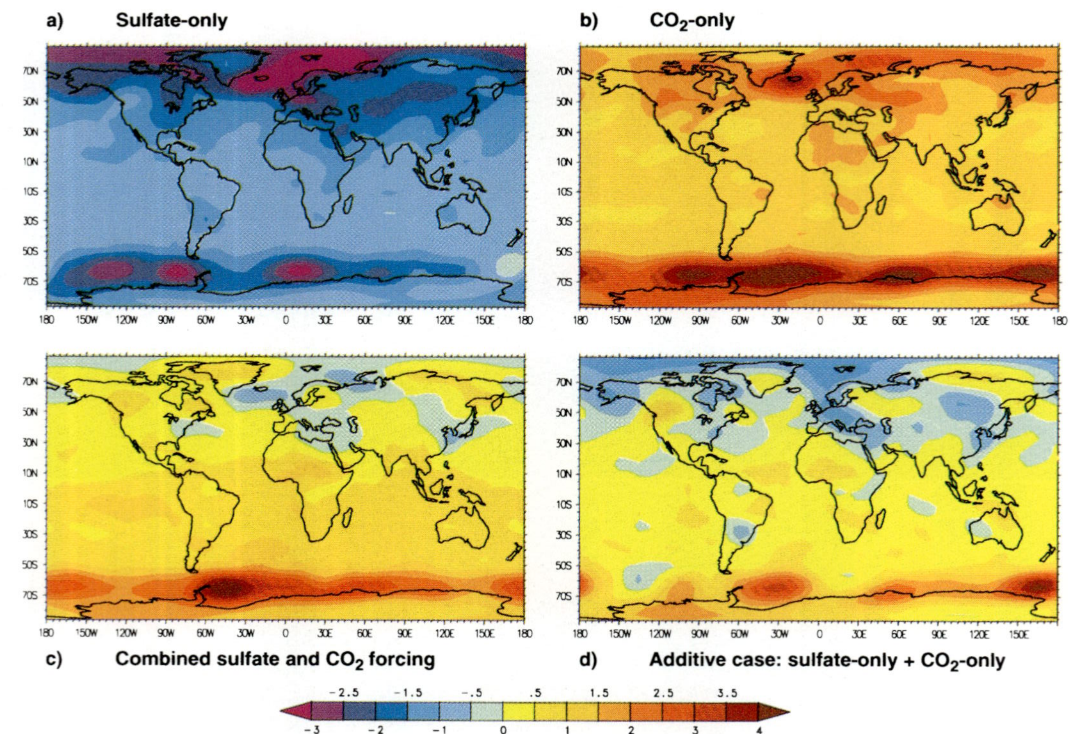

Colour Plate 5.1   The annually averaged surface temperature response to forcing by sulfate only (panel a), CO_2 only (panel b), the combined forcing by CO_2 plus sulfate aerosols (panel c), and the result of adding the fields from the sulfate only and CO_2 only fields (panel d). Differences (in K) are between local temperatures with and without forcing.

ANNUAL MEAN ΔT (GLOBAL-MEAN SUBTRACTED)

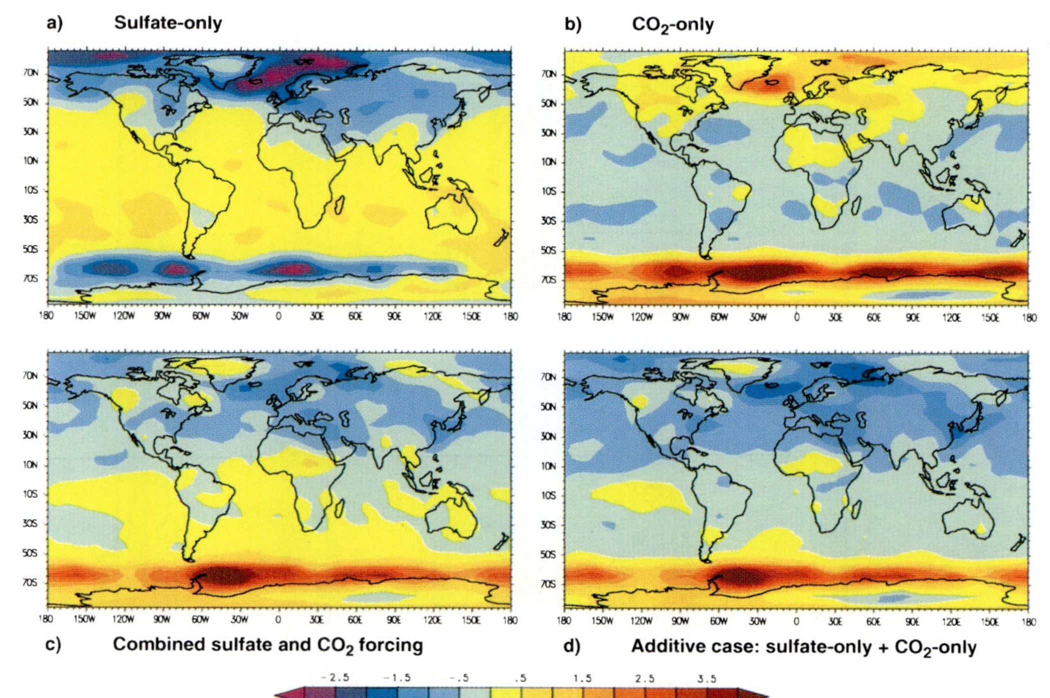

Colour Plate 5.2   The pattern of response for the annual average surface temperature change after subtraction of the global average temperature change. Panels (a) to (d) refer to forcing by sulfate only (panel a), CO_2 only (panel b), the combined forcing by CO_2 plus sulfate aerosols (panel d), and the result of adding the fields from the sulfate only and CO_2 only fields (panel d). Differences (in K) are between local temperatures with and without forcing after subtracting the global mean change.

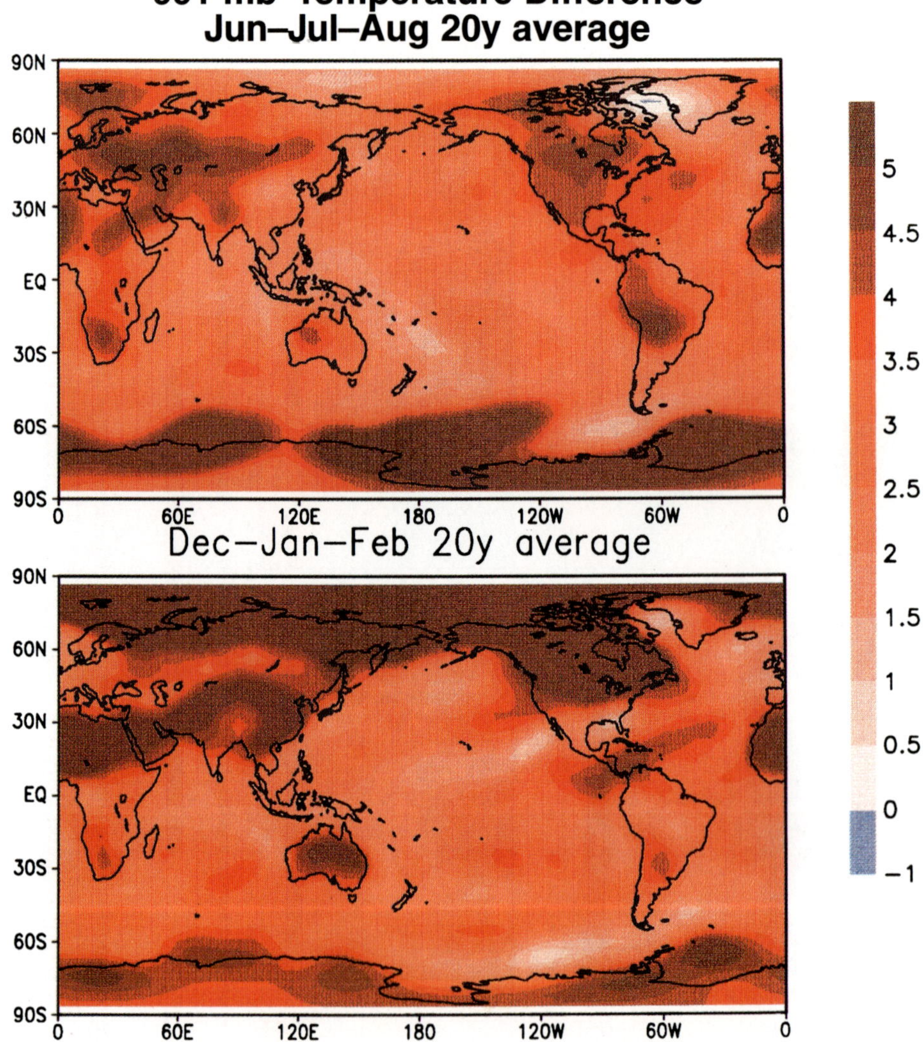

Colour Plate 7.1 The geographical distribution of surface air temperature change between transient and control experiments at the time of CO_2 doubling, 20-year average. The two seasons of December, January and February and June, July and August are shown.

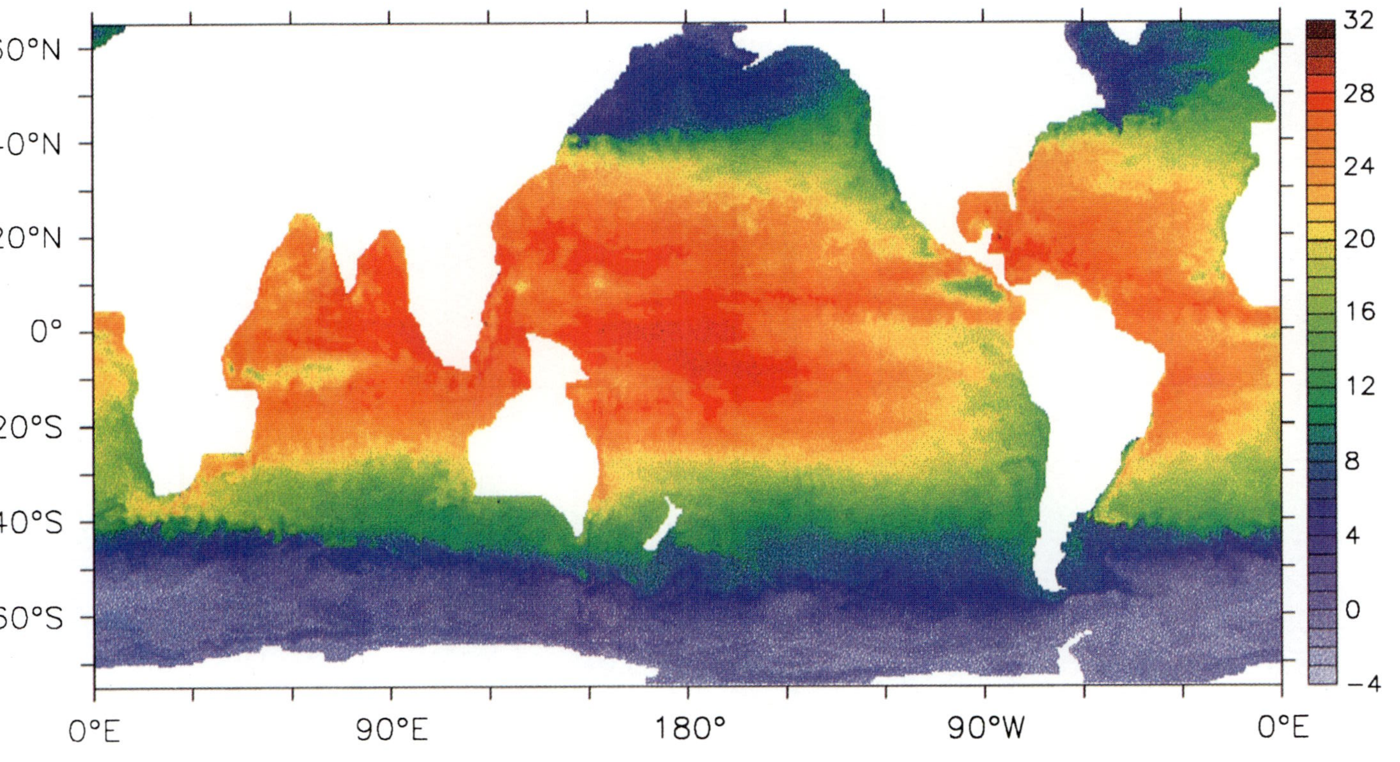

Colour Plate 8.1   Instantaneous snapshot of global ocean temperature at level 2 (37.5 m) for the Parallel Ocean Climate Model run at 0.5 degrees resolution and forced by ECMWF daily winds from 1980–1989. This snapshot corresponds to 1 January 1987 in the model.

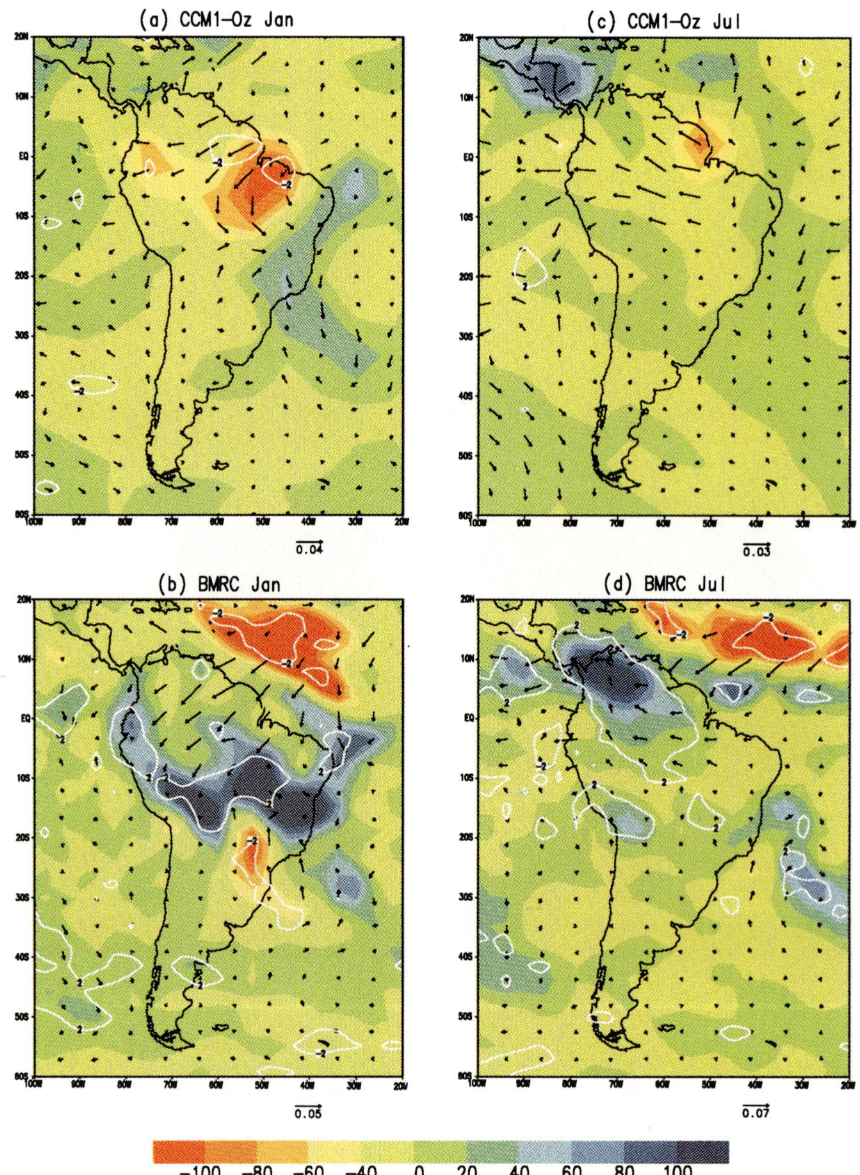

Colour Plate 12.1   Changes in monthly total precipitation (mm month^{-1}) and the low-level moisture advection following deforestation in two model simulations. Shaded areas represent changes in total precipitation and white contour lines are the signal to noise ratio of changes in precipitation with a value of 20. Vectors in all plots show the changes in horizontal moisture advection (kg*m / kg*s) at the lowest model level and the scale of the vectors represents different values as indicated in the individual plots: (a) for CCM1-OZ in January; (b) for BMRC in January; (c) for CCMI-OZ in July; and (d) for BMRC in July.

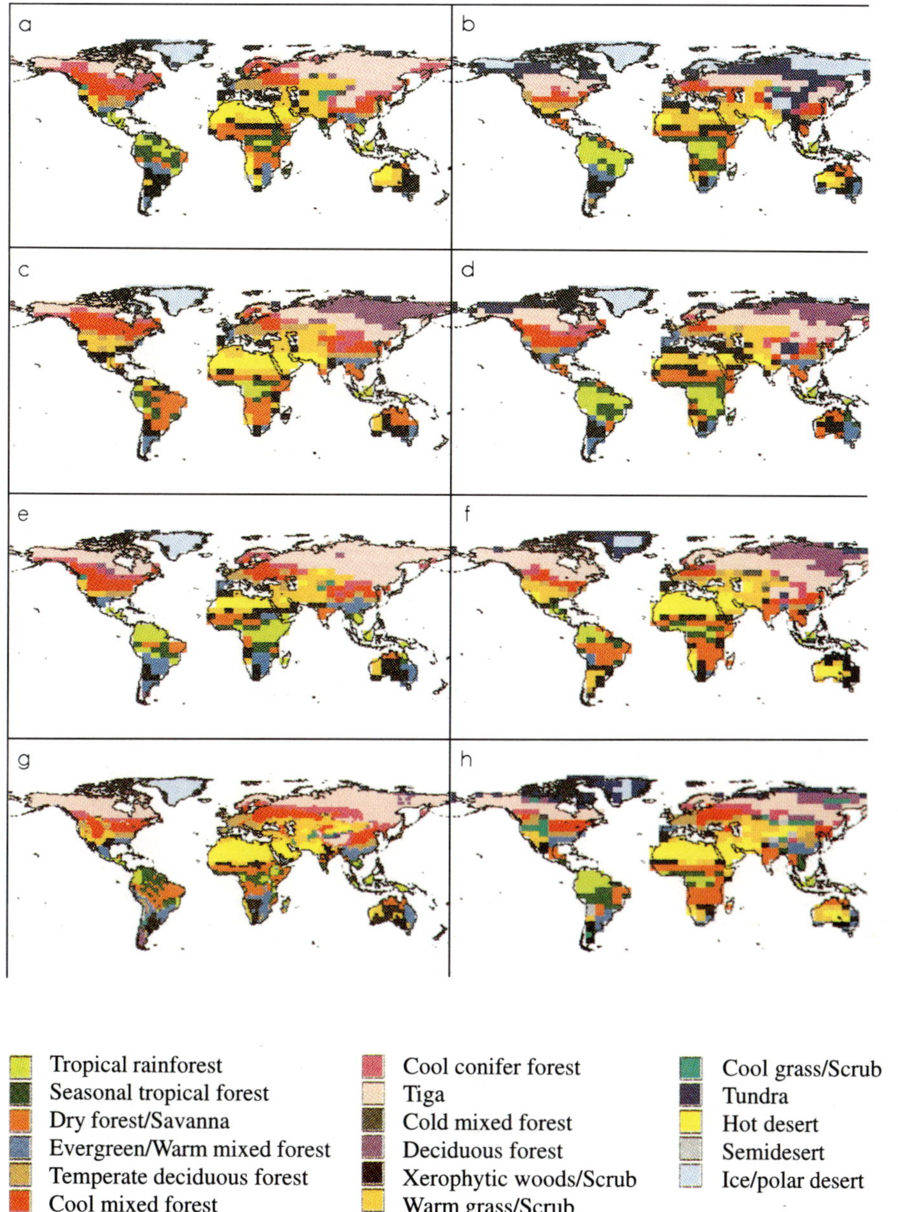

Colour Plate 16.1   Vegetation distribution predicted by the BIOME model using the simulated control climates from: (a) CCM1-OZ, (b) CCM1, (c) CCMO, (d) CCM1W, (e) CCM2 at R15 resolution, (f) BMRC (aggregated to an R15 grid), (g) CCM2 at T63 resolution, and (h) the observed climate datasets from Legates and Willmott (1990) and ISCCP (1992).

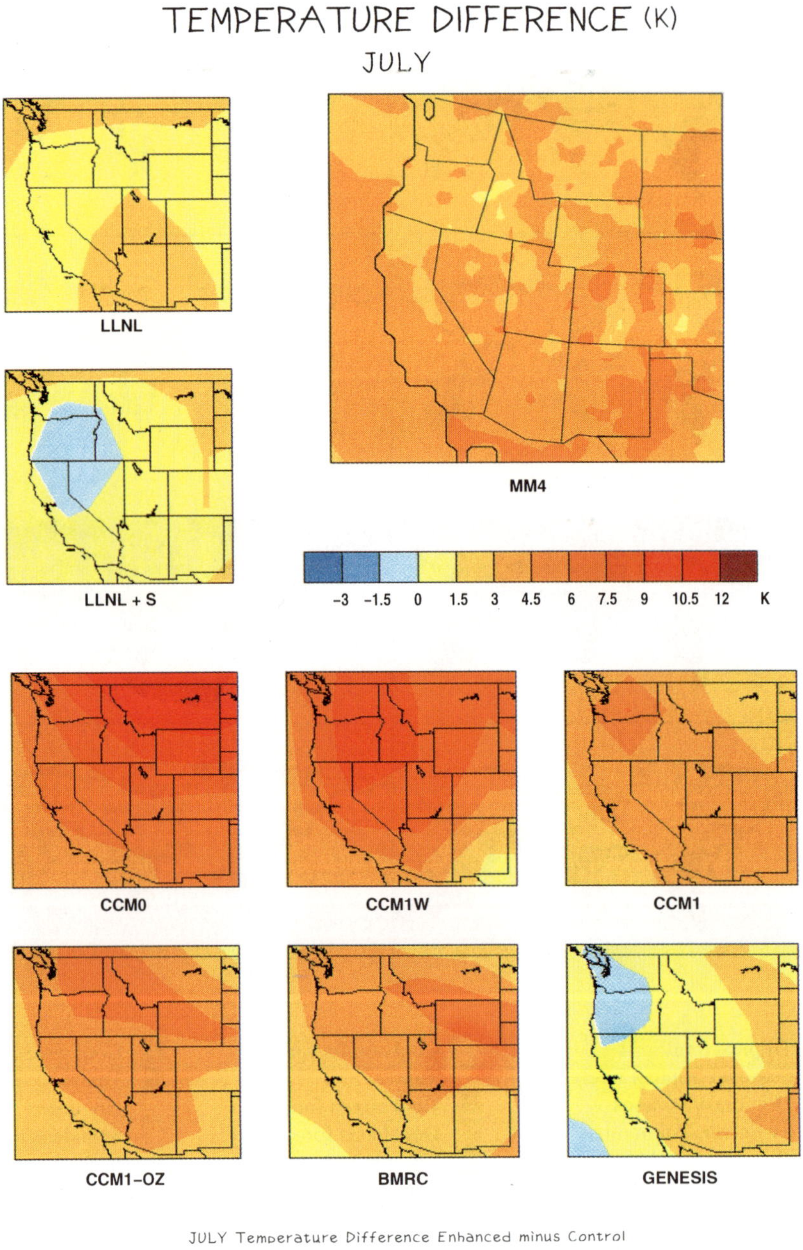

Colour Plate 18.1   July greenhouse warming (surface air temperature change, K) predicted for Western USA by eight MECCA models. The lower six models are as described in Chapter 16. The upper panels show results from the LLNL GCM with (+S) and without sulfate aerosols as described in Chapter 5 and a regional model simulation by MM4.

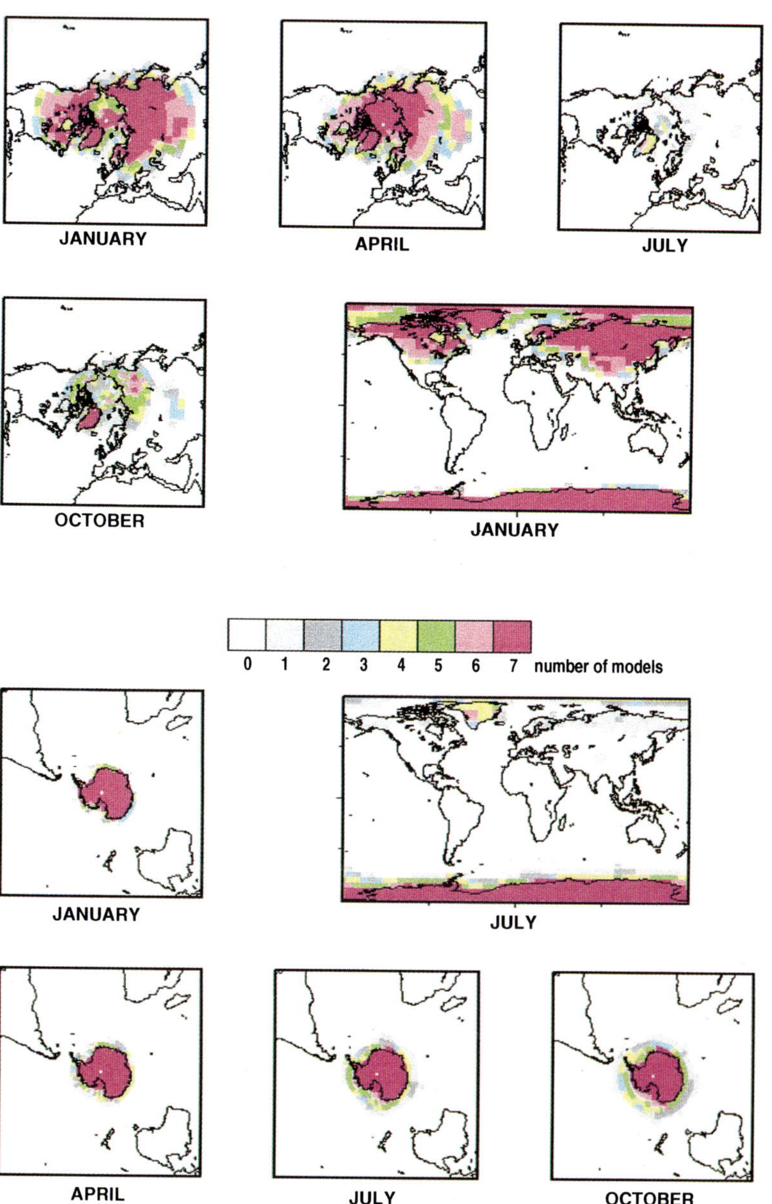

Colour Plate 18.2 Agreement (or confidence) in the greenhouse-climate extent of snow cover in January as depicted by seven MECCA models. Deep pink means that all seven models agree that snow exists in a greenhouse-warmed world. White means that all seven agree that no snow exists.

INDEX